AF686013

Numerical Heat Transfer and Fluid Flow 2022

Numerical Heat Transfer and Fluid Flow 2022

Editors

Artur Bartosik
Dariusz Asendrych

MDPI

Basel • Beijing • Wuhan • Barcelona • Belgrade • Novi Sad • Cluj • Manchester

Editors

Artur Bartosik
Kielce University of Technology
Kielce, Poland

Dariusz Asendrych
Czestochowa University of Technology
Czestochowa, Poland

Editorial Office
MDPI
St. Alban-Anlage 66
4052 Basel, Switzerland

This is a reprint of articles from the Special Issue published online in the open access journal *Energies* (ISSN 1996-1073) (available at: https://www.mdpi.com/journal/energies/special_issues/NHTFF).

For citation purposes, cite each article independently as indicated on the article page online and as indicated below:

Lastname, A.A.; Lastname, B.B. Article Title. *Journal Name* **Year**, *Volume Number*, Page Range.

ISBN 978-3-0365-9853-6 (Hbk)
ISBN 978-3-0365-9854-3 (PDF)
doi.org/10.3390/books978-3-0365-9854-3

© 2023 by the authors. Articles in this book are Open Access and distributed under the Creative Commons Attribution (CC BY) license. The book as a whole is distributed by MDPI under the terms and conditions of the Creative Commons Attribution-NonCommercial-NoDerivs (CC BY-NC-ND) license.

Contents

About the Editors

Artur Bartosik

Artur Bartosik has been appointed as a researcher and academic teacher at the Kielce University of Technology since 1983. In 1993, he established the Centre for Continuing Education at the University and in 2001 the Swietokrzyskie Centre for Innovation and Technology Transfer, which he managed as President of the Board from 2001 to 2007. Currently, he is the head of the Department of Production Engineering and the head of the Industrial Laboratory for Low Emission and Renewable Energy Sources. Together with his team, he works to establish a structure for university-wide multidisciplinary collaboration in terms of energy self-sufficiency. His scientific interest is focused on experiments, modelling, and simulation of single- and two-phase flows, including convective heat transfer. Artur, as a researcher, has spent 2 years at the University of Saskatchewan in Canada and several weeks at Ilmenau Technical University in Germany. He is the Editor-in-Chief of WSEAS Transactions on Fluid Mechanics and an Associate Editor of the Journal of Hydrology and Hydromechanics. He is a member of professional bodies and scientific committees of international conferences. He serves lectures for students on fluid mechanics; heat transfer; fluid-flow machinery; hydrotransport; renewable energy sources; transnational technology transfer; applied fluid mechanics. He continuously participates in industrial projects.

Dariusz Asendrych

Dariusz Asendrych's scientific interests are devoted to experimental and numerical fluid dynamics. Former interests included aerodynamics of jets and jest flames, modelling fibre suspension flows, and chemical absorption of carbon dioxide in gas purification installations. The present scientific fields of interest cover hydrodynamics of two-phase flows in porous media, heat transfer in gas–liquid systems in packed beds, modelling surface wettability, and three-phase oil–water slurry flows. Dr. Asendrych served as a secretary of a Polish Pilot Centre of an EU scientific association ERCOFTAC (1997-2010). He was active in the COST program. He participated in the Actions E36 (2006-2008), E48 (2005-2009), and FP1005 (2011-2015), which were related to the modelling of fibre suspension flows.

energies

MDPI

Editorial

Numerical Heat Transfer and Fluid Flow: New Advances

Artur S. Bartosik

Faculty of Management and Computer Modelling, Department of Production Engineering, Kielce University of Technology, Al. Tysiaclecia P.P. 7, 25-314 Kielce, Poland; artur.bartosik@tu.kielce.pl

1. Introduction

This Special Issue, titled 'Numerical Heat Transfer and Fluid Flow 2022', presents articles addressed to *Energies* and is a continuation of the 2021 edition [1]. The authors of this Special Issue describe solutions of scientific and industrial relevance in a specific field of heat transfer and fluid flow, including technical devices, nanofluids, industrial processes, dedicated perforations or mechanically deformed pipes, the transport of solid particles, etc. These articles also serve as catalysts for future directions and priorities in numerical heat transfer and fluid flow.

Numerical fluid mechanics is based on the governing equations of continuity, momentum, and energy and has been systematically developed since the second half of the twentieth century. Great interest in solving specific engineering problems requires mathematical and numerical modelling, simulations, and experiments on heat exchange and fluid flow for a variety of single- and multiphase flows and boundary conditions. The increase in the efficiency of computational methods and the availability of commercial packages makes it possible to solve complex engineering problems more accurately and faster. These also allow us to analyse the complex phenomena of the dynamic and thermal boundary layers [2].

Formulating an engineering problem of fluid flow and heat transfer requires the prior development of a physical model that includes important features of the phenomenon, such as thermophysical properties, flow geometry, boundary, and initial conditions. The next step is to develop a mathematical model. The mathematical model should start with the general form of the governing equations of continuity, momentum, and energy. As a result of the assumptions made in the physical model, a specific form of the mathematical model is formulated. The set of equations can be solved analytically, which is usually difficult and impractical, or numerically. If numerical methods are considered, direct numerical simulation (DNS), Reynolds-averaged Navier–Stokes equations (RANS), or large eddy simulation (LES) can be taken into account. The DNS method is time-consuming and expensive. The DNS method regards instantaneous parameters, which has a limited application for a large number of engineering problems as engineers are interested mainly in time-averaged parameters. Turbulence simulation using RANS or LES methods has been proven for a variety of engineering applications. Such methods are less time-consuming compared to the DNS method and, therefore, cost less; however, the set of equations requires closure. The problem of closure requires an additional equation or an additional set of equations, such as those proposed for turbulence models. The main requirements for turbulence models include universality, economy, extensionality, and reality [3,4]. The ability to simulate fluid flow and convective–diffusive heat transfer includes velocity-, pressure-, and temperature-dependent variables and remains one of the main challenges in CFD. Researchers are still looking for the best ways to influence the level of turbulence to reduce friction or increase the efficiency of mixing flowing components and heat transfer. We still try to follow nature and implement the best available engineering solutions, which is a great challenge in the development of CFD [5,6].

Considering the intensification of heat exchange, we recognise passive or active methods. Passive methods, such as increasing the area of heat transfer or inserting solid particles

Citation: Bartosik, A.S. Numerical Heat Transfer and Fluid Flow: New Advances. *Energies* **2023**, *16*, 5528. https://doi.org/10.3390/en16145528

Received: 10 July 2023
Revised: 14 July 2023
Accepted: 16 July 2023
Published: 21 July 2023

Copyright: © 2023 by the author. Licensee MDPI, Basel, Switzerland. This article is an open access article distributed under the terms and conditions of the Creative Commons Attribution (CC BY) license (https://creativecommons.org/licenses/by/4.0/).

into a fluid, or increasing the intensity of turbulence by shaping an insert with a dedicated perforation or mechanically deformed pipe, have been studied for several years, and some of them have become commercial solutions [7–10]. Active methods, such as air injection, bubble or vortex generation, or proper pulsation, can lead to an increase in the heat transfer coefficient and, finally, can produce an increase in heat exchange [11–13]. Both methods contribute to changes in the dynamic and thermal boundary layers by increasing the intensity of turbulence, allowing for the better mixing of flowing components.

The articles collated in this Special Issue are useful for researchers, engineers, and students who are focused on fluid flow and heat transfer.

2. Review of New Advances

Efficient numerical methods and fast computers mean that even complex phenomena, including the 3D domain, complex boundary conditions, fluid incompressibility, and changeable thermophysical properties, have the possibility of being solved.

However, in some simplified phenomena, analytical solutions are still valuable and possible. Based on the Navier–Stokes equations, Urbanowicz et al. [14] reviewed analytical models to accelerate the laminar flow of an incompressible Newtonian fluid in a pipe. The authors reviewed the solutions, which included two main approaches, such as an imposed pressure gradient or an imposed flow rate. An extension of this theory to turbulent accelerated pipe flows was also discussed [14]. The authors concluded that the discussed solutions extend the theory of analytical solutions of simplified two-dimensional Navier–Stokes equations and can be used not only to study the behaviour of liquids during accelerated pipe flow, but also to test the accuracy of commercial CFD packages [14].

Nanoparticles suspended in a fluid, called nanofluids, exist broadly in engineering. Nanomaterials possess outstanding optical, electrical, and thermophysical properties and therefore can be found in electronics, medicine, biotechnology, and energy transport systems, of which superconductors are one of those that exploit electrical and thermophysical properties. Nanofluids have been known for many decades and are particularly useful for intensifying heat exchange in engines, heat exchangers, or cooling mechanical devices [15]. Cieslinski [16] reviewed the recent achievements in the numerical modelling of the forced convection of nanofluids in round and smooth tubes and divided them into single- and two-phase approaches. The author synthesised three main challenges for engineers who deal with nanofluids, which are as follows.

- Provide reliable thermophysical properties of nanoparticles, liquids, and nanofluids;
- Provide accurate and reliable methods to calculate the heat transfer coefficient and friction factor;
- Stabilise nanofluids as they tend to aggregate to a larger size, which can result in sedimentation and/or separation.

The author noted that the majority of mathematical models assume a steady heat flow density acting on a chosen pipe length, as well as steady wall temperature. The author recognised 30 validated mathematical models for single-phase flows and 23 for two-phase flows and collected empirical correlations for the Nusselt number for laminar and turbulent flows. Cieslinski found that some researchers, such as Lotfi et al. [17] and Mokmeli and Saffar-Avval [18], observed that the precision of a single-phase approach is similar to that of a two-phase approach, while other researchers, like Maïga [19] for instance, indicated that two-phase models more realistically reproduce the results of measurements. Some results of the simulations emphasised the enhancement of heat transfer caused by the selected solid particles, while others noted the marginal influence. This clearly indicates that the process of nanofluid flow is complex, depends on the boundary layer and solid particles, and is not yet well understood. Cieslinski noted that, in the case of turbulent flow, researchers solve the closure problem primarily by using the standard k-ε turbulence model [16].

Foam is a dispersion of particles in a continuous medium in which the particles are gas bubbles and the medium is a liquid. Bubbles consist of one or more coexisting gas components, mainly air or vapour. The lamellar interfacial length can reach orders of nanometres. Foam is present mainly in chemical and industrial processes [20]. Understanding the foam behaviour and simulating its movement is a difficult task because of the highly dispersed time and length scales where multiphase and multicomponent systems occur. Mobarak et al. [21] proposed an approach to simulate some of the chosen phenomena using the Boltzmann lattice method and discussed the limitations. The authors presented simulations of bubble rise in a partially filled flask-like container that included convective–diffusive heat transfer. The authors stated that they made progress in the direction of the modelling and parameterisation of full-scale industrial rectification columns, where foaming is a critical and often occurring problem [21].

Recently, great interest has been shown in the application of metal foams to high-power batteries, heat and cooling devices, and compact electronic heat sinks [22]. Metal foams can reduce the size and mass of devices because of their low density. Shan et al. [23] studied heat transfer in an electronic radiator filled with metal foam. The authors analysed the effect of several factors, such as flow rate, pores per linear inch, and number of fins, on the heat transfer coefficient. The authors proved that the metal foam reflects a much stronger ability of heat transfer compared to the electronic radiator without the metal foam. They found the optimal number of fins and observed that the higher the number of pores per linear inch, the greater the heat transfer coefficient [23].

The enhancement of heat transfer using passive methods demonstrates an advantage over active methods as they do not require external power. However, passive methods cause greater friction, which requires extra power to the pump, compressor, blower, or ventilator. Considering geometrically complex applications, such as ribbed pipes or cylinders with a wavy leading edge, several factors that affect turbulence and the drag coefficient can be recognised [24]. These are examples of the pitch and shape of the rib, the angle of attack of the rib, the coefficient of channel blockage, etc. [7–10]. Recently, Mousavi et al. [25] described a variety of passive methods used to intensify the heat transfer process. Kugele et al. [26] used the LES approach for a single-started helically ribbed pipe and validated the velocity and heat transfer between the ribs using experimental data obtained by Virgilio et al. [27,28]. The authors considered an acrylic glass with two different helical turbulators inserted into the pipe. The starting points for their mathematical model were continuity, Navier–Stokes equations, and energy equation in 3D form. They simulated 3D water flow in a pipe at a constant Re = 21,000 and Pr = 7.0, assuming constant thermophysical properties. Kugele et al. [26] demonstrated that the LES simulations gave fairly accurate results for the flow field, the level of turbulence, and the local heat transfer coefficient. Measurements and simulations of the Nusselt number in the pipe with and without helical turbulators confirmed a significant increase in the Nusselt number. Simulations of dynamic and thermal boundary layers performed by Kugele et al. [26] shed more light on complex turbulence processes and their influence on heat transfer in helically ribbed pipes.

The components of the gas turbine must be cooled to maintain its mechanical strength for a reliable operation. Cooling efficiency is essential in the region along the edge of the turbine blades and vanes. Lee et al. [29] applied large-eddy simulations to study turbulent flow in a channel of a certain height with a staggered array of pin fins with an appropriate diameter as a function of the heating loads that are relevant for the cooling of turbine blades and vanes. The results of the simulations were validated on the basis of data obtained from direct numerical simulation and experiments. The authors concluded that there are significant changes in the turbulent flow structure caused by heating loads, creating wall jets next to all heated surfaces [29].

Heating the steam pipeline that connects the boiler to the turbine is essential in the turbine and boiler start-up process. Rapid changes in steam temperature can cause high stresses on pipeline and turbine components. Kaczmarski [30] proposed a solution to the inverse heat transfer problem (IHP) in a steam pipeline, taking into account that the turbine manufacturer sets temporary changes in the steam temperature at the turbine input. The author proposed a method to predict the transient steam temperature at the entrance of the pipe based on the measured (known) steam temperature at the exit of the pipe. Taking into account the heat balance equations, the author formulated a set of ordinary differential equations. The author divided the pipeline domain into control volumes in the radial and longitudinal directions. Next, taking into account the boundary and initial conditions, he solved the set of equations using the Runge–Kutta method. The author presented the results of the simulations, which include time-dependent pressure, mass flow rate, wall and steam temperature, and the relative difference between inlet and outlet steam temperatures. The author concluded that his mathematical model can be used to control boiler operation [30]

The incredible increase in the use of renewable energy sources, which are unfortunately sensitive to weather conditions, has led to the development of thermal and electrical storages (accumulators). Thermal accumulators can use fluid, solid, or phase-change materials [31]. Taler et al. [32] built a test rig for a hybrid (electric–water system), for a building in which a ceramic heat accumulator was installed. The ceramic thermal accumulator contains an outer cylindrical shell inside which the ceramic cylinders and electrical resistance heaters were packed. The air flow generated by the fan was transported through the accumulator, which gave heat to the water flowing in the finned heat exchanger, which in turn was the heat source for the central heating system of the building. The calculated exit air temperature T = f(x,t) was compared with the measured data. The results of the simulations indicated good accuracy; however, for the initial air heating period, the differences were significant. The authors reasoned that their mathematical model could help to choose the best accumulator size for specific needs [32].

The global fibre optics market is rapidly developing, mainly in the electronics and telecommunication industries. Fibre strands can be made of glass or plastic and are known as microstructured optical fibres or photonic crystal fibres [33]. Mathematical models that can describe the fibre drawing process are highly desirable. Luzi et al. [34] developed a mathematical model of a fibre drawing process that includes the general equations of continuity, momentum, and energy. Assuming that the flow is incompressible and axially symmetric without a circumferential velocity component and taking into account the evolution equation for the inner and outer surfaces, developed by Fitt et al. [35], they built an asymptotic mathematical model of the unsteady 2D flow of the capillary drawing process. Luzi et al. [34] performed a successful validation of the mathematical model for annular capillaries. The authors performed simulations of the temperature and velocity distributions during the fibre production process. Their study can lead to a better understanding of the fibre drawing process and demonstrate the influence of physical quantities on the production process [34].

The high consumption of oil and gas results in an increase in explorations in deep-depth onshore and offshore territories. It is known that a high formation temperature causes failures in current operating designs [36]. In the drilling process, the drilling mud provokes a cooling effect in the formation. An accurate prediction of the temperature of the mud that flows into the drilling tube and annulus is very desirable [37]. Pioneering work was carried out by Bullard [38] who analytically predicted the temperature distribution using the diffusion equation. Jang et al. [39] proposed a mathematical model of transient heat transfer to compute the radial temperature in the drilling hole, annulus, and formation. The authors used the concept of quantifying thermal disturbance, called a thermally disturbed radius. On the basis of this concept, the authors can predict how long thermal disturbance occurs radially in the formation. Their physical model assumes that the drilling operation system consists of five subsections, such as the drilling hole, drilling pipe, annulus, casing, and reservoir formation. Taking into account some assumptions, the authors developed a mathematical model of unsteady transient heat transfer in a two-dimensional domain to derive the thermally disturbed radius. The authors performed a validation of the mathematical model based on the measurements and obtained satisfactory agreement. As a result of simulations, the dependence of temperature on the depth in the hole, temperature profiles, and thermal disturbance radius based on time and temperature were presented.

The authors reasoned that their mathematical model allows one to predict the temperature of the formulated well [39].

Great interest is expressed in increasing the efficiency of wind, water turbines, and photovoltaic modules to increase their contribution to the total amount of energy produced [40,41]. Hybrid platforms, which use more than one source of renewable energy, are becoming more popular and cost-effective. Torres et al. [42] conducted a study of a hybrid platform, which includes offshore wind and water turbines. The water turbine, which uses tidal power, is installed in a structured tower (together with a wind turbine). The main objective was to calculate the power generated by wind and water turbines [42]. Using the continuity equation and the RANS equations, the set of 3D equations was closed, taking into account the standard k-ε turbulence model. Simulations were performed assuming that the inlet velocity to the domain was the same as the velocity measured on a real platform and at the right height/depth. The results of the simulations were presented as graphs of velocity fields. On the basis of the simulations, the authors were able to calculate the energy production for the hybrid platform over a year [42].

The transport of the solid phase, using water as a carrier liquid, is commonly applied in industry. Solid–liquid transport usually requires more energy compared to a single-phase flow with the same flow rate. It is well known that a small amount of deflocculant results in a decrease in viscosity and, as a consequence, a decrease in wall shear stress. Jaworska-Jozwiak and Dziubinski [43] performed rheological measurements of fine dispersive limestone slurry with and without deflocculant in a wide range of solid concentrations. In the rheological experiments, the authors observed a significant decrease in viscosity and wall shear stress. Next, using the Bernoulli equation for the real liquid, the authors formulated an algebraic mathematical model for the hydrotransport of limestone from a reservoir to a settling tank for the chosen manufacturing enterprise. The authors reasoned that by adding a proper amount of the chosen deflocculant to the limestone slurry, it is possible to reduce the power consumption of the centrifugal pump motor by more than 50% [43].

The high level of water pollution around the world caused by the presence of plastics requires research, which helps to understand the spread of pollution. Plastic particles can settle and spread as sediment or flow freely with the carrier liquid or remain floating [44]. Kevorkijan et al. [45] conducted research on the settling process of plastic particles with diameters between 1 and 10 mm. The authors performed predictions and experiments on 3D turbulent flow in an open channel containing water and plastic particles. Their mathematical model includes CFD computations for liquid flow and the Lagrangian tracking method for plastic particles. The mathematical model dedicated to liquid flow constitutes time-averaged continuity and momentum equations, while the closure problem was solved using the standard k-ε turbulence model. The mathematical model dedicated to plastic particles constitutes the equations for the conservation of mass and Newton's second law. The authors performed simulations using Ansys software for vertical and horizontal velocity profiles, average settling time, and average downstream settling distance of plastic particles, as well as the formation of particle clouds. The authors reasoned that the proposed mathematical model of the settlement of particles in an open channel, which has been validated for turbulent flow, can be used on a larger scale to predict the propagation of mesoplastics in rivers [45].

3. Conclusions

Analysing the articles contributed to this Special Issue, titled *Numerical Heat Transfer and Fluid Flow 2022*, one can say that all articles are applied to specific environmental or engineering problems. The articles are encountered in fluid dynamics and/or heat transfer in machines, production and exploration process, environment, and other related areas of mechanical engineering.

Some articles dealt with simulations, while others presented their own experimental data. The articles, which dealt with simulations, contained a physical model with major assumptions such as the physical properties of the flowing medium, geometry of the flow domain, boundary, and initial conditions, and then a mathematical model. Mathematical models were formulated using conservation laws, such as equations of continuity, N-S, and energy. The authors formulated the set of equations for a variety of applications and solved them numerically, taking into account the convergence criteria and ensuring a mesh-independent solution. All simulations were performed using commercial software, and most of the mathematical models were properly validated.

Through the variety of approaches presented in this Special Issue, titled *Numerical Heat Transfer and Fluid Flow 2022*, the reader can find a description of the physical phenomenon, physical and mathematical models, and experimental data, and can gain a better understanding of the phenomena in a specific engineering application, including the interpretation of computed and measured quantities.

Funding: This research did not receive external funding.

Data Availability Statement: Data are contained within this article.

Acknowledgments: The author thanks the contributors of this Special Issue, titled *Numerical Heat Transfer and Fluid Flow 2022*, for valuable articles and MDPI for the invitation to act as a Guest Editor. Special thanks are extended to the editorial staff of *Energies* for efficient daily cooperation and great participation.

Conflicts of Interest: The author declares that there is no conflict of interest.

References

1. Bartosik, A. Numerical heat transfer and fluid flow: A review of contributions to the special issue. *Energies* **2022**, *15*, 2922. [CrossRef]
2. Patankar, S.V. (Ed.) *Numerical Heat Transfer and Fluid Flow*; Taylor and Frances Inc.: Abingdon, UK, 1980; p. 2014.
3. Spalding, D.B. *Turbulence Models for Heat Transfer*; Report HTS/78/2; Department of Mechanical Engineering, Imperial College of London: London, UK, 1978.
4. Spalding, D.B. *Turbulence Models—A Lecture Course*; Report HTS/82/4; Department of Mechanical Engineering, Imperial College of London: London, UK, 1983.
5. Castelvecchi, D. Mysteries of turbulence unravelled. *Nature* **2017**, *548*, 382–383. [CrossRef] [PubMed]
6. Klebert, P.; Su, B. Turbulence and flow field alterations inside a fish sea cage and its wake. *Appl. Ocean Res.* **2020**, *98*, 102–111. [CrossRef]
7. Zong, Y.; Bai, D.; Zhou, M.; Zhao, L. Numerical studies on heat transfer enhancement by hollow-cross disk for cracking coils. *Chem. Eng. Process. Process Intensif.* **2019**, *135*, 82–92. [CrossRef]
8. Nakhchi, M.E.; Esfahani, J.A. Numerical investigation of heat transfer enhancement inside heat exchanger tubes fitted with perforated hollow cylinders. *Int. J. Therm. Sci.* **2020**, *147*, 106153. [CrossRef]
9. Ahn, K.; Song, J.C.; Lee, J.S. Dependence of conjugate heat transfer in ribbed channel on thermal conductivity of channel wall: An LES Study. *Energies* **2021**, *14*, 5698. [CrossRef]
10. Pandey, L.; Singh, S. Numerical analysis for heat transfer augmentation in a circular tube heat exchanger using a triangular perforated Y-shaped insert. *Fluids* **2021**, *6*, 247. [CrossRef]
11. Hagiwara, Y. Effects of bubbles, droplets or particles on heat transfer in turbulent channel flows. *Flow Turbul. Combust.* **2011**, *86*, 343–367. [CrossRef]
12. Hamed, H.; Mohhamed, A.; Khalefa, R.; Habeeb, O. The effect of using compound techniques (passive and active) on the double pipe heat exchanger performance. *Egypt. J. Chem.* **2021**, *64*, 2797–2802. [CrossRef]
13. Zhao, J.; Zhang, B.; Fu, X.; Yan, S. Numerical study on the influence of vortex generator arrangement on heat transfer enhancement of oil-cooled motor. *Energies* **2021**, *14*, 6870. [CrossRef]
14. Urbanowicz, K.; Bergant, A.; Stosiak, M.; Deptuła, A.; Karpenko, M. Navier-Stokes solutions for accelerating pipe flow—A review of analytical models. *Energies* **2023**, *16*, 1407. [CrossRef]
15. Bhanvase, B.; Barai, D. *Nano Fluid for Heat and Mass Transfer—Fundamentals, Sustainable Manufacturing and Applications*; Academic Press: London, UK, 2021; p. 448. Available online: https://www.elsevier.com/books-and-journals (accessed on 29 April 2021).
16. Cieśliński, J. Numerical modelling of forced convection of nanofluids in smooth, round tubes: A review. *Energies* **2022**, *15*, 7586. [CrossRef]
17. Lotfi, R.; Saboohi, Y.; Rashidi, A.M. Numerical study of forced convective heat transfer of nanofluids: Comparison of different approaches. *Int. Commun. Heat Mass Transf.* **2010**, *37*, 74–78. [CrossRef]
18. Mokmeli, A.; Saffar-Avval, M. Prediction of nanofluid convective heat transfer using the dispersion model. *Int. J. Therm. Sci.* **2010**, *49*, 471–478. [CrossRef]

19. Maïga, S.; Nguyen, C.T.; Galanis, N.; Roy, G. Heat transfer behaviours of nanofluids in a uniformly heated tube. *Superlattices Microstruct.* **2004**, *35*, 543–557. [CrossRef]
20. Cantat, I.; Cohen-Addad, S.; Elias, F.; Graner, F.; Höhler, R.; Pitois, O.; Rouyer, F.; Saint-Jalmes, A. *Foams: Structure and Dynamics*; OUP: Oxford, UK, 2013.
21. Mobarak, M.; Gatternig, B.; Delgado, A. On the simulations of thermal liquid foams using Lattice Boltzmann method. *Energies* **2023**, *16*, 195. [CrossRef]
22. Shen, L.; Xu, S.; Bai, Z.; Wang, Y.; Xie, J. Experimental study on thermal and flow characteristics of metal foam heat pipe radiator. *Int. J. Therm. Sci.* **2021**, *159*, 106572. [CrossRef]
23. Shan, X.; Liu, B.; Zhu, Z.; Bennacer, R.; Wang, R.; Theodorakis, P.E. Analysis of the heat transfer in electronic radiator filled with metal foam. *Energies* **2023**, *16*, 4224. [CrossRef]
24. Ferreira, P.H.; Araújo, T.B.; Carvalho, E.O.; Fernandes, L.D.; Moura, R.C. Numerical investigation of flow past bio-inspired wavy leading-edge cylinders. *Energies* **2022**, *15*, 8993. [CrossRef]
25. Mousavi Ajarostaghi, S.S.; Zaboli, M.; Javadi, H.; Badenes, B.; Urchueguia, J.F. A Review of recent passive heat transfer enhancement methods. *Energies* **2022**, *15*, 986. [CrossRef]
26. Kügele, S.; Mathlouthi, G.O.; Renze, P.; Grützner, T. Numerical simulation of flow and heat transfer of a discontinuous single started helically ribbed pipe. *Energies* **2022**, *15*, 7096. [CrossRef]
27. Virgilio, M.; Mayo, I.; Dedeyne, J.; Geem, K.V.; Marin, G.; Arts, T. Effects of 2-D and 3-D helical inserts on the turbulent flow in pipes. *Exp. Therm. Fluid Sci.* **2020**, *110*, 109923. [CrossRef]
28. Virgilio, M.; Mayo, I.; Dedeyne, J.; Geem, K.V.; Marin, G.; Arts, T. Influence of obstacles on the wall heat transfer for 2D and 3D helically ribbed pipes. *Int. J. Heat Mass Transf.* **2020**, *148*, 119087. [CrossRef]
29. Lee, C.S.; Shih, T.I.P.; Bryden, K.M.; Dalton, R.P.; Dennis, R.A. Strongly heated turbulent flow in a channel with pin fins. *Energies* **2023**, *16*, 1215. [CrossRef]
30. Kaczmarski, K. Identification of transient steam temperature at the inlet of the pipeline based on the measured steam temperature at the pipeline outlet. *Energies* **2022**, *15*, 5804. [CrossRef]
31. Li, P.-W.; Chan, C.L. *Thermal Energy Storage Analyses and Designs*; Academic Press: London, UK, 2017.
32. Taler, D.; Taler, J.; Sobota, T.; Tokarczyk, J. Cooling modelling of an electrically heated ceramic heat accumulator. *Energies* **2022**, *15*, 6085. [CrossRef]
33. Russell, P.S. Photonic-crystal fibers. *J. Light. Technol.* **2006**, *24*, 4729–4749. Available online: https://opg.optica.org/abstract.cfm?URI=OFC-2009-OTuC1 (accessed on 13 July 2023). [CrossRef]
34. Luzi, G.; Lee, S.; Gatternig, B.; Delgado, A. An asymptotic energy equation for modelling thermo fluid dynamics in the optical fibre drawing process. *Energies* **2022**, *15*, 7922. [CrossRef]
35. Fitt, A.D.; Furusawa, K.; Monro, T.M.; Please, C.P.; Richardson, D.J. The mathematical modelling of capillary drawing for holey fibre manufacture. *J. Eng. Math.* **2002**, *43*, 201–227. [CrossRef]
36. Wu, B.; Zhang, X.; Jeffrey, R.G. A model for downhole fluid and rock temperature prediction during circulation. *Geothermics* **2014**, *50*, 202–212. [CrossRef]
37. Dirksen, R. Upgrading formation-evaluation electronics for high-temperature drilling environments. *J. Pet. Technol.* **2011**, *63*, 24–26. [CrossRef]
38. Bullard, E.C. The time necessary for a bore hole to attain temperature equilibrium. *Geophys. J. Int.* **1947**, *5*, 127–130. [CrossRef]
39. Jang, M.; Chun, T.S.; An, J. The transient thermal disturbance in surrounding formation during drilling circulation. *Energies* **2022**, *15*, 8052. [CrossRef]
40. Micallef, D.; van Bussel, G. A Review of urban wind energy research: Aerodynamics and other challenges. *Energies* **2018**, *11*, 2204. [CrossRef]
41. Santamaría, L.; Vega, M.G.; Pandal, A.; Pérez, J.G.; Velarde-Suárez, S.; Oro, J.M.F. Aerodynamic performance of VAWT airfoils: Comparison between wind tunnel testing using a new three-component strain gauge balance and CFD modelling. *Energies* **2022**, *15*, 9351. [CrossRef]
42. Torres, J.P.N.; De Jesus, A.S.; Lameirinhas, R.A.M. How to improve an offshore wind station. *Energies* **2022**, *15*, 4873. [CrossRef]
43. Jaworska-Jozwiak, B.; Dziubinski, M. Effect of deflocculant addition on energy savings in hydrotransport in the lime production process. *Energies* **2022**, *15*, 3869. [CrossRef]
44. Klein, S.; Dimzon, I.K.; Eubeler, J.; Knepper, T.P. Analysis, occurrence, and degradation of microplastics in the aqueous environment. In *Handook of Environmental Chemistry*; Springer: Cham, Switzerland, 2018; Volume 58, pp. 51–67.
45. Kevorkijan, L.; Žic, E.; Lešnik, L.; Biluš, I. Settling of mesoplastics in an open-channel flow. *Energies* **2022**, *15*, 8786. [CrossRef]

Disclaimer/Publisher's Note: The statements, opinions and data contained in all publications are solely those of the individual author(s) and contributor(s) and not of MDPI and/or the editor(s). MDPI and/or the editor(s) disclaim responsibility for any injury to people or property resulting from any ideas, methods, instructions or products referred to in the content.

energies

MDPI

Review

Navier-Stokes Solutions for Accelerating Pipe Flow—A Review of Analytical Models

Kamil Urbanowicz [1,*], Anton Bergant [2,3], Michał Stosiak [4], Adam Deptuła [5] and Mykola Karpenko [6]

1 Faculty of Mechanical Engineering and Mechatronics, West Pomeranian University of Technology in Szczecin, 70-310 Szczecin, Poland
2 Litostroj Power d.o.o., 1000 Ljubljana, Slovenia
3 Faculty of Mechanical Engineering, University of Ljubljana, 1000 Ljubljana, Slovenia
4 Faculty of Mechanical Engineering, Wrocław University of Science and Technology, 50-370 Wrocław, Poland
5 Faculty of Production Engineering and Logistics, Opole University of Technology, 45-758 Opole, Poland
6 Faculty of Transport Engineering, Vilnius Gediminas Technical University, LT-10223 Vilnius, Lithuania
* Correspondence: kamil.urbanowicz@zut.edu.pl

Abstract: This paper reviews analytical solutions for the accelerated flow of an incompressible Newtonian fluid in a pipeline. This problem can be solved in one of two ways according to the (1) imposed pressure gradient or (2) flow rate. Laminar accelerated flow solutions presented in a number of publications concern cases where the two driving mechanisms are described by simple mathematical functions: (a) impulsive change; (b) constant change; (c) ramp change, etc. The adoption of a more complex and realistic description of the pressure gradient or flow rate will be associated with a profound mathematical complexity of the final solution. This is particularly visible with the help of the universal formula derived by several researchers over the years and discussed in this paper. In addition to the solutions strictly defined for laminar flow, an interesting extension of this theory is the theory of underlying laminar flow for the analysis of turbulent accelerated pipe flows (TULF model developed by García García and Alvariño). The TULF model extends the Pai model developed more than 60 years ago, which has been previously used for steady flows only. The discussed solutions extend the theory of analytical solutions of simplified two-dimensional Navier–Stokes equations and can be used not only to study the behavior of liquids during accelerating pipe flow but they can also be used to test the accuracy of commercial CFD codes.

Keywords: Navier–Stokes equations; analytical solution; accelerated flow; pipe flow; duct flow

Citation: Urbanowicz, K.; Bergant, A.; Stosiak, M.; Deptuła, A.; Karpenko, M. Navier-Stokes Solutions for Accelerating Pipe Flow—A Review of Analytical Models. *Energies* **2023**, *16*, 1407. https://doi.org/10.3390/en16031407

Academic Editors: Artur Bartosik and Dariusz Asendrych

Received: 30 November 2022
Revised: 5 January 2023
Accepted: 20 January 2023
Published: 31 January 2023

Copyright: © 2023 by the authors. Licensee MDPI, Basel, Switzerland. This article is an open access article distributed under the terms and conditions of the Creative Commons Attribution (CC BY) license (https://creativecommons.org/licenses/by/4.0/).

1. Introduction

Fluid flow in hydraulic pipes is often time-varying flow. This means that the basic parameters of the flow, i.e., the average value of velocity and pressure in the analyzed pipe cross-section, dynamically change over time. Time-varying flows can be accelerated, decelerated, reverse, pulsating, oscillating and water hammer-type flows. This paper reviews analytical models for accelerating pipe flows.

Starting with the work of Navier [1], the basic hydrodynamic equations (according to the work of Darrigoll [2], these have been re-derived at least four times chronologically by the following well-known researchers: Cauchy, Poisson, Saint-Venant and Stokes), are the subject of numerous studies. Analyzing the literature on analytical solutions of laminar accelerated flows, it can be seen that two groups of flows can exist [3,4]: the first is flows forced by the occurrence of a step pressure gradient change, while the second group of flows is those in which the fluid movement is forced by a step change in the flow rate.

Much earlier, solutions from the first group were analyzed, i.e., when the flow is forced by a change of the pressure gradient along the length of the pipe. The oldest paper found for this review was published by the Italian scientist Roiti [5] in 1871, who worked at the University of Pisa. He studied accelerated flow in a simple vertical water system. The pressure change was obtained by rapidly opening the valve. Apart from experimental research, Roiti's most important achievement was the presentation of an analytical theory of this flow. This referred to solutions previously described by another researcher from the University of Pisa, namely Betti [6], who analyzed heat flows. Another important work was written in 1882 by Gromeka [7]. Unfortunately, due to the fact that it was published only in Russian, it remains known mainly among authors of Russian origin [8–11], who mostly know it from an important reprint of all Gromeka's works in a book form, published on the 100th anniversary of his birth [12]. It was only thanks to the work of Szymański [13] published in 1932 in a respected French mathematical journal (Journal de Mathématiques Pures et Appliquées—still existing and new papers being published) that this solution was noticed and reproduced in many books on the flow of viscous fluids [14–16]. The accelerated solution forced by a rapid, strictly defined pressure gradient was then analyzed by a number of researchers who extended its applicability. Among these studies, the work of Gerbes [17], was the first to apply the Laplace transform to its derivation, should be noted, as well as that of Ito [18], who extended the scope of this solution to make it suitable for the analysis of unsteady transient flows occurring between one steady flow (of the Hagen–Poiseuille type) and another characterized by a higher average velocity.

For short pipes, the entrance length plays an important role. Solutions for this unsteady acceleration problem devoted to the development of the velocity profile along the pipe length and in time have been derived by Atabek [19] and Avula [20]. Fan [21] solved the problem of accelerated flow in rectangular ducts. Solutions in ducts with an arbitrary geometry were the subject of research by Laura [22]. This issue has been recently analyzed by Muzychka-Yovanovich [23,24] as well. In the following years, similar solutions were developed for non-Newtonian fluid flows [25–28] for which the shearing stress is not linearly related to the rate of shearing strain [29].

The second group of models, i.e., those in which the function of velocity or flow rate is defined in the main equation of motion, has been analyzed since 1933, when Vogelpohl's short technical note was published [30]. In his note, Vogelpohl also mentioned the experimental research carried out at the University of Berlin by Prof. Föttinger on accelerated flows. In addition, he derived a formula (using the Whittaker method [31]) for the time-varying pressure gradient for this type of flow (later re-derived by Andersson and Tiseth [32]) for a constant value of average velocity. Other researchers who were interested in this type of flow were Weinbaum and Parker [33], although their interest was mostly focused on the decelerated flow resulting from the sudden closing of the gate valve. Weinbaum and Parker's research was the inspiration for subsequent works, including the work of Andersson and Tiseth [32], which is particularly important from the point of view of this review. It was in this work that for the first time a complete analytical solution for the flow velocity profile was derived for a constant mean velocity scenario that can be the effect of piston movement. It should be mentioned here that Andersson and Kristoffersen [34] had previously analyzed a wide range of flows forced by a pressure gradient (RGS type) as well as the correction to this solution proposed by Otis [35]. The Anderson and Tiseth solution has been improved by Das and Arakeri [36], who presented analytical solutions for a complete piston cycle (acceleration period, motion with constant velocity, deceleration period and final period). The Das and Arakeri solutions for the first two periods (i.e., acceleration and piston work at constant velocity) can be treated as a complete solution when imposing a ramp-type motion. This solution was further refined by Kannaiyan et al. [37] by introducing the possibility of determining the transition from one steady state to another.

The main experimental studies that confirmed the correctness of the solution based on a step change in pressure were carried out by van de Sande et al. [38] and Lefebvre and White [39]. These studies show that laminar flow is maintained during acceleration for a relatively long time—in the paper by van de Sande et al. [38] to Re = 57,500, and in the work of Lefebvre and White [39] to Re > 10^5. This delayed transition from laminar to turbulent flow was also mentioned by Goldberg in a discussion on van de Sande et al. at the Pressure Surges Conference in Canterbury [38]. Goldberg (p. 499 of Pressure Surges 1980 proceedings) wrote about a long distant oil transportation pipeline (105 miles long, twenty inches in diameter) located in the Gulf of Mexico in which, during the start-up, a delayed transition to turbulent flow was detected.

As opposed to pressure-driven flows, the other type of acceleration occurs in systems due to a step change in flow rate, e.g., piston-driven flow. In an early experiment by Kataoka et al. [40], the measured results did not agree well with the theory as an annular jet effect was reported. The research, which indicated the correctness of the theoretical solution for a fixed flow rate, was carried out by a team led by Chaudhury et al. [41]. Recently He et al. [42] have reported that the whole acceleration phase in pipes is a laminar–turbulent bypass transition, so even during the initial flow, some turbulent structures that initially occupy the near wall region are present in the flow.

Although this paper concerns only accelerated Newtonian liquid flows in pipes with circular cross-sections (typical, commonly used in practice, Figure 1), the scope of research is very extensive. Readers interested in further solutions (flows through other cross-sections, accelerated flows of non-Newtonian fluids, etc.), which were developed on the basis of the accelerated flows theory discussed in this review, will be offered additional reading elsewhere.

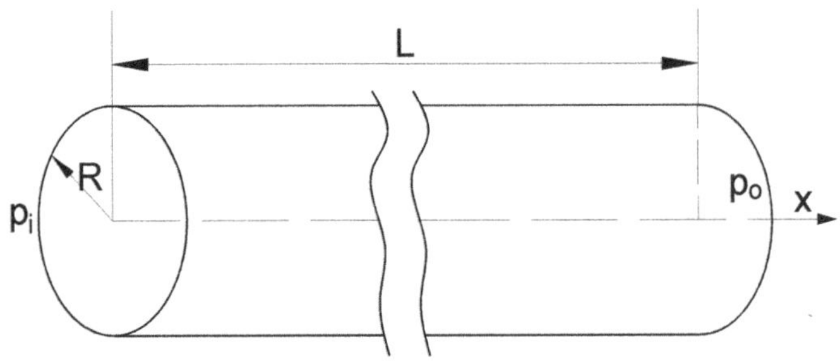

Figure 1. Layout of the pipe considered in the present review.

In Sections 2 and 3 of the present review paper, accelerating laminar flow solutions will be discussed. Laminar flow means that the orderly movement of the fluid along parallel paths occurs in which the fluid elements do not mix with each other, and there is, therefore, a purely viscous mechanism of momentum and energy exchange. Section 4 discusses a turbulent semi-analytical solution. Turbulent flow is understood as a chaotic motion of a fluid. During this flow the elements of the fluid mix with each other, which leads to the intensification of mass, momentum and energy exchange.

2. Accelerated Laminar Flows Driven by Pressure Gradient

2.1. Rapid Instantaneous Increment of Pressure Gradient (Roiti-Gromeka-Szymański Solution)

As mentioned in the introduction, the accelerated flow of liquid in pipes (Figure 1) has been the subject of many studies. All the authors who successfully derived the final solution started from the dynamic equation of motion first derived by Navier [1]:

$$\frac{\partial v}{\partial t} = -\frac{1}{\rho}\frac{\partial p}{\partial x} + \nu\left(\frac{\partial^2 v}{\partial r^2} + \frac{1}{r}\frac{\partial v}{\partial r}\right) \rightarrow \frac{\partial v}{\partial t} = G + \frac{\nu}{r}\frac{\partial}{\partial r}\left(r\frac{\partial v}{\partial r}\right) \tag{1}$$

They assumed the incompressibility of the liquid and that the velocity field is only dependent on the pipe radius and time $v = v(r,t)$. This means that the problem is simplified to isothermal flow in long, constant inner diameter pipelines, where the entrance effects can be neglected. Other boundary conditions are:

(a) flow starts initially from rest:

$$v(r,0) = 0, \tag{2}$$

(b) no-slip condition as viscous fluid in contact with a rigid wall will adhere to the wall due to the effects of viscosity [43]. In the analyzed problem, the solid pipe wall boundary velocity is assumed to be equal to zero:

$$v(R, t) = 0, \tag{3}$$

(c) final velocity profile consistent with the parabolic Hagen–Poiseuille equation:

$$v(r, \infty) = v_{max}\left(1 - \left(\frac{r}{R}\right)^2\right), \tag{4}$$

where: $v_{max} = -\frac{R^2}{4\mu}\frac{\partial p}{\partial x} = \frac{\Delta p}{L}\frac{R^2}{4\mu} = G\frac{R^2}{4\nu}$; $\Delta p = p_o - p_i$, p_o—outlet pressure, p_i—inlet pressure. Maximal velocity occurs at the pipe axis ($r = 0$).

(d) sudden imposition of a pressure gradient (Figure 2a):

$$\begin{cases} G = 0 \; for \; t < 0 \\ G = G_\infty = -\frac{1}{\rho}\frac{\Delta p}{L} \; for \; t \geq 0 \end{cases} \tag{5}$$

The literature review in the Introduction indicates that the final solution was found a relatively long time ago by at least three authors. Roiti [5] studied theoretically and experimentally the unsteady flow of liquid inside a vertically fixed cylindrical pipe, with its gravitational outflow to the atmosphere. He carried out the integration of general formulas using the same method by which his senior colleague Prof. Betti [6] determined the temperature distribution acting on the cylinder. Roiti was the only researcher who, in addition to formulas for the flow velocity, also derived formulas for the displacement of fluid elements during this movement.

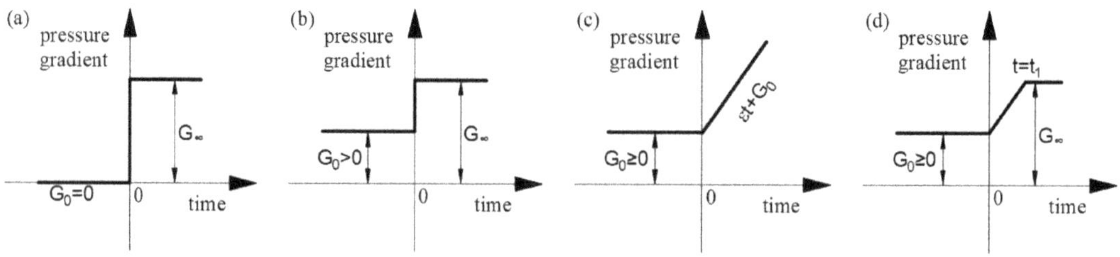

Figure 2. Changes of pressure gradients: (**a**) step change from rest, (**b**) step change from initial to final value, (**c**) linear increase (**d**) ramp change.

Another researcher dealing with this phenomenon, albeit excluding gravitational forces (flow in the horizontal duct), was Gromeka [7]. The work in which he derived the final solution for this type of flow was his doctoral thesis. Gromeka also based his derivation on the analogy existing between some problems related to the movement of viscous fluids and those known from the theory of heat conduction. He carefully used the method of solving heat distributions in circular cylinders proposed by Poisson in his dissertations on "Mémoire sur la distribution de la chaleur dans les corps solides" [44]. In addition to deriving the final formula for the velocity profile (for step gradient change Figure 2a), Gromeka proved that the flow which is forced by a pulsating pressure gradient also for large times tends towards the Hagen–Poiseuille flow. Gromeka also derived formulas for accelerated flow with possible slippage of liquid elements on the walls of the pipe. Both Gromeka's and Roiti's solutions remained unnoticed by the world of science for a long time, likely due to a combination of being published in scientific journals by the researchers' home universities and the slow movement of information at this time. Over a hundred years ago, there was no internet, and materials of this type were not widely available (in contrast to today's open-access papers). Interestingly, Roiti's solution was not quoted in the work of Allievi [45] or that of other Italian (and non-Italian) researchers of unsteady flows, e.g., Fasso [46], Aresti [47].

Fifty years after the publication of Gromeka's thesis, and sixty after the publication of Roiti's work, this interesting topic was addressed by a young Polish researcher, Piotr Szymański, who had just defended his doctoral thesis. Szymański, unaware of the work of his predecessors, during a one-year training course in Paris in 1928, applied the theory of the Fourier–Bessel series to find an analytical solution to this problem. In addition, after deriving the final formula, he conducted a number of mathematical studies on the continuity of its derivatives (using Abel transformations). It is worth mentioning that Szymański in his work [13] analyzed, similarly to Gromeka, the pulsating nature of the flow, and this took place long before Womersley or Uchida's work. In the case of pulsating flow, Szymański analyzed the derivatives of the presented general solution of this problem with the help of the Fourier series theory as well as on the basis of M. Lebesque's theorem. Like Gromeka, however, he did not define the final solution to this problem, which is known today. It is also worth adding here that Szymański's main work was preceded by the publication of a conference paper of the 3rd International Congress of Applied Mechanics held in Stockholm in 1930 [48]. Among the distinguished participants of this congress was Theodore von Kármán.

Due to the fact that the analyzed solution, as can be seen from the above discussion, was derived independently by three researchers, it is suggested to define this solution with the abbreviation RGS derived from the first letters of the surnames of the authors. Leaving aside the tedious details of the derivations, the final solution of the system of Equation (1), assuming a step change in pressure, is the following function:

$$v(r,t)_{RGS} = v_{max}\left[\left[1-\left(\frac{r}{R}\right)^2\right] - 8\sum_{n=1}^{\infty}\frac{J_0\left(\lambda_n\frac{r}{R}\right)}{\lambda_n^3\cdot J_1(\lambda_n)}e^{-\lambda_n^2\frac{\nu}{R^2}t}\right], \tag{6}$$

where λ_n are the nth zero, or root, of a Bessel function $J_0(\lambda_n)$.

An exemplary flow that develops according to the RGS solution described by Equation (6) is presented in Figure 3.

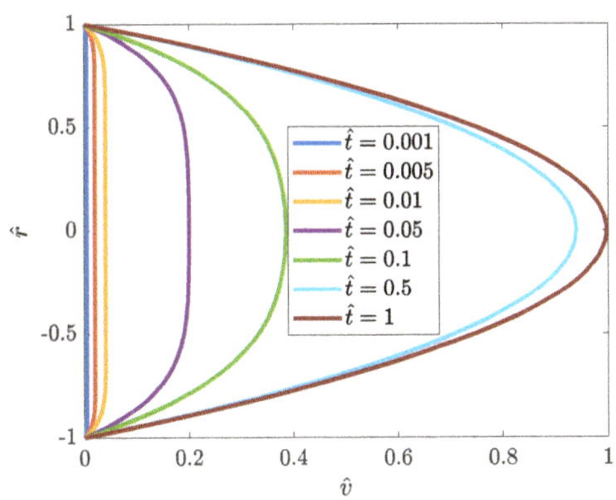

Figure 3. Development of velocity profile in RGS-accelerated flow driven by a step change of pressure gradient.

Gerbes in the 1950s [17] developed the same accelerated solution with help of the Laplace technique; in his work, he also presented a solution for decelerated flow as well as a solution for pulsating flow. The generalization of Equation (6) for the vertical (upward and downward) as well as arbitrary sloping pipes was presented recently by

Urbanowicz et al. [49]. The maximal velocity for this case depends on the pipe slope angle β as follows (Figure 4):

$$v_{max} = \frac{R^2}{4\mu L}(\Delta p - g\rho L \sin\beta). \tag{7}$$

From Equation (7), it follows that in the case of upward vertical flow where $\beta = 90°$: $v_{max} = \frac{R^2}{4\mu L}(\Delta p - g\rho L)$, while in the case of downward vertical flow where $\beta = -90°$, the maximal final velocity has the largest possible value equal to $v_{max} = \frac{R^2}{4\mu L}(\Delta p + g\rho L \sin\beta)$.

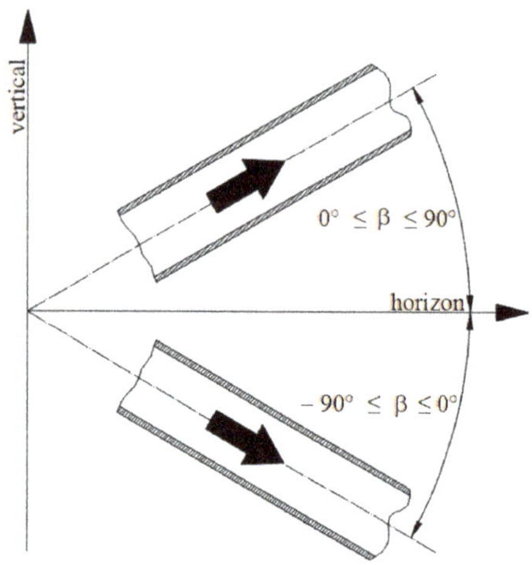

Figure 4. Definition of pipe slope angle (adapted from [49]).

Knowing the solution for velocity distribution at the pipe cross–section (Equation (6)) enables us to calculate the following important parameters describing unsteady accelerated pipe flow:

(a) average flow velocity: $v_m = \frac{2}{R^2}\int_0^R rv(r,t)dr$
(b) wall shear stress: $\tau_w = -\mu \frac{\partial v(r,t)}{\partial r}\Big|_{r=R}$
(c) friction factor: $f = \frac{8\tau_w}{\rho v_m^2}$

The final form for the RGS-type accelerated flow is [49]:

$$\begin{cases} v_{m,\,RGS} = v_\infty\left(1 - 32\sum\limits_{n=1}^{\infty}\frac{e^{-\lambda_n^2\hat{t}}}{\lambda_n^4}\right) \\ \tau_{w,RGS} = \frac{4\mu v_\infty}{R}\left(1 - 4\sum\limits_{n=1}^{\infty}\frac{e^{-\lambda_n^2\hat{t}}}{\lambda_n^2}\right) \\ f_{RGS} = \frac{64}{Re_\infty}\left[1 - 4\sum\limits_{n=1}^{\infty}\frac{e^{-\lambda_n^2\hat{t}}}{\lambda_n^2}\right]\cdot\left[1 - 32\sum\limits_{n=1}^{\infty}\frac{e^{-\lambda_n^2\hat{t}}}{\lambda_n^4}\right]^{-2} \end{cases} \tag{8}$$

Ito [18] generalized the above solution for the case where the initial velocity of the fluid is not zero, which means a steady Hagen–Poiseuille flow is a starting point, and next an instantaneous pressure gradient change takes place from an initial value equal G_0 to a new final value G_∞ (Figure 2b). In this case, the final solution is steady-state Hagen–Poiseuille flow but represented with a higher maximal velocity (in the axial position). The analytical solution obtained for this extended case is:

$$v(r,t)_{RGSI} = \frac{G_\infty(R^2 - r^2)}{4\nu} - \frac{2R^2(G_\infty - G_0)}{\nu}\sum_{n=1}^{\infty}\frac{J_0\left(\lambda_n\frac{r}{R}\right)}{\lambda_n^3\cdot J_1(\lambda_n)}e^{-\lambda_n^2\frac{\nu}{R^2}t} \tag{9}$$

The RGS analytical solution was initially verified experimentally by Letelier and Leutheusser [50] and in more detail by van de Sande et al. [38], Baibikov et al. [8] and Lefebvre and White [39]. In all the mentioned experimental works, good agreement between the RGS solution and the experimental results was indicated. As it is known from the standard textbooks on steady pipe flow, the critical Reynolds number is about 2320 [14]. Below this value laminar flow takes place. For Reynolds numbers of up to 3000, transitional flow occurs, and above that value, turbulent flow can be assumed. Lefebvre and White [39] noticed that the laminar flow persisted in their experimental runs up to very high values of Reynolds numbers in the range between $2{\cdot}10^5$ and $5{\cdot}10^5$. In another study by van de Sande et al. [38], a lower value (Re = 57,500) was found. In both cases, however, it can be seen that the values of the critical Reynolds number exceeded the critical one in steady-state flow. From a paper that is a continuation of the Lefebvre and White study [51], a transitional Reynolds number formula has been proposed $Re_t \approx 450D\left(a/\nu^2\right)^{1/3}$, in which a is acceleration, which was constant during experiments. A different formula based on the Knisely et al. experiments [52] is $Re_t \approx 1.33D\left[\left(a/\nu^2\right)^{1/3}\right]^{1.86}$.

2.2. Other Solutions Driven by a Gradient Change

Ito [18], in addition to extending the RGS solution of Equation (9), has developed the solution for the case when the pressure gradient given to a pipe begins to increase or decrease in proportion to the time (gradient change scenario from Figure 2c):

$$\begin{cases} t < 0 \; then \; -\frac{1}{\rho}\frac{\Delta p}{L} = G_0 \\ t \geq 0 \; then \; -\frac{1}{\rho}\frac{\Delta p}{L} = \varepsilon t + G_0 \end{cases}. \quad (10)$$

The final analytical solution in the Ito (I) case is:

$$v_I = \frac{(\varepsilon t + G_0)\left(R^2 - r^2\right)}{4\nu} - \frac{\varepsilon\left(R^2 - r^2\right)\left(3R^2 - r^2\right)}{64\nu^2} + \frac{2R^4\varepsilon}{\nu^2}\sum_{n=1}^{\infty}\frac{J_0\left(\lambda_n \frac{r}{R}\right)}{\lambda_n{}^5 J_1(\alpha_n)}e^{-\lambda_n{}^2\frac{\nu}{R^2}t}. \quad (11)$$

There is a problem with the above equation regarding how the value of ε function shall be calculated. The analysis of this equation shows that the lower the value of the ε coefficient (representing jerk as its unit is $[\mathrm{m/s^3}]$), the smaller the angle of inclination of the increasing gradient time curve, which will physically mean a larger time required to obtain the selected reference average value of the flow velocity.

The review of analytical solutions in this group revealed that among all derived analytical solutions, there is not one in which the ramp pressure gradient change is gradual, as shown in Figure 2d. Such a solution seems desirable because in practice there are no technical possibilities to change the pressure gradient in an instant way (for example, the valve opening time in the work of van de Sande et al. was about 0.1 s). In real systems, the change of pressure gradient will be strongly related to the valve opening time.

Avula [53] and Avula and Young [54] indicated that changes in velocity during accelerating flows in real systems occur differently than described by the above classical RGS theory. During their research, they experimentally recorded the pressure gradient histories (for selected values of Reynolds numbers—see Figure 5), then they described their mathematical form in an approximate way:

$$f(t_A) = \frac{1}{2}\frac{dp^*}{d\hat{x}} = a_1\left(1 - e^{-a_2 t_A}\right) + a_3 t_A{}^{a_4} e^{-a_5 t_A}. \quad (12)$$

where $a_1, \ldots, a_5$ are constants calibrated with the reference to experimental results and $\hat{x}$, p^* and t_A are the normalized axial coordinate, pressure and time, respectively.

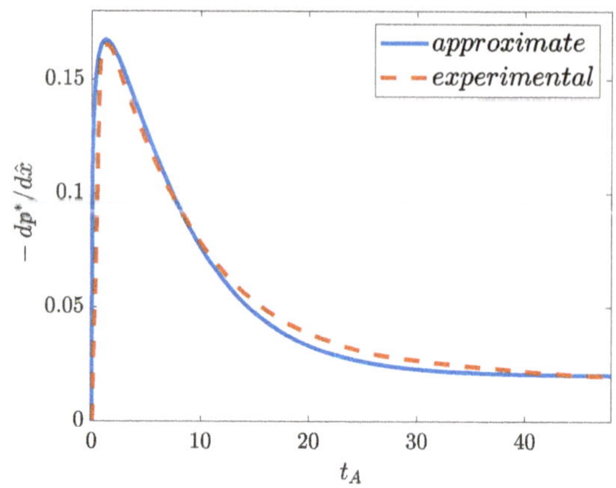

Figure 5. Exemplary variation of pressure gradient with time $R_e \approx 1600$.

The final modified semi-analytical Avula's solution for the velocity profile is:

$$v^* = 2\sum_{n=1}^{\infty} \frac{J_0\left(\eta_n \frac{r}{R}\right)}{\eta_n^3 \cdot J_1(\eta_n)} \left\{ \frac{a_1 Re}{2\eta_n^2}\left(1 - exp\left(-2\frac{\eta_n^2}{Re} t_A\right)\right) + \frac{a_1 Re}{a_2 Re - 2\eta_n^2}\left(e^{-a_2 t_A} - e^{-2\frac{\eta_n^2}{Re} t_A}\right) + a_3 e^{-2\frac{\eta_n^2}{Re} t_A} \int_0^{t_A} u^{a_4} e^{\left(-a_5 u + 2\frac{\eta_n^2}{Re} u\right)} du \right\}, \tag{13}$$

where dimensionless time $t_A = t\frac{v_{m,\infty}}{R}$, $v_{m,\infty} = -\frac{R^2}{8\mu}\frac{\partial p}{\partial x} = \frac{\Delta p}{L}\frac{R^2}{8\mu}$ is the final steady-state mean velocity and η_n is a consecutive zero of a Bessel function of zero order $J_0\left(\sqrt{sRe/2}\right) = 0$.

As the third right-hand side term of this solution (Equation (13)) cannot be evaluated explicitly, this solution needed to be evaluated partially numerically. This numerical solution significantly limits its practical application. A different mathematical form of the function $f(t_A)$ to describe changes in the pressure gradient over time should be found. Such a new function should be integrable to give a complete analytical solution for a wide range of final Reynolds numbers. Illustrative comparisons of the results obtained with the classical RGS model and Avula's solution are shown in Figures 6 and 7.

Figure 6. Velocity profiles comparison of RGS and Avula solutions for relatively: (**a**) small times; (**b**) large times (adapted from [53]).

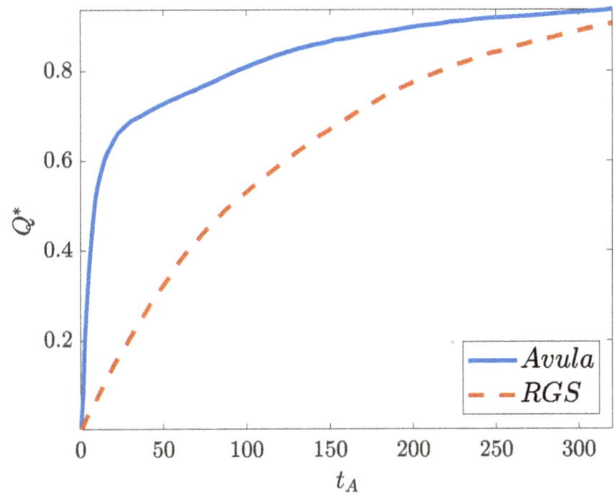

Figure 7. Time variation of discharge.

It can be seen from the comparisons in Figures 6 and 7 that a large discrepancy in the profile dynamics is noticeable for small values of the dimensionless time $t_A < 10$, while for larger values $t_A > 200$, the Avula solution begins to catch up with the classical RGS solution.

A similar solution to the above was analyzed by Smith [55], who in his derivation referred to some of the solutions discussed in the classical textbook "Hydrodynamics" by Dryden et al. [56]. Smith's solution was obtained for the following pressure gradient:

$$\frac{\partial p}{\partial x} = b\mu k^2 e^{-\nu k^2 t}, \tag{14}$$

where k and b are positive constants and μ and ν are the dynamic and kinematic viscosities, respectively. The final Smith's (SM) solution is as follows:

$$v(r,t)_{SM} = b\left\{1 - \frac{J_0(kr)}{J_0(kR)}\right\}e^{-\nu k^2 t} + 2bk^2 \sum_{n=1}^{\infty} \frac{J_0\left(\lambda_n \frac{r}{R}\right)e^{-\lambda_n^2 \frac{\nu}{R^2} t}}{\lambda_n\left(\frac{\lambda_n^2}{R^2} - k^2\right)J_1(\lambda_n)}. \tag{15}$$

When deriving the above solution, it was assumed that the pressure gradient appears at time $t = 0$ and then gradually decays exponentially to zero (and not to a constant value). Such a flow cannot be treated strictly as accelerated as two different periods occur. Firstly, liquid accelerates until the fluid reaches the maximal mean velocity, while after that it starts to decelerate until it comes back to rest again. That is why it is not a subject of the present review. However, this analytical solution is presented and discussed shortly, mainly because it has an interesting feature when the forced decay rate is equal to one of the natural decay rates. It is analogous to the feature experienced in the classical problem regarding oscillations of a linear pendulum in the case when the forcing frequency is equal to the natural frequency (there are two distinct singularities in this final solution).

Other problems with the accelerated pipe flow solution of the RGS type have been reported by Otis [35]. This author noticed that, usually, the pressure gradient does not remain constant but rather it is the total head H_i, which is schematically presented in Figure 8.

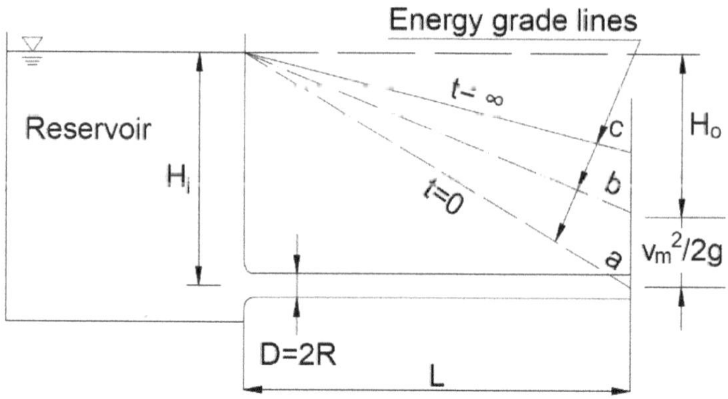

Figure 8. Otis reference drawing indicating time variation of energy grade line (adapted from [35]).

During a flow acceleration, some portion of this head is utilized to establish the kinetic head of the flow. The above is illustrated in Figure 8 in which there are three energy grade lines: (a) just after the instantaneous valve opening, (b) at a later time and (c) after reaching steady-state flow (Hagen–Poiseuille steady state). During flow development, the pressure gradient along the pipe is diminished because of the observed velocity head development. With the above in mind, the main Navier equation becomes non-linear and it is necessary to use numerical methods to model it. Otis [35] presented experimental results (of accelerated flow) of an unnamed researcher and a realistic numerical example of a start-up for SAE 10W-30 oil driven by a head of 1.053 m in a 10-cm diameter pipe with a length of 6.25 m. The main conclusions were that the final steady-flow will occur in less than half the time predicted by the RGS solution. In the same paper, Otis developed a start-up parameter in the following form:

$$M \equiv \frac{G_0 D^4}{2048 \nu^2 \rho L}, \tag{16}$$

where $G_0 = \gamma H_i / L$. Otis concluded that analysis of this parameter revealed that when $M < 1$, the wall shear stress rises monotonically with time, while for $M > 1$, the shear stress overshoots the steady-state value. In addition, Otis stated: "Such behavior resulting from the fact that the time constant for flow start-up decreases with M, whereas the time constant for boundary layer growth is independent of M." Otis's correction was verified numerically by Singh [57], who derived a corrected equation that he did not solve analytically but led to a numerical form suitable for the creation of a computer program and succeed numerical calculations. Patience and Mehrota [58] wrote the Otis start-up parameter in a much simpler form:

$$M \equiv \frac{Re \cdot D}{64L} \tag{17}$$

They also noted that both Szymański [13] and Otis [35] overlooked the hydrodynamic developing region effect in their works. Additionally, Patience and Mehrota proved that in the analyzed comparison start-up cases by Otis, the flow was not fully developed. Their conclusion was based with help of the Fargie and Martin ξ coefficient [59]. Andersson and Kristoffersen also dealt with this problem. In [34], they demonstrated that the start-up flow Otis parameter M defined by Equation (16) is the crucial parameter also when the entrance effects are taken into account. For long pipes ($M = 0$), the Anderson and Kristoffersen numerical solution differs by less than 1% from the exact analytic solution obtained by RGS. For shorter pipes, these authors observed that the start-up period and the resulting steady-state flow rate are significantly reduced due to the entrance region effects. When $M = 0.5$ the start-up time and ultimate flow rate are reduced by about 55% and 40%, respectively, compared with the classical solution of RGS derived for long pipes. In view of the above comments and discussed works, Patience and Mehrota [60] introduced a correction of the method proposed by Otis.

2.3. Universal Solution for Arbitrary Pressure Gradient

The analysis of a number of scientific papers for the purposes of this review revealed that there is a universal solution that can be used to determine the velocity profile for any change in the pressure gradient. Information about this solution will be the topic of this subsection.

Fan in his Ph.D. thesis [61] derived analytical solutions for: (a) impulsive pressure gradient (here the pressure gradient is a function of Dirac delta function); (b) flow under a constant pressure gradient (analogous to RGS solution); (c) flow under a harmonically oscillating pressure gradient; (d) a general solution by superposition integrals and (e) a general solution by the transformation of the governing equation of motion to a homogeneous equation. The latter helped Fan to derive a final analytical solution for an accelerated flow in a pipe with a rectangular cross-section, which will not be discussed due to limited practical applications. The most important solution according to this subsection title is the general solution defined by superposition integral. Since the considered problem is linear, Fan noticed that the solution for any arbitrary $f(t)$ can be obtained by one of the following superposition integrals:

$$v(r,t) = -\int_0^t f(u)\overline{v}_1(r,t-u)du = -\int_0^t f(t-u)\overline{v}_1(r,u)du, \tag{18}$$

where: $\frac{1}{\rho}\frac{\partial p}{\partial x} = f(t)$ and $\overline{v}_1(\hat{r},\hat{t}) = 2\sum_{n=1}^{\infty}\frac{J_0(\hat{r}\lambda_n)}{\lambda_n J_1(\lambda_n)}e^{-\lambda_n^2\hat{t}}$ is the solution that Fan obtained for impulsive pressure gradient, or:

$$v(r,t) = -f(0)\overline{v}_2(r,t) - \int_0^t f'^{(u)}\overline{v}_2(r,t-u)du = -\int_0^t f(u)\frac{\partial\overline{v}_2(r,t-u)}{\partial t}du, \tag{19}$$

where $\overline{v}_2(r,t)$ is the RGS solution $\overline{v}_2(r,t) = \frac{1}{4v}\left[\left[R^2 - r^2\right] - 8R^2\sum_{n=1}^{\infty}\frac{J_0\left(\lambda_n\frac{r}{R}\right)}{\lambda_n^3\cdot J_1(\lambda_n)}e^{-\lambda_n^2\frac{v}{R^2}t}\right]$.

The First Fan integral solution Equation (18) is of simple mathematical structure. The second solution (known as Duhamel's integrals) is based on the more demanding RGS solution Equation (19). Both superposition solutions described by Equations (18) and (19) were derived for the case with the initial flow at rest (mean velocity equal to zero). A solution that takes into account the initial flow was derived by Daneshyar [62]. Daneshyar gives a solution for an arbitrary $f(t)$, using the theory of integral transforms to solve the main Equation (1). He proved after Sneddon [63] that if ξ_n is chosen to be a root of:

$$J_0(R\xi_n) = 0 \tag{20}$$

then

$$\int_0^R r\left(\frac{\partial^2 v}{\partial r^2} + \frac{1}{r}\frac{\partial v}{\partial r}\right)J_0(r\xi_n)dr = -\xi_n^2\overline{v} \tag{21}$$

where

$$\overline{v} = \int_0^R rv(r,t)J_0(r\xi_n)dr \tag{22}$$

is the finite Hankel transform of v. When both sides of Equation (1) are multiplied by $rJ_0(r\xi_n)$ and integrated (within the limits 0 to R), the following ordinary differential equation is obtained for $\overline{v}$:

$$\frac{d\overline{v}}{dt} + \nu\xi_n^2\overline{v} = f(t)\int_0^R rJ_0(r\xi_n)dr = f(t)\frac{R}{\xi_n}J_1(R\xi_n) \tag{23}$$

The solution that satisfies the initial condition $\overline{v}(\xi_n, 0) = \overline{v}_0$ is:

$$\overline{v} = \int_0^t f(u) \frac{R}{\xi_n} J_1(R\xi_n) exp\left[-v\xi_n^2(t-u)\right] du + \overline{v}_0 \tag{24}$$

And, next, with the inversion theorem of finite Hankel transforms, a general solution is found to be:

$$v = v_0 + \frac{2}{R^2} \sum_{n=1}^{\infty} \frac{J_0(r\xi_n)}{[J_1(R\xi_n)]^2} \overline{v} = v_0 + \frac{2}{R} \sum_{n=1}^{\infty} \frac{J_0(r\xi_n)}{\xi_n J_1(R\xi_n)} \int_0^t f(u) exp\left[-v\xi_n^2(t-u)\right] du \tag{25}$$

that is valid for an arbitrary pressure gradient (hidden under $f(t)$ function).

It can be now noticed that substituting for $R\xi_n = \lambda_n$ one gets $\xi_n = \lambda_n / R$, which, when inserted into the above formula, will reduce this solution to the following form:

$$v = v_0 + 2\sum_{n=1}^{\infty} \frac{J_0\left(\frac{r}{R}\lambda_n\right)}{\lambda_n J_1(\lambda_n)} \int_0^t f(u) e^{-\lambda_n^2(t-u)} du, \tag{26}$$

In the special case of the above solution, v_0 can be treated as an initiation laminar flow profile of steady Hagen–Poiseuille type flow.

A careful literature review reveals that universal solutions of this type were derived many times. Chronologically, the earliest form of this solution was presented by Vogepohl in a short note published in 1933 [30]. Then in 1956, Roller, in his master's thesis [64], noticed (similarly to Daneshyar about 15 years later) that the solution of the Navier momentum equation can be obtained with the use of the analogy to the temperature distribution in a cylindrical rod (defining the analogous boundary and initial condition) [63]. Zielke in his Ph.D. thesis [65] rederived Fan's solution in the form of Equation (19) using Laplace transforms. Hersey and Song by using Laplace transforms [66], and Avula with the help of the Cauchy residue and convolution integral theorem [53], derived the same solution but based on different normalized quantities. These solutions were again derived and used with the help of the eigenfunction method in the work of Xiu et al. [67], where the starting flow was analyzed, and in the Sun and Wang paper [68], where the water hammer case (starting from steady Hagen–Poiseuille flow) was examined:

$$\hat{v}(\hat{r}, \hat{x}, \hat{t}) = \sum_{n=1}^{\infty} \frac{16 J_0(\hat{r}\lambda_n)}{\lambda_n^3 J_1(\lambda_n)} e^{-\lambda_n^2 \hat{t}} + \sum_{n=1}^{\infty} \int_0^{\hat{t}} \frac{2 J_0(\hat{r}\lambda_n)}{\lambda_n J_1(\lambda_n)} \left[-\frac{\partial \hat{H}(\hat{x}, u)}{\partial \hat{x}}\right] e^{-\lambda_n^2(\hat{t}-u)} du \tag{27}$$

where: $\hat{x} = x/L$; $\hat{r} = r/R$; $\hat{v} = v/v_0$ and $\hat{H} = \frac{\Delta p R^2}{L\mu v_0}$.

In the above equation, an interesting identity is used (for initial time $\hat{t} = 0$), that can be found in many papers among them the ones discussed in this review i.e., Szymański [13], Gerbes [17], Xiu et al. [67], etc:

$$\sum_{n=1}^{\infty} \frac{8 J_0(\lambda_n \hat{r})}{\lambda_n^3 J_1(\lambda_n)} = (1 - \hat{r}) \text{ for } 0 \le \hat{r} \le 1 \tag{28}$$

The most recent study in which this universal equation was re-derived and examined is Lee's paper published in 2017 in Applied Mathematical Modelling [69]. The motivation of this section is to systematize the derivation of the universal formula discussed above. The list of references is presented in Table 1.

Table 1. A list of works in which a universal formula was derived.

No	Author/s	Year	Reference
1	Vogelpohl	1933	[30]
2	Roller	1956	[64]
3	Fan	1964	[61]
4	Song	1966, 1967	[66,70]
5	Zielke	1966	[65]
6	Avula	1968, 1969	[53,71]
7	Daneshyar	1970	[62]
8	Xiu et al.	1995	[67]
9	Sun and Wang	1995	[68]
10	Lee	2017	[69]

2.4. Comments on the Pressure Gradient Driven Flows

As noticed, the formulas discussed above concerned pipes with circular cross-sections. Readers interested in other solutions, e.g., similar accelerated flow in ducts with different geometry, are referred to other works: the solution to this problem for ducts with a rectangular cross-section was derived in the works of Fan and Erdogan [21,61,72]; start-up flow in an annulus was developed by Müller [73]; for other cross-sections, an intensive study was carried out by Laura [22]; solution for pipes, taking into account the slip of the fluid on the wall [74]; start-up flow in a circular porous pipe [75]; development of unsteady flow at the entrance length of a circular tube starting from rest [19,20,54,76]; the effects of time-dependent viscosity [77–79]; unsteady laminar flow in tubes with a tapered wall thickness [80], etc.

Due to the need to use the zeros of the Bessel function, the presented solutions are a challenge, because it is usually necessary to write a short program in software such as Matlab or Wolfram Mathematica, hence the approximation forms of these formulas discussed by Muzychka-Yovanovich [23,24] and Urbanowicz et al. [4] are also an interesting proposition. It is also worth adding that governing equations of motion, are respectively analogous to heat conduction in a long cylinder with constant thermal conductivity [81]. So all the results presented in this paper for laminar flows are directly transformable to the solution of heat conduction in long cylinders with internal heat generation (simply replacing pressure gradient G by heat generation source term $\frac{\dot{g}}{c_p \cdot \rho}$; kinematic viscosity ν by thermal diffusivity α and velocity field $v(r,t)$ by temperature field $T(r,t)$).

Moss developed [82] a dimensionless flow acceleration parameter that takes a zero value in the case of an impulsively begun flow and non-zero values for exponentially increasing flows. When this parameter is increased beyond a critical value (7.059), the boundary layer never merges. The method developed by Moss is suitable for the application of any stability analyses of unsteady flows, which is very useful to study the physical insights of different flow fields. Pozzi and Tognaccini [83] have extended the analysis of the present accelerating problem in pipes for the thermo-fluid dynamic field arising in an infinite pipe with a circular section when the incompressible fluid is impulsively started from rest by a sudden jump to a constant value of the axial pressure gradient. These authors derived analytical solutions for the temperature field taking additionally into account the dissipation of kinetic energy (Eckert number different from zero) in the relevant case of Prandtl number equal to one. The final solution has been obtained and discussed for four cases depending on the condition imposed on the wall: constant temperature, adiabatic wall, assigned heat loss, and assigned constant flux.

To sum up, the analysis of the presented analytical solutions of accelerated flows in pressure lines forced by a change in pressure gradient shows that:

- the derivation of the analytical formula is missing for the initially linear change of the pressure gradient with subsequent stabilization on a constant value (ramp jump of pressure gradient) (Figure 2d);
- selected experimental studies confirmed the effectiveness of the RGS solution;
- it seems that Avula's solution has great practical potential because the real pressure gradient may have a course similar to that recorded experimentally by Avula. However, further research and work are needed to simplify this solution, firstly by selecting a function representing the gradient that can be integrated (this will make it possible to omit numerical solutions) and secondly, it is necessary to simplify the dimensionless description so that it is not based on the Reynolds number and complex zeros from zero-order Bessel functions with highly complicated arguments;
- there is a certain universal analytical solution that allows the determination of the formula for the flow velocity profile for any function describing the pressure gradient. This solution has been rediscovered many times over the years.

3. Accelerated Laminar Flows Driven by Sudden Imposition of Flow Rate

3.1. Rapid Instantaneous Increment of Flow Rate

In the second group of models, the equation from which the final solution is derived is the very same equation Equation (1) that was discussed in the previous section. The pipe is also treated as long in this case, so that the influence of the formation of the flow profile in the entrance section (additional resistance) can be neglected. The fluid is assumed to be incompressible and the pipe is assumed to be horizontal (ignoring gravitational forces). The original authors of this solution are the Norwegians Helge I. Andersson and Knut L. Tiseth [32]. In this model, a sudden imposition of a constant flow rate is assumed. Such flow can be treated as generated by a piston that is suddenly set in motion with a constant speed. Mathematically, this particular kind of flow is subject to the integral constraint:

$$Q \equiv \int_0^R 2\pi r v dr = constant \tag{29}$$

where Q is the flow rate.

During the start-up of this flow, the uniform motion is set initially into the pipe and the time scale for viscous diffusion is finite and of the order $\frac{R^2}{\nu}$, which means that only an infinitesimally thin viscous layer exists at $t = 0$.

The complete boundary and initial conditions are:

(a) uniform distribution of velocity at the cross-section:

$$v(r,0) = \frac{1}{2} v_{max} \tag{30}$$

(b) unsteady motion is characterized by the no-slip condition at the pipe wall Equation (3);

(c) the velocity profile gradually approaches the steady Hagen–Poiseuille flow solution Equation (4);

(d) an arbitrary velocity scale was defined as the maximum steady state velocity, i.e., $v = v_{max}$ for $r = 0$.

The main equations in the Anderson and Tiseth paper [32] were scaled and presented in dimensionless form. The dimensionless pressure gradient solution (in this case variable in time) tends to four when fluid approaches its steady state after acceleration. The dimensionless velocity solution was split into steady and transient deviation terms made up of two separate functions, one being related to the radial position $S(\hat{r})$ and other to time $T(\hat{t})$. Applying a boundary condition and using the integral of the momentum equation helped Anderson and Tiseth to derive a partial differential equation that was next separated into two ordinary differential equations, the solution of which was found in a form of an infinite Bessel function series:

$$v(r,t)_{AT} = 2v_{\infty}\left[\left(1-\left(\frac{r}{R}\right)^{2}\right)+2\sum_{n=1}^{\infty}\frac{J_0\left(\alpha_n\frac{r}{R}\right)-J_0(\alpha_n)}{\alpha_n^2 J_0(\alpha_n)}\cdot exp\left(-\alpha_n^2\hat{t}\right)\right] \quad (31)$$

where α_n is the nth zero of the Bessel function $J_2(\alpha_n)$ and v_{∞} is the final mean velocity of the flow.

Based on the solution of Equation (31), formulas can be derived for the basic parameters of this type of accelerated unsteady flow [4]:

$$\begin{cases} \tau_{w,AT} = \frac{2\mu v_{\infty}}{R}\left[2+\sum_{n=1}^{\infty}e^{-\alpha_n^2\hat{t}}\right] \\ \left(\frac{\partial p}{\partial x}\right)_{AT} = \frac{-4\mu v_{\infty}}{R^2}\left[2+\sum_{n=1}^{\infty}e^{-\alpha_n^2\hat{t}}\right] \\ f_{AT} = \frac{64}{Re_{\infty}}\left[1+\frac{1}{2}\sum_{n=1}^{\infty}e^{-\alpha_n^2\hat{t}}\right] \end{cases} \quad (32)$$

The course of the AT solution over time is shown in Figures 9 and 10. Figure 9 shows the dynamics of the development of the velocity profile, while Figure 10 compares the change in the maximum velocity value (in the axis of the pipe) obtained with the AT and RGS models. The last comparison shows that in the AT solution, the final profile similar to the Hagen–Poiseuille flow is obtained much faster (for $\hat{t} \approx 0.2$) than with the use of the RGS solution.

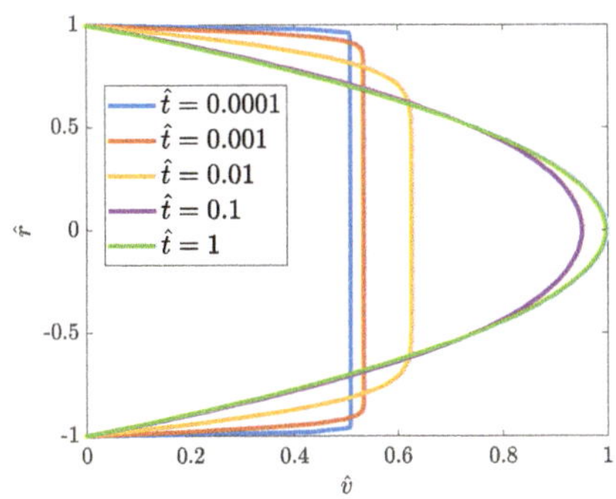

Figure 9. Development of velocity profile in AT accelerated flow driven by an abruptly imposed constant volume flux.

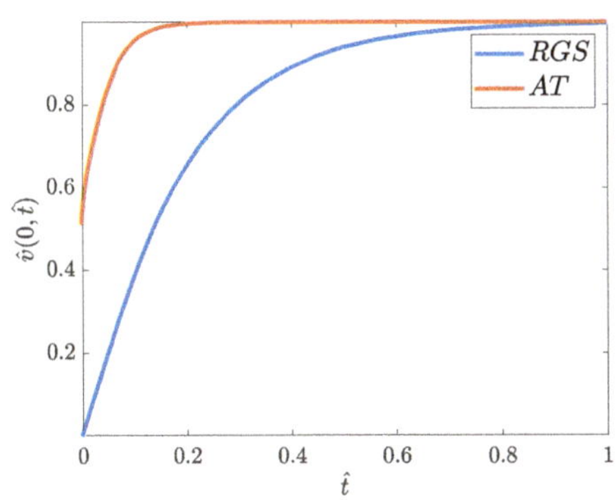

Figure 10. Normalized axial velocity comparison of RGS and AT solutions.

The experimental and numerical validations of this model have been performed by Chaudhury et al. [41]. The conclusion was that the Andersson and Tiseth [32] analytical model is also valid for finite-length tubes at locations beyond the entrance flow development length. Chaudhury et al. additionally write that: "This has been demonstrated by observing the same flow at $x/D = 55$ downstream of the inlet. The developed transient event is insensitive to the position of the piston provided the piston is more than two piston diameters away from the tube entrance. Under these conditions, results apply for constant volume flux start-up flows in physically similar piston pumps. The flow development region is significantly shorter spatially and temporally than in constant pressure gradient-driven flows."

Other experimental runs that are frequently cited were performed by Kataoka et al. [40] Kataoka et al. reported that during their experiment an "annular jet effect" (AJE) was observed. Anderson and Tiseth [32] concluded that this AJE results from the unintentional pressure oscillations induced in the early stage of the start-up period. García García and Alvariño were motivated by these experimental results and this very untypical AJE. They found [84] that in another experimental paper by Maruyama et al. [85], the authors showed the initial stage of an AJE, thus confirming the Kataoka et al. discovery. García García and Alvariño write that: "the origin of the AJE is the result of a partial or local accordion effect, which only involves the flow away from the centreline. During an interval around $\tau \approx 0.02$, the mean velocity is greater in a region midway between the core and the wall. Later, U-profiles, $\tau \gtrsim 0.06$, show a more conventional accordion effect, affecting the complete profile (global deformation). Now, the qualitative difference between early and late turbulence should emerge: the former does not present the accordion effect but AJE, whereas the latter manifests varying degrees of local accordion effects that translate into lone concavities and AJE" and "With slow turbulence, the mid-section experiences a greater increase, even yielding an AJE if the acceleration of first stage is high enough, whereas the core velocity tends to decrease".

Sparrow et al. [76] derived an equation for velocity formulation in a pressure inlet from a reservoir. Interestingly it has the same mathematical form as the AT solution for accelerating pipe flow:

$$v(r,t)_S = 2v_\infty\left[\left(1-\left(\frac{r}{R}\right)^2\right)+\sum_{n=1}^{\infty}\frac{2}{\alpha_n^2}\left\{\frac{J_0\left(\alpha_n\frac{r}{R}\right)}{J_0(\alpha_n)}-1\right\}e^{-\alpha_n^2 X^*}\right] \quad (33)$$

The difference is that in place of AT dimensionless time, the Sparrow et al. analytical solution has the dimensionless distance function X^* calculated as the ratio of a stretched axial coordinate x^* multiplied by the kinematic viscosity divided by the mean velocity multiplied by a square of the pipe radius $X^* = \frac{x^* \nu}{R^2 v_\infty}$.

3.2. Other Flow Rate Solutions and Comments

Das and Arakeri [36] prepared and made experiments in which the motion was generated by a piston. In this case, the pressure is unknown and determined indirectly by the piston motion. Assuming incompressible flow, the piston motion is felt immediately at each cross-section of the pipe and the volume flux at any cross-section corresponds to the volume flux due to piston motion:

$$2\pi\int_0^R rv\,dr = v_p(t)\pi R^2 \quad (34)$$

where v_p is a piston valve. Similarly, as in Anderson and Tiseth and pressure-gradient-driven flows, the main equation is the Navier momentum equation defined by Equation (1). Using the above assumption, the solution obtained for the reverse flow that was firstly accelerated from the rest, then was constant and finally decelerated to rest, as presented in Figure 11, will be investigated below. During the piston acceleration ($0 \le t \le t_0$) the solution is:

$$v_{DA} = \frac{v_p}{t_0}\left[2t\left(1-\left(\frac{r}{R}\right)^2\right) + \underbrace{\frac{R^2}{\nu}\left(\frac{1}{8}\left(1-\left(\frac{r}{R}\right)^4\right) - \frac{1}{6}\left(1-\left(\frac{r}{R}\right)^2\right)\right)}_{T_1}\right] + 2v_p\frac{R^2}{\nu}\sum_{n=1}^{\infty}\frac{J_0(\alpha_n)-J_0\left(\alpha_n\frac{r}{R}\right)}{\alpha_n^3 J_1(\alpha_n)}\cdot\underbrace{\frac{\exp\left(-\alpha_n^2\frac{\nu}{R^2}t\right)}{t_0}}_{T_2} \quad (35)$$

where: v_p—piston velocity during the time t_0 to t_1.

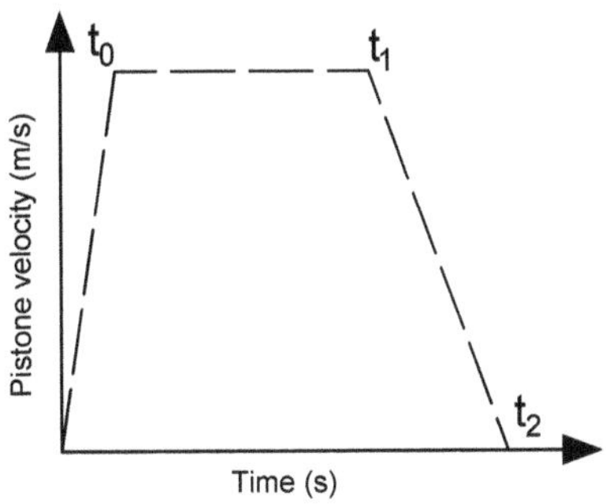

Figure 11. Typical trapezoidal variation of piston velocity (motion) with time.

For a constant piston velocity ($t_0 \le t \le t_1$):

$$v_{DA} = 2v_p\left(1-\left(\frac{r}{R}\right)^2\right) + v_p\frac{2R^2}{\nu}\sum_{n=1}^{\infty}\frac{J_0(\alpha_n)-J_0\left(\alpha_n\frac{r}{R}\right)}{\alpha_n^3 J_1(\alpha_n)}\left(\underbrace{T_2 - \frac{\exp\left(-\alpha_n^2\frac{\nu}{R^2}(t-t_0)\right)}{t_0}}_{T_3}\right) \quad (36)$$

During piston deceleration ($t_1 \le t \le t_2$):

$$v_{DA} = 2v_p\frac{R^2}{\nu}\sum_{n=1}^{\infty}\frac{J_0(\alpha_n)-J_0\left(\alpha_n\frac{r}{R}\right)}{\alpha_n^3 J_1(\alpha_n)}\left(\underbrace{T_3 - \frac{\exp\left(-\alpha_n^2\frac{\nu}{R^2}(t-t_1)\right)}{t_2-t_1}}_{T_4}\right) + 2v_p\frac{t_2-t}{t_2-t_1}\left(1-\left(\frac{r}{R}\right)^2\right) - \frac{R^2}{\nu}\frac{v_p}{t_2-t_1}T_1 \quad (37)$$

and after the piston has stopped ($t_2 \le t \le \infty$):

$$v_{DA} = 2v_p\frac{R^2}{\nu}\sum_{n=1}^{\infty}\frac{J_0(\alpha_n)-J_0\left(\alpha_n\frac{r}{R}\right)}{\alpha_n^3 J_1(\alpha_n)}\left(T_4 + \frac{\exp\left(-\alpha_n^2\frac{\nu}{R^2}(t-t_2)\right)}{(t_2-t_1)}\right) \quad (38)$$

where: α_n—zeros of the second-order Bessel function $J_2(\alpha_n)$, v_p is piston velocity, t_0 is the time of acceleration, t_1 is the time when a stepper motor is switched off and t_2 is the time when piston motion stops.

Solutions for the piston acceleration Equation (35) and constant piston velocity Equation (36) when $t_1 \to \infty$ form an analytical solution for the ramp change of velocity.

Das and Arakeri pointed out after analyzing the flow profiles for the piston deceleration phase that the velocity profiles close to the walls can be in the opposite direction to the core flow. So, this type of unsteady flow does have inflection points and hence can become unstable at relatively low Reynolds numbers.

A comparison of the development of flow profiles in the Anderson and Tiseth solution (continuous lines) and Das and Arakeri ramp change version (the dotted lines obtained with the help of Equations (35) and (36)) is presented in Figure 12. It can be seen that in the final phase from the dimensionless time $\hat{t} = 0.05$ both solutions start to approach each other. The Das and Arakeri solution (DA model) starts from the rest value, not as in the AT model from the non-physical mean velocity $\hat{v}_m = 0.5$.

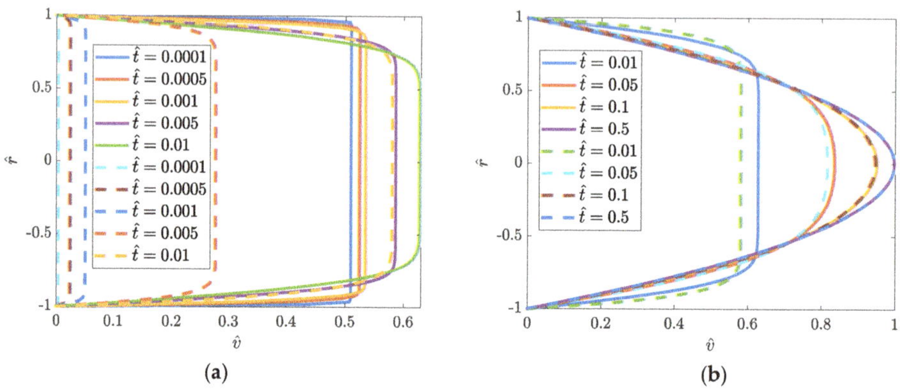

Figure 12. Comparison of Das and Arakeri model (dotted lines; $\hat{t_0} = 0.01$) with Anderson and Tiseth model (solid lines): (**a**) $\hat{t} \leq 0.01$, (**b**) $\hat{t} \geq 0.01$.

Kannaiyan et al. [37] further generalized the AT and DA solutions introducing a condition of double-step changes in the flow rate:

$$Q(t) = v_m(t)A = \begin{cases} 0;\ t \leq 0 \\ Q_i = v_{m,i} \cdot A;\ 0 < t \leq t_c \\ Q_\infty = v_{m,\infty} \cdot A;\ t_c < t \leq \infty \end{cases} \tag{39}$$

where: $v_{m,i}$—initial mean velocity and $v_{m,\infty}$—final mean velocity of flow.

With this assumption, it was possible to derive the solution for acceleration from one steady state (or developing one) to another steady flow. This means that in this solution, compared to the Andersson and Tiseth, and Das and Arakeri, solutions, the flow does not need to start accelerating from rest.

The final solution for the axial velocity profile is:

$$v(r,t)_{KVN} = 2v_{m,\infty}\left[\left(1-\left(\frac{r}{R}\right)^2\right) - \sum_{n=1}^{\infty}\frac{4}{\alpha_n^2}\left[1-\frac{J_0\left(\alpha_n\frac{r}{R}\right)-J_0(\alpha_n)}{J_0(\alpha_n)}\right]\left[v_{m,\infty}+v_{m,i}\left(e^{-\alpha_n^2\frac{\nu}{R^2}t_c}-1\right)\right]e^{-\alpha_n^2\frac{\nu}{R^2}t}\right] \tag{40}$$

The solution for time-dependent pressure gradient in this case takes the following form:

$$\left(\frac{\partial p}{\partial x}\right)_{KVN} = \frac{4\mu v_{m,i}}{R^2}\sum_{n=1}^{\infty}\left[1-e^{-\alpha_n^2\frac{\nu}{R^2}t_c}\right]e^{-\alpha_n^2\frac{\nu}{R^2}t} - \frac{4\mu v_{m,\infty}}{R^2}\left[2+\sum_{n=1}^{\infty}e^{-\alpha_n^2\frac{\nu}{R^2}t}\right] \tag{41}$$

In a recent paper, Kannaiyan et al. [86] analyzed the stability of a laminar pipe flow subjected to a step-like increase in the flow rate.

Summing up the second group of models, in which the flow is forced by a specific change in the flow rate, it can be concluded that:

- the Andersson and Tiseth solution is characterized by a non-physical jump in the flow rate from zero to 0.5. The above results from the fact that the velocity profile changes from a flat plug profile to a parabolic one over time (but maintains constant mean velocity). A similar change takes place at the entrance section from the reservoir to the pipe, as the parabolic profile is here formed along the pipe length in a similar way, which was noticed and described by Sparrow et al. [76];

- the Das and Arrakeri model seems to be correct from a practical point of view; in this model, the velocity profile starts its development from the zero value (Figure 12);
- the solution discussed in this section can be used in practice only in cases where the flow occurs as a result of the piston's motion.

4. Accelerated Turbulent Pipe Flow—TULF Model by García García and Alvariño

In this section, the model that extends from steady to unsteady turbulent pipe flow will be discussed. The model is based on Pai's [87] idea of decomposing the turbulent velocity profile to the sum of the Hagen–Poiseuille parabola and a purely turbulent component. The starting point of García García and Alvariño model (TULF model; TULF is an abbreviation for the theory of underlying laminar flow) is the non-homogeneous dimensionless Reynolds-averaged Navier–Stokes partial differential equation (RANSE) [84,88–90]:

$$\frac{\partial v}{\partial \hat{t}} - \left(\frac{\partial^2 v}{\partial \hat{r}^2} + \frac{1}{\hat{r}}\frac{\partial v}{\partial \hat{r}}\right) = -\frac{\partial \hat{p}}{\partial \hat{x}} - \frac{1}{\hat{r}}\frac{\partial(\hat{r}\sigma)}{\partial \hat{r}} = \Pi(\hat{t}) + \Sigma(\hat{t},\hat{r}) \tag{42}$$

with the following initial and boundary conditions:

$$v(0,\hat{r}) = v_0(\hat{r}),\ v(\hat{t},1) = 0,\ \frac{\partial v(\hat{t},0)}{\partial \hat{r}} = 0 \tag{43}$$

and knowledge of the dimensionless Reynolds-averaged continuity equation:

$$\frac{\partial v}{\partial \hat{x}} = 0 \tag{44}$$

In the above equations $\Pi(\hat{t})$ is the mean (minus) pressure gradient, $\sigma(\hat{t},\hat{r})$ is the Reynolds Shear Stress (RSS) field, satisfying the condition $\sigma(\hat{t},0) = \sigma(\hat{t},1) = 0$ and the function $\Sigma(\hat{t},\hat{r}) = -\frac{1}{\hat{r}}\frac{\partial(\hat{r}\sigma)}{\partial \hat{r}}$ is defined by the authors of this model as the Weighted Reynolds Shear Stress Gradient (WRSSG).

The new semi-analytical solution was obtained by García García and Alvariño [84,88–90] in the following steps:

(a) solving the homogeneous equation (left-hand side of Equation (42)) for the boundary conditions (43). Firstly, it was split up into two ordinary differential equations (separate for variable $\hat{r}$ and $\hat{t}$). Equation with $\hat{r}$ constitutes a classical Sturm–Liouville problem whose solution is the set of normalized eigenfunctions, while the other one (dependent on $\hat{t}$) is a temporal equation. The final general transient velocity field solution takes a form similar to Fan's solution obtained for impulsive pressure gradient (solution $\overline{v}_1(\hat{r},\hat{t})$ mentioned under Equation (18)), the only differences were coefficients determined from initial conditions;

(b) obtaining the solution for the non-homogeneous case with the help of an integrating factor and knowledge of the source functions $\Pi(\hat{t})$ and $\Sigma(\hat{t},\hat{r})$, which were assumed as well-behaved functions. The mean velocity field $v(\hat{t},\hat{r})$ is composed of three terms: v_I—corresponds to transient decay of the mean initial velocity $v_{m,0}$; v_P—defines the unsteady response of the time-dependent mean pressure gradient; v_R—source of velocity field caused by the turbulence's Reynolds shear stress gradient. From [90], it follows that function $\Sigma(\hat{t},\hat{r})$ can be treated as a sum of components of the eigenfunction expansions of $\sigma/\hat{r}$ and $\partial\sigma/\partial\hat{r}$, respectively. In turbulent flows, the analytical form of $\sigma(\hat{t},\hat{r})$ is not known, and must be determined from known reliable experimental data;

(c) using Pai's method [87], the mean velocity field v_I can be composed as a sum of underlying laminar flow $v_{IL}(\hat{t},\hat{r})$ and the pure turbulent component $v_{IT}(\hat{t},\hat{r})$. This allowed the decomposition of the general mean flow equation into the sum of two fields, laminar and turbulent, respectively (see block diagram in Figure 13), as proposed by García García and Alvariño. The temporal evolution of v_L and v_T are different, such a situation generates asynchronism (distortion) of the mean velocity field of unsteady flows resulting in many phenomena noticed earlier only in the experimental research. Many of them have been discussed and explained in the authors publications [84,89,90];

(d) the mean pressure gradient (MPG) was assumed to undergo a linear change until reaching its final value. The MPG change time can be controlled directly by an experimenter (associated with the mean valve-aperture time), and therefore, it can be assumed that it will remain virtually unchanged in all realizations of the flow. The second source term, the RSS field, is strictly connected with turbulence. A linear homotopy transition of this field is assumed. It follows that RSS is modeled similarly to MPG, but the slope can be assumed differently ($\Delta\hat{t}$ versus $\hat{t}_2$ will be discussed later on) Such a simplified ramp approach has one degree of freedom and singularities which can cause visible unrealistic peaks of final solutions, however, as shown in [84,90], the quality of the modeling compliance can be considered satisfactory. In starting flow from rest the RSS must evolve from a zero value to some constant value related to a steady-state final flow. To model the final state, the model of Pai is used [87], which gives acceptable results for moderate Reynolds numbers (it is a limitation of this semi-analytical model). The Pai model helps to define the initial and final flow in simple polynomial form.

The terms ($\Pi_0, \Pi_\infty; \chi_0, \chi_\infty$ and q_0, q_∞) that govern RSS, mean velocity and WRSSG are defined in [84] as Spatial Degrees of Freedom (SDoF) that need to be calculated for the initial and final times. They define the radial dependence of the relevant flow fields. Coefficient χ_0 is the initial centreline turbulent dissipation being the ratio $v_L(\hat{t}=0,\hat{r}=0)/v(\hat{t}=0,\hat{r}=0)$ of underlying laminar flow to the mean velocity at the centerline of the pipe. The turbulence field can be switched off if one assumes $\chi_0 = 1$. The other initial coefficient q_0 is a best-fitting integer power. The final values of the coefficients χ_∞ and q_∞ need to be defined from the experimental velocity profile of the final turbulent flow.

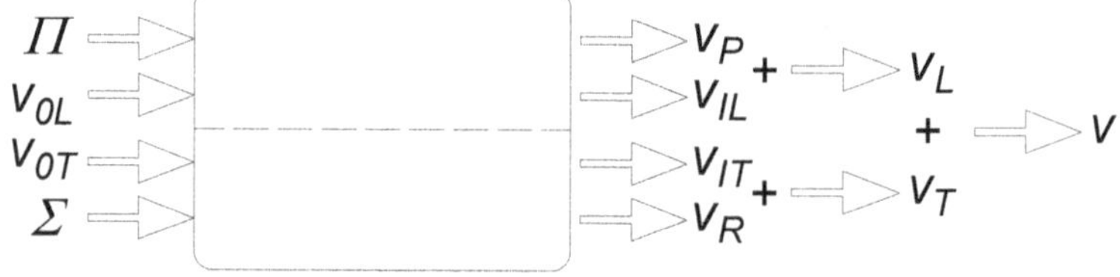

Figure 13. TULF model component-wise block diagram of detailed velocity field decomposition.

If the accelerated flow is started from rest then the initial values of RSS, velocity field and WRSSG are: $\sigma_0(\hat{r}) = 0$; $v_0(\hat{t}=0,\hat{r}) = 0$ and $\Sigma_0(\hat{r}) = 0$, while the final values can be calculated from the following equations [84]:

$$\begin{Bmatrix} \sigma(\hat{t},\hat{r}) \\ \Sigma(\hat{t},\hat{r}) \end{Bmatrix} = \begin{cases} 0 & \text{if } \hat{t} < \hat{t}_0, \\ 2\eta^2\left(\frac{3}{2}-\eta\right)\begin{Bmatrix} \sigma_\infty(\hat{r}) \\ \Sigma_\infty(\hat{r}) \end{Bmatrix} & \text{if } \hat{t}_0 \le \hat{t} \le \hat{t}_2, \\ \begin{Bmatrix} \sigma_\infty(\hat{r}) \\ \Sigma_\infty(\hat{r}) \end{Bmatrix} & \text{if } \hat{t} > \hat{t}_2. \end{cases} \tag{45}$$

where $\eta = \frac{\hat{t}-\hat{t}_0}{\hat{t}_2-\hat{t}_0} \in [0,1]$, $\hat{t}_0$ is the mean time of transition to turbulence; $\hat{t}_2$ is the mean turbulence-settling time. These times are not easily controlled by any experimenter (the values may change from one experiment to the other). They must be defined in the TULF model too, they are external parameters that cannot be determined from analytical theory. The relation between mentioned times $\hat{t}_0$, $\hat{t}_2$ and $\Delta\hat{t}$ defined as Temporal Degrees of Freedom (TDoF) can be a source of division of the flows into four classes [84]:

(I) $\hat{t}_0 < \Delta\hat{t}$—here turbulence on average begins before the increase in MPG is over (early transition to turbulence);
(II) $\hat{t}_0 > \Delta\hat{t}$—the turbulence on average begins after the MPG becomes constant (late transition to turbulence);
(III) $\hat{t}_2 - \hat{t}_0 < \Delta\hat{t}$—the turbulence's increase rate is faster than that of MPG with an offset $\hat{t}_0$ (fast turbulence evolution);
(IV) $\hat{t}_2 - \hat{t}_0 > \Delta\hat{t}$—the turbulence increases slower than the MPG, even after subtracting the offset (slow turbulence evolution).

The above four types of accelerated pipe flows are the subject of intensive research discussed in [84];

(e) the use of formulas discussed in earlier points made it possible to determine the final solution for the terms $v_L(\hat{t},\hat{r})$ and $v_T(\hat{t},\hat{r})$ for the analyzed case of accelerated flow for appropriate ranges of dimensionless time.

The final solution [84] for $v_L(\hat{t},\hat{r})$ is:

(a) for $0 \le \hat{t} \le \Delta\hat{t}$:

$$v_L(\hat{t},\hat{r}) = \frac{12\Pi(\hat{t})}{\Delta\hat{t}^3}\sum_{n=1}^{\infty}\frac{J_0(\lambda_n\hat{r})}{\lambda_n^9 J_1(\lambda_n)}\left[-\frac{\lambda_n^6\hat{t}^3}{3} + \lambda_n^4\left(\frac{\lambda_n^2\Delta\hat{t}}{2}+1\right)\hat{t}^2 + \left(\lambda_n^2\Delta\hat{t}+2\right)\left(1-\lambda_n^2\hat{t}-e^{-\lambda_n^2\hat{t}}\right)\right] \tag{46}$$

(b) while for $\hat{t} > \Delta\hat{t}$:

$$v_L(\hat{t},\hat{r}) = \frac{12\Pi(\hat{t})}{\Delta\hat{t}^3}\sum_{n=1}^{\infty}\frac{J_0(\lambda_n\hat{r})}{\lambda_n^9 J_1(\lambda_n)}\left[\left(2-\lambda_n^2\Delta\hat{t}\right)e^{-\lambda_n^2(\hat{t}-\Delta\hat{t})} - \left(2+\lambda_n^2\Delta\hat{t}\right)e^{-\lambda_n^2\hat{t}} + \frac{\lambda_n^6\Delta\hat{t}^3}{6}\right] \tag{47}$$

And the final solution [84] for $v_T(\hat{t},\hat{r})$:

(a) for $0 \le \hat{t} \le \hat{t}_0$:

$$v_T(\hat{t},\hat{r}) = 0 \tag{48}$$

(b) for $\hat{t}_0 < \hat{t} \le \hat{t}_2$:

$$\begin{aligned} v_T(\hat{t},\hat{r}) = {} & \frac{\Pi(\hat{t})}{(\hat{t}_2-\hat{t}_0)^2}\frac{q(\chi-1)}{\chi(q-1)}\sum_{n=1}^{\infty}\frac{J_0(\lambda_n\hat{r})}{\lambda_n^6(J_1(\lambda_n))^2}\Big[{}_1F_2\Big(q;q+1,1;-\tfrac{\lambda_n^2}{4}\Big) \\ & -\tfrac{2J_1(\lambda_n)}{\lambda_n}\Big]\Big\{3\Big[\big((\hat{t}-\hat{t}_0)\lambda_n^2-1\big)^2+1-2e^{-\lambda_n^2(\hat{t}-\hat{t}_0)}\Big] \\ & -\tfrac{2}{(\hat{t}_2-\hat{t}_0)\lambda_n^2}\Big[(\hat{t}-\hat{t}_0)^3\lambda_n^6-3(\hat{t}-\hat{t}_0)^2\lambda_n^4+6\Big((\hat{t}-\hat{t}_0)\lambda_n^2-1+e^{-\lambda_n^2(\hat{t}-\hat{t}_0)}\Big)\Big]\Big\}\infty \end{aligned} \tag{49}$$

(c) for $\hat{t} > \hat{t}_2$:

$$\begin{aligned} v_T(\hat{t},\hat{r}) = {} & \frac{\Pi(\hat{t})}{(\hat{t}_2-\hat{t}_0)^2}\frac{q(\chi-1)}{\chi(q-1)}\sum_{n=1}^{\infty}\frac{J_0(\lambda_n\hat{r})}{\lambda_n^6(J_1(\lambda_n))^2}\Big[{}_1F_2\Big(q;q+1,1;-\tfrac{\lambda_n^2}{4}\Big)-\tfrac{2J_1(\lambda_n)}{\lambda_n}\Big]\Big\{\lambda_n^4(\hat{t}_2-\hat{t}_0)^2 \\ & -6e^{-\lambda_n^2(\hat{t}-\hat{t}_0)}\Big(\tfrac{2}{(\hat{t}_2-\hat{t}_0)\lambda_n^2}+1\Big)+6e^{-\lambda_n^2(\hat{t}-\hat{t}_2)}\Big(\tfrac{2}{(\hat{t}_2-\hat{t}_0)\lambda_n^2}-1\Big)\Big\} \end{aligned} \tag{50}$$

In Equations (49) and (50), ${}_1F_2(a;b,c;x)$ is the Generalized Hypergeometric function, which generally is calculated in the following way:

$${}_1F_2(a;b,c;x) = \sum_{n=1}^{\infty}\frac{(a)_n}{(b)_n(c)_n}\frac{x^n}{n!} \tag{51}$$

In Equation (51), $(z)_n$ is the Pochhammer's symbol, which is defined as:

$$(z)_n = z(z+1)(z+2)\cdots(z+n-1) = \prod_{k=1}^{n}(z+k-1) = \frac{\Gamma(z+n)}{\Gamma(z)},\ (z)_0 = 1,\ z \in \mathbb{N} \quad (52)$$

The complete accelerated solution is a sum of laminar and turbulent components:

$$v(\hat{t},\hat{r}) = v_L(\hat{t},\hat{r}) + v_T(\hat{t},\hat{r}) \quad (53)$$

An example of simulation with this model for a case of accelerated pipe flow from rest to $Re = 56677$ is presented in Figure 14. Input data for this early and slow class of accelerated flow are: $\Pi(\hat{t}) = 4.0912{\cdot}10^6$; $\Delta\hat{t} = 0.006$; $q = 45$; $\chi = 29.3758$; $\hat{t}_0 = 0.0042$; $\hat{t}_2 = 0.012$.

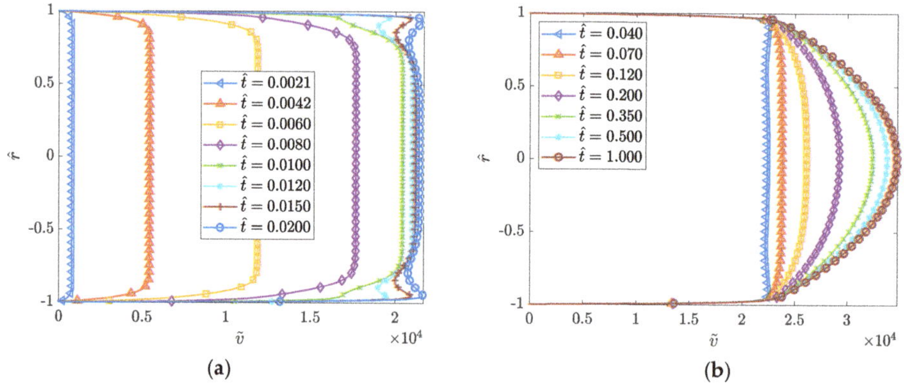

Figure 14. Results of calculation using the García García and Alvariño TULF model: (**a**) early stage; (**b**) late stage.

In summary, the TULF solution presented by García García and Alvariño is an interesting addition to the theory of analytical methods for solving this type of accelerated flow. With its help, it was possible to theoretically justify many phenomena occurring during accelerated flows (e.g., the hyperlaminar jet effect [91], the lone concavity [92], or the annular jet effect [40,85,93]), which until now had only been observed experimentally. These findings show the role of analytical solutions in the continuous scientific progress regarding the issues of transient liquid flows through pressurized pipes. They were derived assuming a pressure gradient change as a flow generator. This means that an equivalent solution could also be derived for forced flow by means of a piston.

5. Conclusions

Two types of accelerated flows described in the literature were analyzed: (1) acceleration as a result of a change in pressure gradient (occurs in systems where a change in pressure occurs as a result of a sudden opening of the gate valve) and (2) acceleration resulting from the assumption of a specific change in the flow rate (such a situation occurs in hydraulic systems with forced movement due to the displacement of piston elements). A survey of the literature shows that the standard 1932 Szymański's solution has been independently derived much earlier by at least two scientists: (1) the Italian Roiti in 1871 and the Russian Gromeka in 1882, hence it should be quoted as the RGS solution, as adopted in this work from the letters of the authors' surnames. An interesting solution, which is the prototype of the 1992 Andersson and Tiseth solution, was derived in 1933 by Vogelpohl. This combines the universal solution for any change in the pressure gradient discussed in detail in Section 2.3 of this work with solutions from the second group, i.e., flows forced by a jump in the flow rate. Vogelpohl succeeded in determining the pressure gradient function such that the average value of the flow velocity is kept constant (identical to that obtained almost sixty years later by Anderson and Tiseth). All the analytical solutions discussed in this work are of great practical value, as they give the opportunity to accurately analyze the variability in velocity profiles, mean velocity values (or pressure gradients), shear stresses on the pipe wall, dynamic friction coefficients, etc. Thanks to this, they are perfect for verifying commercial programs in the field of CFD (computational fluid dynamics). The completed review of known analytical models of accelerated flow indicates that Telionis was right when he wrote in his book that "impulsive fluid motions do not exist in reality". In real life the motion of a piston is never instantaneous as assumed in the Anderson and Tiseth solution; similarly, an instantaneous pressure gradient never takes place in real systems (as an effect of the valve opening time). So real flow is always somewhere between these two solutions defined and discussed in this paper. In the case of accelerated flows resulting from the impact of the pressure gradient at the inlet and outlet of the conduit, it seems that further research is necessary to show the real time variability of pressure gradient functions in laminar flows. This will determine whether it can actually be described by a function similar to the one observed in experimental studies by Avula in 1968, if so, it will still be necessary to choose a function that will give a complete analytical solution (and not, as now, the integral of this function should be determined numerically). There is a need to summarize all scientific papers describing experimental studies of accelerated flows—most are focused on the transition between laminar and turbulent flow. Experimental studies strictly concerning only laminar flows ($Re < 2320$) are almost non-existent in the literature.

A drawback of the presented models is their mathematical complexity (infinite series). They are based on Bessel functions and their zeros. For these zeros no analytical formula has been developed to date, hence they must be determined numerically. Bessel functions, on which all discussed solutions of velocity profile are based, are still simulation problems, especially for very small and large arguments. To overcome this problem in the computer programs used to carry out sample comparisons presented in several figures, this function was expanded in a Taylor series for small arguments, and asymptotic formulas for large arguments were used. According to this approach, this solution should be approximated with much simpler functions to be more useful in practice.

A significant strength of the presented analytical solutions is the possibility to define other related flow parameter-time dependences: mean velocity (in pressure-gradient-driven flows), the pressure gradient (in flow-rate-driven flows), friction factor and wall shear stress. These parameters are very helpful for a better understanding of the accelerated flow characteristics.

The semi-analytical solution for turbulent flow (TULF model developed by García García and Alvariño) is the most complex of all the analyzed solutions. Here, the turbulent velocity field is dependent on Bessel functions, their zeros and a generalized hypergeometric function. There is no easy and straightforward way to define the required spatial degrees of freedom coefficients (Π, χ and q—defining respectively: mean pressure gradient, initial centreline turbulent dissipation and best-fitting integer power). However, it is possible to theoretically investigate and better understand the behavior of experimentally discovered phenomena such as: the lone concavity, annular jet effect and hyperlaminar jet effect. This solution would be useful to study the behavior of other related phenomena not discovered experimentally yet. Finally, Table 2 summarizes the advantages and disadvantages of the reviewed analytical solutions.

Table 2. Short summary of the reviewed analytical solutions valid for laminar accelerating pipe flow (except TULF model).

		Imposed Pressure Gradient	
Solution	**Equation/s**	**Advantages**	**Disadvantages**
RGSI [5,7,13,18]	(9)	able to calculate the unsteady velocity profile for instantaneous pressure gradient changes (not necessary from rest)	in original form valid only for horizontal infinite long pipes—Urbanowicz's correction needed for sloping pipe
I [18]	(11)	showed what would be the mathematical solution of first stage of ramp change of the pressure gradient	uselessness in practise, as the pressure gradient should stabilize as in unsolved case of ramp change
Avula [53]	(13)	promising results that more faithfully illustrate the real history of the velocity-profile changes in real systems	complicated normalization (Re number dependence) used by Avula, which ended up with complicated zeros of the Bessel function. Not fully analytically solvable, as a result of the applied approximation form of the pressure gradient
general [30,61,62]	(26)	useful for general mathematical form of pressure gradient (suggested assumptions of pressure gradients that will give analytical solution of this integral)	convolution integral of pressure gradient and exponential function need to be solved (analytically—suggested, or numerically)
TULF [84,88–90]	(46)–(50)	only one developed for turbulent flow, theoretically possible investigation of experimentally discovered phenomena (the lone concavity, annular jet effect, hyperlaminar jet effect)	no easy and straightforward way to define the required spatial degrees of freedom coefficients, use of generalized hypergeometric function needed
		Imposed Flow Rate	
AT [32]	(31)	valid for finite-length tubes at locations beyond the entrance flow development length	assumes unphysical jump (change) of axial velocity from 0 to 0.5 value
DA [36]	(35) and (36)	based on a more realistic ramp change of flow rate assumption (unphysical velocity jump excluded)—the results are more consistent with experiments in early stage of acceleration	two different solutions defined for piston acceleration and constant piston velocity. Unclear rule to select time t_0 defining finish of the piston acceleration period
KVN [37]	(40)	generalized AT and DA solutions, gives the possibility to derive the solution for acceleration from one steady state (or developing one) to other steady flow (a condition of double-step changes of the flow rate introduced)	non-linear increase phase assumed, which gives unrealistic jump of the axial velocity in calculations

In the present review, it was also noted that, to date, there is a lack of an analytical solution as well as experimental studies of the accelerated flow of Newtonian fluid in plastic conduits (HDPE, ABS, PVC, PP, PB). In these conduits, the water hammer theory implies that the additional damping is due to the delayed strains that occur during this type of flow. The influence of these deformations is taken into account in the equation of continuity and not momentum, so the question remains whether in accelerated flows the same profiles and development times will be obtained in pipes with identical parameters (diameter, roughness, wall thickness) but made of two different materials: plastic and metal.

Author Contributions: Conceptualization, K.U., A.B. and M.S.; methodology, K.U. and M.K.; software, K.U., M.K. and A.D.; validation, K.U., A.D. and M.S.; formal analysis, K.U., A.D. and M.S.; investigation, K.U., A.B, M.S. and M.K.; resources, K.U.; data curation, K.U. and A.D.; writing—original draft preparation, K.U. and A.B.; writing—review and editing, K.U., A.B., A.D. and M.S.; visualization, K.U. and M.K.; supervision, K.U. and A.B.; project administration, K.U. and M.S.; funding acquisition, K.U. and A.B. All authors have read and agreed to the published version of the manuscript.

Funding: A. Bergant gratefully acknowledges the support of the Slovenian Research Agency (ARRS) conducted through the research project L2-1825 and programme P2-0162.

Data Availability Statement: Codes generated during the study and experimental data are available from the corresponding author by request.

Conflicts of Interest: The authors declare no conflict of interest.

Nomenclature

A	pipe cross-sectional area (m^2)
a	acceleration (m/s^2)
b, k	positive constants in Smith solution
$a_1, \ldots, a_5$	calibrated constants of Avula's pressure gradient field
c_P	specific heat capacity (J/(kgK))
D	pipe internal diameter (m)
f	friction factor (-)
G	normalized pressure gradient (m/s^2)
g	acceleration due to gravity (m/s^2)
$\dot{g}$	volumetric heat source term (W/m^3)
H	pressure head (m)
$\hat{H}$	dimensionless head (-)
L	pipe length (m)
M	Otis start-up parameter (-)
p	pressure (Pa)
p^*	dimensionless Avula's pressure $p^* = \frac{\Delta p}{2\rho U^2}$ (-)
$\hat{p}$	normalized pressure (-)
Q	flow rate (m^3/s)
Q^*	normalized flow rate (-)
q	best fitting integer TULF model coefficient (-)
R	pipe internal radius (m)
Re	Reynolds number $\mathrm{Re} = \frac{Dv_m}{\nu}$ (-)
r	radial coordinate (m)
$\hat{r}$	normalized radial coordinate (-)
s	Laplace complex variable (s^{-1})
T	temperature field (°C)
$T_1, \ldots, T_4$	terms in Das-Arakeri solution
t	time (s)
t_A	dimensionless Avula's time (-)
$\hat{t}$	dimensionless time $\hat{t} = t\frac{\nu}{R^2}$ (-)
u	dummy variable in convolutions integrals (s)
v	velocity field (m/s)
v_m	mean velocity (m/s)
$\hat{v}$	dimensionless velocity $\hat{v} = v/v_{max}$ (-)
$\widetilde{v}$	dimensionless García García and Alvariño velocity $\widetilde{v} = v{\cdot}R/\nu$ (-)
v^*	dimensionless Avula's velocity $v^* = v/U$ (-)
v_0	initial velocity (m/s)
X^*	dimensionless distance function (-)
x	axial coordinate (m)
x^*	stretched axial coordinate (m)
$\hat{x}$	normalized axial coordinate $\hat{x} = x/R$ (-)

α	thermal diffusivity coefficient (m^2/s)
α_n	successive zeros of the Bessel function $J_2(\alpha_n)$ (-)
β	pipe slope angle (°)
γ	specific weight of the liquid (N/m^3)
$\Delta\hat{t}$	dimensionless time interval of linear MPG growth (mean valve aperture time) (-)
ε	jerk coefficient (m/s^3)
η	time scale coefficient in TULF model (-)
η_n	successive zeros of the Bessel function $J_0\left(\sqrt{sRe/2}\right)$ (-)
λ_n	successive zeros of the Bessel function $J_0(\lambda_n)$ (-)
μ	dynamic viscosity of liquid (Pa·s)
ν	kinematic viscosity of liquid (m^2/s)
ξ_n	successive roots fulfilling relation $J_o(R\xi_n) = 0$ (m^{-1})
Π	normalized mean pressure gradient in TULF solution (-)
ρ	density of liquid (kg/m^3)
Σ	weighted Reynolds shear stress gradient (-)
σ	Reynolds shear stress field in TULF solution (-)
τ_w	wall shear stress (Pa)
χ	turbulent dissipation coefficient (-)
Subscripts	
0	initial
m	mean
o	outlet
p	piston
t	transitional
i	inlet
w	wall
∞	final
Acronyms	
AT	Anderson and Tiseth solution
DA	Das and Arakeri solution
I	Ito solution
KVN	Kannaiyan–Varathalingarajah–Natarajan solution
MPG	mean pressure gradient
RANSE	Reynolds-averaged Navier–Stokes equation
RGS	Roiti–Gromeka–Szymański solution
RGSI	Roiti–Gromeka–Szymański-Ito solution
RSS	Reynolds shear stress
S	Sparrow solution
SM	Smith solution
TULF	theory of underlying laminar flow
WRSSG	weighted Reynolds shear stress gradient

References

1. Navier, C.L.M.H. Mémoire sur les lois du mouvement des fluides. *Mémoires L'académie R. Sci. L'institut Fr.* **1823**, *6*, 389–440.
2. Darrigol, O. Between Hydrodynamics and Elasticity Theory: The First Five Births of the Navier-Stokes Equation. *Arch. Hist. Exact Sci.* **2002**, *56*, 95–150. [CrossRef]
3. Letelier S, M.F.; Leutheusser, H.J. Unified Approach to the Solution of Problems of Unsteady Laminar Flow in Long Pipes. *J. Appl. Mech.* **1983**, *50*, 8–12. [CrossRef]
4. Urbanowicz, K.; Firkowski, M.; Bergant, A. Comparing analytical solutions for unsteady laminar pipe flow. In Proceedings of the 13th International Conference on Pressure Surges, Bordeaux, France, 14–16 November 2018; pp. 311–326.
5. Roiti, A. Sul movimento dei liquidi. *Ann. Della Sc. Norm. Super. Pisa—Cl. Sci.* **1871**, *1*, 193–240.
6. Betti, E. Alcune Determinazioni delle Temperature Variabili di un Cilindro. Tipografia dei FF; Nistri: Pisa, Italy, 1868.
7. Gromeka, I.S. On a theory of the motion of fluids in narrow cylindrical tubes. *Uch. Zap. Kazan. Inst.* **1882**, *112*. (In Russian)

8. Baibikov, B.S.; Oreshkin, F.; Prudovskii, A.M. Frictional resistance in the case of accelerated flow in a tube. *Fluid Dyn.* **1981**, *16*, 749–751. [CrossRef]
9. Ovsyannikov, V.M. Calculation of accelerated motion of fluid in a tube. *Fluid Dyn.* **1981**, *16*, 770–772. [CrossRef]
10. Logov, I.L. Frictional resistance to accelerated flow in a tube. *Fluid Dyn.* **1984**, *18*, 978–983. [CrossRef]
11. Loitsyanskii, L.G. *Mechanics of Liquids and Gases*, 2nd revised ed.; International Series of Monographs in Aeronautics and Astronautics, Division II: Aerodynamics; Jones, R.T., Jones, W.P., Eds.; Pergamon Press: Oxford, UK, 1966; Volume 6, pp. 17–18.
12. Gromeka, I.S. *Collected Works*; Izd. AN SSSR: Moscow, Russia, 1952.
13. Szymański, P. Quelques solutions exactes des équations de l'hydrodynamique du fluide visqueux dans le cas d'un tube cylindrique. *J. Math. Pures Appliquées* **1932**, *11*, 67–108.
14. Schlichting, H.; Gersten, K. *Boundary Layer Theory*, 9th ed.; McGraw-Hill: New York, NY, USA, 2017; p. 140.
15. White, F.M. *Viscous Fluid Flow*, 3rd ed.; McGraw-Hill: New York, NY, USA, 2006; p. 125.
16. Telionis, O.P. *Unsteady Viscous Flows*; Springer: New York, NY, USA, 1981; pp. 91–92.
17. Gerbes, W. Zur instationären, laminaren Strömung einer inkompressiblen, zähen Flüssigkeit in kreiszylindrischen Rohren. *Z. Angew. Phys.* **1951**, *3*, 267–271.
18. Ito, H. Theory of Laminar Flow through a Pipe with Non-Steady Pressure Gradients. *Trans. Jpn. Soc. Mech. Eng.* **1952**, *18*, 101–108. [CrossRef]
19. Atabek, H.B. Development of flow in the inlet length of a circular tube starting from rest. *ZAMP* **1962**, *13*, 417–430. [CrossRef]
20. Avula, X.J.R. Analysis of suddenly started laminar flow in the entrance region of a circular tube. *Appl. Sci. Res.* **1969**, *21*, 248–259. [CrossRef]
21. Fan, C.; Chao, B.-T. Unsteady, laminar, incompressible flow through rectangular ducts. *J. Appl. Math. Phys. (ZAMP)* **1965**, *16*, 351–360. [CrossRef]
22. Laura, P.A.A. Unsteady, laminar, incompressible flow through ducts of arbitrary, doubly connected cross section. *Rev. De La Unión Matemática Argent.* **1976**, *27*, 197–206.
23. Muzychka, Y.; Yovanovich, M. Compact models for transient conduction or viscous transport in non-circular geometries with a uniform source. *Int. J. Therm. Sci.* **2006**, *45*, 1091–1102. [CrossRef]
24. Muzychka, Y.; Yovanovich, M. Unsteady viscous flows and Stokes's first problem. *Int. J. Therm. Sci.* **2010**, *49*, 820–828. [CrossRef]
25. Chen, C.-I.; Yang, Y.-T. Unsteady unidirectional flow of an Oldroyd-B fluid in a circular duct with different given volume flow rate conditions. *Heat Mass Transf.* **2004**, *40*, 203–209. [CrossRef]
26. Nazar, M.; Mahmood, A.; Athar, M.; Kamran, M. Analytic solutions for the unsteady longitudinal flow of an oldroyd-b fluid with fractional model. *Chem. Eng. Commun.* **2012**, *199*, 290–305. [CrossRef]
27. Wang, X.; Xu, H.; Qi, H. Transient magnetohydrodynamic flow and heat transfer of fractional Oldroyd-B fluids in a microchannel with slip boundary condition. *Phys. Fluids* **2020**, *32*, 103104. [CrossRef]
28. Rahaman, K.; Ramkissoon, H. Unsteady axial viscoelastic pipe flows. *J. Non-Newtonian Fluid Mech.* **1995**, *57*, 27–38. [CrossRef]
29. Gerhart, P.M.; Gerhart, A.L.; Hochstein, J.I. *Munson, Young, and Okiishi's Fundamentals of Fluid Mechanics*, 8th ed.; John Wiley and Sons: Hoboken, NJ, USA, 2016; p. 16.
30. Vogelpohl, G. Über die Ermittlung der Rohreinlaufströmung aus den Navier-Stokesschen Gleichungen. *ZAMM J. Appl. Math. Mech./Z. Angew. Math. Mech. Vorträge Hauptversamml. Würzburg Ges. Angew. Math.* **1933**, *13*, 422–449. [CrossRef]
31. Whittaker, E.T. On the numerical solution of integral-equations. *Proc. R. Soc. London. Ser. A* **1918**, *94*, 367–383. [CrossRef]
32. Andersson, H.I.; Tiseth, K.L. Start-up flow in a pipe following the sudden imposition of a constant flow rate. *Chem. Eng. Commun.* **1992**, *112*, 121–133. [CrossRef]
33. Weinbaum, S.; Parker, K.H. The laminar decay of suddenly blocked channel and pipe flows. *J. Fluid Mech.* **1975**, *69*, 729–752. [CrossRef]
34. Andersson, H.; Kristoffersen, R. Start-up of laminar pipe flow. In Proceedings of the AIAA/ASME/SIAM/APS 1st National Fluid Dynamics Congress, Cincinnati, OH, USA, 25–28 July 1988; pp. 1356–3805. [CrossRef]
35. Otis, D.R. Laminar Start-Up Flow in a Pipe. *J. Appl. Mech.* **1985**, *52*, 706–711. [CrossRef]
36. DAS, D.; Arakeri, J.H. Transition of unsteady velocity profiles with reverse flow. *J. Fluid Mech.* **1998**, *374*, 251–283. [CrossRef]
37. Kannaiyan, A.; Varathalingarajah, T.; Natarajan, S. Analytical solutions for the incompressible laminar pipe flow rapidly subjected to the arbitrary change in the flow rate. *Phys. Fluids* **2021**, *33*, 043601. [CrossRef]
38. Van de Sande, E.; Belde, A.P.; Hamer, B.J.G.; Hiemstra, W. Velocity profiles in accelerating pipe flows starting from rest. In Proceedings of the 3rd International Conference on Pressure Surges, Canterbury, UK, 25–27 March 1980; pp. 1–14, paper A1.
39. Lefebvre, P.J.; White, F.M. Experiments on Transition to Turbulence in a Constant-Acceleration Pipe Flow. *J. Fluids Eng.* **1989**, *111*, 428–432. [CrossRef]
40. Kataoka, K.; Kawabata, T.; Miki, K. The start-up response of pipe flow to a step change in flow rate. *J. Chem. Eng. Jpn.* **1975**, *8*, 266–271. [CrossRef]
41. Chaudhury, R.A.; Herrmann, M.; Frakes, D.H.; Adrian, R.J. Length and time for development of laminar flow in tubes following a step increase of volume flux. *Exp. Fluids* **2015**, *56*, 22. [CrossRef]
42. He, K.; Seddighi, M.; He, S. DNS study of a pipe flow following a step increase in flow rate. *Int. J. Heat Fluid Flow* **2016**, *57*, 130–141. [CrossRef]
43. Deville, M.O. *An Introduction to the Mechanics of Incompressible Fluids*; Springer: Cham, Switzerland, 2022; p. 20. [CrossRef]

44. Poisson, S.D. Mémoire sur la distribution de la chaleur dans les corps solides. *J. L'ecole Polytech.* **1823**, *19*, 249–403.
45. Allievi, L. Teoria generale del moto perturbato dell'acqua nei tubi in pressione (colpo d'ariete). *Il Politec.—G. Dell'ingegnere Archit Civ. Ed Ind. (Fasc.)* **1903**, *33*, 360–371.
46. Faccò, C.A. Avviamento del moto di una corrente liquida in un tubo di sezione costante: Influenza delle resistenze. *Rendiconti Inst. Lomb.—Acad. Sci. E Lett.* **1956**, *90*, 305–342.
47. Aresti, G. Sul moto di un fluido viscoso, incompressible, lungo un tubu cylindrico (rotondo). *Rend. Semin. Della Fac. Sci dell'Università Cagliari* **1934**, *4*, 91–93.
48. Szymański, P. Sur l'écoulement non permanent du fluide visqueux dans le tuyau. In Proceedings of the III Congrès International de Mécanique Appliquée, Stockholm, Sweden, 24–29 August 1930; pp. 249–254.
49. Urbanowicz, K.; Tijsseling, A.S.; Firkowski, M. Comparing convolution-integral models with analytical pipe- flow solutions *J. Phys. Conf. Ser.* **2016**, *760*, 012036. [CrossRef]
50. Letelier S, M.F.; Leutheusser, H.J. Skin Friction in Unsteady Laminar Pipe Flow. *J. Hydraul. Div.* **1976**, *102*, 41–56. [CrossRef]
51. Lefèbvre, P.J.; White, F.M. Further Experiments on Transition to Turbulence in Constant-Acceleration Pipe Flow. *J. Fluids Eng.* **1991**, *113*, 223–227. [CrossRef]
52. Knisely, C.W.; Nishihara, K.; Iguchi, M. Critical Reynolds Number in Constant-Acceleration Pipe Flow from an Initial Steady Laminar State. *J. Fluids Eng.* **2010**, *132*, 091202. [CrossRef]
53. Avula, X.J.R. A Combined Method for Determining Velocity of Starting Flow in a Long Circular Tube. *J. Phys. Soc. Jpn.* **1969**, *27*, 497–502. [CrossRef]
54. Avula, X.J.R.; Young, D.F. Start-up Flow in the Entrance Region of a Circular Tube. *ZAMM-J. Appl. Math. Mech./Z. Angew. Math. Mech.* **1971**, *51*, 517–526. [CrossRef]
55. Smith, S.H. Classroom Note: Time-Dependent Poiseuille Flow. *SIAM Rev.* **1997**, *39*, 511–513. [CrossRef]
56. Dryden, H.L.; Murnaghan, F.D.; Bateman, H. *Hydrodynamics*; Dover: New York, NY, USA, 1956.
57. Singh, T. Incipient flow of elastico-viscous fluid in a pipe. *Eng. Comput.* **1992**, *9*, 81–91. [CrossRef]
58. Patience, G.S.; Mehrotra, A.K. Discussion: "Laminar Start-Up Flow in a Pipe". *J. Appl. Mech.* **1987**, *54*, 243–244. [CrossRef]
59. Fargie, D.; Martin, B.W. Developing laminar flow in a pipe of circular cross-section. *Proc. R. Soc. Lond. Ser. A* **1971**, *321*, 461–476. [CrossRef]
60. Patience, G.S.; Mehrotra, A.K. Laminar start-up flow in short pipe lengths. *Can. J. Chem. Eng.* **1989**, *67*, 883–888. [CrossRef]
61. Fan, C. Non-Steady, Viscous, Incompressible Flow in Cylindrical and Rectangular Conduits (with Emphasis on Periodically Oscillating Flow). Ph.D Thesis, University of Illinois, Champaign, IL, USA, 1964.
62. Daneshyar, H. Development of unsteady laminar flow of an incompressible fluid in a long circular pipe. *Int. J. Mech. Sci.* **1970**, *12*, 435–445. [CrossRef]
63. Sneddon, I.N. *Fourier Transforms*; McGraw-Hill: New York, NY, USA, 1951.
64. Roller, J.E. Unsteady Flow in a Smooth Pipe after Instantaneous Opening of a Downstream Valve. Master's Thesis, Georgia Institute of Technology, Atlanta, GA, USA, 1956.
65. Zielke, W. Frequency-Dependent Friction in Transient Pipe Flow. Ph.D. Thesis, University of Michigan, Ann Arbor, MI, USA, 1966.
66. Hershey, D.; Song, G. Friction factors and pressure drop for sinusoidal laminar flow of water and blood in rigid tubes. *AIChE J.* **1967**, *13*, 491–496. [CrossRef]
67. Xiu, W.; Sun, J.G.; Sha, W.T. Transient flows and pressure waves in pipes. *J. Hydrodyn. Ser. B* **1995**, *2*, 51–59.
68. Sun, J.G.; Wang, X.Q. Pressure transient in liquid lines. In Proceedings of the ASME/JSME Pressure Vessels and Piping Conference, Honolulu, HI, USA, 23–27 July 1995.
69. Lee, Y. Analytical solutions of channel and duct flows due to general pressure gradients. *Appl. Math. Model.* **2017**, *43*, 279–286. [CrossRef]
70. Song, G. Determination of Friction Factors for the Pulsatile Laminar Flow of Water and Blood in Rigid Tubes. Ph.D. Thesis, University of Cincinnati, Cincinnati, OH, USA, 1966.
71. Avula, X.J.R. Unsteady Flow in the Entrance Region of a Circular Tube. Ph.D. Thesis, Iowa State University, Ames, IA, USA, 1968.
72. Erdogan, M.E. On the flows produced by sudden application of a constant pressure gradient or by impulsive motion of a boundary. *Int. J. Non-Linear Mech.* **2003**, *38*, 781–797. [CrossRef]
73. Müller, W. Zum Problem der Anlaufströmung einer Flüssigkeit im geraden Rohr mit Kreisring- und Kreisquerschnitt. *ZAMM-J. Appl. Math. Mech./Z. Angew. Math. Mech.* **1936**, *16*, 227–238. [CrossRef]
74. Avramenko, A.A.; Tyrinov, A.I.; Shevchuk, I.V. An analytical and numerical study on the start-up flow of slightly rarefied gases in a parallel-plate channel and a pipe. *Phys. Fluids* **2015**, *27*, 042001. [CrossRef]
75. Kuznetsov, A.V.; Avramenko, A.A. Start-Up Flow in a Channel or Pipe Occupied by a Fluid-Saturated Porous Medium. *J. Porous Media* **2009**, *12*, 361–367. [CrossRef]
76. Sparrow, E.M.; Lin, S.H.; Lundgren, T.S. Flow Development in the Hydrodynamic Entrance Region of Tubes and Ducts. *Phys. Fluids* **1964**, *7*, 338. [CrossRef]
77. Vardy, A.E.; Brown, J.M. Influence of time-dependent viscosity on wall shear stresses in unsteady pipe flows. *J. Hydraul. Res.* **2010**, *48*, 225–237. [CrossRef]
78. Vardy, A.E.; Brown, J.M.B. Laminar pipe flow with time-dependent viscosity. *J. Hydroinform.* **2011**, *13*, 729–740. [CrossRef]

79. Daprà, I.; Scarpi, G. Unsteady Flow of Fluids with Arbitrarily Time-Dependent Rheological Behavior. *J. Fluids Eng.* **2017**, *139*, 051202. [CrossRef]
80. Wiens, T.; Etminan, E. An Analytical Solution for Unsteady Laminar Flow in Tubes with a Tapered Wall Thickness. *Fluids* **2021**, *6*, 170. [CrossRef]
81. Cengel, Y.A. *Heat Transfer a Practical Approach*, 2nd ed.; Mcgraw-Hill: New York, NY, USA, 2002; p. 70.
82. Moss, E.A. Laminar pipe flows accelerated from rest. *N&O J.* **1991**, 7–14.
83. Pozzi, A.; Tognaccini, R. The effect of the Eckert number on impulsively started pipe flow. *Eur. J. Mech. B Fluids* **2012**, *36*, 120–127. [CrossRef]
84. García, F.J.G.; Alvariño, P.F. On the influence of Reynolds shear stress upon the velocity patterns generated in turbulent starting pipe flow. *Phys. Fluids* **2020**, *32*, 105119. [CrossRef]
85. Maruyama, T.; Kato, Y.; Mizushina, T. Transition to turbulence in starting pipe flows. *J. Chem. Eng. Jpn.* **1978**, *11*, 346–353. [CrossRef]
86. Kannaiyan, A.; Natarajian, S.; Vinoth, B.R. Stability of a laminar pipe flow subjected to a step-like increase in the flow rate. *Phys. Fluids* **2022**, *34*, 06410. [CrossRef]
87. Pai, S. On turbulent flow in circular pipe. *J. Frankl. Inst.* **1953**, *256*, 337–352. [CrossRef]
88. García García, F.J. Transient Discharge of a Pressurised Incompressible Fluid through a Pipe and Analytical Solution for Unsteady Turbulent Pipe Flow. Ph.D. Thesis, Higher Polytechnic College—University of A Coruña, A Coruña, Spain, 2017. Available online: https://hdl.handle.net/2183/18502 (accessed on 30 November 2022).
89. García, F.J.G.; Alvariño, P.F. On an analytic solution for general unsteady/transient turbulent pipe flow and starting turbulent flow. *Eur. J. Mech. B Fluids* **2018**, *74*, 200–210. [CrossRef]
90. García, F.J.G.; Alvariño, P.F. On an analytical explanation of the phenomena observed in accelerated turbulent pipe flow. *J. Fluid Mech.* **2019**, *881*, 420–461. [CrossRef]
91. Annus, I.; Koppel, T.; Sarv, L.; Ainola, L. Development of Accelerating Pipe Flow Starting from Rest. *J. Fluids Eng.* **2013**, *135*, 111204. [CrossRef]
92. Kurokawa, J.; Morikawa, M. Accelerated and Decelerated Flows in a Circular Pipe: 1st Report, Velocity Profile and Friction Coefficient. *Bull. JSME* **1986**, *29*, 758–765. [CrossRef]
93. Viola, J.P.; Leutheusser, H.J. Experiments on Unsteady Turbulent Pipe Flow. *J. Eng. Mech.* **2004**, *130*, 240–244. [CrossRef]

Disclaimer/Publisher's Note: The statements, opinions and data contained in all publications are solely those of the individual author(s) and contributor(s) and not of MDPI and/or the editor(s). MDPI and/or the editor(s) disclaim responsibility for any injury to people or property resulting from any ideas, methods, instructions or products referred to in the content.

energies

MDPI

Review

Numerical Modelling of Forced Convection of Nanofluids in Smooth, Round Tubes: A Review

Janusz T. Cieśliński

Faculty of Mechanical and Ocean Engineering, Gdańsk University of Technology, Narutowicza 11/12, 80233 Gdansk, Poland; jcieslin@pg.edu.pl

Abstract: A comprehensive review of published works dealing with numerical modelling of forced convection heat transfer and hydrodynamics of nanofluids is presented. Due to the extensive literature, the review is limited to straight, smooth, circular tubes, as this is the basic geometry in shell-and-tube exchangers. Works on numerical modelling of forced convection in tubes are presented chronologically in the first part of the article. Particular attention was paid to the method of the solution of governing equations, geometry of the heating section, and boundary conditions assumed. Influence of nanoparticles on heat transfer and flow resistance are discussed. Basic information is summarized in tabular form, separately for single-phase approach and two-phase models. The second part of the article contains the correlation equations proposed in the presented papers for the calculation of the Nusselt (Nu) number or heat transfer coefficient, separately for laminar and turbulent flow. Details of the type of nanofluids, the concentration of nanoparticles, and the Reynolds (Re) number range are also presented. Finally, advantages and disadvantages of individual numerical approaches are discussed.

Keywords: forced convection; nanofluids; straight; smooth; round tubes; heat transfer; flow resistance

Citation: Cieśliński, J.T. Numerical Modelling of Forced Convection of Nanofluids in Smooth, Round Tubes: A Review. *Energies* **2022**, *15*, 7586. https://doi.org/10.3390/en15207586

Academic Editor: Gianpiero Colangelo

Received: 13 September 2022
Accepted: 10 October 2022
Published: 14 October 2022

Publisher's Note: MDPI stays neutral with regard to jurisdictional claims in published maps and institutional affiliations.

Copyright: © 2022 by the author. Licensee MDPI, Basel, Switzerland. This article is an open access article distributed under the terms and conditions of the Creative Commons Attribution (CC BY) license (https://creativecommons.org/licenses/by/4.0/).

1. Introduction

Nanofluids are one of the techniques for heat transfer enhancement [1,2]. The goal is obvious: to reduce the heat transfer surface area, thus reducing the consumption of materials and energy necessary for the manufacture of heat exchangers. However, in order for nanofluids to find practical applications, several problems must be solved. The first and foremost is to provide engineers with accurate and reliable methods to calculate heat transfer coefficients and friction factors. The second problem is the thermophysical properties of nanofluids, the determination of which is not easy [3,4]. The third problem that has so far held back the practical application of nanofluids is their stability [5,6]. Due to the costs of experimental research and the long design and construction time of measuring stands, as well as tedious measurements, numerical methods are an indispensable approach that allow for quick and precise assessment of new technologies. However, it should be remembered that each numerical work should be verified experimentally or, if possible, by an analytical solution. There are two main approaches in the modeling of nanofluid flows, e.g., [7–11]. Due to the size of nanoparticles (similar to the dimensions of liquid molecules), it is assumed that the resulting mixture forms a homogeneous liquid, the properties of which result from the properties of the base liquid and solid particles. Hence, the classical methods of continuum mechanics are used to solve the set of governing equations. In the second approach, a nanofluid is treated as two phase solid-liquid mixture.

This paper presents state of the art in the field of numerical modeling of the forced convection of nanofluids in straight, smooth, round tubes.

2. Methods

Maïga et al. [12] considered the problem of nanofluid flow inside a uniformly heated tube and applied a single-phase approach. In general, it was observed that the inclusion of nanoparticles considerably enhanced heat transfer for both the laminar and turbulent regimes. Such improvement of heat transfer becomes more pronounced with the increase of the nanoparticle concentration (NPC). On the other hand, the presence of particles has produced adverse effects on the wall friction that also increases with the NPC. In [13], Maïga et al. studied the laminar flow of nanofluids inside a uniformly heated tube. It was shown that the presence of nanoparticles induced drastic effects on the wall shear stress (WSS) that increased appreciably with the NPC increase. In [14], Maïga et al. studied the turbulent flow of nanofluids inside a uniformly heated tube. A single-phase approach was used to solve the system of non-linear and coupled governing equations, and the k-ε model was employed in order to model the turbulence. It was found that HTC increases with NPC increase. Heris et al. [15] investigated the laminar flow of nanofluid in an isothermal tube. A dispersion model with constant thermophysical properties of nanofluids were assumed. Heat transfer enhancement was detected with NPC increase. Moreover, the Nu number decreased with nanoparticle diameter increase for the given NPC. Behzadmehr et al. [16] studied the turbulent flow of nanofluids in a uniformly heated tube. Probably for the first time, a mixture model was applied to study nanofluid behavior. It was found that the mixture model better interpreted experimental results than the single-phase approach. Bianco et al. [17] studied laminar flow of nanofluids for a single value of the Re number in a uniformly heated tube. A single-phase approach and the discrete particles model were employed with either constant or temperature-dependent properties. Heat transfer enhancement was observed with NPC increase. In [18], Bianco et al. investigated developing the laminar flow of nanofluid in a circular, uniformly heated tube. A single-phase approach and two-phase model (discrete particles model) were employed with either constant or temperature-dependent properties. The maximum difference in the average HTC between single-phase and two-phase models was about 11%. Heat transfer enhancement increases with the NPC increase, but it is accompanied by increasing WSS. Higher HTC and lower WSS were detected in the case of temperature dependent models. In Figure 1 results of numerical calculations by use of single-phase (1Phase) approach and DPM model with constant (con.) and variable (var.) thermophysical properties of nanofluid are shown.

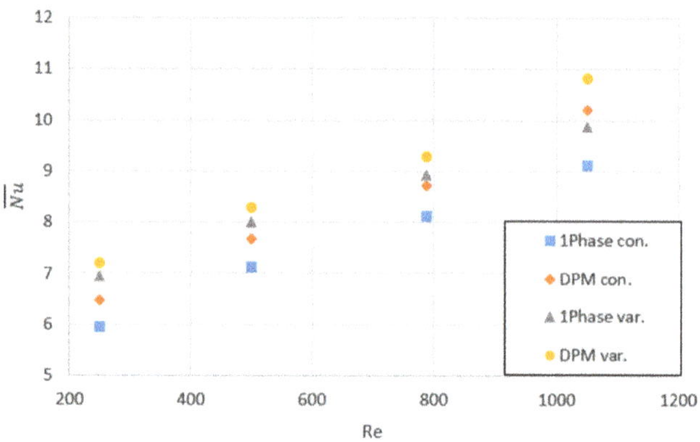

Figure 1. $\overline{\text{Nu}}$-Re relationship for water–γAl_2O_3 (1%) nanofluid—data source [18].

He et al. [19] studied the heat transfer of nanofluids flowing through a straight tube under laminar flow conditions by using both the single-phase method and combined Euler–Lagrange method. The results show significant enhancement of heat transfer of nanofluids particularly in the entrance region. The results suggest that the HTC is more affected by the thermal conductivity than by the viscosity and the Brownian force; the lift force and the thermophoretic force play very small role. Namburu et al. [20] analyzed the turbulent flow of nanofluids flowing through a circular tube under a constant heat flux condition. The nanofluids were considered as a conventional single-phase fluid. The thermophysical properties are dependent upon the temperature and NPC. It was found that nanofluids containing smaller diameter nanoparticles have higher viscosity and Nu numbers. Fard et al. [21] studied the laminar convective heat transfer of nanofluids in a circular tube under a constant wall temperature condition. Single-phase and two-phase models were used for the prediction of temperature, flow field, and calculation of HTC. The results showed that HTC increases with an increase in NPC, and the heat transfer enhancement increases with Peclet (Pe) number. The two-phase model shows better agreement with experimental measurements. Based on the results of the CFD simulation, it was concluded that the two-phase approach gives better predictions for heat transfer rate compared to the single-phase model. Lotfi et al. [22] studied the forced convection of a nanofluid in a horizontal uniformly heated tube. A two-phase Eulerian model was implemented to study a flow field. A single-phase model and two-phase mixture model were also used. The comparison of calculated results with experimental values shows that the mixture model is more precise. It is illustrated that the single-phase model and the two-phase Eulerian model underestimates the Nu number in comparison with the experimental data. Figure 2 shows a comparison of numerical calculations by use of single-phase (1Phase) approach and two two-phase models, namely a mixture model (Mixture) and Eulerian (E) model.

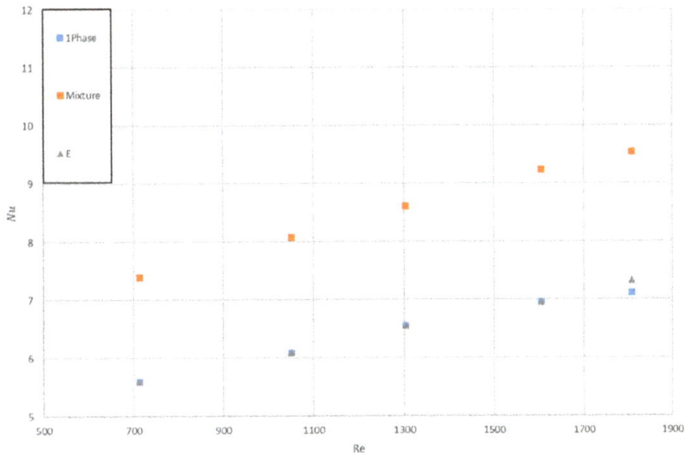

Figure 2. Nu-Re relationship for water–Al_2O_3 (1%) nanofluid and x/D = 63—data source [22].

Mokmeli and Saffar-Avval [23] studied the laminar flow of nanofluids in uniformly heated and isothermal tubes. A single-phase approach and dispersion model were used to simulate the effect of nanoparticle diameter and NPC on heat transfer. It was established that HTC increases with the decrease of nanoparticles, and the single-phase model, contrary to the dispersion model, predicts the decrease of the Nu number against NPC. It was shown that dispersion model better reproduces experimental data. Ebrahimnia-Bajestan et al. [24] analyzed the effects of NPC, particle diameter, Brownian motion, Re number, type of nanoparticles and base fluid on the HTC and pressure drop (PD) of nanofluids during laminar flow at constant heat flux boundary condition. A single-phase model was used to simulate nanofluid behavior. It was established that HTC increases with NPC and aspect ratio increase, and decreases with nanoparticle diameter increase. Moraveji et al. [25] studied the influence of nanoparticle diameter on HTC in the developing region of a tube using the single-phase nanofluid model. Laminar flow with a constant heat flux boundary condition was examined. It was established that HTC deteriorated with an increase in the axial location and particle diameter. Bianco et al. [26] studied the turbulent flow of nanofluids in uniformly heated tube using the single-phase approach and mixture model. Constant thermophysical properties of nanofluids with standard k-ε turbulence model were applied. The authors emphasized the importance of correctly determining the properties of nanofluids. Bayat and Nikseresht [27] studied the effect of type of nanofluid on PD and heat transfer during laminar flow in a uniformly heated tube. The single-phase approach with temperature and NPC dependent properties of nanofluids was applied. The most important conclusion is that it is possible to enhance heat transfer with lower WSS by the selection of a proper nanofluid. Akbari et al. [28] applied a single-phase and three different two-phase models (VOF, mixture, Eulerian) to analyze the laminar convection of nanofluids in a horizontal tube with uniform wall heat flux. The predictions by the three two-phase models were essentially the same. The two-phase models give closer predictions of the HTC to the experimental data than the single-phase model. In [29] Akbari et al. used the same numerical approaches as in [28] but this time studied the turbulent flow of nanofluids in a horizontal tube with uniform wall heat flux. It was established that the predictions by the single-phase model and by the two-phase models favors the single-phase approach. Since the single-phase model is also simpler to implement and requires less computer memory and CPU time, it was concluded that it is more appropriate for the conditions under study. Figure 3 shows comparison of numerical calculations by use of single-phase (1Phase) approach and three two-phase models, namely Eulerian (E) model, mixture model (Mixture), and volume of fluid model (VOF).

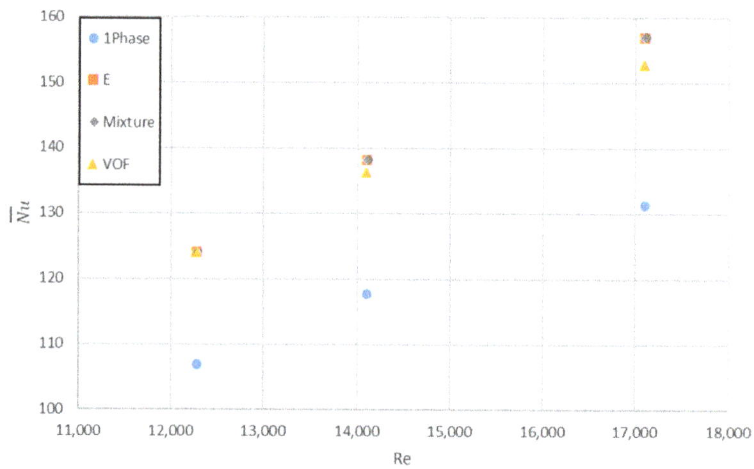

Figure 3. $\overline{\text{Nu}}$-Re relationship for water–Al_2O_3 (1%) nanofluid—data source [29].

Alvarino et al. [30] studied the effect of the Brownian motion and thermophoretic diffusion on heat transfer during laminar flow of nanofluid in uniformly heated tube. A single-phase approach was considered. It was established that the heat transfer enhancement by the nanofluid had to be attributed to its thermophysical properties rather than to another transport mechanism. Tahir and Mital [31] investigated a developing laminar flow of nanofluid in a uniformly heated tube by use of discrete phase modeling (DPM) and a Euler–Lagrangian approach. The fluid was treated as a continuous medium and the flow field was solved based on Navier–Stokes equations. The nanoparticles were individually tracked in a Lagrangian reference frame and their trajectories were determined using particle force balance. Using this approach, a good match was obtained between the numerical model and the experimental results reported in the literature. Based on statistical analysis, it was established that almost all of the variation in the HTC can be explained due to changes in the three independent variables. The Re number is the most significant variable impacting the HTC, while NPC is the least significant. The HTC linearly increases with both Re number and NPC, but shows non-linear parabolic decrease with an increase in particle size. In addition, the three variables only weakly interact with each other in terms of their impact on the average HTC. Balla et al. [32] analyzed the effect of NPC on heat transfer and PD of nanofluids during laminar flow in a uniformly heated tube. It was established that the Nu number, HTC, and PD strongly increase with the increase of NPC. In [33] Bayat and Nikseresht studied influence of temperature and NPC on heat transfer and PD during turbulent flow of nanofluid in a uniformly heated tube. A single-phase approach with temperature and nanoparticle concentration dependent properties of nanofluids was applied. It was concluded that application of nanofluids in a turbulent flow regime results in substantially higher PD and pumping power (PP) compared to base fluid for the same Re number. Moraveji and Esmaeili [34] investigated the laminar flow of nanofluid in uniformly heated tube. A single-phase model and discrete phase approach with selected properties of nanofluids was calculated as temperature dependent. The HTC increase with NPC increase was observed. There was no significant difference between the results for the single-phase model and discrete model. Davarnejad et al. [35] studied the influence of NPC on heat transfer during the laminar flow of nanofluids. A single-phase approach with properties of nanofluids calculated for the mean temperature was applied. Constant heat transfer coefficient was assumed as a boundary condition at the tube wall. It was shown that HTC increases with NPC and Pe number increase. In [36] Davarnejad et al. analyzed the influence of nanoparticle size and NPC on heat transfer during laminar flow in a uniformly heated tube. A single-phase approach with properties of nanofluids calculated for the mean temperature was applied. It was shown that HTC increases with nanoparticle size decrease. Kayaci et al. [37] studied the influence of NPC on heat transfer during the turbulent flow of nanofluids in an isothermally heated tube. A single-phase approach with the k-ε turbulence model was applied. A slight increase of HTC with NPC increase was observed. Hejazian and Moraveji [38] conducted an investigation on turbulent nanofluid flow inside an isothermal tube. A single-phase approach and mixture model with the k-ε turbulence model were tested. It was established that Nu number increases with NPC increase. Second, the mixture model better fits experimental data than the single-phase approach. Göktepe et al. [39] tested the ability of the single-phase model and two-phase models (Eulerian–Eulerian and Eulerian–mixture) to reproduce published experimental HTC and friction factors (FF) for laminar flow at the developing region of uniformly heated tube. It was established that two-phase models predict HTC and FF more accurately than single-phase models. Among single-phase models, the dispersion model that uses velocity gradient to define dispersion conductivity, was found to be more effective compared to the other. Moreover, the Eulerian–Eulerian model is recommended for cases when no prior experimental data is available. Figure 4 shows a comparison of numerical calculations by the use of two single-phase models with differently assumed dispersion conductivity (1PhaseD1, 1PhaseD2) and two two-phase models, namely Eulerian–mixture model (E–Mixture) and Eulerian–Eulerian model (E–E).

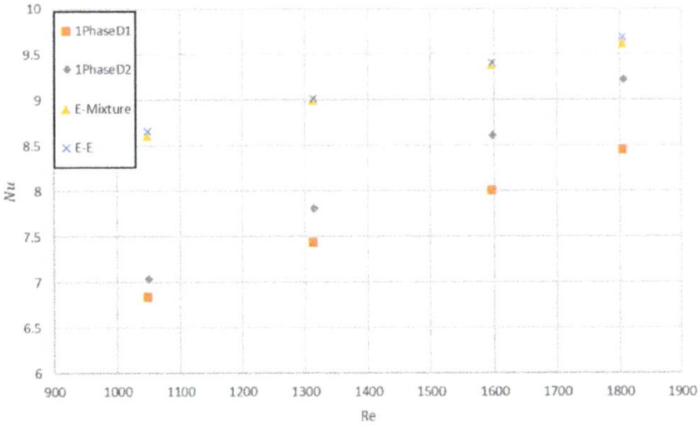

Figure 4. Nu-Re relationship for water–Al_2O_3 (1.6%) nanofluid and x/D = 116—data source [39].

Saha and Paul [40] studied the influence of nanoparticle size, NPC, and Brownian motion on heat transfer and WSS during the turbulent flow of a nanofluid inside a uniformly heated tube. The single-phase approach with k-ε turbulence model was applied to solve governing equations. It was established that heat transfer is more affected by the diameter of nanoparticles and Brownian motion than the thermal conductivity of a nanofluid Bianco et al. [41] studied the influence of NPC on the turbulent flow of nanofluids in tubes with a constant wall temperature. A mixture model of nanofluid with k-ε turbulence model was adopted. Although Nu number increases with NPC increase, PD increases more, so the application of nanofluids is questionable. In [42] Saha and Paul applied a Eulerian–Eulerian multi-phase mixture model to analyze the turbulent flow of nanofluids in a uniformly heated tube. It was found that heat transfer performance determined by the multi-phase mixture model is better than the single-phase model [40]. Aghaei et al. [43] simulated the effect of NPC and nanoparticle diameter on the turbulent behavior of nanofluids during flow in tube with a constant temperature. A mixture model with k-ε turbulence model was applied. It was shown that there is an optimum value for the Nu number depending on the NPC. Moreover, the increase of nanoparticle diameter results in a Nu number decrease Minea [44] studied the turbulent flow of nanofluids in a long tube with an isothermal entrance zone and a uniformly heated section of the developed region. A single-phase approach with the k-ε turbulence model has been applied. It was established that HTC increases monotonically with NPC. Although PP increases with NPC in three case studies, the opposite trend was determined in one case study. Nasiri-lohesara [45] stated that the enhancement of heat transfer with NPC increased during turbulent nanofluid flow inside a tube-in-tube heat exchanger. However, the HTC rate of increase was less than the WSS increase. A mixture model of nanofluid with k-ε turbulence model was applied. Ehsan and Noor [46] established a substantial enhancement of heat transfer during the flow of nanofluid in a rough tube compared to a smooth tube. Moreover, it was shown that there is an optimum value of NPC for which the PP is less than the base liquid. A single-phase approach with an SST k-ω turbulence model was applied. Mahdavi et al. [47] tested different turbulence models by use of a mixture model and DPM implemented in ANSYS-FLUENT code to study the behavior of nanofluids flow in uniformly heated tubes of various lengths and diameters. According to Mahdavi et al. the realizable and standard k-ε models provided the same results in most of the simulations. The Reynolds stress model (RSM) overestimates PD compared with the other k-ε models, while the renormalization group (RNG) model overestimates HTC. The DPM model is recommended for the discussed cases, however, as it was stressed, the number of particles plays a key role in the simulations. Purohit et al. [48] studied the laminar flow of nanofluids inside a uniformly heated tube by the use of the single-phase approach. It was established that HTC and WSS increase monotonically with NPC increase. Jahanbin [49] used single-phase and dispersion models to study the laminar flow of nanofluids in a uniformly heated tube. Moreover, in both models, constant and temperature dependent properties of nanofluids were tested. It was established that dispersion model shows better agreement with the existing experimental data than the single-phase model. Elahmer et al. [50] studied the laminar flow of nanofluids in a tube with a wall loaded with constant or periodic heat flux. A single-phase approach was applied. Regardless of the method of heating the wall of the tube, HTC increased with NPC increase. Albojamal and Vafai [51] conducted simulations of laminar nanofluid flow in a tube with a uniformly heated wall. A single-phase model, the Lagrangian–Eulerian model (DPM), and the mixture model were adopted with both constant and temperature-dependent properties of nanofluids. It was observed that the DPM model overestimated the HTC. The mixture model predicted unreasonable heat transfer enhancement, particularly for high NPC. The single-phase model displayed a very good agreement with the experimental data. It was established that HTC increases with NPC and Re number increase, but it is punished by PD and WSS increases. Finally, it was concluded that the proposed single-phase model reproduces the experimental data with sufficient accuracy and there is no need for two-phase models. In Figure 5, results of numerical calculations by the use of the single-phase (1Phase) approach and two two-phase models, namely DPM and mixture, with constant (con.) and variable (var.) thermophysical properties of nanofluid, are shown.

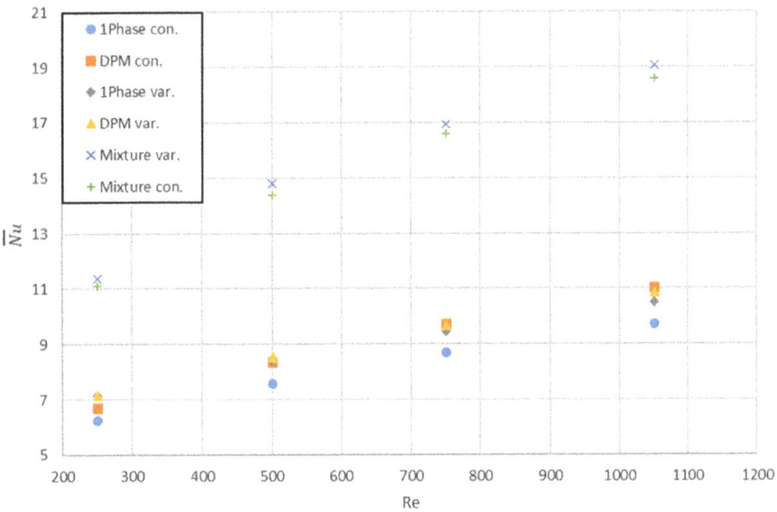

Figure 5. $\overline{\text{Nu}}$-Re relationship for water–Al_2O_3 (4%) nanofluid—data source [51].

Rashidi et al. [52] used single-phase model, VOF model, mixture model, and Eulerian model to study entropy generation during turbulent flow of nanofluids in a uniformly heated tube. The k-ε turbulence model was applied. It was established that the entropy generation were very similar for the single-phase and mixture models. Rabby et al. [53] studied the laminar flow of nanofluids in a uniformly heated tube by use of the single-phase approach. It was determined that a substantial HTC increase is accompanied by a small FF increase with NPC increase. Boertz et al. [54] conducted a numerical simulation of turbulent nanofluid flow in a tube with a uniformly heated wall by use of the single-phase approach with an SST k-ω turbulence model. It was established that the Nu number increases significantly with NPC increase; however, FF and PP also increase. Kristiawan et al. [54] used a Eulerian approach to study laminar and turbulent flows of nanofluid in a uniformly heated tube. For turbulent flow, the k-ε turbulence model was applied. It was determined that the HTC of nanofluid increases considerably compared to the base liquid in both laminar and turbulent flow. According to [55] the Eulerian approach should be applied for nanofluids with higher NPC. Sajjad et al. [56] studied the laminar flow of nanofluid in a uniformly heated tube by use of the single-phase model with temperature-dependent properties. It was established that an increase in local HTC is more distinct in the entrance region. Moreover, the addition of nanoparticles results in a higher average HTC, but is punished with higher PD with NPC increase. Minea et al. [57] conducted an inter-comparative study on the 3D laminar flow of nanofluids in uniformly heated tube by the use of four single-phase models and two mixture models. It was established that numerical results overpredict the experimental data and the two-phase approach is more appropriate to simulate nanofluid behavior. Onyiriuka and Ikponmwoba [58] applied a mixture model to study the laminar flow of nanofluids inside a uniformly heated tube. Constant properties of the tested bio-nanofluid were assumed. It was established that HTC increases with NPC increase. Jamali and Toghraie [59] studied the transition from laminar to turbulent flow of nanofluid in an isothermal tube by use of the single-phase approach. The effect of the nanoparticle diameter, NPC, and type of nanoparticles was studied. It was established that the addition of nanoparticles does not affect the onset of transition. Fadodun et al. [60] studied the turbulent flow of nanofluids in a uniformly heated tube by the use of the single-phase approach. The k-ε turbulence model was applied with temperature-dependent properties of nanofluid. Heat transfer enhancement with simultaneous PD increase was observed for NPC increase. Saeed and Al-Dulaimi [61] applied the single-phase approach and four different sets of thermophysical properties to study the laminar flow of a nanofluid in a uniformly heated tube. It was determined that both temperature-dependent and temperature-independent models of thermophysical properties of nanofluids can correctly reproduce the behavior of nanofluids during laminar flow. Uribe et al. [62] studied the 3D laminar and turbulent flows of nanofluid in a uniformly heated tube by the use of the single-phase approach. It was established that HTC increases with NPC increase, while thermal boundary thickness decreases with NPC increase. It was stated that there is no need to apply two-phase models to describe nanofluid behavior. Taskesen et al. [63] studied the influence of channel geometry (circular, square, triangular, and rectangular) on the thermo-hydraulic behavior of nanofluid during laminar flow. A single-phase model with temperature-dependent properties was applied. Heat transfer enhancement was observed with an NPC increase, and a circular cross-section was superior over other geometries. Yildiz and Aktürk [64] used a single-phase approach to study the 3D turbulent flow of a nanofluid inside a uniformly heated tube. The standard k-ε turbulence model with temperature-dependent properties was applied. A substantial increase of Nu number and HTC was recorded. However, FF increases markedly with NPC increase. In Tables 1 and 2 details of the studies in which the single-phase approach and two-phase models were applied are presented, respectively.

Table 1. Numerical studies dealing with single-phase approach.

Executors	Approach	Nanofluid	Geometry/BCon	Re Number/ NPC	Result
Maïga et al. [12]	Single-phase	water–γAl_2O_3	D = 10 mm L = 1000 mm q = const.	$10^4 < Re < 5 \times 10^4$ $0 < \varphi_v$ [%] < 10	AHTCE (60%) AWSSI (180%)
Maïga et al. [13]	Single-phase	water–γAl_2O_3 EG–γAl_2O_3	D = 10 mm L= 1000 mm q = const. t_w = const.	$250 < Re < 10^3$ (Water) $6.31 < Re < 631$ (EG) $0 < \varphi_v$ [%] < 10	AHTCE (180%) AWSSI (1207%) for EG–γAl_2O_3
Maïga et al. [14]	Single-phase	water–γAl_2O_3	D = 10 mm L= 1000 mm q = const.	$10^4 < Re < 5 \times 10^5$ $0 < \varphi_v$ [%] < 10	FDHTCE (75%) FDWSSI (581%)
Heris et al. [15]	Dispersion model	water–Al_2O_3 water–CuO water–Cu	D = 5 mm L = 1000 mm t_w = const.	2500 < P e< 6500 $0.5 < \varphi_v$ [%] < 5	ANuNE (100%)
Namburu et al. [20]	Single-phase	EG-water (60:40)/ CuO, Al_2O_3, SiO_2	D = 10 mm L= 800 mm q = const	$10^4 < Re < 10^5$ $0 < \varphi_v$ [%] < 6	AHTCE (75%)
Mokmeli and Saffar-Avval [23]	• Single-phase • Dispersion model	water–γAl_2O_3 ATF-G water–Al_2O_3	D = 4.5 mm L = 970 mm D = 4.57 mm L = 457 mm D = 6 mm L = 1000 mm q = const. t_w = const.	$19.67 < Re < 1810$ $1 < \varphi_m$ [%] < 1.6	Single-phase model → NuND (31%) Dispersion model → NuNE (15%)

Table 1. *Cont.*

Executors	Approach	Nanofluid	Geometry/BCon	Re Number/ NPC	Result
Ebrahimnia-Bajestan et al. [24]	Single-phase	EG-water (60:40) water; Al_2O_3, CuO, CNT, TNT	D = 4.57 mm L = 2000 mm q = const.	500 < Re < 1460 $0 < \varphi_v$ [%] < 6	HTCE (22%) PDI (3880%)
Moraveji et al. [25]	Single-phase	water–Al_2O_3	D = 4.75 mm L = 970 mm q = const.	500 < Re < 2500 $1 < \varphi_m$ [%] < 6	HTCE (20%)
Bayat and Nikseresht [27]	Single-phase	Water-Al_2O_3 EG- Al_2O_3 EG-water (60:40)/Al_2O_3 EG-water (30:70)/Al_2O_3	D = 4.57 mm L = 2000 mm q = const.	100 < Re < 2000 $0 < \varphi v$ [%] < 9	AHTCE (42%) ASSI (500%)
Alvarino et al. [30]	Single-phase	water–Al_2O_3	D = 4.5 mm L = 970 mm q = const.	750 < Re < 1750 $0 < \varphi_v$ [%] < 1.6	HTCE (12%)
Balla et al. [32]	Single-phase	water–Al_2O_3 water–CuO water–TiO_2	D = 10 mm L = 2000 mm q = const.	100 < Re < 1000 $0 < \varphi_v$ [%] < 4	HTCE (25%)
Bayat and Nikseresht [33]	Single-phase	EG-water (60:40)/Al_2O_3	D = 10 mm L = 1000 mm q = const.	$10^4 < Re < 10^5$ $1 < \varphi_v$ [%] < 10	AHTCE (70%) PDI (230%)
Davarnejad et al. [35]	Single-phase	water–Al_2O_3	D = 6 mm L= 1000 mm h = const.	700 < Re < 2050 $0 < \varphi_v$ [%] < 2.5	HTCE (25%)
Davarnejad et al. [36]	Single-phase	water–Al_2O_3	D = 6 mm L = 1000 mm q = const.	700 < Re < 2050 $0 < \varphi_v$ [%] < 2.5	AHTCE (7%)
Kayaci et al. [37]	Single phase	water–TiO_2	D = 8.13 mm L = 1500 mm t_w = const.	7233 < Re < 13,340 $0.2 < \varphi_v$ [%] < 1	HTCE (3%) PDI (6.5%) FFI (5%) WSSI (5%)
Saha and Paul [40]	Single-phase	water–Al_2O_3 water–TiO_2	D = 19 mm L = 1000 mm q = const.	$10^4 < Re < 10^5$ $0 < \varphi_v$ [%] < 6	ANuNE (60%) AWSSI (1363%)
Minea [44]	Single-phase	water–Al_2O_3	D= 12 mm L = 8760 mm q = const.	$10^7 < Re < 3 \times 10^7$ $0 < \varphi_v$ [%] <4	HTCE (26%) PPI (16%) PPD (10%)
Ehsan and Noor [46]	Single-phase	water–Al_2O_3	D = 3 mm L = 300 mm q = const.	$10^4 < Re < 3 \times 10^4$ 1–5%vol.	HTCE (39%) PPI (203%)
Purohit et al. [48]	Single-phase	water–Al_2O_3 water–ZrO_2 water–TiO_2	L = 1000 mm q = const.	1150 < Re < 1900 $0.5 < \varphi_v$ [%] < 2	HTCE (18%) WSSI (3%)
Jahanbin [49]	• Single phase • Dispersion model	water– Al_2O_3	D = 10 mm L = 1000 mm q = const.	250 < Re < 1250 $1 < \varphi_m$ [%] < 4	HTCE (49%) WSSI (156%)

Table 1. *Cont.*

Executors	Approach	Nanofluid	Geometry/BCon	Re Number/ NPC	Result
Elahmer et al. [50]	Single-phase	EG-CNT EG-CNT-Ag	D = 10 mm q = const. q-periodic	Pe = 10^3 and Pe = 2 × 10^3 0 < $\varphi_{v,CNT}$ [%] < 10 $\varphi_{v,Ag}$ [%] = 1	HTCE (67%)
Rabby et al. [53]	Single-phase	water–Al_2O_3	D = 5 mm L = 750 mm q = const.	100 < Re < 1400 1 < φ_v [%] < 5	HTCE (32%) NuNE (15%) FFI (2%)
Boertz et al. [54]	Single-phase	EG/W (60:40) $_{wt}$–SiO_2	D= 3.14 mm L= 1168 mm q = const.	6 × 10^3 < Re < 1.2 × 10^4 2 < φ_v [%] < 10	ANuNE (15%) FFI (13%)
Sajjad et al. [56]	Single-phase	EG/W (40:60) $_{wt}$–GO	D = 4.5 mm L = 2000 mm q = const.	400 < Re < 2000 0.01 < φ_m [%] < 0.1	HTCE (13%) PDI (112%)
Jamali and Toghraie [59]	Single-phase	water–Al_2O_3 water–CuO	D = 8.14 mm L= 7400 mm t_w= const.	500 < Re < 13,000 0<φ_v [%] <4	AHTCE (14%)
Fadodun et al. [60]	Single-phase	water–Al_2O_3	D = 20 mm L = 2000 mm q = const.	5000 < Re < 15,000 1 < φ_v [%] < 5	AHTCE PDI (values not specified)
Saeed and Al-Dulaimi [61]	Single-phase	water–TiO_2	D = 4 mm L = 2000 mm q = const.	Re = 900 0.6 < φ_v [%] < 1.18	HTCE (28%)
Uribe et al. [62]	Single-phase	water–CuO water–Al_2O_3 water–Fe_2O_3	D =25.4 mm L = 400 mm q = const.	10^3 < Re < 11 × 10^3 0.01 < φ_v [%] < 0.05	HTCE (18%)
Taskesen et al. [63]	Single-phase	water–Fe_3O_4	D =16 mm L = 1500 mm q = const.	500 < Re < 2000 1 < φ_v [%] < 5	ANuNE (4%) PDI (50%)
Yildiz and Aktürk [64]	Single-phase	water–Al_2O_3	D = 17 mm L = 350 mm q = const.	4000 < Re < 9000 1 < φ_v [%] < 4	HTCE (53%) ANuNE (32%) FFI (41%)

Table 2. Numerical studies dealing with two-phase approach.

Executors	Approach	Nanofluid	Geometry/BCon	Re Number/ NPC	Result
Behzadmehr et al. [16]	• Mixture model • Single-phase	water–Cu	0≤ x/D ≤ 100 q = const.	10,515 < Re < 22,540 φ_v = 1%	ANuNE (15%) FC ≈ const.
Bianco et al. [17]	• Discrete-phase • Single-phase	water–Al_2O_3	D = 10 mm L = 1000 mm q = const.	Re = 1050 1 < φ_v [%] < 4	HTCE (20%) WSSI (54%)
Bianco et al. [18]	• Discrete-phase • Single-phase	water–γAl_2O_3	D = 10 mm L = 1000 mm q = const.	250 < Re < 1050 1 < φ_v [%] < 4	NuNE(17%) WSSI (101%)
He et al. [19]	• Euler-DPM • Single-phase	water–TiO_2	D = 4 mm L = 2000 mm q = const.	Re = 900, Re = 1500 0.24 < φ_v [%] < 1.18	HTCE (14%)

Table 2. *Cont.*

Executors	Approach	Nanofluid	Geometry/BCon	Re Number/ NPC	Result
Fard et al. [21]	• Two-phase (CFX code) • Single-phase	water–Cu water–CuO water–Al_2O_3	D = 6 mm L = 1000 mm t_w = const.	700 < Re < 2050 $0 < \varphi_v$ [%] < 3	AHTCE (54%) for CuO
Lotfi et al. [22]	• Mixture model • Eulerian model • Single-phase	water–Al_2O_3	D = 45 mm L = 970 mm q = const.	700 < Re < 1800 $2 < \varphi_v$ [%] < 7	NuND (46%)
Bianco et al. [26]	• Mixture model • Single-phase	water–Al_2O_3	D = 10 mm L = 1000 mm q = const.	10^4 < Re < 10^5 $0 \leq \varphi_m$ [%] < 6	HTCE (20%)
Akbari [28]	• VOF • Mixture model • Eulerian model • Single-phase	water–Al_2O_3	D = 4.5 mm L = 1500 mm q = const	Re = 1050 and Re = 1600 φ_v [%] < 2	E model → AHTCD (15%) Single-phase model → AHTCD (44%)
Akbari [29]	• VOF • Mixture model • Eulerian model • Single-phase	water–Al_2O_3 water–Cu	D = 19 mm L = 1500 mm D = 10 mm L = 800 mm q = const.	9.7×10^4 < Re < 17.7×10^4 $1 < \varphi_v$ [%] < 2	ANuNE (213%) for VOF and Mixture model ANuNE (0%) for single-phase model FFI (13%)
Tahir and Mital [31]	Discrete-phase model	water–Al_2O_3	D = 10 mm L = 1000 mm q = const.	250 < Re < 1250 $1 < \varphi_v$ [%] < 4	AHTCE (25%)
Moraveji and Esmaeili [34]	• Discrete-phase • Single-phase	water–Al_2O_3	D = 10 mm L = 1000 mm q = const.	250 < Re < 1050 $0 < \varphi_v$ [%] < 4	AHTCE (32%)
Hejazian and Moraveji [38]	• Mixture model • Single-phase	water–TiO_2	D = 5 mm L = 1200 mm t_w = const.	4.8×10^3 < Re < 30.5×10^3 $0.05 < \varphi_v$ [%] < 0.25	ANuNE(52%)
Göktepe et al. [39]	• Eulerian–mixture model • Eulerian–Eulerian • Single-phase • Dispersion model	water–Al_2O_3	D = 4.5 mm L = 2000 mm q = const.	1050 < Re < 1810 $0.6 < \varphi_v$ [%] < 1.6	HTCD or HTCE depending on the model and (x/D) value FFI (2%)
Bianco et al. [41]	Mixture model	water–Al_2O_3	D = 10 mm L = 1000 mm tw = const.	20,103 < Re < 105 0 < φv [%] < 4	HFI (33%) PPI (450%)
Saha and Paul [42]	Eulerian–Eulerian model	water–Al_2O_3 water–TiO_2	D = 19 mm L = 1000 mm q = const.	10^4 < Re < 10^5 $0 < \varphi_v$ [%] < 6	NuNE (62%) WSSI (1335%) FCI
Aghaei et al. [43]	Mixture model	water–Al_2O_3	D = 20 mm L = 1000 mm t_w = const.	10^4 < Re < 10^5 $0 < \varphi_v$ [%] < 4	NuN ≈ const. for $\varphi_v \leq 0.02$ NuND (4%) for $\varphi_v \leq 0.02$ PDI (17%) FCI (5%)
Nasiri-lohesara [45]	Mixture model	water–γAl_2O_3	D = 10 mm L = 650 mm	2×10^4 < Re < 5×10^4 $0 < \varphi_v$ [%] < 6	HTCE (32%) WSSI (203%)
Mahdavi et al. [47]	• Mixture model • Discrete-phase	water–Al_2O_3 water–ZrO_2 water–TiO_2 water–SiO_2	D = 10.26 mm L = 3000 mm D = 3.5 mm L = 600 mm D = 9.4 mm L = 3000 mm D = 10.6 mm L = 1500 mm q = const., q = 0	3500 < Re < 63,000 $0.2 < \varphi_v$ [%] < 3.6	**Mixture model** • k-εRNG-overpredicts ExHTC (12%) • k-εS & R and RSM fit ExHTC **DPM** k-εRNG, k-εS&R, and RSM overpredict ExHTC (1.8%)

Table 2. *Cont.*

Executors	Approach	Nanofluid	Geometry/BCon	Re Number/ NPC	Result
Albojamal and Vafai [51]	• Mixture model • Discrete-Phase • Single-phase	water–Al_2O_3	D = 10 mm L = 1000 mm q = const.	$250 < Re < 1460$ $1 < \varphi_v$ [%] < 4	HTCE: • Single-phase (19%) • DPM (23%) • Mixture (90%) WSSI: • Single-phase (103%) • DPM (184%) • Mixture (80%)
Rashidi et al. [52]	• VOF • Mixture model • Eulerian model • Single-phase	water–TiO_2	D = 8.13 mm L = 1500 mm q = const.	$9 \times 10^3 < Re < 21 \times 10^3$	NHTC fits ExHTC
Kristiawan et al. [55]	Eulerian model	water–TiO_2	D = 5 mm L = 2000 mm q = const.	$500 < Re < 1200$ $4 \times 10^3 < Re < 14 \times 10^3$ $0.24 < \varphi_v$ [%] < 1.18	HTCE: • Laminar (20%) • turbulent (22%)
Minea et al. [57]	• Mixture model • Single-phase	water–TiO_2	D = 8 mm L = 2000 mm q = const.	$1 < \varphi m$ [%] < 2.5	TD (9%)
Onyiriuka and Ikponmwoba [58]	Mixture model	water–mango leaves	D = 19 mm L = 1000 mm q = const.	$250 < Re < 1250$ $1 < \varphi_v$ [%] < 3	AHTCE (12%)

3. Heat Transfer Correlation Equations

As seen in Tables 3 and 4, a limited number of numerically developed correlation equations have been proposed. It is necessary to stress the very limited range of applications of the presented correlations.

Table 3. Correlation equations for laminar flow of nanofluids.

Authors	Equation	Range
Maïga et al. [13]	$\overline{Nu} = 0.086Re^{0.55}Pr^{0.5} q = \text{const.}$ $\overline{Nu} = 0.28Re^{0.35}Pr^{0.36}\ t_w = \text{const.}$	$Re < 1000$ $6 \leq Pr \leq 753$ $0 \leq \varphi_v$ [%] ≤ 10
Moraveji et al. [25]	$Nu = 2.03\varphi_v{}^{0.06}(x/D)^{-0.37}Re^{0.293}Pr^{0.6}$	$500 < Re < 2300$ $6.8 < Pr < 11.97$ $1 < \varphi_m$ [%] < 6
Tahir and Mital [31]	$\overline{h} = 399 - 278(d_p/100) + 568(Re/1250) + 8(\varphi_v/4) + 180(d_p/100)^2 - 100(Re/1250)^2 - 72(d_p/100)(Re/1250)$	$250 < Re < 150$ water–Al_2O_3 $1 < \varphi_v$ [%] < 4
Moraveji and Esmaeili [34]	$Nu = 0.716Re^{0.314} \cdot Pr^{0.6}\ \varphi_v{}^{0.3}$	$250 < Re < 1050$ water–Al_2O_3 $0 < \varphi_v$ [%] < 4
Davarnejad et al. [35]	$Nu = 0.18665\varphi_v{}^{-0.00728}(x/D)^{0.1036}Re^{0.368718}Pr^{0.3992}$	$700 < Re < 2050$ water–Al_2O_3 $0 < \varphi_v$ [%] < 2.5
Taskesen et al. [63]	$\overline{Nu} = 0.873747Re^{0.312881} + 4.98564\varphi_v$	$500 < Re < 2000$ water–Fe_3O_4 $1 < \varphi_v$ [%] < 5

Table 4. Correlation equations for turbulent flow of nanofluids.

Authors	Equation	Range
Maïga et al. [13]	$\overline{Nu} = 0.085Re^{0.71}Pr^{0.35}$	$10^4 < Re < 5 \times 10^5$ $6.6 < Pr < 13.9$ $0 \leq \varphi_v\ [\%] \leq 10$
Hejazian and Moraveji [38]	$\overline{Nu} = 0.00218Re^{1.0037}Pr^{0.5}(1+\varphi_v)^{154.6471}$	$4800 < Re < 30{,}500$ $5.5 < Pr < 5.59$ $0 \leq \varphi_v\ [\%] \leq 0.25$
Saha and Paul [40]	water–Al_2O_3 $Nu = 0.01272Re^{0.85861}Pr^{0.42986}\left(\frac{d_f}{d_p}\right)^{-0.0017}$ water–TiO_2 $Nu = 0.01259Re^{0.85926}Pr^{0.43020}\left(\frac{d_f}{d_p}\right)^{-0.00068}$	$10^4 < Re < 10^5$ $8.45 < Pr < 20.29$ $4\% < \varphi_v < 6\%$ $10 < d_p\ [nm] < 40$
Saha and Paul [42]	water–Al_2O_3 $Nu = 0.01260Re^{0.85589}Pr^{0.44709}\left(\frac{d_f}{d_p}\right)^{-0.00176}$ water–TiO_2 $Nu = 0.01518Re^{0.84071}Pr^{0.44083}\left(\frac{d_f}{d_p}\right)^{-0.00534}$	$10^4 < Re < 10^5$ $8.45 < Pr < 20.29$ $2\% < \varphi_v < 6\%$ $10 < d_p\ [nm] < 40$
Aghaei et al. [43]	$\overline{Nu} = 0.5277 \cdot 10^{-2}Re + 0.1625 \cdot 10^5\varphi_v^2 + 0.3066 \cdot 10^3\varphi_v + 0.3084 \cdot 10^2$	$10^4 < Re < 10^5$ $Pr = 6.13$ $0.001 < \varphi_v < 0.04$ $25 < d_p\ [nm] < 100$
Fadodun et al. [60]	$\log_e \overline{Nu} = 4.23 + 0.13Re + 0.0063\varphi - 0.036T_{in}$	$10^4 < Re < 10^5$ water–Al_2O_3 $0.01 < \varphi_v < 0.05$ $20 < d_p\ [nm] < 100$ $290 < T\ [K] < 330$

4. Conclusions

As shown in Table 1, a single-phase approach was used in the majority of works. The conclusions of the researchers who used single-phase models and two-phase models regarding the comparison of both approaches are different. Some of them [22,23] observed that the accuracy of the single-phase model is similar to the two-phase models. Others [11,12] indicate that the two-phase models more faithfully reproduce the results of experimental research than the single-phase approach. It was stressed that for nanofluids, the number of particles in the computational domain, even for a very small NPC, is very large. Therefore, due to limitations of the software abilities, the Lagrangian–Eulerian approach is extremely difficult to implement.

Several investigators [19,30,31,34,39,47,51,57] implemented mechanisms influencing convective heat transfer of nanofluids previously considered by Buongiorno [65]. According to [19,30], Brownian motion and thermophoresis have a negligible impact on heat transfer during the forced convection of nanofluids. Moreover, Albojamal and Vafai [51] demonstrated that the Brownian motion, thermophoresis, virtual mass, pressure gradient, and Saffman lift force do not significantly influence mean HTC. It was shown in [47] that gravity, virtual mass, pressure gradient, Brownian, and lift forces have no impact on mean flow field. As stated in [57], gravity force has to be included in all numerical models.

There is also no consensus among researchers regarding the assumption of constant or temperature-dependent thermophysical properties of nanofluids. In [33,34,61], a greater accuracy of the model using temperature-dependent thermophysical properties was observed, compared to the model where the thermophysical properties were constant.

Generally, the standard k-ε turbulence model was used to model turbulent flows of nanofluids [14,16,20,26,29,33,37,40,55,60,62]. Only a few studies used other models of turbulence [4–8].

Virtually all the results of numerical studies indicate the intensification of the heat transfer of nanofluids compared to base liquids, regardless of the type of flow (laminar or turbulent), although the results of the experimental studies are not so clear, e.g., [66–68].

There is also full agreement that the addition of nanoparticles increases flow resistance (PD, FF, and WSS).

Many researchers point to the need to intensify research on the thermophysical properties of nanofluids, as they determine the accuracy of the single-phase model first of all. In this context, however, one must bear in mind the limitations of experimental research [69].

The heat transfer correlations proposed in the literature have a very limited range of applicability. No numerically developed correlation for determination of the friction factor of nanofluids flowing inside a horizontal smooth tube has been presented.

Future works should concern a benchmark solution for laminar and turbulent flow of nanofluids similar to that of Buongiorno et al. in relation to thermal conductivity [70]. It is necessary to conduct a comparative study of the influence of divergent thermophysical properties of base liquids on the final results concerning the influence of nanoparticles on heat transfer, as was carried out in [57].

Funding: This research received no external funding.

Conflicts of Interest: The authors declare no conflict of interest.

Nomenclature

a	Thermal diffusivity	(m^2/s)
d_f	Base fluid molecule diameter	(m)
d_p	Particle diameter	(m)
D	Inside diameter of tube	(m)
$\overline{h}$	Average heat transfer coefficient	($W/(m^2\ K)$)
k	Thermal conductivity	($W/(m\ K)$)
$Nu = \frac{hD}{k}$	Local Nusselt number	(-)
$\overline{Nu} = \frac{\overline{h}D}{k}$	Average Nusselt number	(-)
$Pe = RePr$	Peclet number	(-)
$Pr = \frac{\upsilon}{a}$	Prandtl number	(-)
q	Heat flux	(W/m^2)
$Rr = \frac{uD}{\nu}$	Reynolds number	(-)
u	Velocity	(m/s)
x	Axial coordinate	(m)
Greek symbols		
υ	Kinematic viscosity	(m^2/s)
φ	Nanoparticle concentration	(-)
Subscripts		
In	Inlet	
m	Mass	
nf	Nanofluid	
p	Particle	
v	Volume	
w	Wall	
Abbreviations		
1Phase con.	Single phase model with constant properties	
1PhaseD1	Single phase model with thermal dispersion effect	
1PhaseD2	Single phase model with thermal dispersion effect	
1Phase var.	Single phase model with variable properties	
AHTCD	Average heat transfer coefficient enhancement	
AHTCE	Average heat transfer coefficient enhancement	
ANuNE	Average nu number enhancement	

ASSI	Average shear stress increase
ATF	Automatic transmission fluid
AWSSI	Average wall shear stress increase
BCon	Boundary conditions
CFD	Computational fluid dynamics
CNT	Carbon nanotubes
CPU	Computational processing unit
DPM	Discrete phase model
DPM con.	Discrete phase model with constant properties
DPM var.	Discrete phase model with variable properties
E	Eulerian
E-E	Eulerian–Eulerian model
E-Mixture	Eulerian–mixture model
EG	Ethylene glycol
ExHTC	Experimental heat transfer coefficient
FC	Friction coefficient
FCI	Friction coefficient increase
FF	Friction factor
FFI	Friction factor increase
FDHTCE	Fully developed heat transfer coefficient enhancement
FDWSSI	Fully developed wall shear stress increase
G	Graphite
GO	Graphene oxide
HFI	Heat flux increase
HTC	Heat transfer coefficient
HTCE	Heat transfer coefficient enhancement
HTCD	Heat transfer coefficient deterioration
HTNI	Heat transfer negligible impact
FF	Friction factor
FFI	Friction factor increase
NHTC	Numerical heat transfer coefficient
NPC	Nanoparticle concentration
NuND	Nusselt number deterioration
NuNE	Nusselt number enhancement
PD	Pressure drop
PDD	Pressure drop decrease
PDI	Pressure drop increase
PP	Pumping power
PPD	Pumping power decrease
PPI	Pumping power increase
RNG	Re-normalization group
RSM	Reynolds stress model
TD	Temperature decrease
TNT	Titanate
VOF	Volume of fluid
WSS	Wall shear stress
WSSI	Wall shear stress increase

References

1. Webb, R.L. *Principles of Enhanced Heat Transfer*; John Wiley&Sons, Inc.: New York, NY, USA, 1994.
2. Mousa, M.H.; Miljkovic, N.; Nawaz, K. Review of heat transfer enhancement techniques for single phase flows. *Renew. Sust. Energ. Rev.* **2021**, *137*, 110566. [CrossRef]
3. Gupta, M.; Singh, V.; Kumar, R.; Said, Z. A review on thermophysical properties of nanofluids and heat transfer applications. *Renew. Sust. Energ. Rev.* **2017**, *74*, 638–670. [CrossRef]

4. Angayarkanni, S.A.; Philip, J. Review on thermal properties of nanofluids: Recent developments. *Adv. Colloid Interface Sci.* **2015**, *225*, 146–176. [CrossRef] [PubMed]
5. Ilyas, S.U.; Pendyala, R.; Marneni, N. Stability of nanofluids. Topics in Mining, Metallurgy and Materials Engineering. In *Engineering Applications of Nanotechnology. From Energy to Drug Delivery*; Korada, V.S., Hamid, N.H.B., Eds.; Springer: Berlin/Heidelberg, Germany, 2017. Available online: https://www.springerprofessional.de/engineering-applications-of-nanotechnology/11992454 (accessed on 1 July 2022).
6. Abdullah, M.; Malik, S.R.; Iqbal, M.H.; Sajid, M.M.; Shad, N.A.; Hussain, S.Z.; Razzaq, W.; Javed, Y. Sedimentation and stabilization of nano-fluids with dispersant. *Colloids Surf. A Physicochem. Eng. Asp.* **2018**, *554*, 86–92. [CrossRef]
7. Kamyar, A.; Saidur, R.; Hasanuzzaman, M. Application of Computational Fluid Dynamics (CFD) for nanofluids. *Int. J. Heat Mass Transf.* **2012**, *55*, 4104–4115. [CrossRef]
8. Kumar, S.; Chakrabarti, S. A Review: Enhancement of Heat Transfer with Nanofluids. *Int. J. Eng. Res. Technol.* **2014**, *3*, 549–557.
9. Kakaç, S.; Pramuanjaroenkij, A. Single-phase and two-phase treatments of convective heat transfer enhancement with nanofluids-A state-of-the-art review. *Int. J. Therm. Sci.* **2016**, *100*, 75–97. [CrossRef]
10. Mahian, O.; Kolsi, L.; Amani, M.; Estellé, P.; Ahmadi, G.; Kleinstreuer, C.; Marshall, J.S.; Siavashi, M.; Taylor, R.A.; Niazmad, H.; et al. Recent advances in modeling and simulation of nanofluid flows-Part I: Fundamentals and theory. *Phys. Rep.* **2019**, *790*, 1–48. [CrossRef]
11. Mahian, O.; Kolsi, L.; Amani, M.; Estellé, P.; Ahmadi, G.; Kleinstreuer, C.; Marshall, J.S.; Taylor, R.A.; Abu-Nada, E.; Rashidi, S.; et al. Recent advances in modeling and simulation of nanofluid flows-Part II: Applications. *Phys. Rep.* **2019**, *791*, 1–59. [CrossRef]
12. Maïga, S.; Nguyen, C.T.; Galanis, N.; Roy, G. Heat transfer behaviours of nanofluids in a uniformly heated tube. *Superlattices Microstruct.* **2004**, *35*, 543–557. [CrossRef]
13. Maïga, S.; Palm, S.J.; Nguyen, C.T.; Roy, G.; Galanis, N. Heat transfer enhancement by using nanofluids in forced convection flows. *Int. J. Heat Fluid Flow.* **2005**, *26*, 530–546. [CrossRef]
14. Maïga, S.; Nguyen, C.T.; Galanis, N.; Roy, G.; Maré, T.; Coqueux, M. Heat transfer enhancement in turbulent tube flow using Al_2 O_3 nanoparticle suspension. *Int. J. Numer. Methods Heat Fluid Flow* **2006**, *16*, 275–292. [CrossRef]
15. Heris, S.Z.; Esfahany, M.N.; Etemad, G. Numerical Investigation of Nanofluid Laminar Convective Heat Transfer through a Circular Tube. *Numer. Heat Transf. A Appl.* **2007**, *52*, 1043–1058. [CrossRef]
16. Behzadmehr, A.; Saffar-Avval, M.; Galanis, N. Prediction of turbulent forced convection of a nanofluid in a tube with uniform heat flux using a two phase approach. *Int. J. Heat Fluid Flow* **2007**, *28*, 211–219. [CrossRef]
17. Bianco, V.; Manca, O.; Nardini, S. Numerical investigation of transient single phase forced convection of nanofluids in circular tubes. *WIT Trans. Eng. Sci.* **2008**, *61*, 3–12.
18. Bianco, V.; Chiacchio, F.; Manca, O.; Nardini, S. Numerical investigation of nanofluids forced convection in circular tubes. *Appl. Therm. Eng.* **2009**, *29*, 3632–3642. [CrossRef]
19. He, Y.; Men, Y.; Zhao, Y.; Lu, H.; Ding, Y. Numerical investigation into the convective heat transfer of TiO_2 nanofluids flowing through a straight tube under the laminar flow conditions. *Appl. Therm. Eng.* **2009**, *29*, 1965–1972. [CrossRef]
20. Namburu, P.K.; Das, D.K.; Tanguturi, K.M.; Vajjha, R.S. Numerical study of turbulent flow and heat transfer characteristics of nanofluids considering variable properties. *Int. J. Therm. Sci.* **2009**, *48*, 290–302. [CrossRef]
21. Fard, M.H.; Esfahany, M.N.; Talaie, M.R. Numerical study of convective heat transfer of nanofluids in a circular tube two-phase model versus single-phase model. *Int. Commun. Heat Mass Transf.* **2010**, *37*, 91–97. [CrossRef]
22. Lotfi, R.; Saboohi, Y.; Rashidi, A.M. Numerical study of forced convective heat transfer of Nanofluids: Comparison of different approaches. *Int. Commun. Heat Mass Transf.* **2010**, *37*, 74–78. [CrossRef]
23. Mokmeli, A.; Saffar-Avval, M. Prediction of nanofluid convective heat transfer using the dispersion model. *Int. J. Therm. Sci.* **2010**, *49*, 471–478. [CrossRef]
24. Ebrahimnia-Bajestan, E.; Niazmand, H.; Duangthongsuk, W.; Wongwises, S. Numerical investigation of effective parameters in convective heat transfer of nanofluids flowing under a laminar flow regime. *Int. J. Heat Mass Transf.* **2011**, *54*, 4376–4388. [CrossRef]
25. Moraveji, M.K.; Darabi, M.; Haddad, S.M.H.; Davarnejad, R. Modeling of convective heat transfer of a nanofluid in the developing region of tube flow with computational fluid dynamics. *Int. Commun. Heat Mass Transf.* **2011**, *38*, 1291–1295. [CrossRef]
26. Bianco, V.; Manca, O.; Nardini, S. Numerical investigation on nanofluids turbulent convection heat transfer inside a circular tube. *Int. J. Therm. Sci.* **2011**, *50*, 341–349. [CrossRef]
27. Bayat, J.; Nikseresht, A.H. Investigation of the different base fluid effects on the nanofluids heat transfer and pressure drop. *Heat Mass Transf.* **2011**, *47*, 1089–1099. [CrossRef]
28. Akbari, M.; Galanis, N.; Behzadmehr, A. Comparative analysis of single and two-phase models for CFD studies of nanofluid heat transfer. *Int. J. Therm. Sci.* **2011**, *50*, 1343–1354. [CrossRef]
29. Akbari, M.; Galanis, N.; Behzadmehr, A. Comparative assessment of single and two-phase models for numerical studies of nanofluid turbulent forced convection. *Int. J. Heat Fluid Flow.* **2012**, *37*, 136–146. [CrossRef]
30. Alvarino, P.F.; Jabardo, J.M.S.; Arce, A.; Galdo, M.I.L. Heat transfer enhancement in nanofluids. A numerical approach. *J. Phys. Conf. Ser.* **2012**, *395*, 012116. [CrossRef]

31. Tahir, S.; Mital, M. Numerical investigation of laminar nanofluid developing flow and heat transfer in a circular channel. *Appl. Therm. Eng.* **2012**, *39*, 8–14. [CrossRef]
32. Balla, H.H.; Abdullah, S.; Zulkifli, R. Effect of oxides nanoparticles materials on the pressure loss and heat transfer on nanofluids in circular pipes. *J. Appl Sci.* **2012**, *12*, 1396–1401. [CrossRef]
33. Bayat, J.; Nikseresht, A.H. Thermal performance and pressure drop analysis of nanofluids in turbulent forced convective flows. *Int. J. Therm. Sci.* **2012**, *60*, 236–243. [CrossRef]
34. Moraveji, M.K.; Esmaeili, E. Comparison between Single-Phase and Two-Phases CFD Modeling of Laminar Forced Convection Flow of Nanofluids in Circular Tube under Constant Heat Flux. *Int. Commun. Heat Mass Transf.* **2012**, *39*, 1297–1302. [CrossRef]
35. Davarnejad, R.; Barati, S.; Zakeri, M. Simulation of Convective Heat Transfer of a Nanofluid in a Circular Cross-section. *Int. J. Eng.* **2013**, *26*, 571–576. [CrossRef]
36. Davarnejad, R.; Barati, S.; Kooshki, M. CFD simulation of the effect of particle size on the nanofluids convective heat transfer in the developed region in a circular tube. *Springerplus* **2013**, *2*, 192. [CrossRef] [PubMed]
37. Kayaci, N.; Balcilar, M.; Tabatabaei, M.; Celen, A.; Yıldız, O.; Dalkilic, A.S.; Wongwises, S. Determination of the Single-Phase Forced Convection Heat Transfer Characteristics of TiO_2 Nanofuids Flowing in Smooth and Micro-Fin Tubes by Means of CFD and ANN Analyses. *Curr. Nanosci.* **2013**, *9*, 61–80.
38. Hejazian, M.; Moraveji, M.K. A Comparative Analysis of Single and Two-Phase Models of Turbulent Convective Heat Transfer in a Tube for TiO_2 Nanofluid with CFD. *Numer. Heat Transf. A Appl.* **2013**, *63*, 795–806. [CrossRef]
39. Göktepe, S.; Atalık, K.; Ertürk, H. Comparison of single and two-phase models for nanofluid convection at the entrance of a uniformly heated tube. *Int. J. Therm. Sci.* **2014**, *80*, 83–92. [CrossRef]
40. Saha, G.; Paul, M.C. Numerical analysis of the heat transfer behaviour of water based Al_2O_3 and TiO_2 nanofluids in a circular pipe under the turbulent flow condition. *Int. Commun. Heat Mass Transf.* **2014**, *56*, 96–108. [CrossRef]
41. Bianco, V.; Manca, O.; Nardini, S. Performance analysis of turbulent convection heat transfer of Al_2O_3 water-nanofluid in circular tubes at constant wall temperature. *Energy* **2014**, *77*, 403–413. [CrossRef]
42. Saha, G.; Paul, M.C. Heat transfer and entropy generation of turbulent forced convection flow of nanofluids in a heated pipe. *Int. Commun. Heat Mass Transf.* **2015**, *61*, 26–36. [CrossRef]
43. Aghaei, A.; Sheikhzadeh, G.A.; Dastmalchi, M.; Forozande, H. Numerical investigation of turbulent forced-convective heat transfer of Al_2O_3–water nanofluid with variable properties in tube. *Ain Shams Eng. J.* **2015**, *6*, 577–585. [CrossRef]
44. Minea, A.A. Simulation of Nanofluids Turbulent Forced Convection at High Reynolds Number: A Comparison Study of Thermophysical Properties Influence on Heat Transfer Enhancement. *Flow Turbul Combust.* **2015**, *94*, 555–575. [CrossRef]
45. Nasiri-lohesara, M. Heat Transfer Enhancement and Hydrodynamic Characteristics of Nanofluid in Turbulent Flow Regime. *J. Energy* **2015**, *2015*, 814717. [CrossRef]
46. Ehsan, M.M.; Noor, S. Study of Heat Transfer Performance and Pumping Power Improvement of Nanofluid Through a Rough Circular Tube. *Int. Res. J. Eng. Technol.* **2016**, *3*, 1709–1717.
47. Mahdavi, M.; Sharifpur, M.; Meyer, J.P. Simulation study of convective and hydrodynamic turbulent nanofluids by turbulence models. *Int. J. Therm. Sci.* **2016**, *110*, 36–51. [CrossRef]
48. Purohit, N.; Purohit, V.A.; Purohit, K. Assessment of nanofluids for laminar convective heat transfer: A numerical study. *Eng. Sci. Technol. Int. J.* **2016**, *19*, 574–586. [CrossRef]
49. Jahanbin, A. Comparative Study on Convection and Wall Characteristics of Al_2O_3—Water Nanofluid Flow inside a Miniature Tube. *Eng. J.* **2017**, *20*, 169–181. [CrossRef]
50. Elahmer, M.; Abboudi, S.; Boukadida, N. Nanofluid effect on forced convective heat transfer inside a heated horizontal tube. *Int. J. Heat Technol.* **2017**, *35*, 874–882. [CrossRef]
51. Albojamal, A.; Vafai, K. Analysis of single phase, discrete and mixture models, in predicting nanofluid transport. *Int. J. Heat Mass Transf.* **2017**, *108*, 225–237. [CrossRef]
52. Rashidi, M.M.; Nasiri, M.; Shadloo, M.S.; Yang, Z. Entropy Generation in a Circular Tube Heat Exchanger Using Nanofluids: Effects of Different Modeling Approaches. *Heat Transf. Eng.* **2017**, *38*, 853–866. [CrossRef]
53. Rabby, I.I.; Amin, S.A.M.S.; Rahman, S.; Islam, A.K.M.S. Numerical Investigation of Laminar Convective Heat Transfer and Friction Factor of a Pipe by Using Al_2O_3-Water Nanofluid. In Proceedings of the International Conference on Mechanical, Industrial and Energy Engineering, Khulna, Bangladesh, 23–24 December 2018.
54. Boertz, H.; Baars, A.J.; Cieśliński, J.T.; Smolen, S. Numerical Study of Turbulent Flow and Heat Transfer of Nanofluids in Pipes. *Heat Transf. Eng.* **2018**, *39*, 241–251. [CrossRef]
55. Kristiawan, B.; Santoso, B.; Wijayanta, A.T.; Aziz, M.; Miyazaki, T. Heat Transfer Enhancement of TiO_2/Water Nanofluid at Laminar and Turbulent Flows: A Numerical Approach for Evaluating the Effect of Nanoparticle Loadings. *Energies* **2018**, *11*, 1584. [CrossRef]
56. Sajjad, M.; Kamran, M.S.; Shaukat, R.; Zeinelabdeen, M.I.M. Numerical investigation of laminar convective heat transfer of graphene oxide/ethylene glycol-water nanofluids in a horizontal tube. *Eng. Sci. Technol. Int. J.* **2018**, *21*, 727–735. [CrossRef]
57. Minea, A.A.; Buonomo, B.; Burggraf, J.; Ercole, D.; Karpaiya, K.R.; Di Pasqua, A.; Sekrani, G.; Steffens, J.; Tibaut, J.; Wichmann, N.; et al. NanoRound: A benchmark study on the numerical approach in nanofluids' simulation. *Int. Commun. Heat Mass Transf.* **2019**, *108*, 104292. [CrossRef]

58. Onyiriuka, E.J.; Ikponmwoba, E.A.A. Numerical investigation of mango leaves-water nanofluid under laminar flow regime *Niger. J. Technol.* **2019**, *38*, 348–354. [CrossRef]
59. Jamali, M.; Toghraie, D. Investigation of heat transfer characteristics in the developing and the developed flow of nanofluid inside a tube with different entrances in the transition regime. *J. Therm. Anal. Calorim.* **2020**, *139*, 685–699. [CrossRef]
60. Fadodun, O.G.; Amosun, A.A.; Salau, A.O.; Olaloye, D.O.; Ogundeji, J.A.; Ibitoye, F.I.; Balogun, F.A. Numerical investigation and sensitivity analysis of turbulent heat transfer and pressure drop of Al_2O_3/H_2O nanofluid in straight pipe using response surface methodology. *Arch. Thermodyn.* **2020**, *41*, 3–30.
61. Saeed, F.R.; Al-Dulaimi, M.A. Numerical investigation for convective heat transfer of nanofluid laminar flow inside a circular pipe by applying various models. *Arch. Thermodyn.* **2021**, *42*, 71–95.
62. Uribe, S.; Zouli, N.; Cordero, M.E.; Al-Dahhan, M. Development and validation of a mathematical model to predict the thermal behaviour of nanofluids. *Heat Mass Transf.* **2021**, *57*, 93–110. [CrossRef]
63. Taskesen, E.; Tekir, M.; Gedik, E.; Arslan, K. Numerical investigation of laminar forced convection and entropy generation of Fe_3O_4/water nanofluids in different cross-sectioned channel geometries. *J. Therm. Eng.* **2021**, *7*, 1752–1767. [CrossRef]
64. Yildiz, M.; Aktürk, A. Numerical Investigation on Heat Transfer and Hydraulic Performance of Al_2O_3–Water Nanofluid as a Function of Reynolds Number and Flow Velocity. *Int. J. Heat Mass Transf.* **2021**, *11*, 535–547. [CrossRef]
65. Buongiorno, J. Convective Transport in Nanofluids. *Int. J. Heat Trans.* **2006**, *128*, 240–250. [CrossRef]
66. Williams, W.; Buongiorno, J.; Hu, L.W. Experimental Investigation of Turbulent Convective Heat Transfer and Pressure Loss of Alumina/Water and Zirconia/Water Nanoparticle Colloids (Nanofluids) in Horizontal Tubes. *J. Heat Trans.* **2008**, *130*, 042412. [CrossRef]
67. Sommers, A.D.; Yerkes, K.L. Experimental investigation into the convective heat transfer and system-level effects of Al_2O_3-propanol nanofluid. *J. Nanopart Res.* **2009**, *12*, 1003–1014. [CrossRef]
68. Sahin, B.; Manay, E.; Akyurek, E.F. An Experimental Study on Heat Transfer and Pressure Drop of CuO-Water Nanofluid. *J. Nanomater.* **2015**, *16*, 790839. [CrossRef]
69. Zhang, X.; Li, J. A review of uncertainties in the study of heat transfer properties of nanofluids. *Heat Mass Transf.* **2022**, 1–33. [CrossRef]
70. Buongiorno, J.; Venerus, D.C.; Prabhat, N.; McKrell, T.J.; Townsend, J.; Christianson, R.J.; Tolmachev, Y.V.; Keblinski, P.; Hu, L.-W.; Alvarado, J.L.; et al. A benchmark study on the thermal conductivity of nanofluids. *J. Appl. Phys.* **2009**, *106*, 094312. [CrossRef]

energies

Article

On the Simulations of Thermal Liquid Foams Using Lattice Boltzmann Method

Mohammad Mobarak [1,*], Bernhard Gatternig [1,2,3] and Antonio Delgado [1,3]

[1] Institute of Fluid Mechanics, Friedrich-Alexander-Universität Erlangen-Nürnberg, 91058 Erlangen, Germany
[2] Process Engineering and Circular Economy, University of Applied Sciences Weihenstephan-Triesdorf, 91746 Weidenbach, Germany
[3] German Engineering Research and Development Center, LSTME Busan, Busan 1276, Republic of Korea
* Correspondence: mohammad.mobarak@fau.de; Tel.: +49-9131-85-29484

Abstract: Liquid foams exist in a wide variety of chemical and industrial processes, and they can contaminate the end-product and cause time and economical losses. Understanding and simulating foam is not a straightforward task, due to the highly dispersed time and length scales where the physical phenomena occur. Surfactants' or proteins' length scales are far beyond the capability of macroscopic and even mesoscopic numerical fluid solvers, yet the macroscales are still required to be resolved. Meanwhile, the lattice Boltzmann method (LBM) has gained much attention and success as a mesoscopic approach which can deal with complex multiphase multicomponent systems. The aim of this study is to implement LBM to simulate liquid foams while considering the accompanying thermal effects. A coupled multiphase multicomponent thermal flow model and its selected add-ons from the literature are tuned and explained, limitations and future suggestions are fairly discussed. Validations and a final study case are shown as an example for the proposed model and its applicability in thermal liquid foams. Finally, a delicate treatment to back couple the effect of temperature on the surface tension is proposed, hence considering one aspect of the Marangoni effect. Initial results show promising behavior, which can be material for future investigations.

Keywords: lattice Boltzmann method; multiphase multicomponent flows; heat transfer; liquid foam

Citation: Mobarak, M.; Gatternig, B.; Delgado, A. On the Simulations of Thermal Liquid Foams Using Lattice Boltzmann Method. *Energies* **2023**, *16*, 195. https://doi.org/10.3390/en16010195

Academic Editors: Artur Bartosik and Dariusz Asendrych

Received: 14 November 2022
Revised: 19 December 2022
Accepted: 20 December 2022
Published: 24 December 2022

Copyright: © 2022 by the authors. Licensee MDPI, Basel, Switzerland. This article is an open access article distributed under the terms and conditions of the Creative Commons Attribution (CC BY) license (https://creativecommons.org/licenses/by/4.0/).

1. Introduction

Foams exist in several chemical and industrial processes, and they often delay and obstruct the production. Foam is a complex multiphase multicomponent system, and it often consists of one or more coexisting gas components (air and vapor) in the bubble and liquid phase represented in the lamella (water). The lamellar interfacial length can reach orders of nanometers depending on whether the stabilizing agents are surfactants or proteins [1]. The lamella of protein foams normally have higher length scales than surfactants' ones [2]. The repulsive forces between the surfactants or protein molecules in the lamellar interface often induce high positive values of disjoining pressure, which is the reason for stable foams [3]. Quantification of these disjoining pressure isotherms is often undertaken by thin film pressure balance technique, where the thickness of the thin film is controlled and measured with the film pressure until reaching the Black Newton film and, hence, rupture [4].

Thermal effects and gradients add more physical complexities to the foam system, such as evaporation–condensation, viscosity variations and the local variation of surface tension, leading to the Marangoni effect [5]. These complex phenomena interact with each other. For example, one of the most foam-dynamics influencing phenomena is the liquid drainage through the lamella [2]. However, temperature variation can highly affect the surface tension, leading to an induced flow, which may be in an unfavorable direction opposing the drainage one, hence leading to more stable foam [6]. There also exists the coarsening or Ostwald ripening in foams, where gas diffuses from smaller bubbles to larger bubbles [1]. This process is normally more important on the larger time scales, when compared to the drainage and convective mass flow.

Modelling and simulation of these interacting processes are not straightforward, although some aspects can be tackled using the mesoscopic lattice Boltzmann method (LBM), which has proven its capability of modelling complex multiphase and thermal flows [7]. The idea of the current work is to propose an approach to simulate some of these phenomena and discuss the current limitations based on the literature and current experience.

For the multiphase models in LBM, the two most famous methods are the free-energy [8,9] and the pseudopotential Shan-Chen (SC) [10,11] approaches. The limitation of the high density ratio and viscosity ratios are always issues for both methods. The adopted approach in this work is the SC model, since it is possible to extend it to approach the simulation of liquid foams by reaching positive values of disjoining pressure. The original SC model can also reach higher stable density ratios up to 100 and can be extended to reach ratios up to 1000, though with a modified equations of state (EOS). In the work by Yuan and Schaefer [12], they studied several equations of state, where Peng-Robinson (P-R) EOS [13] yielded the highest stable density ratio and the least spurious currents, which is a well-known issue in multiphase models. The higher viscosity ratios issue can generally be stabilized by switching from the standard LBM single relaxation time (SRT) collision operator to either the two or multiple relaxation time (TRT or MRT) [14]. SC model can also reach comparable and even slightly higher surface tensions than the free-energy approach [14].

There also exists the free-surface (FS) LBM approach, where the dense component is considered while the other one is neglected; this somehow resembles the Volume of Fluid (VOF) approach in LBM. In the work by Körner et al. [15] and later by Ataei et al. [16], much success has been gained with reaching stable foam by introducing linearly variable repulsive forces as a function of the lamellar distance. Interfacial curvature estimation [17] is required for the FS-LBM in general, in addition to ray tracing algorithms for the lamellar distance evaluation. Nevertheless, the presented work is mainly focused on the SC LBM approach, where neither the tracking nor the curvature estimation is required, not to mention the lamellar distance evaluation.

The SC LBM is based on modelling the molecular interaction between phases and components to reach the phase separation "bottom-up approach" [18]. The original SC can be extended to model both multiphase and multicomponent system, while being able to tune the density ratio and surface tension separately [19]. In order to simulate foams, the surfactants effects shall be implemented into these interaction forces. This can be carried out by using the mid-range interaction (repulsion) in addition to the short-range one [20,21]. This benefit emerges from the nature of the SC LBM, modeling the molecular interactions themselves. This can lead to a positive disjoining pressure, hence delaying bubbles coalescence and obtaining foam! It is worth mentioning that even using a mesoscale method such as LBM, it is practically impossible to resolve the length scales of surfactants molecules; that is why only their forces' effects are modelled.

In the literature, there also exists the model presented by Chen et al. [22] and Nekovee et al. [23], which is based on introducing two extra sets of LBM distribution functions in addition to the SC LBM system, representing the surfactants species (position) and their own dipolar orientation, although, in the current work, we preferred avoiding this approach for foam simulations for three reasons. First, the model adds much more complexity to the already saturated system of LBM distribution functions. Second, relating the relaxation time, hence the diffusivity of the surfactants position and orientation, to the physical ones could not be seen to be possible. On the other hand, in the mid-range interaction model, the resulting disjoining pressure can be evaluated and, hence, there might be a way to relate it to the experimental isotherms. Third, it is reported in the work from Mukherjee et al. [24] that the model by itself could not inhibit the bubbles coalescence.

Thermal phase change from liquid to gas and vice versa using LBM has been proposed by Gong and Cheng [25] and further enhanced by Li et al. [26]. The model is based on coupling the temperature field by including a source term responsible for the enthalpy of phase change with the SC multiphase model, which is modified by the temperature dependent P-R EOS. However, in addition to the stability issues which are encountered, maintaining both the thermodynamic consistency for phase change and obtaining positive disjoining pressure, which is an essential property of foam formation and stability, has to our knowledge not been developed in the literature.

In the literature, the Enthalpy-based LBM to model phase change for liquid to solid states also exists [27,28], which might be a compromise to investigate the phase change in extremely stable foam by only tracking the interface front between liquid and gas phases accordingly with the temperature field.

The current work aims to propose a recipe for model combination in the scope of SC LBM, in order to simulate the thermal effects in liquid foam without taking into account the phases change. The convective diffusive heat transfer and thermal effects consideration while simulating the flow field in liquid foam is the key innovation in this work. Coarsening or Ostwald ripening will not be discussed in this work, since controlling the mass diffusivity of gas in liquids using the SC model separately is numerically impossible.

The paper is arranged as follows: the LBM and SC model are briefly proposed, showing the multiphase–multicomponent simulation parameters, the implemented forcing scheme and the TRT. The mid-range interaction model is proposed and explained. The temperature field and how it is implemented is shown. Units and non-dimensional groups to change from lattice to physical units are discussed. For the following section, a model verification is carried out by relating the interaction numerical parameters to the physical ones through the Young–Laplace test, where surface tension and density ratio (using SC EOS) are tuned and evaluated for different bubble sizes. Rayleigh–Taylor instability (RTI) is shown as a validation case for the implemented multiphase–multicomponent model and compared with numerical test cases from the literature. Simulation samples from the bubble rise diagram by Clift et al. [29] are presented in brief as a further qualitative validation for different Reynolds and bonds numbers, showing the model capability of capturing the interfacial behavior for different bubble regimes. In the application section, the developed model is applied for a case study of one-way coupled heat transfer in dynamic liquid foam while using the mid-range interaction model. Finally, a new technique to consider the variable surface tension due to temperature gradients is proposed, hence suggesting a technique to account for the Marangoni effect in thermal foams using SC LBM. A summary and conclusion section is shown to finalize the whole presented model, discuss the achievement fairly and suggest future work in order to reach the final aim of simulating the foam in industrial scale processes.

2. Materials and Methods

2.1. Lattice Boltzmann Method

The lattice Boltzmann method is based on the well-known Boltzmann Equation, where particles' motion and interaction in space and time are represented by the particle distribution function f, as shown in Equation (1) [30].

$$\frac{\partial f}{\partial t} + \hat{v} \cdot \nabla f = Q \tag{1}$$

where t, $\hat{v}$ and Q are time, particle microscopic velocity and the collision operator. Q is originally found in a complex integral form which can be approximated using the Bhatnagar–Gross–Krook (BGK) relaxation, namely the single relaxation time (SRT) [31]. It represents the particles' relaxation from non-equilibrium to the local equilibrium state during collision, through the time τ as shown in Equation (2) [30], where the equilibrium distribution function f^{equ} can be formulated in the scope of the kinetic theory of gases using the Maxwell–Boltzmann distribution.

$$Q = -\frac{1}{\tau}[f - f^{equ}] \tag{2}$$

Discretizing the particle distribution function into a subspace of discrete lattice velocities i, mimicking the degrees of freedom of where the particles can propagate in space, the distribution function can be shown as $f_i(\hat{x}, t)$. This can be explained as the probability of finding a particle in position vector $\hat{x}$ at time t with a microscopic speed in the direction i, using the explicit time integration, which when based on the trapezoidal rule [32] maintains the second order accuracy of the LBM in time [14]. The lattice Boltzmann equation (LBE) finally reads as:

$$f_i(\hat{x} + \hat{e}_i\Delta t, t + \Delta t) = f_i(\hat{x}, t) - \frac{1}{\tau}(f_i(\hat{x}, t) - f_i^{equ}(\hat{x}, t)) \tag{3}$$

The above equation carries the core procedure of the LBM, showing the propagation of particles from one lattice position to the other and the collision operation, i.e., relaxation [33]. The equilibrium distribution function is based on the Hermite series expansion of the Maxwell–Boltzmann distribution with the truncation up to the second order macroscopic velocity, which is valid for low lattice Mach numbers and can be written as:

$$f_i^{equ} = w_i\rho\left[1 + 3\frac{\hat{e}_i \cdot \hat{u}}{c^2} + \frac{9}{2}\frac{(\hat{e}_i \cdot \hat{u})^2}{c^4} - \frac{3}{2}\frac{\hat{u} \cdot \hat{u}}{c^2}\right] \tag{4}$$

where $\hat{e}_i$, w_i, c, ρ and $\hat{u}$ are the lattice unit vectors, weighting parameter, lattice speed (lattice spacing/lattice time step size $\Delta x/\Delta t = 1$), macroscopic density and velocity, respectively.

The macroscopic density and velocity can be evaluated from the 0th and 1st moment of the distribution function, respectively, as follows:

$$\rho = \sum_{i=0}^{8} f_i\ ;\ \ \hat{u} = \frac{1}{\rho}\sum_{i=0}^{8} f_i\hat{e}_i \tag{5}$$

The common D2Q9 is used here representing a 2D case and 9 lattice directions. The weighting parameters for the D2Q9 according to the short-range interaction from Figure 1 (presented in red) are listed in Table 1. Although LBM in accordance with the SC model (Section 2.2) is easily represented in 3D, the interest in this work is limited to the 2D case. Taking into consideration that the 2D mid-range interaction model (Section 2.5) requires the information from 24 neighboring nodes, the 3D case would require 92 in order to maintain the same 8th order of isotropy [21], which means much higher computational cost.

Table 1. Weighting parameter w_i values for the D2Q9 lattice.

Direction $\hat{e}_i$	w_i
0	4/9
1, 2, 3, 4	1/9
5, 6, 7, 8	1/36

The LBM relaxation time can be related to the fluid kinematic viscosity through the Chapman–Enskog expansion, and comparing it with the incompressible Navier–Stokes equation leads to the following equation:

$$\tau = 3v_{LU} + 0.5 \tag{6}$$

where v_{LU} stands for the kinematic viscosity in lattice unis (LU). The Mach number M, lattice speed of sound c_s and static pressure p based on the isothermal ideal equation of state can be presented, respectively, as follows:

$$M = \frac{||\hat{u}||}{c_s},\ \ c_s = \frac{c}{\sqrt{3}},\ \ p = \rho {c_s}^2 \tag{7}$$

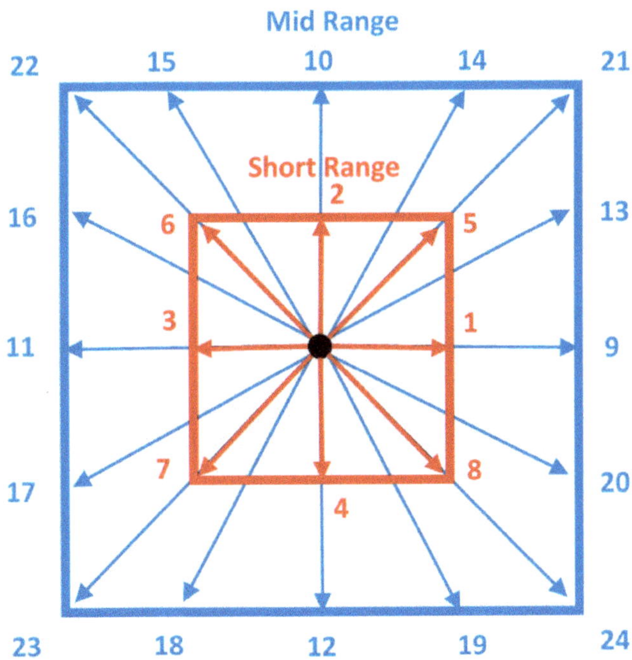

Figure 1. D2Q9 Lattice for short range nodes (red only). D2Q25 Lattice for short- and mid-range nodes (red and blue).

2.2. *Shan-Chen Multiphase Multicomponent Model*

The main benefit of the Shan-Chen [10,11] multiphase multicomponent model is that it does not dictate tracking any interface, since the separation automatically occurs due to the repulsive and attraction forces between components and phases, respectively. The SC is a pseudopotential method based on the potential function ψ as shown in Equation (8), which leads to the non-ideal equation of state as shown in Equation (9). The potential function can be written in two cases, and case A is reported to be more stable than case B [19] although case B will be used in a specific case as explained in Section 2.5.

$$\psi^{(\sigma)} = \begin{cases} \rho_o \left[1 - e^{\frac{-\rho^{(\sigma)}}{\rho_O}} \right] & case\ \ A \\ \rho^{(\sigma)} & case\ \ B \end{cases} \tag{8}$$

$$p = c_s{}^2 \sum_{\sigma} \left(\rho^{(\sigma)} + \frac{1}{2} G_{\sigma\sigma}\, \psi^{(\sigma)2} \right) + \frac{c_s{}^2}{2} \sum_{\sigma,\,\tilde{\sigma}} G_{\sigma\tilde{\sigma}}\, \psi^{(\sigma)}\, \psi^{(\tilde{\sigma})} \tag{9}$$

where $\rho_o, \sigma, \tilde{\sigma}$ and G represent the reference density which is taken as unity, the respected component, opposite components indices and the interaction parameter between phases and components, respectively. Here, both the interaction between phases and components are included. Only two components are considered in this work, 1 and 2. Setting the value of $G_{11} = 0$, $G_{12} = G_{21} > 0$, and $G_{22} < 0$ will lead to the presence of the two components with the coexistence of the component 2 in both phases. Tuning the two non-zero parameters will lead to controlling both the density ratio and surface tension separately.

The local SC interaction force for each component with respect to the neighboring nodes will then be presented as:

$$\hat{F}_{SC}{}^{(\sigma)}(\hat{x}) = -\psi^{(\sigma)}(\hat{x})\left(G_{\sigma\tilde{\sigma}}\sum_{i=0}^{8} w_i\,\psi^{(\tilde{\sigma})}(\hat{x}+c_i\Delta t)\,\hat{e}_i \;+\; G_{\sigma\sigma}\sum_{i=0}^{8} w_i\,\psi^{(\sigma)}(\hat{x}+c_i\Delta t)\,\hat{e}_i\right) \quad (10)$$

For the short range interactions, the potential function $\psi(x + c_i\Delta t)$ requires the density information from the first level of neighbors, which are 8 for the case of D2Q9.

Two LBE are then required for both components with two sets of distribution functions, which can be written as:

$$f_i{}^{(\sigma)}(\hat{x}+\hat{e}_i\Delta t,\;\; t+\Delta t) = f^{(\sigma)}{}_i(\hat{x},t) - \frac{1}{\tau^{(\sigma)}}\left(f_i{}^{(\sigma)}(\hat{x},t) - f_i{}^{equ^{(\sigma)}}(\hat{x},t)\right) + S_i{}^{(\sigma)}(\hat{x},t) \quad (11)$$

where $S_i{}^{(\sigma)}(\hat{x},t)$ is a source term that incorporates the interaction and any other external forces based on the selected forcing scheme shown in Section 2.3, while $\tau^{(\sigma)}$ is responsible for the kinematic viscosity of each component. The equilibrium distribution function for each component can be written as:

$$f_i{}^{equ^{(\sigma)}} = w_i\rho^{(\sigma)}\left[1 + 3\frac{\hat{e}_i\cdot\hat{u}'}{c^2} + \frac{9}{2}\frac{(\hat{e}_i\cdot\hat{u}')^2}{c^4} - \frac{3}{2}\frac{\hat{u}'\cdot\hat{u}'}{c^2}\right] \quad (12)$$

where $\rho^{(\sigma)}$ and $\hat{u}'$ are the density of each component evaluated as $\rho^{(\sigma)} = \sum_{i=0}^{8} f_i{}^{(\sigma)}$ and the corrected macroscopic velocity of the mixture, which is shown in Equation in 14 in Section 2.3.

2.3. Forcing Scheme

The adopted forcing scheme is the one by Guo et al. [34], as it gave the most stable simulations, lowest interfacial spurious currents and, hence, the highest stable interaction parameter when compared to the simple method by Buick and Greated [35] or the velocity shifting method, which was adopted in the original SC model [10,11]. The scheme by Guo et al. [34] is originally evaluated while considering the discrete lattice effects in order to recover the macroscopic continuity and momentum with external force. It was also reported by Yu and Fan [36] that the model by Guo can produce a τ-independent (viscosity independent) surface tension.

The evaluation of the source term can be written in the terms of the SC interaction force as follows:

$$S_i{}^{(\sigma)} = \left(1 - \frac{1}{2\,\tau^{(\sigma)}}\right)R_i{}^{(\sigma)};\;\; R_i{}^{(\sigma)} = \frac{w_i}{c_s{}^2}\left(F_{SC_\alpha}{}^{(\sigma)}\,e_{i\alpha} + \frac{e_{i\alpha}e_{i\beta} - c_s{}^2\delta_{\alpha\beta}}{c_s{}^2}F_{SC_\alpha}{}^{(\sigma)}\,u'\beta\right) \quad (13)$$

where α, β = 1, 2 for the 2D case, while the corrected macroscopic velocity of the mixture $\hat{u}'$ shall written as a function of the interaction (or any external forces) as follows:

$$\hat{u}' = \frac{1}{\rho}\sum_{\sigma}\left(\sum_{i=0}^{8} f_i{}^{(\sigma)}\hat{e}_i + \frac{\hat{F}_{SC}{}^{(\sigma)}}{2}\right);\;\; \rho = \sum_{\sigma}\rho^{(\sigma)} \quad (14)$$

Any external gravity or wall forces can be also added directly to the SC interaction forces.

2.4. Two Relaxation Time

Two phase flows always suffer from instabilities, especially when reaching the limiting cases of interaction parameters which are essential in order to tune density ratio and the surface tension together. Additionally, high viscosity ratio is a very high source of instability. This will always lead to the two relaxation time (TRT), which gives an extra degree of freedom for the simulations, yet in a simple way. This is available by splitting the distribution functions to symmetric ${f_i^{(\sigma)}}^+ = \left(f_i^{(\sigma)} + f_{-i}^{(\sigma)}\right)/2$ and asymmetric parts ${f_i^{(\sigma)}}^- = \left(f_i^{(\sigma)} - f_{-i}^{(\sigma)}\right)/2$, where $-i$ stands for the opposite lattice velocity direction. Each of the new distribution functions is relaxed with two separated relaxation times ${\tau^{(\sigma)}}^+$ and ${\tau^{(\sigma)}}^-$, and controlled through the magic parameter Λ, as shown in Equation (15). It is reported that the value of $\Lambda = 1/4$ provides the highest stability [14].

$$\Lambda = \left({\tau^{(\sigma)}}^+ - 0.5\right)\left({\tau^{(\sigma)}}^- - 0.5\right); \quad {\tau^{(\sigma)}}^+ = 3{v_{LU}}^{(\sigma)} + 0.5 \tag{15}$$

while the source term according to Guo et al. [34] shall also be modified as follows [37,38]:

$$ {S_i^{(\sigma)}}_{TRT} = \left(1 - \frac{1}{2{\tau^{(\sigma)}}^+}\right){R_i^{(\sigma)}}^+ + \left(1 - \frac{1}{2{\tau^{(\sigma)}}^-}\right){R_i^{(\sigma)}}^- \tag{16}$$

where ${R_i^{(\sigma)}}^+ = \left(R_i^{(\sigma)} + R_{-i}^{(\sigma)}\right)/2$ and ${R_i^{(\sigma)}}^- = \left(R_i^{(\sigma)} - R_{-i}^{(\sigma)}\right)/2$ following same idea of splitting the distribution functions. Apart from the collision and forcing steps, the algorithm is exactly the same as the SRT.

2.5. Mid-Range Interaction Model

A positive value of disjoining pressure Π is a major macroscopic property of stable thin films, due to the existence of repulsive forces between protein or surfactant molecules [3]. The disjoining pressure can be related to the film tension γ^f as follows:

$$\gamma^f = 2\gamma - \int_{\infty}^{l} \Pi dl + \Pi l \tag{17}$$

where γ and l are the surface tension and the film thickness, respectively.

The D2Q25 model is used in order to reach positive values of disjoining pressure, hence stable films and foam [20,21]. The neighboring nodes, far till $\sqrt{8}$ from the central one, are considered as represented in Figure 1 (both red and blue colors). This will yield to a new interaction parameter responsible for the repulsive forces between the same component, which can be called $\widetilde{G}_{11} = \widetilde{G}_{22} > 0$. The extra SC force shall be added to Equation (10), and can be written as:

$$\hat{\widetilde{F}}_{SC}{}^{(\sigma)}(\hat{x}) = -\,\psi^{(\sigma)}(\hat{x})\widetilde{G}_{\sigma\sigma}\sum_{\widetilde{i}=0}^{24} \widetilde{w}_{\widetilde{i}}\,\psi^{(\sigma)}(\hat{x} + c_{\widetilde{i}}\Delta t)\,\hat{e}_{\widetilde{i}} \tag{18}$$

where $\hat{e}_{\widetilde{i}}$ and $\widetilde{w}_{\widetilde{i}}$ represent the mid-range lattice unit vectors and the weighting parameter as shown in Table 2.

Table 2. Weighting parameter $\widetilde{w}_{\widetilde{i}}$ values for the D2Q25 lattice.

Direction $\hat{e}_{\widetilde{i}}$	$\widetilde{w}_{\widetilde{i}}$
0	247/420
1, 2, 3, 4	4/63
5, 6, 7, 8	4/135
9, 10, 11, 12	1/180
13, 14, ..., 20	2/945
21, 22, 23, 24	1/15,120

The full non-ideal static pressure (EOS) shall then be modified using the scope of Equation (9) as follows:

$$p = {c_s}^2 \sum_{\sigma} \left(\rho^{(\sigma)} + \frac{1}{2} \left(G_{\sigma\sigma} + \tilde{G}_{\sigma\sigma} \right) {\psi^{(\sigma)}}^2 \right) + \frac{{c_s}^2}{2} \sum_{\sigma,\tilde{\sigma}} G_{\sigma\tilde{\sigma}}\, \psi^{(\sigma)}\, \psi^{(\tilde{\sigma})} \tag{19}$$

2.6. Thermal Field

Temperature field can be added in LBM as a third set of the distribution function, in addition to the two from the two fluid components. The LBE for the temperature distribution function $h_i(\hat{x}, t)$ can be written following the laws of diffusion and convection as follows:

$$h_i(\hat{x} + \hat{e}_i \Delta t, t + \Delta t) = h_i(\hat{x}, t) - \frac{1}{\tau_T}\left(h_i(\hat{x}, t) - {h_i}^{equ}(\hat{x}, t)\right) \tag{20}$$

where τ_T and ${h_i}^{equ}(\hat{x}, t)$ are the temperature relaxation time and the equilibrium temperature distribution function. Both can be written as follows:

$$\tau_T = 3\alpha_{LU} + 0.5 \tag{21}$$

$${h_i}^{equ}(\hat{x},\, t) = w_i T \left[1 + 3\frac{\hat{e}_i \cdot \hat{u}'}{c^2} \right] \tag{22}$$

where α_{LU}, T and $\hat{u}'$ are thermal diffusivity in lattice units, the macroscopic temperature evaluated from the zeroth moment of the temperature distribution function ($T = \sum_{i=0}^{8} h_i$) and the macroscopic mixture velocity imported from the flow field, respectively. It is reported in the work by Guo et al. [39] that truncating the equilibrium temperature distribution function after the first order velocity term still recovers the convective diffusive temperature equation. Although it is a common practice to use a lower resolution lattice for the temperature field (e.g., D2Q4) than for the flow field [39], in the current work the typical D2Q9 as for the flow field is used for the three sets of distribution functions.

Originally, this can be seen as a one-way coupled temperature field with the flow one, although the relation to the surface tension can be locally back-coupled using a locally variable interaction parameter, and will be shown in Section 4.2.

It is worth mentioning that the implementation of the TRT is only essential for the flow field, while for the temperature field SRT is sufficient since the stability issue is not critical. If TRT was to be used, the asymmetric part of the relaxation time would be the one responsible for the thermal diffusion, not the symmetric one as in the kinematic viscosity case according to the scope of Equation (15).

2.7. Units and Non-Dimensional Groups

In order to maintain the physical similarity between the numerical simulations in lattice units, the following non-dimensional groups shall be introduced:

1. Reynolds Number:

$$\mathrm{Re} = \frac{\sqrt{gL}L}{v} \tag{23}$$

where g and L are the gravity and the characteristic length, respectively. Characteristic length will be considered as the domain width or bubble diameter depending on the study case.

2. Atwood Number:

$$\mathrm{At} = \frac{\rho_l|_2 - \rho|_1}{\rho_l|_2 + \rho|_1} \tag{24}$$

where $\rho_l|_2$ and $\rho|_1$ are the liquid phase of the high density component 2 and the low density component 1. It is worth mentioning that, for the short-range interactions, the Atwood number (density ratio) main controller is the interaction parameter G_{22}.

3. Bonds Number:

$$\text{Bo} = \frac{\rho g L^2}{\gamma} \tag{25}$$

where ρg and γ represent the force field applied on the fluid component and the surface tension, whose evaluation with respect to the SC LBM will be discussed in Section 3.1, respectively. It is worth mentioning that, for the short-range interactions, the surface tension main controller is the interaction parameter $G_{12} = G_{21}$.

4. Prandtl Number:

To relate the flow and the thermal relaxation times, the Prandtl number shall be introduced for the two fluid components, as follows:

$$\text{Pr} = \frac{v}{\alpha} \tag{26}$$

3. Model Verification and Validation

3.1. Young–Laplace Test

In order to adjust the surface tension with the SC interaction parameters G, hence making the connection between the numerical interaction forces and the physical macroscopic property, the Young–Laplace test is always used as a key for model verification. A liquid or gas bubble is initiated in a fully periodic computational domain with a reasonable density for both components accordingly with the chosen interaction parameters. The bubble is then left to reach the steady state, the pressure inside and outside the bubble are evaluated according to Equation (9) or (19), bubble diameter (radius) is measured and the 2D surface tension can be evaluated according to the Young–Laplace equation as follows:

$$\Delta p = p_{inner} - p_{outer} = \frac{\gamma}{R} \tag{27}$$

where p_{inner}, p_{outer} and R are the inner and outer static pressure and bubble diameter, respectively. The adopted criteria to estimate the interface was half the density of the heavier component liquid phase.

The common procedure is to perform several tests for each combination of interaction parameters and evaluate the surface tension according to the linear regression of the Laplace pressure Δp and $1/R$. Figure 2 shows two values of interaction parameters for single phase-two components with density ratio = 1. The two lines represent the two limiting cases: separation between components and numerical stability. The chosen interaction parameters equivalent to the resulting surface tensions in LU are shown on the figure legends. Figure 3 also shows the two limiting cases of surface tension for a two phase two component system with density ratio ≈ 10 between the liquid phase of the high density component 2 and the low density component 1 (At ≈ 0.8). The chosen combination of interaction parameters which led to the resulting density ratio and surface tension in LU are shown on the legend. For the higher surface tension case in Figure 3, due to a stability issues, TRT with magic parameter $\Lambda = 1/4$ was required.

For both Figures 2 and 3, the cases of lower surface tension show a non-zero density inside the bubble and a higher interfacial thickness. This is an inherent behavior of the SC model, since it is a diffuse interface approach and does not offer perfectly immiscible components [14]. Density contours in LU are shown for the four Young–Laplace test samples.

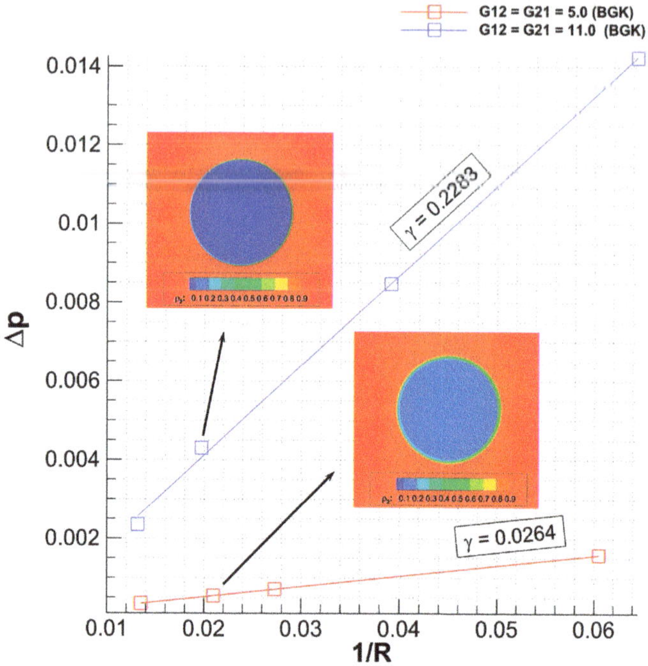

Figure 2. Young–Laplace test for the two limiting surface tension cases for density ratio ≈ 1 (single phase two components). Surface tension and density contours are presented in LU.

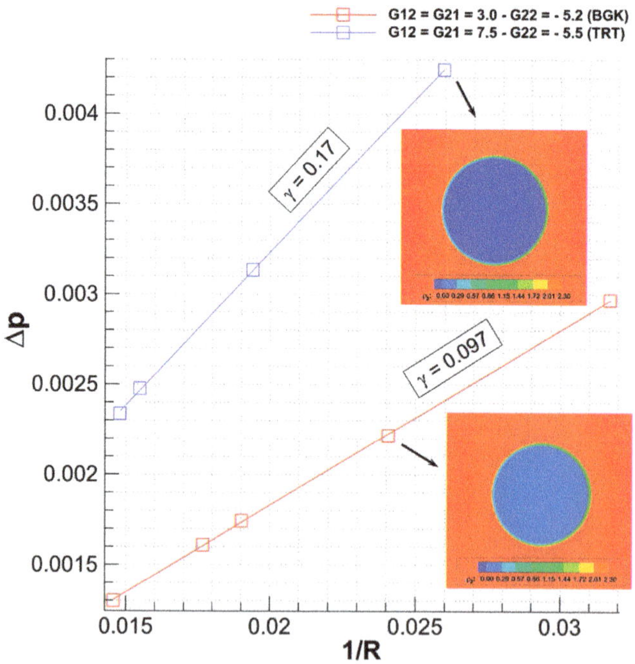

Figure 3. Young–Laplace test for the two limiting surface tension cases and density ratio ≈ 10 (two phase two component). Surface tensions and density contours are presented in LU.

3.2. Rayleigh–Taylor Instability

As a model validation, Rayleigh–Taylor Instability is investigated for two different resolutions: 256 × 1024 and 1024 × 4096. As shown in Figure 4, the computational domain consists of a heavier liquid on top impinging the bottom lighter one, with an initial single mode wave of an amplitude of 10% of the domain width. Vertical boundaries are set as periodic. The upper and lower boundaries are set as no-slip walls. Reynolds, Atwood and Bonds numbers are set as 256, 0.5 and 100, respectively. The model shows fair qualitative agreement with the work by He et al. [40] and Chiappini et al. [41]. It is worth mentioning that the surface tension was neglected in both of their works and, therefore, the Bonds numbers are not similar for all cases. Nevertheless, the Bonds number in the shown simulations is fixed for both domain resolutions. The two cases show good mesh independency except for the last time stamp, which is as issue reported and also encountered by [40,41] for both low and high Reynolds numbers, although the last time stamp of the low-resolution case is close to the one from [41], except that the bubble separation appeared slightly before our presented case.

(a)

(b)

Figure 4. RTI with time plotted on the same density contour levels (Re = 256, At ≈ 0.5, Bo ≈ 100). (**a**) Resolution: 256 × 1024. (**b**) Resolution: 1024 × 4096.

The selected SC interaction parameters for reaching the required density ratio and surface tension are: $G_{12} = G_{21} = 3.15$, $G_{22} = -3.4$ and $G_{22} = -3.4$. The kinematic viscosities of both components are kept the same ($v_{component\ 1} = v_{component\ 2}$). The adopted wall boundary condition is based on the extrapolation scheme by Chen et al. [42], where ghost nodes are introduced behind the boundaries and unknown distribution functions are extrapolated from the inner fluid nodes. Other boundary treatments, such as the non-equilibrium extrapolation method from Guo et al. [43], showed less stability and sudden phase changes close to the walls, which is why it was avoided in the current work. Other types of boundary treatments could be material for future investigations.

3.3. Bubble Rise Diagram

Some samples from the well-known bubble rise diagram by Clift et al. [29] were numerically investigated in order to qualitatively validate the presented SC LBM model for different Reynolds and Bonds numbers. Periodic boundary condition is set on the domain's edges, while selected time stamps from each case are shown in Figure 5. The Atwood number is set as zero, while the SC interaction parameters are $G_{12} = G_{21} = 6.0$ and $G_{11} = G_{22} = 0$. The selected points have the following Bonds and Reynolds numbers:

1. Bo = 100, Re = 500
2. Bo = 100, Re = 100
3. Bo = 10, Re = 100
4. Bo = 10, Re = 10

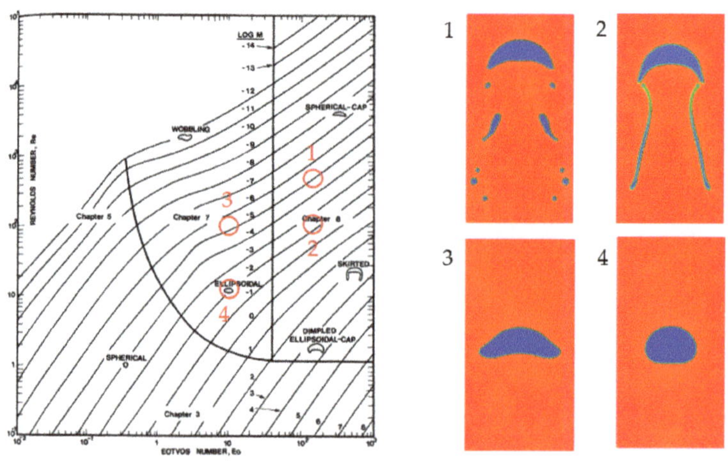

(a) (b)

Figure 5. (**a**) Shape regimes for bubble rise after Clift et al. [29]. (**b**) Four simulation samples from different points on the diagram.

The results have very good qualitative agreement with the shape regimes for bubble rise, showing the complex interfacial deformation behaviors. In Case 1, with the highest Reynolds and Bonds number, edges of the bubble are quickly split to smaller bubbles and only the cap remains. In Case 2, with lower Reynolds number, the bubble skirts and the splitting is less severe. In Case 3, with lower Bonds number, the bubble is connected, though it starts wobbling along its motion. In Case 4, with the lowest Reynolds and Bonds number, the bubble shows a slight spherical shape.

All the simulations in this section were carried out using the TRT with $\Lambda = 1/4$ for stability issues that occurred with SRT at high Reynolds and Bonds numbers. It is Worth mentioning that the Reynolds number in this simulation is based on $\sqrt{2gR}$ according to Equation (23) instead of the terminal velocity as in [29], which will yield to ~5% difference between both Reynolds numbers according to the terminal velocity formula by Davis and Taylor [44].

4. Application

4.1. Thermal Field in Liquid Foam

For the study case of thermal foam, randomly generated bubbles are initiated at the bottom of a partially filled flask-like domain and left to rise due to the gravitational effect (buoyancy) as shown in Figure 6. No slip boundary condition is set on the four walls. The thermal field model is implemented, where Dirichlet boundary condition is set for the three and bottom (source) walls. All the simulation parameters are shown in Table 3. The D2Q25 interaction model is implemented to obtain stable foam, while the SC potential function for the interaction between the two components are set using the case *B* from Equation (8), following the scope of work by Dollet et al. [45] and Fei et al. [46], as it provided much higher bubble stability than case *A*. For the other two same-component interactions, case *B* is still used. The surface tension is re-estimated for this case following the same procedure as in Section 3.1, to evaluate the Bonds number. Although the interaction parameters presented by Fei et al. [46] to reach the highest positive disjoining pressure were $G_{12} = G_{21} = 3.0$, $G_{11} = G_{22} = -8.0$ and $\widetilde{G}_{11} = \widetilde{G}_{22} = 7.0$, the required repulsive forces for this bubble rising case between the same components seemed to be higher than the attraction ones in order to maintain stable bubbles.

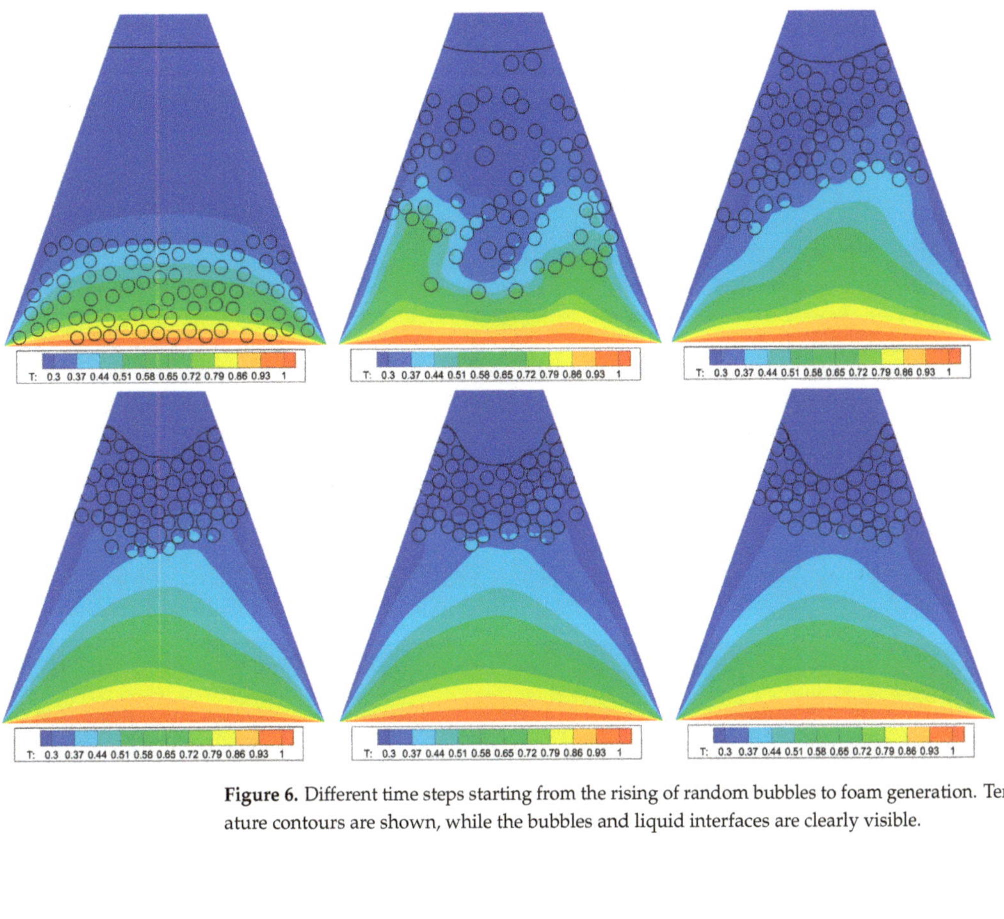

Figure 6. Different time steps starting from the rising of random bubbles to foam generation. Temperature contours are shown, while the bubbles and liquid interfaces are clearly visible.

Table 3. Simulation parameters for the thermal foam case study.

Simulation Parameter	Value
Resolution	1000 × 1000
Re	10
Bo	0.1
$\mathrm{Pr}_{component\ 1}$	0.72
$\mathrm{Pr}_{component\ 2}$	6.0
$G_{12} = G_{21}$	5.45
$G_{11} = G_{22}$	−2.0
$\tilde{G}_{11} = \tilde{G}_{22}$	5.0
θ *	20°
T_{source}/T_{ref}	1.0
T_{walls}/T_{ref}	0.25
$R_{initial}$	20 LU
$v_{component\ 1} = v_{component\ 2}$	-

*: Cone angle.

The simulation time evolution shows the generation of foam layers from rising bubbles as shown in the figure. Before imposing the bubbles' flow, a pre-time loop is used to reach the steady state for the thermal field. A one-way coupled temperature field is also shown, where the thermal convection due to the bubbles' motion and rearrangement in the lamella is clearly visible. Thermal diffusion inside bubbles is shown to be much slower than in the liquid due to the different Prandtl number. It is not possible to reach this kind of stable foam using solely the short-range interaction model. The bubbles' deformation without or with much delayed coalescence, either with the upper interface or between bubbles themselves, are obvious through the shown interfaces. Contact angle is not considered in this work, which will affect the upper interface.

This setting of the simulation shows a promising tool to investigate the thermal and flow field in thermal foam, where a wide range of length scales are resolved, starting from the lamellar film and two bubbles' interaction to a full-scale domain with a cluster of foam as presented. If the model is extended to include the phase change in foam, one can expect a breakthrough in the field of simulation and optimization of industrial rectification columns, where foam formation is a critical and an often-encountered issue.

4.2. Locally Variable Surface Tension Due to Temperature

Surface tension gradient in lamella due to temperature can induce flow in an unfavorable direction which is opposite to the drainage one (Marangoni effect); this affects the foam stability. The situation arises when the temperature on the top surface is lower than bottom, leading to higher surface tension on the top than in the bottom. Hence, flow advection can occur in the direction opposing the gravity, as shown in Figure 7 [6].

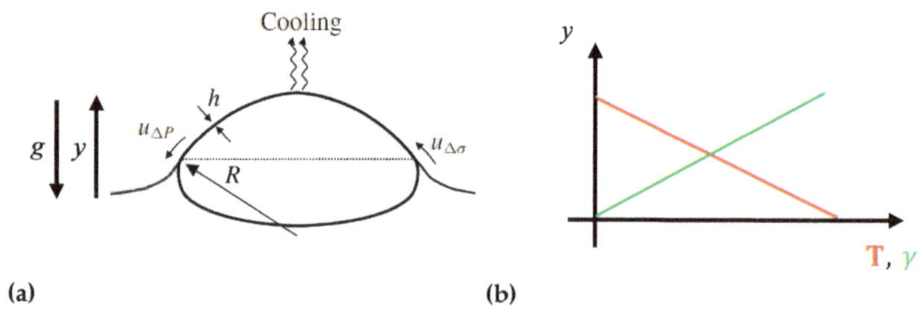

Figure 7. (**a**) Bubble with lamella showing the drainage direction and Marangoni convection, after Poulain et al. [6]. (**b**) Representation of surface tension variation due to temperature gradient.

A relation between surface tension and temperature can be written as:

$$\gamma(T) = \gamma_{ref} - \frac{\partial \gamma}{\partial T}\left(T - T_{ref}\right) \tag{28}$$

where γ_{ref} is a reference surface tension value at a reference temperature, T_{ref} = 25 °C , for example. A proposed way to back-couple the effect of temperature on the flow field, specifically on the surface tension, can be carried out through the SC interaction parameter. As a first approximation, Equation (28) can thereby be translated to the interaction parameter as shown in Equation (29). The relation can be argued to be mathematically plausible, since, practically, the relation between $G_{\sigma\tilde{\sigma}}$ and γ is almost linear. Additionally, the wide range of surface tension which can be reached as shown is Figures 2 and 3 offers the possibility of surface tension tuning even for high gradients ($\frac{\partial \gamma}{\partial T}$).

$$G_{\sigma\tilde{\sigma}}(T) = G_{\sigma\tilde{\sigma}_{ref}} - \frac{\partial G_{\sigma\tilde{\sigma}}}{\partial T}\left(T - T_{ref}\right) \tag{29}$$

Samples from two simulations are shown in Figure 8, with a spatially variable surface tension and with constant surface tension. The domain is assumed to have linear temperature variation following the schematic on Figure 7b, with a bubble rising from bottom to top until reaching the interface. The case with variable surface tension showed a slightly delayed coalescence, though uninhibited, without using the mid-range interaction model. Flow vectors show flow in a direction opposing the drainage one, as shown in Figure 8a when compared to Figure 8b. Generally, when imposing the surface tension gradient on the bubble, it tends to have slower rising velocity. This can explain, for this situation, the direction of the Marangoni convection.

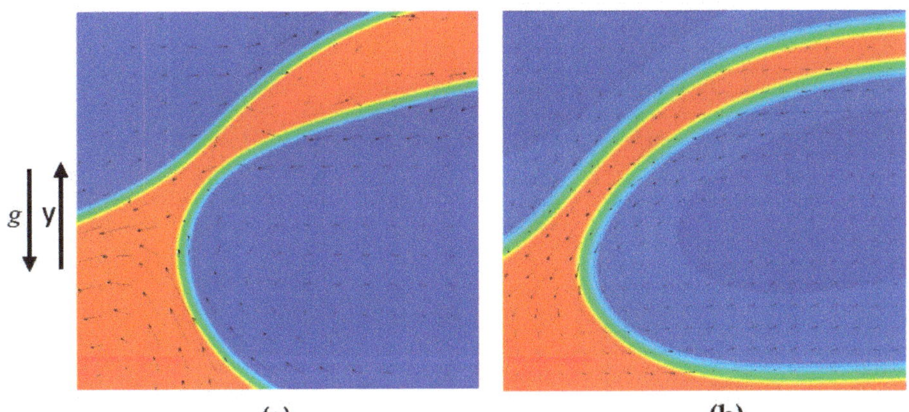

Figure 8. (**a**) Bubble rise simulation case with locally variable surface tension. (**b**) Bubble rise simulation with constant surface tension.

5. Summary and Conclusions

In this work, a literature review was presented regarding the different possible approaches and add-ons required to simulate a thermal multiphase multicomponent system with a complexity such as in thermal liquid foam using the lattice Boltzmann method (LBM). The choice of each ingredient in the model recipe was explained and justified. All the model details and parameters were shown and clarified.

The Shan-Chen (SC) LBM model was explained, considering both short and mid-range interactions, which was a necessity to reach a stable liquid foam and inhibit coalescence, implying the existence of a positive disjoining pressure. The selected forcing scheme and the Two Relaxation Time were presented. The inclusion of the thermal field as a third set of LBM distribution function was shown. The required non-dimensional groups were stated

Later, the Young–Laplace test was shown for two different cases of density ratios The Rayleigh–Taylor Instability test was shown as a validation case for the implemented multiphase–multicomponent model and compared with numerical results from the literature. Four chosen samples from the well-known shape regimes diagram for bubble rise were simulated and presented as a further validation for different Reynolds and Bonds numbers

Finally, a case study for bubble rise in a partially filled flask-like container was simulated, where random bubbles are initiated and left to rise, reaching to the top. Coalescence was significantly inhibited and, hence, liquid foam is generated from these rising bubbles Foam dynamics and bubbles' rearrangements were shown with a very interesting and realistic behavior. Convective diffusive heat transfer is included, showing the effect of the flow field on the thermal one. All the simulation parameters and non-dimensional groups were fairly presented. A novel technique to implement the local variation in surface tension due to temperature gradients in the SC LBM model was proposed. Promising findings using this approach were briefly shown, which can lead after deeper quantification and validation to a unique and delicate way to include the Marangoni effect in the SC LBM model.

The chosen recipe of approaches has succeeded in simulating thermal liquid foam and tracking its dynamics and rearrangements. Although all the presented non-dimensional groups are used, one missing ring is to quantify the disjoining pressure and tune the interaction parameters to reach the experimental physical values and isotherms, which have not been found in the literature yet. In the latest work by Ataei et al. [16], a qualitative validation of the whole numerical simulation of foam with experimental data was also not reachable. The mentioned reason was the unsimilar initial conditions, which is of course an agreeable justification in addition to the unmatched quantity of disjoining pressure. In general, this is to be expected since the numerical simulation of liquid foam as a whole system is far away from maturity.

In order to simulate the full system, the relationship between the effect of surfactants or protein molecules in the microscales, single or two bubbles scale and a full scale of a foam column must be all considered. The current proposed model can be seen as a big step towards the final aim, which is the modelling and parametrization of full-scale industrial rectification columns, where foaming is a critical and an often-occurring issue.

Author Contributions: Validation, B.G. and M.M.; writing—original draft preparation, M.M.; writing—review and editing, A.D., B.G. and M.M; visualization, M.M.; supervision, A.D. and B.G; project administration, A.D. and B.G; funding acquisition, A.D. All authors have read and agreed to the published version of the manuscript.

Funding: This research is funded by the DFG (Deutsche Forschungsgemeinschaft), FEI (Forschungskreis der Ernährungsindustrie), AiF (Arbeitsgemeinschaft industrieller Forschungsvereinigungen) and the Federal Ministry of Economics and Technology. This research is part of the DFG/AiF project (Cluster 6, CV 6): Industrial Joint Research (IGF) "Physikalisch basiertes Management störender Schäume in Produktionsanlagen: Prävention, Inhibierung und Zerstörung". The authors would like to acknowledge the financial support by Friedrich-Alexander-Universität Erlangen-Nürnberg and the Universitätsbibliothek for this publication.

Acknowledgments: The authors gratefully acknowledge the scientific support and HPC resources provided by the Erlangen National High Performance Computing Center (NHR@FAU) of the Friedrich-Alexander-Universität Erlangen-Nürnberg (FAU). The hardware is funded by the German Research Foundation (DFG).

Conflicts of Interest: The authors declare no conflict of interest.

References

1. Cantat, I.; Cohen-Addad, S.; Elias, F.; Graner, F.; Höhler, R.; Pitois, O.; Rouyer, F.; Saint-Jalmes, A. *Foams: Structure and Dynamics*; OUP: Oxford, UK, 2013.
2. Saint-Jalmes, A. Physical chemistry in foam drainage and coarsening. *Soft Matter* **2006**, *2*, 836–849. [CrossRef] [PubMed]
3. Exerowa, D.; Gochev, G.; Platikanov, D.; Liggieri, L.; Miller, R. *Foam Films and Foams: Fundamentals and Applications*; CRC Press: Boca Raton, FL, USA, 2018.
4. Bergeron, V.; Radke, C.J. Equilibrium measurements of oscillatory disjoining pressures in aqueous foam films. *Langmuir* **1992**, *8*, 3020–3026. [CrossRef]
5. Stevenson, P. *Foam Engineering: Fundamentals and Applications*; John Wiley & Sons: Oxford, UK, 2012.
6. Poulain, S.; Villermaux, E.; Bourouiba, L. Ageing and burst of surface bubbles. *J. Fluid Mech.* **2018**, *851*, 636–671. [CrossRef]
7. Chen, S.; Doolen, G.D. Lattice Boltzmann method for fluid flows. *Annu. Rev. Fluid Mech.* **1998**, *30*, 329–364. [CrossRef]
8. Swift, M.R.; Osborn, W.R.; Yeomans, J.M. Lattice Boltzmann simulation of nonideal fluids. *Phys. Rev. Lett.* **1995**, *75*, 830. [CrossRef]
9. Swift, M.R.; Orlandini, E.; Osborn, W.R.; Yeomans, J.M. Lattice Boltzmann simulations of liquid-gas and binary-fluid systems. *Phys. Rev. E* **1996**, *54*, 5041. [CrossRef]
10. Shan, X.; Chen, H. Lattice Boltzmann model for simulating flows with multiple phases and components. *Phys. Rev. E* **1993**, *47*, 1815. [CrossRef]
11. Shan, X.; Chen, H. Simulation of nonideal gases and liquid-gas phase transitions by the lattice Boltzmann equation. *Phys. Rev. E* **1994**, *49*, 2941. [CrossRef]
12. Yuan, P.; Schaefer, L. Equations of state in a lattice Boltzmann model. *Phys. Fluids* **2006**, *18*, 042101. [CrossRef]
13. Peng, D.Y.; Robinson, D.B. A new two-constant equation of state. *Ind. Eng. Chem. Fundam.* **1976**, *15*, 59–64. [CrossRef]
14. Krüger, T.; Kusumaatmaja, H.; Kuzmin, A.; Shardt, O.; Silva, G.; Viggen, E.M. *The Lattice Boltzmann Method*; Springer International Publishing: Cham, Switzerland, 2017.
15. Körner, C.; Thies, M.; Hofmann, T.; Thürey, N.; Rüde, U. Lattice Boltzmann model for free surface flow for modeling foaming. *J. Stat. Phys.* **2005**, *121*, 179–196. [CrossRef]
16. Ataei, M.; Shaayegan, V.; Costa, F.; Han, S.; Park, C.; Bussmann, M. LBfoam: An open-source software package for the simulation of foaming using the Lattice Boltzmann Method. *Comput. Phys. Commun.* **2021**, *259*, 107698. [CrossRef]
17. Thies, M. Lattice Boltzmann Modeling with Free Surfaces Applied to Formation of Metal Foams. Ph.D. Dissertation, Friedrich-Alexander-Universitaet Erlangen-Nuernberg, Erlangen, Germany, 2005.
18. Sukop, M.C. *Lattice Boltzmann Modeling Lattice Boltzmann Modeling*; Springer: Berlin/Heidelberg, Germany, 2006.
19. Chen, L.; Kang, Q.; Mu, Y.; He, Y.L.; Tao, W.Q. A critical review of the pseudopotential multiphase lattice Boltzmann model: Methods and applications. *Int. J. Heat Mass Transf.* **2014**, *76*, 210–236. [CrossRef]
20. Falcucci, G.; Ubertini, S.; Succi, S. Lattice Boltzmann simulations of phase-separating flows at large density ratios: The case of doubly-attractive pseudo-potentials. *Soft Matter* **2010**, *6*, 4357–4365. [CrossRef]
21. Sbragaglia, M.R.L.S.K.; Benzi, R.; Biferale, L.; Succi, S.; Sugiyama, K.; Toschi, F. Generalized lattice Boltzmann method with multirange pseudopotential. *Phys. Rev. E* **2007**, *75*, 026702. [CrossRef] [PubMed]
22. Chen, H.; Boghosian, B.M.; Coveney, P.V.; Nekovee, M. A ternary lattice Boltzmann model for amphiphilic fluids. *Proc. R. Soc. Lond. Ser. A Math. Phys. Eng. Sci.* **2000**, *456*, 2043–2057. [CrossRef]
23. Nekovee, M.; Coveney, P.V.; Chen, H.; Boghosian, B.M. Lattice-Boltzmann model for interacting amphiphilic fluids. *Phys. Rev. E* **2000**, *62*, 8282. [CrossRef] [PubMed]
24. Mukherjee, S.; Berghout, P.; Van den Akker, H.E. A lattice boltzmann approach to surfactant-laden emulsions. *AIChE J.* **2019**, *65*, 811–828. [CrossRef]
25. Gong, S.; Cheng, P. A lattice Boltzmann method for simulation of liquid–vapor phase-change heat transfer. *Int. J. Heat Mass Transf.* **2012**, *55*, 4923–4927. [CrossRef]
26. Li, Q.; Zhou, P.; Yan, H.J. Improved thermal lattice Boltzmann model for simulation of liquid-vapor phase change. *Phys. Rev. E* **2017**, *96*, 063303. [CrossRef]
27. Huang, R.; Wu, H.; Cheng, P. A new lattice Boltzmann model for solid–liquid phase change. *Int. J. Heat Mass Transf.* **2013**, *59*, 295–301. [CrossRef]
28. He, Y.L.; Liu, Q.; Li, Q.; Tao, W.Q. Lattice Boltzmann methods for single-phase and solid-liquid phase-change heat transfer in porous media: A review. *Int. J. Heat Mass Transf.* **2019**, *129*, 160–197. [CrossRef]
29. Clift, R.; Grace, J.R.; Weber, M.E. *Bubbles, Drops, and Particles*; Academic Press: London, UK, 1978.
30. Succi, S. *The Lattice Boltzmann Equation: For Fluid Dynamics and Beyond*; Oxford University Press: Oxford, UK, 2001.
31. Koelman, J.M.V.A. A simple lattice Boltzmann scheme for Navier-Stokes fluid flow. *Europhys. Lett.* **1991**, *15*, 603. [CrossRef]
32. Wilde, D.; Krämer, A.; Küllmer, K.; Foysi, H.; Reith, D. Multistep lattice Boltzmann methods: Theory and applications. *Int. J. Numer. Methods Fluids* **2019**, *90*, 156–169. [CrossRef]
33. Hussein, M.A. On the Theoretical and Numerical Development of Lattice Boltzmann Models for Biotechnology and Its Applications. Ph.D. Dissertation, Technical University of Munich, Munich, Germany, 2010.
34. Guo, Z.; Zheng, C.; Shi, B. Discrete lattice effects on the forcing term in the lattice Boltzmann method. *Phys. Rev. E* **2002**, *65*, 046308. [CrossRef]
35. Buick, J.M.; Greated, C.A. Gravity in a lattice Boltzmann model. *Phys. Rev. E* **2000**, *61*, 5307. [CrossRef]

36. Yu, Z.; Fan, L.S. Multirelaxation-time interaction-potential-based lattice Boltzmann model for two-phase flow. *Phys. Rev. E* **2010**, *82*, 046708. [CrossRef]
37. Seta, T.; Rojas, R.; Hayashi, K.; Tomiyama, A. Implicit-correction-based immersed boundary–lattice Boltzmann method with two relaxation times. *Phys. Rev. E* **2014**, *89*, 023307. [CrossRef]
38. Postma, B.; Silva, G. Force methods for the two relaxation times lattice Boltzmann. *Phys. Rev. E* **2020**, *102*, 063307. [CrossRef]
39. Guo, Z.; Shi, B.; Zheng, C. A coupled lattice BGK model for the Boussinesq equations. *Int. J. Numer. Methods Fluids* **2002**, *39*, 325–342. [CrossRef]
40. He, X.; Chen, S.; Zhang, R. A lattice Boltzmann scheme for incompressible multiphase flow and its application in simulation of Rayleigh-Taylor instability. *J. Comput. Phys.* **1999**, *152*, 642–663. [CrossRef]
41. Chiappini, D.; Bella, G.; Succi, S.; Toschi, F.; Ubertini, S. Improved lattice Boltzmann without parasitic currents for Rayleigh-Taylor instability. *Commun. Comput. Phys.* **2010**, *7*, 423. [CrossRef]
42. Chen, S.; Martinez, D.; Mei, R. On boundary conditions in lattice Boltzmann methods. *Phys. Fluids* **1996**, *8*, 2527–2536. [CrossRef]
43. Guo, Z.; Zheng, C.; Shi, B. An extrapolation method for boundary conditions in lattice Boltzmann method. *Phys. Fluids* **2002**, *14*, 2007–2010. [CrossRef]
44. Davies, R.M.; Taylor, G.I. The mechanics of large bubbles rising through extended liquids and through liquids in tubes. *Proc. R. Soc. Lond. Ser. A Math. Phys. Sci.* **1950**, *200*, 375–390.
45. Dollet, B.; Scagliarini, A.; Sbragaglia, M. Two-dimensional plastic flow of foams and emulsions in a channel: Experiments and lattice Boltzmann simulations. *J. Fluid Mech.* **2015**, *766*, 556–589. [CrossRef]
46. Fei, L.; Scagliarini, A.; Montessori, A.; Lauricella, M.; Succi, S.; Luo, K.H. Mesoscopic model for soft flowing systems with tunable viscosity ratio. *Phys. Rev. Fluids* **2018**, *3*, 104304. [CrossRef]

Disclaimer/Publisher's Note: The statements, opinions and data contained in all publications are solely those of the individual author(s) and contributor(s) and not of MDPI and/or the editor(s). MDPI and/or the editor(s) disclaim responsibility for any injury to people or property resulting from any ideas, methods, instructions or products referred to in the content.

energies

MDPI

Article

Analysis of the Heat Transfer in Electronic Radiator Filled with Metal Foam

Xiaofang Shan [1], Bin Liu [1,*], Zongsheng Zhu [1], Rachid Bennacer [2,*], Rounan Wang [1] and Panagiotis E. Theodorakis [3]

1 Tianjin Key Lab of Refrigeration Technology, Tianjin University of Commerce, Tianjin 300134, China
2 LMPS, ENS Paris-Saclay, CentraleSupélec, Université Paris-Saclay, CNRS, 91190 Gif-sur-Yvette, France
3 Institute of Physics, Polish Academy of Sciences, Al. Lotników 32/46, 02-668 Warsaw, Poland
* Correspondence: lbtjcu@tjcu.edu.cn (B.L.); rachid.bennacer@ens-paris-saclay.fr (R.B.)

Abstract: The performance of an electronic radiator filled with metal foam with a porosity of 96% was studied. The effect of the factors including the flow rates, the pores per linear inch (PPI) and the numbers of fins was analyzed. The results show that the electronic radiator with metal foam reflects a stronger ability of the heat transfer compared to the electronic radiator without metal foam. With the increase in the flow rate between 10 L/h and 60 L/h, the heat transfer coefficient of both of the two electronic radiators will be improved, but it is also dependent on the number of fins. In this study, we find that the heat transfer coefficient first increases and then decreases with the number of fins. The optimum number is three. As for the effect of the PPI, the higher the PPI, the larger the heat transfer coefficient, while the pressure drop always increases with the flow rates' increase, the pores per linear inch (PPI) and the numbers of fins.

Keywords: metal foam; electronic radiator; enhancement of heat transfer; pressure drop

Citation: Shan, X.; Liu, B.; Zhu, Z.; Bennacer, R.; Wang, R.; Theodorakis, P.E. Analysis of the Heat Transfer in Electronic Radiator Filled with Metal Foam. *Energies* **2023**, *16*, 4224. https://doi.org/10.3390/en16104224

Academic Editors: Marco Marengo, Artur Bartosik and Dariusz Asendrych

Received: 25 January 2023
Revised: 10 May 2023
Accepted: 17 May 2023
Published: 20 May 2023

Copyright: © 2023 by the authors. Licensee MDPI, Basel, Switzerland. This article is an open access article distributed under the terms and conditions of the Creative Commons Attribution (CC BY) license (https://creativecommons.org/licenses/by/4.0/).

1. Introduction

With the rapid development of electronic technology, miniaturization and intelligence have become important directions in the development of contemporary equipment. The size of the electronic devices used in industrial production equipment is gradually decreasing, and the chip achieves a continuous increase in the operating frequency as well as integration density. This leads to a rapid increase in the heat flow density of the chip, and its temperature directly affects the stability of the performance. As a result, higher demands are placed on the thermal design of electronic chips. To ensure the operational stability of electronic devices, higher requirements for the reliability of electronic equipment are required. Through a high standard of thermal design work arrangements, reasonable cooling methods are chosen to ensure that the heat escaping from the device is removed from the surface more quickly and that the temperature around the component is always within the safe operating temperature range.

In recent years, metal porous materials have been used due to their unique properties and the combination of structural and functional material properties [1–3]. Widely used in biological, medical [4–6], aerospace [7] and industrial applications [8–12], the high demand for green materials in various fields has driven the development of metal foam. Porous metal foam structures with high specific surface area, high permeability properties and high mechanical strength are being explored as an alternative material to conventional heat exchangers. In the field of heat transfer [13–16], metal foam holds great promise for applications in multifunctional heat exchangers [17–19], cooling systems [20–23], high-power batteries, compact electronic heat sinks [23–28] and so on. Porous metal foam can significantly reduce the size and mass of equipment when used in heat transfer equipment due to their light mass and low density. They have great potential for application in industrial production as well as in related fields such as high technology, gaining more attention and importance.

The use of metal foam as a new compact heat exchanger has been investigated by numerical simulations. Chen et al. [13] presented a three-dimensional numerical simulation to reveal the flow and heat transfer characteristics of a new tube bundle design covered with metal foam. The results showed that tube bundles covered with metal foam with low porosity and low pore density have a strong advantage over bare tube bundles. For example, Kotresha et al. [16] discussed a numerical simulation of a metal foam heat exchanger system performed by a commercial software. The aim is to improve the thermal performance of the heat exchanger by reducing the pressure drop and increasing the heat transfer rate to the maximum.

Metal foam has also been investigated experimentally. Kim et al. and Shen et al. [14,15] found that the use of metal foam significantly reduces the overall thermal resistance compared to a conventional finned heat pipe heat sink. Hsieh et al. [29] experimentally investigated the effects of porosity, pore density (PPI) and air flow rate on the heat transfer characteristics of the aluminum foam heat sinks and found that the increase in porosity and pore density enhanced the non-local thermal equilibrium phenomenon. Liu et al. [30], based on the Reynolds number range of the equivalent spherical diameter of the foam from 32 to 1289, found that the porosity range was 0.87 to all seven types of the aluminum foams, and an empirical Equation was developed to relate the unexpected pressure drop to the unexpected flow rate. Dukhan et al. [31] presented heat transfer measurements within a rectangular block of commercially available aluminum foam subjected to constant heat flow on one side. The temperature profile decays in an exponential fashion as the distance from the heat base increases. Boomsma and Poulikakos [32] experimentally showed that varying the fluid conductivity has a relatively small effect on increasing the effective heat transfer rate.

The heat transfer performance of metal foam under forced convection and natural convection conditions has also been investigated by researchers. Shih et al. [33,34] demonstrated that under impact jet flow conditions for all values of jet–jet spacing, an increase in the pore density was accompanied by an increase in the heat transfer. Bhattacharya et al. [35] investigated forced convection heat transfer in a new finned metal foam heat sink, showing that when fins were added to the metal foam, the heat transfer was significantly enhanced, and the heat transfer coefficient increased with the number of fins until the addition of more fins would lead to the interference with the thermal boundary layer and retard heat transfer. Shen et al. [15] conducted a systematic study and analysis of the thermal and flow characteristics at different air speeds and thermal powers through experiments. It was found that the introduced metal foam significantly reduced the overall thermal resistance by 25.5% compared to a conventional finned heat pipe heat sink.

This paper focuses on the implementation of enhanced heat transfer analysis for metal-filled radiator components with 96% porosity and 10 PPI pore density of copper foam, aluminum foam and 20 PPI pore density of copper foam. The heat transfer performance of the devices at different flow rates and at certain flow conditions is clarified, and the heat transfer geometrical parameters of the metal foam-filled radiators and their heat transfer performance are obtained.

2. Materials and Methods

2.1. Heat Accounting for Radiators

The heat of the hot and cold fluids is exchanged through the solid walls. First, the hot fluid conducts the heat to the solid walls; next, it is conducted via one side of the solid walls toward the other side, and then the heat is transferred to the cold fluid via the walls. Overall, the heat exchange can be distinguished as a heat giving–heat conducting–heat giving process acting in tandem.

If a hot fluid flows in a metal foam or radiator, a cold fluid passes through the metal foam or radiator, and convection occurs to dissipate the heat. Based on the heat balance principle, it is clear that when the ducts are well insulated, then the amount of heat discharged per unit time of the hot flow and the amount of heat absorbed per unit time of the cold flow are equal.

For a radiator with hot and cold fluids flowing against each other, the heat balance equation reads as follows:

$$Q = q_{m,c}(H_{c,2} - H_{c,1}) = q_{m,h}(H_{h,1} - H_{h,2}). \tag{1}$$

Here, $q_{m,c}$ and q_{mhh} indicate the mass flow rate of the cold and hot fluids, respectively; H_c and H_h correspond to the enthalpy per unit of cold and hot fluid (the subscripts c for cold and h for hot) and the subscripts "1" and "2" correspond to the inlet and outlet side of each unit.

If there is no phase change in either the hot or cold fluid inside the radiator, and the specific heat coefficient at constant pressure.of the fluid, c_p, does not change due to a change in the temperature, then Equation (1) can be expressed as follows:

$$Q = q_{m,c}c_{p,c}(t''_1 - t'_1) = q_{m,h}c_{p,h}(t'_2 - t''_2) \tag{2}$$

For overall heat transfer, the heat transfer equation required is as follows:

$$Q = kA\Delta t_m. \tag{3}$$

Here, A is the contact area; Δt_m represents the average temperature difference between the two ends where the fluid is located, i.e.,

$$\Delta t_m = \frac{\Delta t_h - \Delta t_c}{ln(\Delta t_c \Delta t_h)}(\Delta t_c = t''_1 - t'_1; \Delta t_h = t'_2 - t''_2) \tag{4}$$

While K is the heat transfer coefficient, the introduction of Equation (3) into Equation (1) and the association of Equations (1) and (2) lead to the following:

$$k = \frac{q_{m,c}c_{p,c}(t''_c - t'_c)}{A\Delta t_m} = \frac{q_{m,h}c_{p,h}(t'_h - t''_h)}{A\Delta t_m}. \tag{5}$$

According to this Equation, it is clear that determining the inlet and outlet temperatures of the two fluids and the flow rate of the radiator suffices to obtain the total heat transfer coefficient, K.

During actual operation, the heat dissipation as well as the heat absorption achieved by the hot and cold fluids are not in an absolutely equal state and need to be based on the heat balance theory to obtain their relative error in heat balance ΔQ as follows:

$$\Delta Q = \frac{(Q_h - Q_c)}{Q_c} \times 100\%. \tag{6}$$

Here, Q_h denotes fluid heat release and Q_c denotes fluid heat absorption; in the case of $\Delta Q < 5\%$, it can be assumed that the system is in a state of thermal equilibrium.

Forced convection is applied to the fluid in the radiator. The factors that will have an effect on the heat transfer coefficient correspond to the internal diameter of the radiator channel D, the fluid flow rate v, the fluid viscosity μ, the fluid density ρ, the fluid heat capacity c_p, the fluid thermal conductivity k, etc. The expression is as follows:

$$h = f(l, D, \rho, \mu, c_p, k, v). \tag{7}$$

Implementing the measure analysis based on Buckingham's π theorem results in the following:

$$h = f(l, D, \rho, \mu, c_p, k, v), \tag{8}$$

$$\pi_2 = \frac{Dv\rho}{\mu} = Re, \tag{9}$$

$$\pi_3 = \frac{\mu c_p}{k} = Pr. \tag{10}$$

Therefore, Equation (7) can be related as follows:

$$Nu = f_1(Re, Pr). \tag{11}$$

From heat transfer and fluid mechanics, the Nusselt and Prandtl numbers for the flow of a work mass in an electronic heat sink component are known.

Where d_e is the equivalent diameter of the flow channel or the so-called hydraulic diameter, h denotes the convection heat transfer coefficient and λ denotes the thermal conductivity of water. ρ and μ denotes fluid density and viscosity; c_p denotes constant pressure heat capacity.

2.2. Experimental Materials

This paper focuses on the experimental study of copper foam fin radiators with 10 PPI and 20 PPI (as shown in Figure 1). The experimental components used the following dimensions: length 75 mm, width 45 mm and height 17 mm of copper foam radiator, with a base material thickness of 3 mm. The copper foam was cut to the required size using a cutting machine, cleaned with alcohol to remove oil and dust from the surface and then welded to the substrate. The porosity of the copper foam was 96% and the pore density was 10 PPI and 20 PPI, respectively, while the porosity of the aluminum foam was 96% and the pore density was 10 PPI (as shown in Figure 1). The physical properties are shown in Table 1.

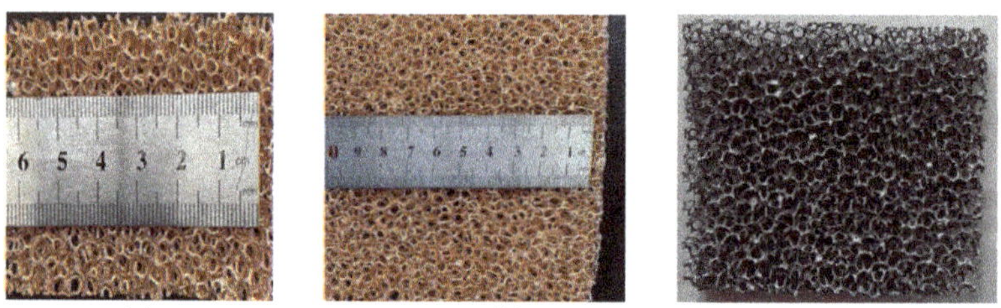

Figure 1. Copper foam and aluminum foam. (**a**) Copper foam of 10 PPI. (**b**) Copper foam of 20 PPI. (**c**) Aluminum foam of 10 PPI.

Table 1. Physical properties of copper foam and aluminum foam.

Materials \ Physical Properties	Pore Density (PPI)	ε	Average Pore Diameter (mm)	Density (g/cm^3)	Thermal Conductivity (W/m·K)	Specific Surface Area (g/m^2)
Copper foam	10	0.96	2.54	0.26	46.21	375
	20	0.96	1.27	0.32	62.56	624
Aluminum foam	10	0.96	2.54	0.09	26.74	375

This paper investigates a copper foam fin radiator at 100 W heating power. The water flow rate was from 10 to 60 L/h and the fins were aligned at 90 degrees to the direction of the water flow. The purpose of this arrangement is to increase the contact surface. The materials filled for the experiments had a pore density of 10 PPI copper metal foam, 10 PPI aluminum metal foam and 20 PPI copper metal foam, as shown in Table 2. The model diagram of the electronic heat sink is shown in Figure 2.

Table 2. Type of material used in the experiment.

Radiator Number	Filling Material Number			
	10 PPI Copper Foam	10 PPI Aluminum Foam	20 PPI Copper Foam	Unfilled
Finless	1	2	3	4
Two fins	1	2	3	4
Triple fins	1	2	3	4
Five fins	1	2	3	4

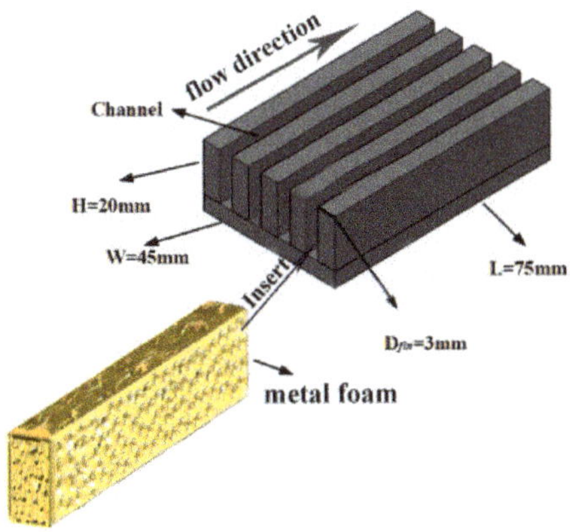

Figure 2. Model diagram of electronic radiator.

2.3. Experimental Setup

The setup contains instruments such as a miniature water pump, an electronic pressure regulator, a paperless recorder, an electrical parameter measuring instrument, a differential pressure transmitter, a constant temperature water bath and a rotameter. A flow chart of the apparatus is shown in Figure 3.

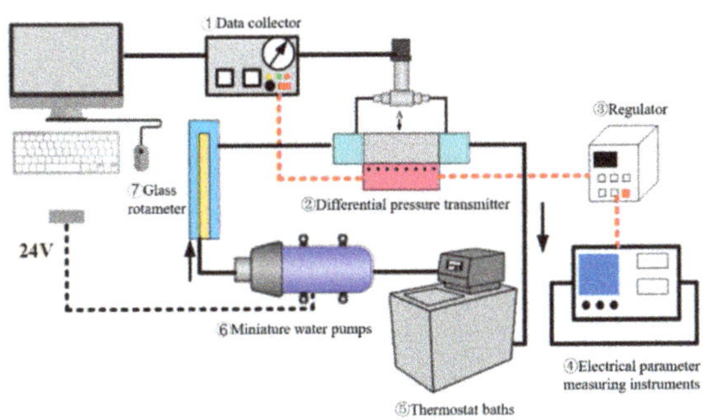

Figure 3. Schematic diagram of experimental process. ① Data collector ② Differential pressure transmitter ③ Regulator ④ Electrical parameter measuring instruments ⑤ Thermostat bath ⑥ Miniature water pumps ⑦ Glass rotameter.

The specific parameters of the main instruments and equipment required in the experiment are as follows: armored thermocouples; miniature water pump, Degree New Weicheng Technology Co., Zhuhai, China; silicon-controlled electronic voltage regulator, Shanghai Steadyford Electric Co., Shanghai, China; paperless recorder, Yokogawa Corporation, Japan; electric parameter measuring instrument, Hangzhou Yuanfang Photoelectric Information Co., Hangzhou, China; differential pressure transmitter, Foshan Yizhong Sensing Instrument Co., Foshan, China; constant temperature water bath, Nanjing Xian'ou

Instrument Manufacturing Co., Nanjing, China; glass rotor flowmeter, Leerda Instruments Co., Portland, OR, USA.

The length of the electronic heat sink was 75 mm and the cross-sectional area was 45 mm × 20 mm. The metal foam of different pore densities were filled into the electronic heat sink. The experimental process is as follows. Water with a constant temperature (20 °C) is fed into the test section through a rotameter. In the test section and the fins (as shown in Figure 4), the heating plate transfers its own heat to the electronic heat sink through thermal conduction and then exchanges heat with the water. Finally, the heat is discharged from the outlet with the flow of water.

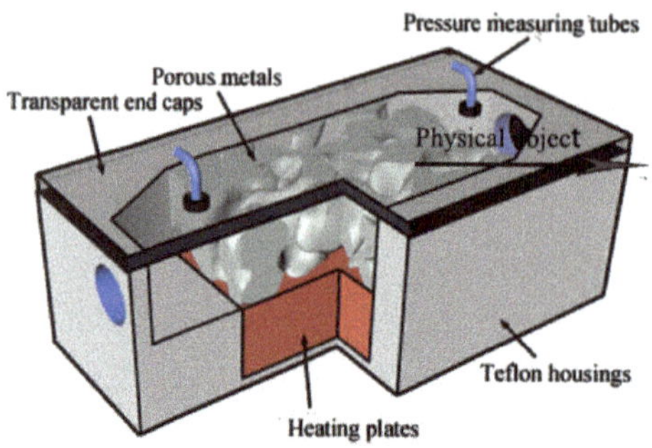

(a)

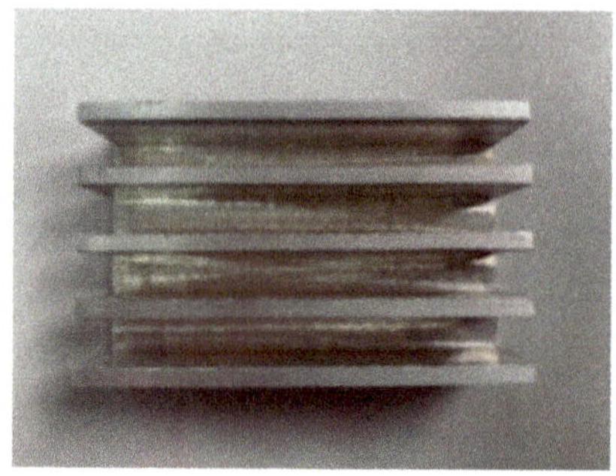

(b)

Figure 4. Test section renderings and the five fins. (**a**) Test section renderings. (**b**) Physical object of five fins.

3. Results and Discussion

3.1. Data Processing Content

This experiment uses the Wilson graphical method to process the data, and thus, derive the convective heat transfer coefficient. The main data to be measured as well as processed are as follows:

(1) To investigate the total heat transfer performance of a heat sink with a metal foam filling.
(2) To investigate the heat transfer performance and to obtain the Nu–Re relationship curves.
(3) To investigate the resistance performance, obtaining the relationship curves between f-Re, Δp-v.
(4) To evaluate the comprehensive enhanced heat transfer performance of the metal foam-filled electronic heat sinks.

3.2. Data Processing Steps

The experiment requires that we record the raw data and process them as follows.

The average temperature, feature size and feature velocity are calculated, and the corresponding physical parameters are obtained.

(1) The average temperature t_0 for convective heat transfer flowing in the channels of the electronic heat sink parts is as follows:

$$t_0 = (t_1 + t_2 + t_3 + \cdots + t_n)/n. \tag{12}$$

where t_0 is the average temperature of the radiator surface; $t_1, t_2 \ldots t_n$ are the radiator surface temperature, °C.

(2) Feature dimensions refer to standard-like geometric dimensions. Convective heat transfer within the electronic heat sink is dimensioned as a microchannel. This experiment is a rectangular cross section, so the equivalent diameter d_e is as follows:

$$d_e = 4\frac{S}{U} = 2\frac{ab}{a+b}. \tag{13}$$

where d_e is the equivalent diameter of the flow channel or the so-called hydraulic diameter; S is the effective cross-sectional area of the flow channel also known as the cross-flow cross-sectional area, m^2; U is the electronic heat sink parts section perimeter, and fluid contact with the solid wall perimeter, m and a, b are the length and width of the micro-channel cross-section, m.

(3) The characteristic velocity in a heat exchanger can be obtained by taking the flow rate and the cross-flow area. The ratio between the two is the characteristic velocity.

In order to obtain the total convective heat transfer Q during the experiment, the heat transfer coefficient h needs to be determined. The total heat was obtained using Equation (2) and averaged when the heat absorbed by the cold fluid was in equilibrium with that of the hot fluid ($\Delta Q < 5\%$). The log-average temperature difference is calculated using Equation (3) and the total heat transfer coefficient K is calculated from Equation (4).

$$Q = qA = h\Delta TA = \rho vc_p(t_{out} - t_{in}). \tag{14}$$

Known from fluid mechanics, the equation for the resistance coefficient of a medium flowing through a system is as follows:

$$f = \frac{\Delta p}{\frac{1}{2}\rho v^2} \cdot \frac{d_e}{l} \tag{15}$$

where Δp is the pressure drop in the experimental section measured by the differential pressure transmitter in the experiment; v is the average flow rate of the tube and l is the distance between the static pressure measurement in the experimental section (110 mm for this experiment).

The resistance coefficients for each group of experiments were fitted using the least squares method as shown in (18) as follows:

$$f = C_f Re^x \tag{16}$$

3.3. Data Processing Results and Performance Curve Analysis

3.3.1. Analytical Comparison of the Total Heat Transfer Coefficient K Versus Flow Rate v

As shown in Figure 5, the heat transfer in the foam-filled electronic heat sink is significantly higher than that of the unfilled heat sink. In the case of a foam-filled electronic heat sink, the heat transfer differs for different pore densities of the metal foam. As the pore density increases, the contact area between the metal foam and the fluid increases, and the total heat transfer area of the heat sink increases. On the other hand, the heat transfer of an electronic heat sink is also related to the metal foam it is filled with. The thermal conductivity of the copper metal foam is higher than that of the aluminum metal foam. Hence, the heat transfer is higher in the same model for the heat sinks filled with copper metal foam material at the same speed conditions.

Equation (16) was used to obtain the relation curve of the heat transfer coefficient K and v of the radiator. As shown in Figure 6, the convective heat transfer coefficient increases by around 31.57% when the number of electronic radiator fins is increased from no fins to two fins. When the number of fins is increased from two to three, the convective heat transfer coefficient increases by about 26.08%. When the number of fins is increased from three to five, the convective heat transfer coefficient decreases, as shown in Figure 7. This is due to an increase in the number of fins accompanied by a decrease in fin spacing and interference in the thermal boundary layer between adjacent fins, resulting in a delay in heat transfer. As a result, there is a maximum number of fins that allows the radiator to achieve optimum heat transfer. As can be seen from the graph, the overall heat transfer coefficient is best when the number of fins is three.

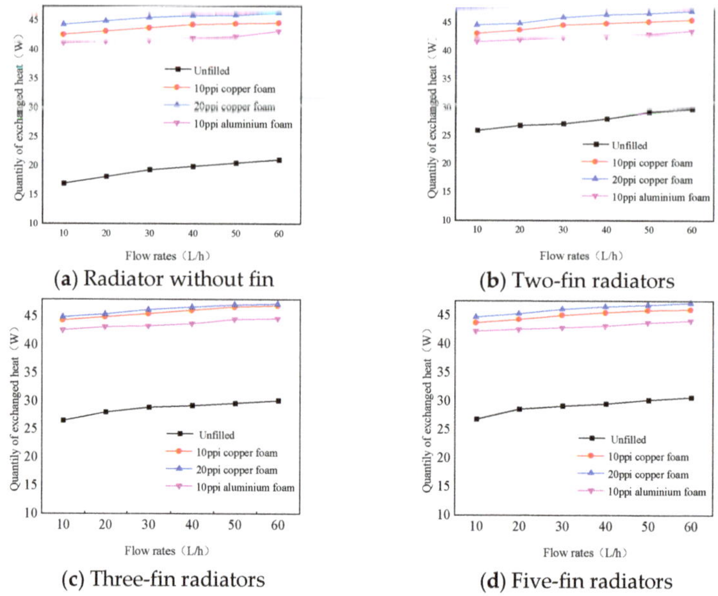

Figure 5. The relationship between heat transfer Q and v of radiator.

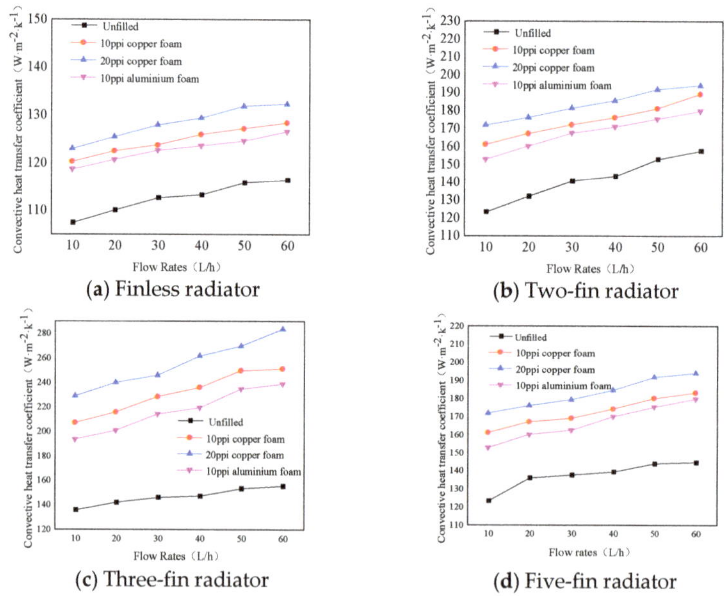

Figure 6. Relation curve of heat transfer coefficient K and v of radiator.

On the other hand, the convective heat transfer coefficients with and without metal foam filling are very different for the same number of finned electronic heat exchangers. The heat transfer coefficient is significantly higher for the foam-filled type of electronic heat sink than for the unfilled type. Moreover, the convective heat transfer coefficient is also related to the pore density of the filled metal foam, which is related to the length of the pore diameter. The greater the pore density, the smaller the corresponding pore length, and the smaller the diameter of the metal foam skeleton. When the porosity of the metal foam is certain, the contact area between the fluid flowing through the metal foam and the solid also increases, thus strengthening the heat transfer capacity of the electronic heat sink. At the same time, the greater the density of the pores, the greater the disruption of the skeleton within the metal foam and the greater the disturbance caused to the fluid flow, which, in turn, enhances the convective heat transfer intensity. Based on the experimental results, it can be concluded that the highest convective heat transfer coefficient is obtained for copper metal foam with a filled pore density of 20 PPI at a fin number of three.

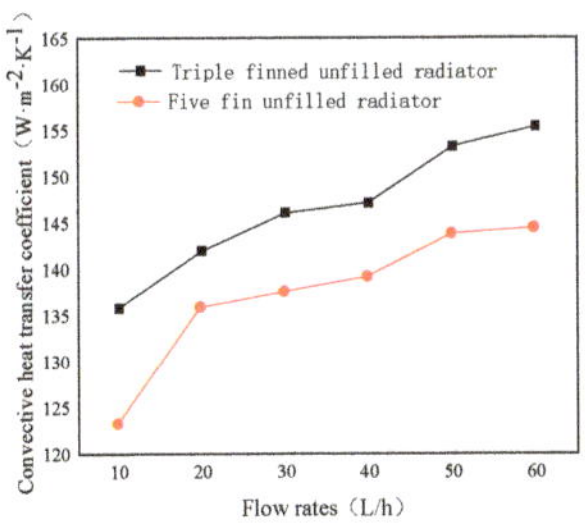

Figure 7. Curve of k versus v for radiators with different numbers of fins.

Equation (2) was used to obtain the heat transfer rate of the radiator. In Figure 8, the heat exchange increasing rate decreases with the increasing flow rate. This is due to the fact that the greater the fluid flow velocity, the greater the degree of turbulence. This results in a consequent decrease in the heat transfer rate. As can be seen from Figure 8, the greatest increase in the heat transfer is in the finless radiator. This is due to the specific surface area of the metal foam itself. The larger the area of contact between the fluid and the solid, the larger the heat transfer area will be, ultimately increasing the heat transfer effect.

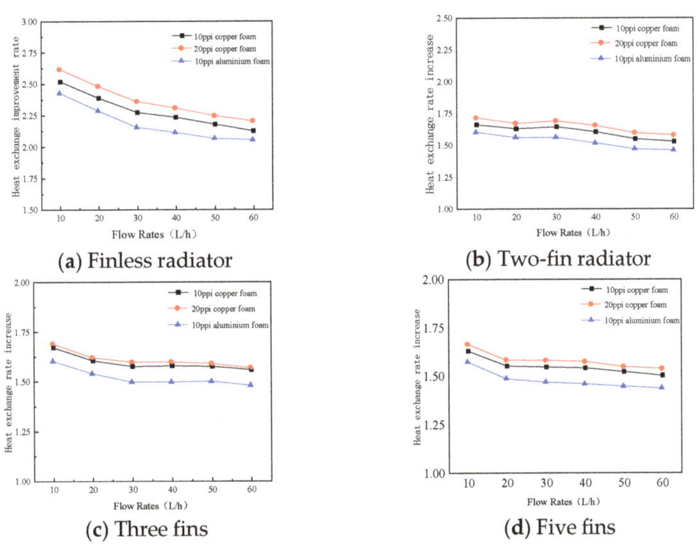

Figure 8. Heat transfer rate of radiator.

Electronic heat sink components with different numbers of pieces are filled with materials of the same porosity and different porosity densities. The rate of increase in the convective heat transfer coefficient gradually decreases when the fluid flow rate is in the range from 10 to 30 L/h. The convective heat transfer coefficient improvement rate gradually increases when the fluid flow rate is in the range of 30–60 L/h. When the fluid flow rate is in the range of 30–60 L/h, the convective heat transfer coefficient improvement rate gradually increases. This indicates that after filling the electronic heat sink components with metal foam, the effect of the enhanced heat transfer is more pronounced at low turbulence and less effective at high turbulence. This is because the main effect of the filled metal foam is to create a cyclonic flow in the core part of the flow. At low flow rates, the metal foam promotes a degree of turbulence and thus enhances heat transfer more effectively. At high flow rates, however, turbulence is sufficiently developed, and the enhanced effect of filling with metal foam on the heat transfer in the flow is reduced.

3.3.2. Analytical Comparison of Nusselt Number Nu with Re and Pressure Drop Δp with v

As shown in Figure 9, the pressure drop across the different models of radiators increases with increasing inlet flow, and the trend of increase can be seen as approximately linear. This is due to the fact that at low inlet flow rates, the dominant flow pattern within the radiator is the laminar flow. As the inlet flow rate increases, the flow pattern within the radiator gradually changes to turbulent flow. The filling of the metal foam increases the flow disturbance within the radiator. This ultimately leads to an increase in the resistance to fluid flow, which, in turn, increases the pressure drop between the inlet and the outlet. For the same inlet flow rate, the pressure drop in the electronic radiator increases with the number of fins contained. However, at a certain inlet flow rate, the difference in pressure drop between different types of electronic heat sink is not significant. This is because the pressure drop generated by the fluid flowing through the electronic heat sink depends primarily on the metal foam itself. Moreover, the increase in the number of fins in practice results in the decrease in the fin spacing, thus, in effect, changing the width of the contact cross-section of the fluid in the direction of movement. This ultimately leads to an uneven velocity distribution of the fluid at the inlet. The pressure drop generated by the electronic heat sink also depends on the pore density of the filled metal foam. The greater the pore density, the more chaotic the metal foam's own skeleton structure becomes, which, in turn, generates greater pressure losses.

Equation (11) was used to obtain the relation curve between the Nu and Re of the radiator. As shown in Figure 10, the Nusselt number of the electronic heat sink filled with metal foam is significantly higher than the Nusselt number of the electronic heat sink not filled with metal foam. This indicates that the heat transfer capability of the heat sink filled with metal foam is enhanced. The heat transfer strength of each type of heat sink also differs under different conditions of the Reynolds number. As it can be seen from Figure 10, the Nusselt number increases with the increasing Reynolds number. Additionally, the Nusselt number increases with the increasing pore density at different pore densities, while it increases and then decreases with the number of fins contained in the electronic heat sink. As can also be concluded from Figure 10 that the Nusselt number reaches a maximum when the number of fins is three. This trend is consistent with that shown by the convective heat transfer coefficient.

Equation (16) was used to obtain the radiator resistance coefficient f and Re relation curve. As shown in Figure 11, the frictional drag coefficient gradually decreases as the Reynolds number increases. As $Re > 3000$, the trend of the decreasing frictional drag coefficient becomes progressively slower. The coefficient of frictional resistance and the Reynolds number follow the same trend for electronic heat sinks with and without metal foam filling. The drag coefficient of the electronic heat sink with metal foam filling is significantly higher than that of the electronic heat sink without metal foam filling, due to the structure of the filling metal foam itself.

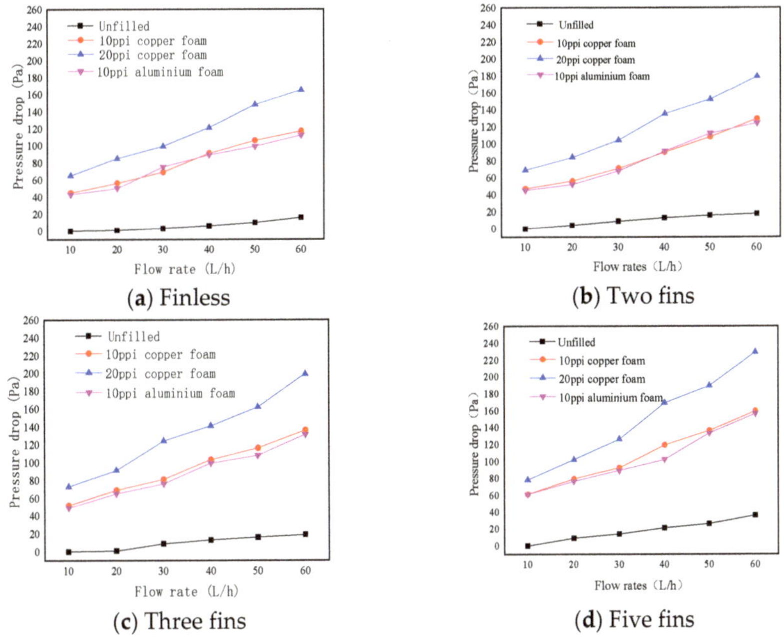

Figure 9. Radiator pressure drop Δp and v curve.

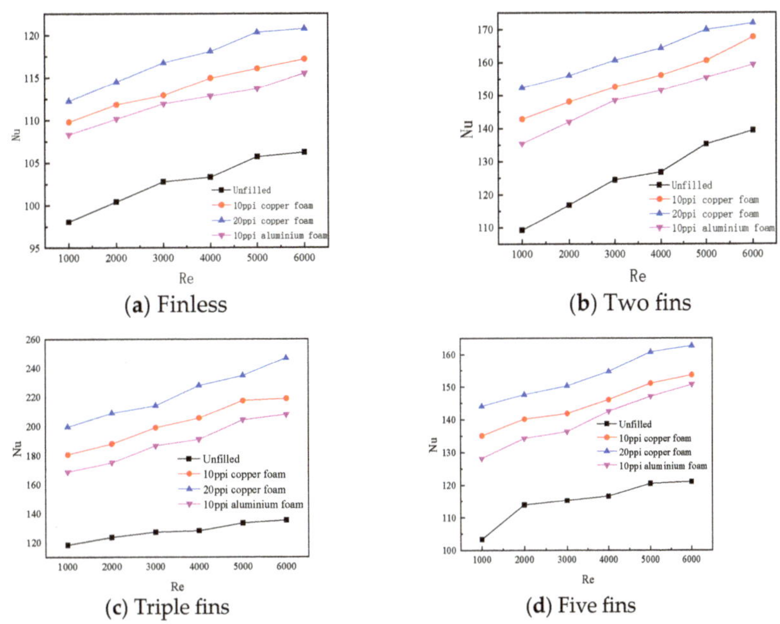

Figure 10. Relation curve between Nu and Re of radiator.

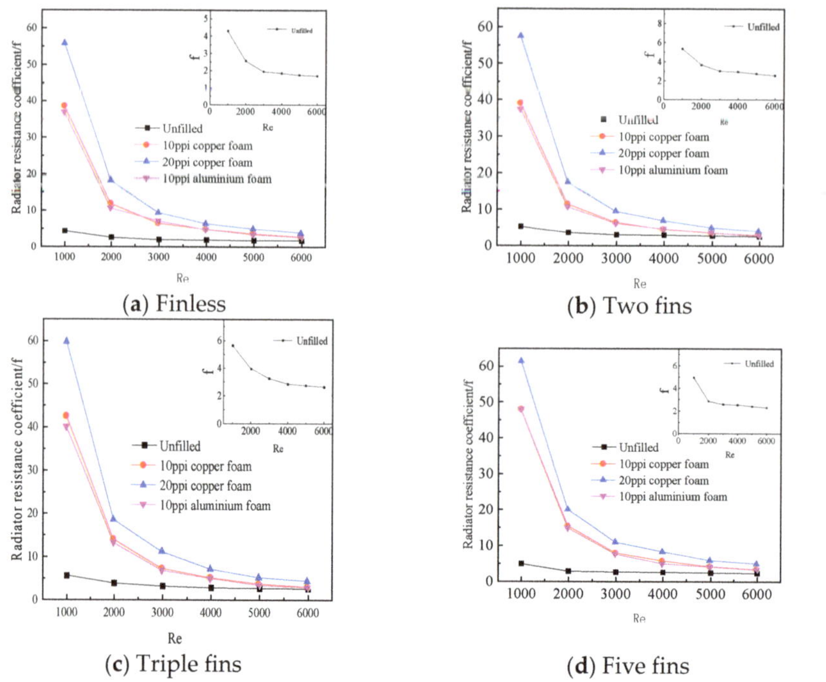

Figure 11. Radiator resistance coefficient of *f* and Re relation curve.

4. Conclusions

Our experimental studies of electronic heat sinks filled with and without metal foam led to the following conclusions.

Within the specified flow rate range (10 L/h to 60 L/h), the convective heat transfer coefficients of both unfilled and filled electronic heat sinks gradually increase as the fluid flow rate increases. However, the convective heat transfer coefficient of the filled electronic heat sink is significantly higher than that of the unfilled electronic heat sink. As the number of fins in an electronic heat sink increase, the convective heat transfer coefficient tends to increase and then decrease. Additionally, the convective heat transfer coefficient is the highest when the number of fins is three.

For metal foam-filled electronic heat sinks, the rate of increase in the heat transfer coefficient gradually decreases as the flow rate increases. The rate of increase in the convective heat transfer coefficient decreases when the fluid flow rate is in the range from 10 to 30 L/h. The rate of increase in the convective heat transfer coefficient gradually increases when the fluid flow rate is between 30 and 60 L/h.

Within the specified range of the fluid flow, the pressure drop of the filled metal foam radiator gradually increases with the increase in the fluid flow. Under the same fin conditions, the greater the pore density of the filled metal foam, the greater the pressure drop. However, the effect on the pressure drop is similar when the metal foam is filled with different materials under the same fins. As the number of electronic heat sink fins increases, the pressure drop tends to rise for both the filled and unfilled metal foam.

Within the specified flow range, the Nusselt number gradually increases with the increasing Reynolds number, considering the effect of *Re*. The trend is the same for different pore densities and different radiator fins. The Nusselt number reaches its maximum value when the number of electronic heat sink fins is three. Therefore, the heat transfer capacity of the heat sink reaches its maximum when the number of fins of the electronic heat sink is three.

Author Contributions: Conceptualization, B.L. and R.B.; Formal analysis, B.L. and Z.Z.; Investigation, X.S. and R.W.; Writing—original draft, X.S.; Writing—review & editing, B.L., R.B. and P.E.T. All authors have read and agreed to the published version of the manuscript.

Funding: This research received no external funding.

Data Availability Statement: Not applicable.

Conflicts of Interest: The authors declare no conflict of interest.

Nomenclature

ε	Metal foam porosity (vol %)
$q_{m,c}$	Mass flow of cold fluid (kg·s^{-1})
$q_{m,h}$	Mass flow of hot fluid (kg·s^{-1})
H_c	Enthalpy per unit of cold fluid (J·kg^{-1})
H_h	Enthalpy per unit of thermal fluid (J·kg^{-1})
$c_{p,c}$	Constant pressure heat capacity of cold fluid (J/(kg·K))
$c_{p,h}$	Constant pressure heat capacity of hot fluid (J/(kg·K))
t''_c	Temperature of the cold fluid at the outlet (°C)
t'_c	Temperature of the cold fluid at the inlet (°C)
t'_h	Temperature of the hot fluid at the inlet (°C)
t''_h	Temperature of the hot fluid at the outlet (°C)
A	Contact area (cm^3)
ρ	Fluid density (kg/m^3)
μ	Fluid viscosity coefficient (Pa·s)
Δt_m	Logarithmic mean temperature difference (°C)
Q_h	Fluid heat release (W)
Q_c	Fluid heat absorption (W)
S	Effective cross-sectional area of the flow channel (m^2)
U	Cross-sectional circumference of electronic heat sink parts (m)
a.b	Length and width of the micro-channel cross-section (m)
ΔQ	Relative error in thermal equilibrium
Nu	Nusselt number
Re	Reynolds number
Pr	Prandtl number
d_e	Equivalent diameter of the flow channel (m)
λ	Thermal conductivity of water W/(m·K)
f	Coefficient of frictional resistance
l	Experimental section pressure measurement hole spacing (m)
Δp	Experimental section pressure drop (Pa)
v	Fluid flow velocity (m/s)
k	Fluid thermal conductivity (W m^{-2} K^{-1})
R^2	Correlation coefficient

References

1. Garcia-Moreno, F.; Mukherjee, M.; Solórzano, E.; Banhart, J. Metal foams–towards microcellular materials. *Int. J. Mater. Res.* **2010**, *101*, 1134–1139. [CrossRef]
2. Clyne, T.; Markaki, A.; Tan, J.-C. Mechanical and magnetic properties of metal fibre networks, with and without a polymeric matrix. *Compos. Sci. Technol.* **2005**, *65*, 2492–24999. [CrossRef]
3. Qin, J.; Chen, Q.; Yang, C.; Huang, Y. Research process on property and application of metal porous materials. *J. Alloys Compd.* **2016**, *654*, 39–44. [CrossRef]
4. Wu, S.; Liu, X.; Yeung, K.W.K.; Liu, C.; Yang, X. Biomimetic porous scaffolds for bone tissue engineering. *Mater. Sci. Eng. R Rep.* **2014**, *80*, 1–36. [CrossRef]
5. Singh, R.; Lee, P.D.; Dashwood, R.J.; Lindley, T.C. Titanium foams for biomedical applications: A review. *Mater. Technol.* **2010**, *25*, 127–136. [CrossRef]
6. Bansiddhi, A.; Sargeant, T.; Stupp, S.; Dunand, D. Porous NiTi for bone implants: A review. *Acta Biomater.* **2008**, *4*, 773–782. [CrossRef]
7. Ding, C.; Zhang, C.; Ma, L.; Sharma, A. Numerical investigation on melting behaviour of phase change materials/metal foam composites under hypergravity conditions. *Appl. Therm. Eng.* **2022**, *207*, 118153. [CrossRef]

8. Zhao, C. Review on thermal transport in high porosity cellular metal foams with open cells. *Int. J. Heat Mass Transf.* **2012**, *55* 3618–3632. [CrossRef]
9. Deng, F.; Li, X.; He, L.; Feng, Y. Experimental evaluation of metal foam for sand control. *J. Pet. Sci. Eng.* **2019**, *176*, 1152–1160 [CrossRef]
10. Brothers, A.; Dunand, D. Amorphous metal foams. *Scr. Mater.* **2006**, *54*, 513–520. [CrossRef]
11. Liu, Y.; He, X.; Deng, C. Self-Piercing Riveting of Metal Foam Sandwich Structures. *Mater. Trans.* **2017**, *58*, 1532–1537. [CrossRef]
12. Xu, C.; Mao, Y.; Hu, Z. Tonal and broadband noise control of an axial-flow fan with metal foams: Design and experimental validation. *Appl. Acoust.* **2017**, *127*, 346–353. [CrossRef]
13. Chen, K.; Wang, X.; Chen, P.; Wen, L. Numerical simulation study on heat transfer enhancement of a heat exchanger wrapped with metal foam. *Energy Rep.* **2022**, *8*, 103–110. [CrossRef]
14. Kim, S.Y.; Paek, J.W.; Kang, B.H. Thermal performance of aluminum-foam heat sinks by forced air cooling. *IEEE Trans. Compon. Packag. Technol.* **2003**, *26*, 262–267.
15. Shen, L.; Xu, S.; Bai, Z.; Wang, Y.; Xie, J. Experimental study on thermal and flow characteristics of metal foam heat pipe radiator *Int. J. Therm. Sci.* **2021**, *159*, 106572. [CrossRef]
16. Kotresha, B.; Gnanasekaran, N. Numerical Simulations of Fluid Flow and Heat Transfer through Aluminum and Copper Metal Foam Heat Exchanger—A Comparative Study. *Heat Transf. Eng.* **2019**, *41*, 637–649. [CrossRef]
17. Chen, T.; Shu, G.; Tian, H.; Zhao, T.; Zhang, H.; Zhang, Z. Performance evaluation of metal-foam baffle exhaust heat exchanger for waste heat recovery. *Appl. Energy* **2020**, *266*, 114875. [CrossRef]
18. Zhao, C.; Lu, W.; Tassou, S. Thermal analysis on metal-foam filled heat exchangers. Part II: Tube heat exchangers. *Int. J. Heat Mass Transf.* **2006**, *49*, 2762–2770. [CrossRef]
19. Kaur, I.; Singh, P. Critical evaluation of additively manufactured metal lattices for viability in advanced heat exchangers. *Int. J. Heat Mass Transf.* **2021**, *168*, 120858. [CrossRef]
20. Saeedan, M.; Ziaei-Rad, M.; Afshari, E. Metal foam as a turbulent flow distributor in the cooling channels of a PEM fuel cell—A numerical study. *Phys. Scr.* **2019**, *94*, 064002. [CrossRef]
21. Shi, J.; Zheng, G.; Chen, Z.; Dang, C. Experimental study of flow condensation heat transfer in tubes partially filled with hydrophobic annular metal foam. *Int. J. Heat Mass Transf.* **2019**, *136*, 1265–1272. [CrossRef]
22. Wang, Y.; Yu, Y.; Jing, Z.; Wang, C.; Zhou, G.; Zhao, W. Thermal performance of lithium-ion batteries applying forced air cooling with an improved aluminium foam heat sink design. *Int. J. Heat Mass Transf.* **2021**, *167*, 120827. [CrossRef]
23. Hossain, M.S.; Shabani, B. Metal foams application to enhance cooling of open cathode polymer electrolyte membrane fuel cells. *J. Power Sources* **2015**, *295*, 275–291. [CrossRef]
24. Saeedan, M.; Afshari, E.; Ziaei-Rad, M. Modeling and optimization of turbulent flow through PEM fuel cell cooling channels filled with metal foam- a comparison of water and air cooling systems. *Energy Convers. Manag.* **2022**, *258*, 115486. [CrossRef]
25. Alhusseny, A.; Al-Aabidy, Q.; Al-Zurfi, N.; Nasser, A.; Aljanabi, M. Cooling of high-performance electronic equipment using graphite foam heat sinks. *Appl. Therm. Eng.* **2021**, *191*, 116844. [CrossRef]
26. Qi, C.; Chen, T.; Tu, J.; Yan, Y. Effects of metal foam on exergy and entropy of nanofluids in a heat sink applied for thermal management of electronic components. *Int. J. Energy Res.* **2020**, *44*, 10628–10651. [CrossRef]
27. Li, Y.; Gong, L.; Xu, M.; Joshi, Y. Hydraulic and thermal performances of metal foam and pin fin hybrid heat sink. *Appl. Therm. Eng.* **2020**, *166*, 114665. [CrossRef]
28. Li, Y.; Gong, L.; Xu, M.; Joshi, Y. Thermal Performance Analysis of Biporous Metal Foam Heat Sink. *J. Heat Transf.* **2017**, *139*, 052005. [CrossRef]
29. Hsieh, W.; Wu, J.; Shih, W.; Chiu, W. Experimental investigation of heat-transfer characteristics of aluminum-foam heat sinks. *Int. J. Heat Mass Transf.* **2004**, *47*, 5149–5157. [CrossRef]
30. Liu, J.; Wu, W.; Chiu, W.; Hsieh, W. Measurement and correlation of friction characteristic of flow through foam matrixes. *Exp. Therm. Fluid Sci.* **2006**, *30*, 329–336. [CrossRef]
31. Dukhan, N.; Chen, K.-C. Heat transfer measurements in metal foam subjected to constant heat flux. *Exp. Therm. Fluid Sci.* **2007**, *32*, 624–631. [CrossRef]
32. Boomsma, K.; Poulikakos, D. On the effective thermal conductivity of a three-dimensionally structured fluid-saturated metal foam. *Int. J. Heat Mass Transf.* **2001**, *44*, 827–836. [CrossRef]
33. Shih, W.H.; Chiu, W.C.; Hsieh, W.H. Height Effect on Heat-Transfer Characteristics of Aluminum-Foam Heat Sinks. *J. Heat Transf.* **2006**, *128*, 530–537. [CrossRef]
34. Singh, P.; Nithyanandam, K.; Mahajan, R.L. An experimental and numerical investigation of forced convection in high porosity aluminum foams subjected to jet array impingement in channel-flow. *Int. J. Heat Mass Transf.* **2019**, *149*, 119107. [CrossRef]
35. Bhattacharya, A.; Mahajan, R.L. Finned Metal Foam Heat Sinks for Electronics Cooling in Forced Convection. *J. Electron. Packag.* **2002**, *124*, 155–163. [CrossRef]

Disclaimer/Publisher's Note: The statements, opinions and data contained in all publications are solely those of the individual author(s) and contributor(s) and not of MDPI and/or the editor(s). MDPI and/or the editor(s) disclaim responsibility for any injury to people or property resulting from any ideas, methods, instructions or products referred to in the content.

energies

MDPI

Article

Numerical Investigation of Flow Past Bio-Inspired Wavy Leading-Edge Cylinders

Paulo Henrique Ferreira *, Tiago Barbosa de Araújo, Eduardo Oliveira Carvalho, Lucas Dantas Fernandes and Rodrigo Costa Moura

Aeronautical Engineering Division, Instituto Tecnológico de Aeronáutica (ITA), São José dos Campos 12228-900, SP, Brazil; tiagotba@ita.br (T.B.d.A.); eduardoc@ita.br (E.O.C.); lucasdantasf29@gmail.com (L.D.F.); moura@ita.br (R.C.M.)
* Correspondence: paulohf@ita.br

Abstract: A numerical investigation is proposed to explore the flow past a novel wavy circular cylinder as a passive flow control, whose shape is determined by a sinusoidal function applied to its leading edge line, similar to studies with wavy leading-edge airfoils. The latter are motivated by the wavy-shaped tubercles found in the flippers of humpback whales, which are believed to improve their maneuverability. Our attempt is, therefore, to assess the effects of leading-edge waviness now on a simpler and canonical geometry: circular cylinders. The present work relies on iLES simulations conducted with Nektar++ at a Reynolds number of 3900. Besides the straight cylinder, two wavy geometries are assessed, which are determined by a single wavelength of 37.5% for two amplitudes, 3% and 11%, based on the mean diameter of the wavy cylinder. Our results showed that, contrary to what is usually the case with traditional wavy cylinders at similar Reynolds numbers, waviness caused a reduction in the near-wake recirculation length and an increase in the mean near-wake turbulent kinetic energy compared to the straight cylinder. This was followed by a reduction in base pressure (up to about 36%) leading to a rise in lift oscillations and also to a significant increase in the mean drag coefficient of up to about 28%. An attempt to detail the flow phenomena is provided, evidencing the emergence of counter-rotating pairs of streamwise vortices between peaks. It is argued that the differences observed in recirculation length, turbulent kinetic energy, and force coefficients start even prior to the formation of these coherent structures and end up with interactions with the near wake.

Keywords: wavy cylinder; wavy leading-edge airfoil; spectral element method; continuous Galerkin method; implicit large-eddy simulation

Citation: Ferreira, P.H.; de Araújo, T.B.; Carvalho, E.O.; Fernandes, L.D.; Moura, R.C. Numerical Investigation of Flow Past Bio-Inspired Wavy Leading-Edge Cylinders. *Energies* **2022**, *15*, 8993. https://doi.org/10.3390/en15238993

Academic Editors: Artur Bartosik and Dariusz Asendrych

Received: 3 October 2022
Accepted: 21 November 2022
Published: 28 November 2022

Publisher's Note: MDPI stays neutral with regard to jurisdictional claims in published maps and institutional affiliations.

Copyright: © 2022 by the authors. Licensee MDPI, Basel, Switzerland. This article is an open access article distributed under the terms and conditions of the Creative Commons Attribution (CC BY) license (https://creativecommons.org/licenses/by/4.0/).

1. Introduction

Many researchers have exhaustively studied flow over bluff bodies, especially cylinders, such as Roshko [1], Williamson [2], and Zdravkovich [3]. Whether due to the simplicity of the geometry or the intriguing complexity of the emerging flow structures, cylinders have always aroused this interest over time. Additionally, this geometry has always been present in everyday life, in different engineering applications, e.g., flow over buildings, wind turbines, and oil risers, which reinforces its importance, especially by knowing they suffer the action of a phenomenon called vortex-induced vibration (VIV). In a simplified way, this is an oscillating effect induced by the interaction of pressure fluctuations created by a vortex-shedding wake on the body itself, which can even compromise its structural integrity [4].

In the search for solutions to mitigate or suppress phenomena such as VIV, or simply to reduce drag or the mean lift fluctuations (i.e., RMS of lift forces), many studies have been carried out with passive flow control mechanisms. In other words, those mechanisms do not have an active device, but work by changing the body geometry itself, e.g., a surface undulation, like the one that will be presented here, or the addition of some fixed external structure to the body surface, such as the well-known vortex generator [5], that helps to improve stall characteristics.

Despite also having a circular cross-section, different from the planform of the typical existing wavy cylinders, the present work proposes a new bio-inspired planform cylinder, as shown in Figure 1, here named the "wavy leading-edge cylinder". The inspiration comes from works on wavy leading-edge wings, whose study was first motivated by the morphological observations made by Fish and Battle [6], in 1995, about humpback whales. This study hypothesized that the presence of some protuberances (also called tubercles) almost regularly distributed on the leading-edge surface of the humpback whale flippers acted like a flow control mechanism responsible for the distinct hydrodynamic characteristics of these animals, which are known to be excellent hunters, and are able to perform complex movements with great ease and under high angles of attack.

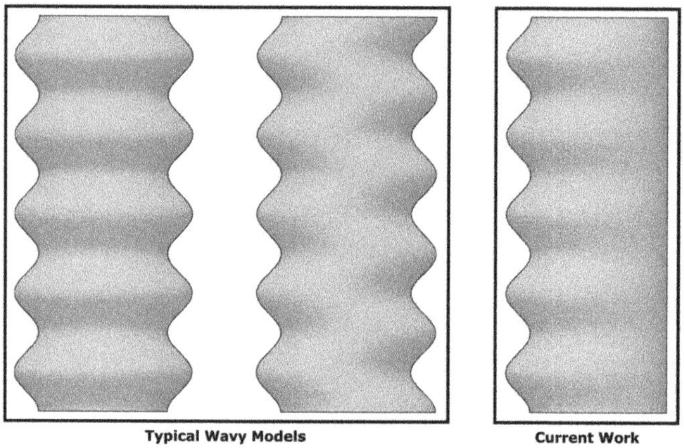

Figure 1. Types of wavy cylinders (in each cylinder, the leading edge appears on the left and the trailing edge on the right). The right-most cylinder shows the waviness adopted in the current work, whereas the other two cylinders show the typical patterns of waviness considered in the literature.

Different studies [7–12] mimicking those tubercles have tried to understand how this geometry allowed to improve maneuverability and whether they could enhance the aerodynamic performance of wings. Those studies have evidenced interesting phenomena and promising results, such as three-dimensional effects like the emergency of longitudinal counter-rotating vortex pairs (CVP), the formation of three-dimensional laminar separation bubbles (LSB), the delay of boundary layer detachment, a more evenly distributed surface pressure, an aerodynamic efficiency improvement (by reduction of drag and/or increase of lift), and so on.

After this brief overview of wavy leading-edge airfoils, we turn to the theme of this work. The proposed waviness modification was created by giving volume to a geometry after sweeping a circular area through a planform constrained by a sinusoidal function curve at the leading-edge line and a straight line at the trailing-edge line, as shown in Figure 2. This is a computational part of a research that also includes an experimental part [13] for the same geometries, but at higher Reynolds numbers. As will be discussed in the present work, the applied waviness resulted in a drag increase at the current (lower) Reynolds of 3900, showing there is also an important Reynolds effect at play.

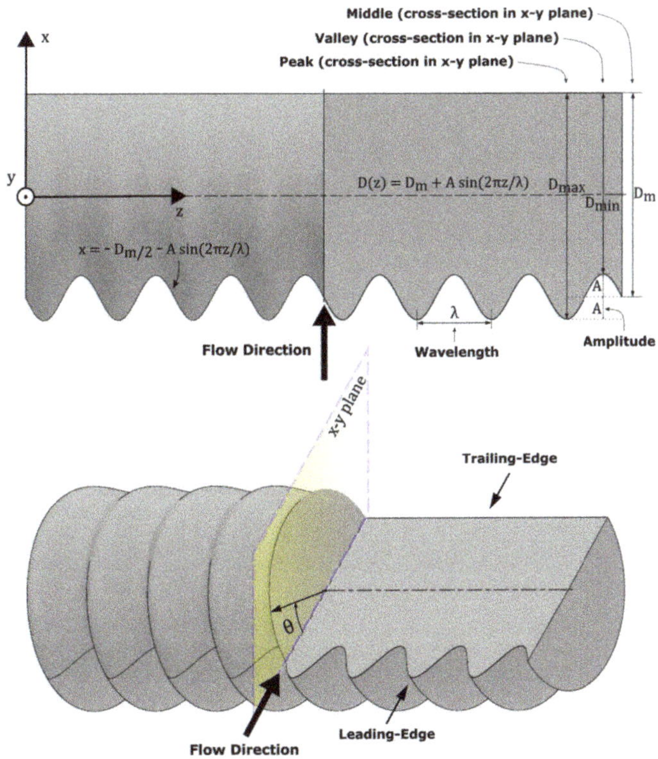

Figure 2. Sectional views—cutting plane (**top**) and three-dimensional view (**bottom**)—of the new flow control geometry proposed: the wavy leading-edge cylinder geometry. The illustrative figure presents the main parameters and constructive function in the x–z plane.

Here, simulations were conducted first to validate the numerical model for the straight cylinder at Re = 3900 and then to investigate two wavy geometries at the same Reynolds, namely two amplitudes for a single wavelength. At this Reynolds number, the straight cylinder is well within the subcritical regime, in which the boundary layer is laminar and transition occurs along the separating shear layer (via Kelvin–Helmholtz instability) leading to a turbulent wake featuring the well-known von-Kármán vortex street. In the near wake region, the interaction between the transitional shear layer and the wake creates three-dimensional turbulent structures that evolve in both streamwise and spanwise directions.

Focusing on wavy cylinders, Ahmed and Bays-Muchmore [14] were among the first authors to carry out experiments with transverse flow over a wavy cylinder. In this case, they measured surface pressure in a wind tunnel for a Reynolds number of 10,000 and performed dye visualizations in a water tunnel for Reynolds numbers of 5000, 10,000, and 20,000. They tested a traditional sinusoidal wavy planform with symmetric waviness in the spanwise direction for a set of wavy models with wavelength (λ) ratio to the mean diameter D_m of $\lambda/D_m = 1.2, 1.6, 2.0$, and 2.4, and a fixed amplitude ratio of $A/D_m = 0.1$. They found that waviness reduces overall drag and observed that the sectional (2D) drag coefficients were greater at peaks (larger diameter section) than at valleys (smaller diameter section). Additionally, they found that separation was anticipated on valleys and delayed on peaks. Note that the frontal pressure typically changes along the spanwise direction, also influencing differences in drag between sections. In addition, the separation line maintained a curvature proportional to the wavy geometry applied, following a wavy pattern. Additionally, from flow topology analysis, they inferred the possible formation of a pair of longitudinal vortices in between adjacent peaks.

Zhang et al. [15] carried out experiments in a water channel at a Reynolds number of 3000 in order to investigate the near wake behavior of a wavy cylinder through a particle image velocimetry (PIV) measurement technique. From the transverse vorticity results, they inferred the formation of structures supposed to be a pair of counter-rotating streamwise vortices at both sides of each peak. These vortices showed an action to suppress the development of both large-scale spanwise vortices and regular vortex shedding. This effect indicated a possible reason for the reduction in the average turbulent kinetic energy (TKE) over the near-wake region, especially after the valley regions. Therefore, this consequential lower TKE also helped to justify lower values of drag coefficient compared to the straight cylinder.

Lam and Lin [16] conducted a large-eddy simulation (LES) to study the flow past wavy cylinders at a Reynolds number of 3000. They tested a combination of parameters for λ/D_m varying from 1.136 to 3.333, and two amplitude ratios, $A/D_m = 0.091$ and 0.152. The most notable result was a reduction of about 18% in drag when compared to the straight cylinder, for $\lambda/D_m = 1.9$ and $A/D_m = 0.152$. In addition, waviness caused a reduction or even a suppression of RMS lift coefficients with no significant changes in Strouhal number (St). Furthermore, they observed that waviness created a difficulty for the vortex sheet to roll up, which ended up moving downstream the vortex formation region. This three-dimensional free shear layer effect helped them to explain the less intense pressure fluctuations over the surface and the lower values of turbulent kinetic energy in the wake (compared to the "two-dimensional" wake of the straight cylinder), leading to higher base pressure values and lower drag coefficients.

Lin and Yu-fen [17] conducted a large-eddy simulation study, with an experimental validation of the straight model in a wind tunnel, to investigate the effect of waviness at a Reynolds number of 3000. They tested a single combination of wavelength and amplitude ratios, namely $\lambda/D_m = 1.5$ and $A/D_m = 0.15$. Their conclusions suggested that waviness imposed a difficulty for the formation of a vortex street, also with less intense pressure fluctuations and smaller values of turbulent kinetic energy over different spanwise cross-sections in the wake. Those effects led to a reduction in both mean drag and RMS lift coefficients. Again, the separation was anticipated in the valleys and delayed in the peaks.

Lam et al. [18] performed a large-eddy simulation and an experimental validation with laser Doppler anemometry (LDA) and load cell measurements in water and wind tunnels, to investigate flow past a yawed wavy cylinder at a Reynolds number of 3900. They tested a single combination of wavelength and amplitude ratios, $\lambda/D_m = 6.0$ and $A/D_m = 0.15$, respectively, with yaw angle varying from 0° to 60°. The results of the flow with yaw angle = 0° (or unyawed), which is the main interest for the present work, showed drag and RMS lift coefficients reductions of up to 14% and 80%, respectively. Power spectral density (PSD) revealed a lower St, namely 0.184 against 0.208 from the straight cylinder. Moreover, pressure and velocity distributions confirmed spanwise periodic and repetitive wake structures, following the pattern imposed by the wavy geometry.

Zhao et al. [19] implemented a scale-adaptive simulation to understand the physics of the flow past a wavy cylinder at Reynolds number 8000 for several configurations. They varied λ/D_m from 3 to 7 for $A/D_m = 0.091$ and 0.152. They confirmed results already obtained by other authors, with reductions in Cd, St, and Cl_{RMS} of up to 30%, 50%, and 92%, respectively, compared to the straight case. Besides associating the reduction of lift fluctuations to the presence of longitudinal vortices, they visualized for certain configurations a change in the pattern of surface streamlines after the separation line (curve). Lastly, they also observed longer mean recirculation lengths, which means spanwise vortex formation moved downstream, which led to a base pressure increase and corresponding drag reduction.

Bai et al. [20] investigated flow past wavy cylinders in a water tunnel at a Reynolds number of 3000 through a time-resolved PIV (TR-PIV) technique. They tested a single model of $\lambda/D_m = 0.152$ and $A/D_m = 6$. PSD results revealed two different vortex-shedding frequencies associated with peak and valleys sections, respectively, of $f = 0.164$ and

$f = 0.214$, the last one very close to the Strouhal number expected for a straight cylinder at this Reynolds number ($St = 0.21$), indicating that larger structures such as von-Kármán vortices still dominate flow. An analysis based on proper orthogonal decomposition (POD) in the spanwise direction corroborated this by showing that the first two dominant modes exhibit the spanwise vortex shedding in valleys and peaks and together concentrate almost 45% of the POD energy. Despite that, near the wake in the streamwise direction, the most energetic POD modes were associated with the presence of the streamwise vortices, which tend to suppress the coherence of spanwise vortices. In addition, they observed that mean recirculation lengths extend further downstream, more in peaks than in valleys, as well as when compared to the straight cylinder.

Karthik et al. [21] studied the flow past a wavy cylinder using LES along with the Ffowcs, Williams and Hawking's (FWH) acoustic analogy at Reynolds number 97,300. They tested all combinations of parameters for $\lambda/D_m = 1.0, 1.5, 2.0$, and 2.5, and $A/D_m = 0.05$, 0.1, 0.15, and 0.2. Drag and RMS lift reduced in all wavy cases, with the greatest drag reduction for $\lambda/D_m = 2.0$, and the minimum RMS lift for $\lambda/D_m = 1.5$, among all amplitudes. The parameters for the optimum case for both drag and sound emission were $\lambda/D_m = 1.892$ and $A/D_m = 0.134$, with a reduction of 33.21% and 6.331 dB compared to the straight case, respectively. They associated these results with a mean increase in base pressure distribution and a reduction in average vorticity in the near wake.

Finally, a recent relevant study was conducted by Zhang et al. [22], in which LES was used for the comparison of different models at Reynolds number 3900, namely for wavelength ratios varying from 1.8 to 5.0 and amplitude ratios from 0.05 to 0.2. They characterized flow structures developed in the wake as rib-like vortices and three-dimensional vortex lines caused by waviness along the span. Besides that, they also identified longitudinal vortices, and evidenced their similarities with some vortex generator mechanisms. Based on the analysis of two components of the transport equation for vorticity, namely stretching and turning terms, they inferred two vortex formations, which they called primary and secondary vortices. The primary seems to induce the secondary one. The first is dominant while the second tends to disappear with increasing wavelengths. A correlation between longer recirculation lengths and drag reduction is observed for all their wavy cylinders. Drag reduction was found for greater amplitudes and/or smaller wavelengths. Analysis of surface streamlines revealed a flow from the peak to valley before the separation line (curve). The pressure at the stagnation position (leading edge) at peaks and valleys is about the same as the straight cylinder, while the middle section presents a reduction in static pressure, especially for the models with largest amplitude.

Despite being a topic already explored in the literature, others wavy cylinders do not have the same inspiration as the present work. In addition to the geometric transformation applied by the undulation being different, the parameters (wavelength and amplitude) also differ from those typically selected in the wavy cylinder literature. In addition to the presentation of the results, an attempt is also made to establish a correlation that explains the emerging phenomena.

As will be seen later, even though the results were different from what was expected based on the wavy airfoils, since it is a recent and developing topic, the understanding of the phenomenon behind this bio-inspired waviness in a simpler and canonical geometry, such as the circular cylinder, may be useful for other authors to be able to better describe the flow over wavy airfoils or others wavy geometries.

Regarding the wavy cylinders already existing in the literature, it was not expected to find the same results, since they are similar geometries, but not the same, even so, apparently, the phenomenology appears to be common to both. Regarding a possible application of this new geometry, which has a preferential or fixed direction, we can mention rotary wings, such as those found in helicopters, and rotate blades in wind turbines.

This paper is organized as follows. In Section 2, the methodology of the study is presented along with the definition of the numerical domain, the description of the meshes employed and boundary conditions adopted. In Section 3, the numerical results are presented and discussed, starting with a validation of the numerical method as per the results obtained for the straight cylinder, followed by a discussion of the main characteristics found for the wavy models and a detailed assessment of three-dimensional flow structures. Finally, Section 4 summarizes the conclusions of the study.

2. Materials and Methods

The present study made use of the high-fidelity implicit large-eddy simulation (iLES) approach based on spectral/hp element methods (SEM) [23]. Broadly speaking, this approach relies on numerical dissipation in lieu of a traditional turbulence model, no subgrid-scale model is employed, and has become very popular in recent decades. In this context, iLES has also been called under-resolved direct numerical simulation (uDNS) [24]. For a detailed study on why and how to use SEM-based iLES/uDNS, the reader is referred to references [25–36].

The SEM adopted was the high-order continuous Galerkin (CG) method coupled with a novel stabilization scheme particularly suited for the simulation of transitional and turbulent flows [37,38]. This technique is called gradient-jump penalty (GJP) and has even been proven superior to modern versions of the well-known spectral vanishing viscosity (SVV) operator [39]. Simulations were conducted with the Nektar++ open-source code [40,41], a mature platform that has been successfully employed multiple times for iLES/uDNS [42–45].

The adopted unsteady CG solver [46] is based on a spectral/hp element method that provides a continuous Galerkin projection as a linear combination of a set of a particular polynomial basis functions Φ_n, such that:

$$u^\delta = \sum_{n \in N} \Phi_n \hat{u}_n \,, \tag{1}$$

where $\hat{u}_n$ are the coefficients of our solution vector $\hat{\mathbf{u}}$ to the approximate solution u^δ in the reduced finite-dimensional function space $U_\delta \subset U$ for a linear differential equation type:

$$\mathbf{L}(\mathbf{u}) = f \,, \tag{2}$$

which can be replaced by our incompressible Navier–Stokes equations to be solved, i.e.,

$$\frac{\partial \mathbf{u}}{\partial t} = \mathbf{N}(\mathbf{u}) - \nabla p + \nu \mathbf{L}(\mathbf{u}) \,, \tag{3}$$

$$\nabla \cdot \mathbf{u} = 0 \,, \tag{4}$$

where:

$$\mathbf{N}(\mathbf{u}) = -(\mathbf{u} \cdot \nabla)\mathbf{u} \,, \tag{5}$$

$$\nu \mathbf{L}(\mathbf{u}) = \nu \nabla^2 \mathbf{u} \,. \tag{6}$$

They are solved along with the appropriate boundary conditions described later on. The pressure is solved by the following velocity correction scheme:

$$\int_\Omega \nabla p^{n+1} \cdot \nabla \phi d\Omega = \int_\Omega \nabla \cdot \left(\frac{-\hat{\mathbf{u}}}{\Delta t} \right) d\Omega + \int_\Gamma \phi \left[\frac{\hat{\mathbf{u}} - \gamma_0 \bar{\mathbf{u}}^{n+1}}{\Delta t} - \nu (\nabla \times \nabla \times \mathbf{u})^* \right] \cdot \mathbf{n} \, dS \,, \tag{7}$$

where Ω is the domain and Γ is the boundary domain, with a backward approximation of the time derivative, such that:

$$\hat{\mathbf{u}} = \mathbf{u}^{\oplus} + \Delta t \, \mathbf{N}^* \,, \tag{8}$$

where $(*)$ denotes the time extrapolation, and $(\oplus)$ the backward differencing. It uses a second-order implicit–explicit (IMEX) time-integration scheme. The velocity is the solution of the Helmholtz problem given by:

$$\frac{\gamma_0 \mathbf{u}^{n+1} - \hat{\mathbf{u}}}{\Delta t} = -\nabla p^{n+1} + \nu \mathbf{L}(\mathbf{u}^{n+1}) \,. \tag{9}$$

2.1. Computational Domain and Baseline Mesh

The simulation domain is initially based on a two-dimensional mesh of quadrilaterals, as shown in Figure 3, which is then extruded along the spanwise direction in a way that allows for the desired waviness to be incorporated. The dimensions of the domain, i.e., the distances upstream and downstream of the body, the height and the width were based on dimensions previously established as sufficient to avoid blockage, with a blockage ratio (diameter/height) less than 3.4%, and allow the proper development of the wake [47,48].

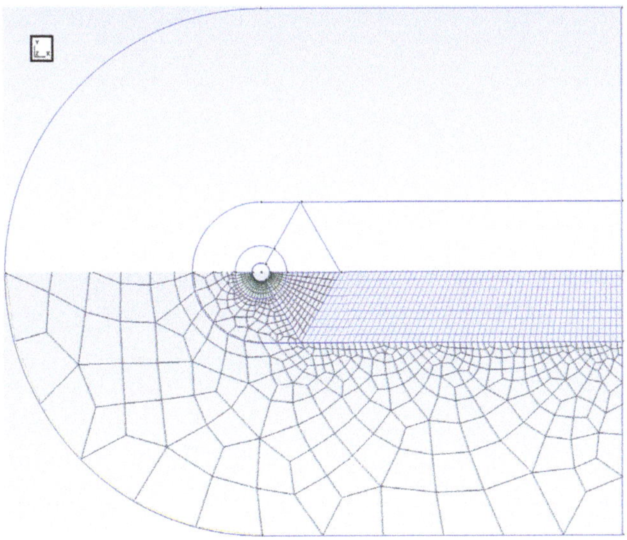

Figure 3. Two-dimensional mesh. It is worth noting that this mesh has not yet received p-type refinement (N_p = 2), that is, in this case, in each direction the elements still will be "divided" in half, then increasing its resolution.

Note the mesh is of a structured pattern around the cylinder and along the wake, whereas elsewhere it is of an unstructured pattern. In SEM, the solution inside each element is represented through a polynomial expansion. Here, the element-wise polynomial order has been chosen as $P = 2$ (parabolic solution per element). Note that CG's nominal order of accuracy is $P + 1$, hence all the solutions presented here are nominally third-order accurate.

The circular cylinder is placed at the center of the domain (origin of Cartesian coordinates) and has a diameter D of unit size. The outermost boundary is made up of a semi-circle (inflow boundary of radius $15D$) and a rectangle whose downstream closure coincides with the outflow boundary (at $x \approx 21D$). The intermediate part of the mesh, as also shown in Figure 3, has another semi-circular section (of radius $4D$) that connects to two symmetric triangular sections. These exist as means for the mesh to transition into the wake region, which extends downstream until the outflow boundary.

Around the cylinder, there is an outer circle (of radius $1.5D$) and an inner circle (of radius $D/2 + \delta_{BL}$) placed very close to the cylinder's surface, as shown in Figure 4. The latter is devoted to capturing the (laminar) boundary layer of the simulation, whereby its thickness is adjusted based on the cylinder's Reynolds number Re_D according to the estimate

$$\delta_{BL} \sim D\,(Re_D)^{-1/2}\,, \qquad (10)$$

which was found to approximately match the boundary layer thickness next to its separation point. This innermost circular layer is subdivided into 5 equispaced layers that are deemed sufficient to resolve the cylinder's laminar boundary layer, given the high-order nature of the elements within the equispaced layers. For example, when using $P = 2$ element-wise polynomials in the solution, the innermost circular region will have $2 \times 5 = 10$ DOF (degrees of freedom) across the radial direction.

It is also worth mentioning that, in terms of resolution power, a SEM-type DOF is somewhat superior to the DOF of classical CFD schemes, as can be seen in dispersion/diffusion analyses [25,27]. In fact, these considerations apply to all the elements in the domain, which is why the mesh in Figures 3 and 4 may seem too coarse for those unfamiliar with SEM. If the mesh was shown with "subelements" representing the element-wise polynomial DOFs, it would look much finer.

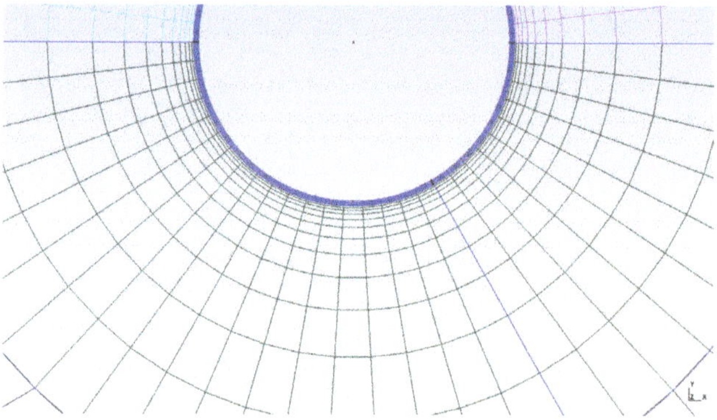

Figure 4. Close-up view of the mesh next to the cylinder. It is worth noting that this mesh has not yet received p-type refinement (N_p = 2), that is, in this case, in each direction the elements still will be "divided" in half, increasing their DOF.

From the edge of the boundary layer region, the radial mesh size grows geometrically toward the outer circle around the cylinder, when the radial mesh size achieves the same mesh spacing employed horizontally along the wake ($\approx$0.25D). Along the cylinder's wall, in the azimuthal direction, 54 equispaced elements (or angular subdivisions) are used to cover the whole circle. This also defines the azimuthal size of the mesh along the outer circle, which in turn defines the vertical mesh size along the wake ($\approx$0.35D).

As mentioned previously, the regions within the cylinder's outer circle, along with the two symmetric triangular regions plus the wake region follow a structured mesh pattern. The other regions of the domain have a mesh with an unstructured pattern. All meshes have been generated with GMSH [49], an open-source finite element mesh generator. A third-order mesh representation was used in all the cases, i.e., one order above the polynomial order adopted for the numerical solution near the surface, as per typical SEM guidelines for simulating flows around curved geometries.

2.2. Mesh Generation

The 2D mesh described in the previous section is the same for all the cases simulated. To obtain the actual three-dimensional meshes, the baseline 2D mesh is extruded along the spanwise direction over a distance $L_z = 3D$, which is typically found to be sufficiently large according to the literature. Waviness is then incorporated in the 3D mesh through a simple mapping transformation, which shifts mesh coordinates on the original surface of the cylinder to the desired case-dependent wavy surface. The mapping is such that the original trailing edge is left untouched, whereas the leading edge becomes wavy according to a sine function of a given wavelength and amplitude (Figure 2). This transformation also shifts vertical mesh coordinates, so that any cross-section (normal to the spanwise direction) slicing the wavy cylinder marks an actual circle in its new surface, albeit one of variable diameter size.

To avoid invalid elements, the mapping intensity is made to decrease linearly with the radial distance, from the cylinder's surface to the outer circle in the intermediate mesh region, where the mapping intensity becomes zero. Hence, only the region within the cylinder's outer circle (recall Figure 3) is affected by the mapping. The mesh coordinates in GMSH's original file are then updated and Nektar++ is able to load it just as a normal GMSH file. A view of a mesh with 11% amplitude waviness is shown in Figure 5.

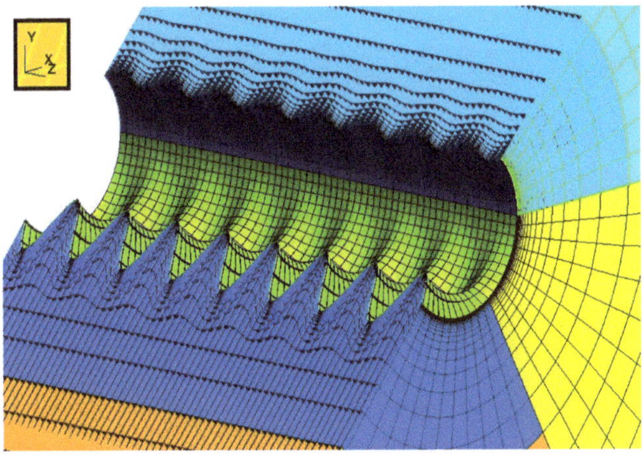

Figure 5. Cross-sectional view of a mesh showing the mapping pattern for $\lambda = 0.375D_m$, and waviness amplitude 11% (the line of contact between the green and dark blue regions is the cylinder's trailing edge).

2.3. Boundary and Initial Conditions

The following boundary conditions were imposed on the boundary surfaces shown in Figure 6:

- Viscous wall (no-slip and impermeability) conditions at the cylinder surface, along with Neumann condition for the pressure ($\partial p/\partial n = 0$, where n denotes the unit vector normal to the relevant surface);
- Periodic condition connecting the opposite surfaces in the spanwise direction (back and front planes in Figure 6);
- Dirichlet boundary condition for the velocity ($u = 1, v = 0, w = 0$) at the inlet surface, along with Neumann condition for the pressure ($\partial p/\partial n = 0$); these same conditions were applied for the outermost top and bottom domain surfaces shown in Figure 6;
- Nektar's high-order outflow boundary condition at the outlet surface, which enforces a Dirichlet condition for the pressure ($p = 0$) while allowing for velocity fluctuations across the outlet.

In order to facilitate the vortex shedding formation and, consequently, speed up the transient state, the following initial conditions given by simple arbitrary functions with respect to the domain coordinates were imposed:

$$u = 1.0 - \frac{0.1\,(x-1.0)}{(x-1.0)^2 + y^2 + 0.1}; \; v = \frac{0.1\,y}{(x-1.0)^2 + y^2 + 0.1}; \; w = 0; \; p = 0\,. \quad (11)$$

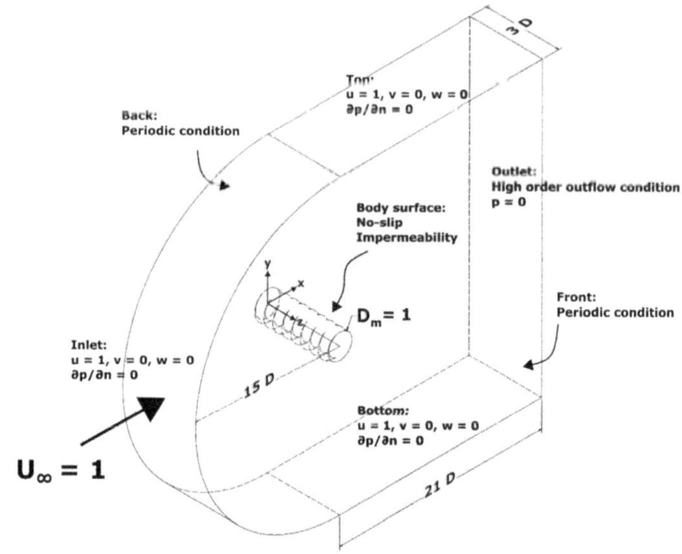

Figure 6. Computational domain (representative, not to scale) showing the surfaces on which boundary conditions are applied.

2.4. Simulation Details and Cases

Three cases have been simulated in the present work, including the straight model, as listed in Table 1. The two wavy models have the same wavelength, $\lambda/D = 0.375$, and undulation amplitudes of 3% and 11%. In all simulations, the initial 100 D/U_∞ time units were considered as numerical transient, in which the turbulent wake was left to accommodate across the domain. This time span corresponds to approximately 20 vortex shedding cycles and also allows for a number of large vortices to first leave the domain through the outlet boundary. The cases were then further simulated in order for the flow statistics to be collected, up to a total of 290 D/U_∞ time units for a good statistical convergence.

Table 1. Geometry parameters for the models simulated.

Case	Type	A/D_m	λ/D_m
Baseline	Straight	-	-
A03	Wavy	0.03	0.375
A11	Wavy	0.11	0.375

We note that, according to Lam and Lin [16], who performed LES of several wavy configurations, cases with smaller wavelengths end up requiring less time for statistics' convergence, provided $\lambda/D < 1$, due to the spanwise periodicity imparted on the wake structures (modulating the spanwise length of the largest wake structures). Therefore, the time required for the convergence of the wavy cases' statistics is expected to be less than that of the straight cylinder case.

To remind the reader, by keeping the same common position on the trailing-edge line, the cross section of the wavy cylinder is always a circle, and its diameter at any position along span (z coordinate) is given by:

$$D(z) = D_m + A\ \sin\left(\frac{2\pi z}{\lambda}\right), \tag{12}$$

Therefore, it is trivial to conclude that the mean diameter of the wavy cylinder is equal to D_m, which is unitary by construction. This mean diameter is used as characteristic length in the calculation of flow parameters such as the Reynolds number. Also, it is important to name two special cross sections that will serve as references for comparisons with the straight cylinder, and between the wavy ones: first, the smallest cross section diameter as "Valley", and second, the largest one as "Peak".

Table 2 presents the computational costs exclusively referring to the simulation of each case, without considering preliminary simulations to evaluate the domain and mesh resolution.

Table 2. Computational cost.

Case	Time Step	CPU Time (h)
Baseline	7.2×10^{-4}	898
A03	5.4×10^{-4}	1321
A11	1.8×10^{-4}	1980

After the simulation results are generated by Nektar++, mean flow fields and instantaneous force values are mainly processed using both post-processing tools already available in Nektar++ and some in-house scripts written for MATLAB®. The in-house scripts assist in processing and plotting the spatial flow field averages over consecutive spanwise wavelengths (given the spatial flow periodicity in this direction), the power spectral density, the graph generation of temporal forces distributions, and coefficients distributions around circumferential positions. Visit, an open-source CFD viewer, is used to visualize the iso-surfaces of the mean streamwise velocity and flow streamlines.

2.5. Mesh Dependence Study

A grid independence study was performed with four different mesh refinements for the smooth cylinder case, by evaluating the convergence of some reference variables (Cd and C_{RMS}). Without loss of generality, the resolution was varied only via h-type refinement, the p-type refinement is the same in all cases, $N_p = 2$. Therefore, the total DOF can be calculated by multiplying N_{Total} by 2^3. The results in Table 3 exhibited a suitable convergence of such variables, with little difference between the values of the medium grid, currently selected for this work, and the fine grid. After confronting it with experimental and computational results from the literature, as shown in the results section, the selection of the medium grid was considered adequate, and the same resolution was assigned to the wavy cases since the parameters of the wavy geometric modification are relatively small and produce a geometry smoothly close to the straight cylinder.

Table 3. Grid independence study: summary of tested cases with different mesh resolutions. The p-type refinement is the same in all cases, $N_p = 2$. Relative to h-type refinement, N_{x-y} is the number of quadrilateral elements in x–y plane, N_z is the number of divisions in spanwise direction, and N_{Total} is the total number of hexahedral elements.

Case	N_{Total}	N_{x-y}	N_z	Cd	Cl_{RMS}
Fine	156.080	3902	40	0.939	0.046
Medium	92.832	2901	32	0.946	0.052
Coarse 1	49.024	1532	32	1.054	0.195
Coarse 2	20.944	1309	16	1.054	0.203

3. Results and Discussion

3.1. Numerical Validation

First of all, the results for the straight model, which is the baseline for this study, were compared to those available in the literature in Table 4. We note that our implicit LES results are closer to the DNS of Beaudan and Moin [50] than to the classic LES of Kravchenko and Moin [51]. The good agreement of these main flow variables serves as a primary validation for the numerical methodology (including the mesh) employed here. Values of lift and drag coefficients over time for the straight case are shown in Figure 7a. The Strouhal number was extracted from the main peak value of the PSD curve, see Figure 8a, calculated from the temporal history of Cl (Figure 7a). The separation angle θ_s was estimated considering the criterion of $C_f = 0$ applied to the curve of skin friction coefficient of the baseline model given in Figure 9.

Table 4. Comparison and validation of results for the straight cylinder at Reynolds 3900.

Author	Nature	Cd	$-C_{pb}$	St	θ_s
Norberg [52]	Experimental	0.98	0.84	0.21	-
Ong and Wallace [53]	Experimental	0.99	-	0.21	86
Lourenco and Shih [54] cited in [50]	Experimental	0.99	0.88	0.22	89
Beaudan and Moin [50]	DNS	0.96	0.89	0.216	85.3
Tremblay [55]	DNS	1.03	0.93	0.215	86.5
Kravchenko and Moin [51]	Classic LES	1.04	0.93	0.21	88
Present work	Implicit LES	0.95	0.84	0.226	86

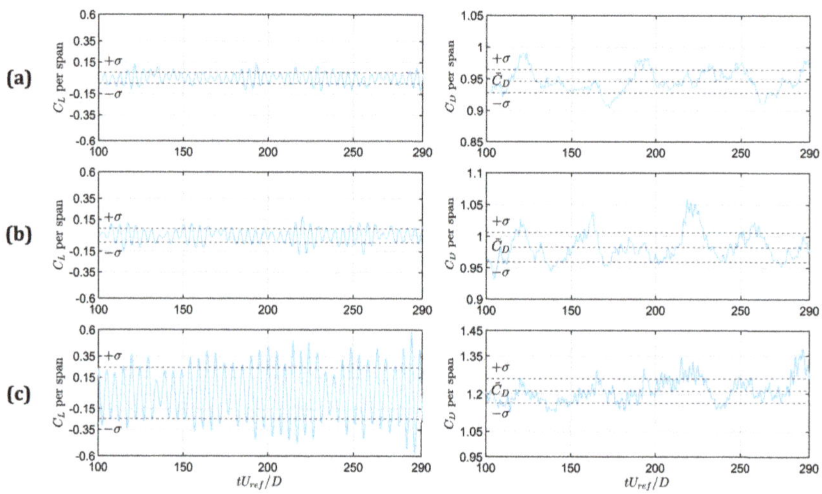

Figure 7. Histories of lift coefficient (**left**) and drag coefficient (**right**) for the baseline (**a**), A03 (**b**), and A11 (**c**) models.

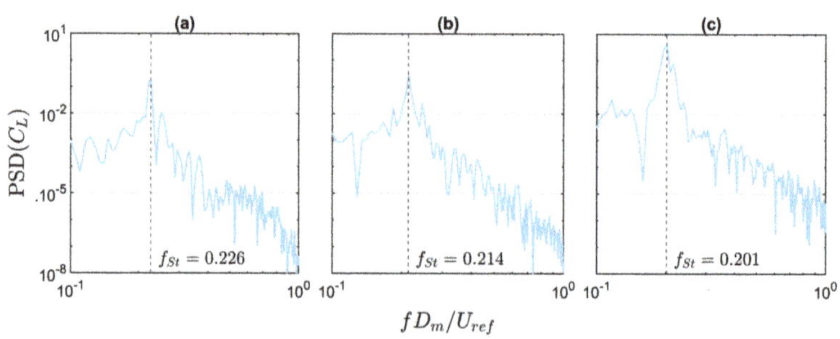

Figure 8. Power spectral density of lift fluctuations for the baseline (**a**) and wavy cases A03 (**b**) and A11 (**c**).

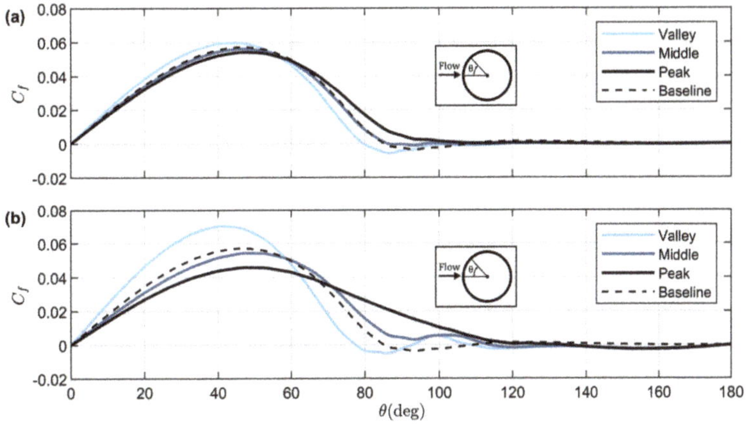

Figure 9. Sectional mean C_f distributions for A03 (**a**) and A11 (**b**). A dashed curve denotes the baseline C_f.

In addition, the Cl_{RMS} value, see Table 5, is in the range of typical values found for this Reynolds number, which is around 0.08 for $Re = 3900$ [56]. Nevertheless, this Reynolds number is still reasonably close to the range in which the so-called lift crisis takes place ($260 < Re < 1600$), where Cl_{RMS} has a large behavioral variation, in which Cl_{RMS} values drop dramatically from ≈0.5 to 0.045, that is, a more significant variation could be observed. Notwithstanding, as shown in Table 4, all mean values of Cd and Cl_{RMS} are in line with experimental and computational results, as well as the Strouhal number, the separation angle and the base pressure coefficient.

Table 5. Summary of relevant non-dimensional coefficients for the simulated cases.

Case	St	Cl_{RMS}	Cd	Cd Valley	Cd Peak	Cd Middle
Baseline	0.226	0.052	0.946	0.946	0.946	0.946
A03	0.214	0.066	0.983	0.941	1.024	0.977
A11	0.201	0.245	1.212	1.068	1.316	1.184

3.2. Changes in Drag, Lift, and Shedding Frequency

We now turn to discuss how the relevant non-dimensional coefficients change as leading-edge undulation is applied. A summary of these quantities is given in Table 5, featuring the non-dimensional shedding frequency or Strouhal number St, lift coefficient RMS values Cl_{RMS}, the mean overall drag coefficient Cd, and also three sectional (2D) drag coefficients. The latter represent the drag of sections containing a valley, a peak, and those in the middle (for which the diameter matches the original, baseline one). The mean values for the overall drag coefficient and the RMS values of the lift coefficient were obtained from the time histories of these two quantities, which can be seen in Figure 7.

The main effects of leading-edge waviness gathered from Table 5 and Figure 7 are an increase in Cd and Cl_{RMS}, especially for case A11, and a slight decrease in the primary shedding frequency value St. We note that the latter have been evaluated from PSD plots of the lift fluctuation history, which are shown in Figure 8. It is also noteworthy that St values decrease gradually as the undulation amplitude is made larger, whereas the increase in Cd and Cl_{RMS} is more sudden, rising significantly from A03 to A11. This is related to a marked reduction in the base pressure coefficient C_{pb}, as will be discussed below in connection to Figure 10.

As made clear in the introduction, the typical effect of undulation in traditional wavy cylinders is to decrease the mean Cd. In order to help unveil why the opposite is being observed in the case of leading-edge waviness, the surface distributions of friction coefficient C_f and pressure coefficient C_p is discussed below. These distributions are shown in Figures 9 and 10, respectively.

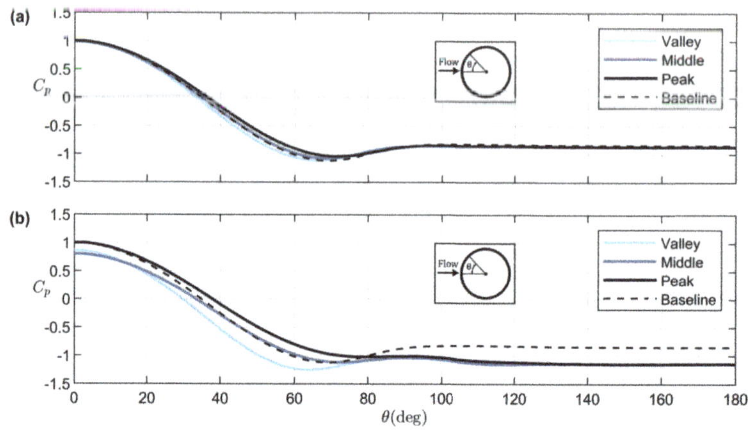

Figure 10. Sectional mean C_p distributions for A03 (**a**) and A11 (**b**). A dashed curve denotes the baseline C_p.

The C_f distributions in Figure 9 show that the curves representing middle sections are not much different from the reference C_f curve of the baseline cylinder (dashed). For the other sections (peak and valley), each C_f curve changes in such a symmetrical way that the overall friction drag for those sections is not significantly altered. This indicates that the changes in overall *Cd* due to undulation are primarily caused by changes in pressure distribution. Still, important information that can be extracted from C_f plots has to do with flow detachment. The mean separation location can be defined from the point in which the C_f curve crosses zero (reversal of velocity profile near the surface). Moreover, the inclination (steepness) of the curve at this point of crossing typically indicates how abrupt the separation is. Hence, by looking at Figure 9, one can conclude that separation is anticipated at valleys and delayed at peaks, these effects being stronger at larger undulation amplitudes. Moreover, it seems that separation at valleys can be rather abrupt (especially for case A11), whereas separation at peaks seems to be much smoother. This is in fact the case, as will be discussed in connection to the iso-surfaces of mean streamwise velocity at near zero value, to be shown later, where vigorous boundary layer detachments are observed when separation occurs at the valleys.

Now, turning to the C_p distributions of Figure 10, one can see why case A03 has no significant change in *Cd* when compared to the baseline cylinder, namely, there is no significant change in pressure distribution, even along the spanwise direction. However, when it comes to case A11, the situation changes dramatically, especially with regard to the base pressure (C_{pb}) which becomes significantly reduced. This is likely the primary cause for the significant rise in mean *Cd* for case A11. For some reason, the base pressure seems to follow the pressure at the first location of separation (valley section), whose detachment occurs slightly before $\theta = 80$ degrees, see Figure 9. Curiously, this does not happen with case A03, since a pressure recovery occurs from its first separation location until $\theta \approx 100$ degrees, after which the base pressure remains constant. It seems that for case A11, something takes between $80 < \theta < 100$ that prevents this pressure recovery. This will be further investigated in the next section through the analysis of the coherent structures that form in connection to the separation along valleys.

Lastly, it is worth mentioning that Figure 10 explains why the sectional drag at valleys is smaller than that at peaks, see Table 5. Essentially, undulation decreases the minimum pressure value on the C_p curve of valleys, imparting a suction on the frontal part of the section and therefore reducing its drag. The opposite effect is observed for the C_p

distribution of the peak section, which then contributes the most to the increased overall mean drag Cd. It is believed that the C_p reduction along valleys is caused by a Venturi-like effect due to the constriction of fluid between to peaks.

3.3. Flow Field Analysis and Coherent Structures

We start by looking at two-dimensional plots of mean spanwise vorticity ω_z for the cylinders' three relevant sections, as shown in Figure 11. Note that these plots show the correct size of each cylinder's section, with valleys having smaller diameters and peaks having larger ones — except for the baseline cylinder, whose plots in the top row of Figure 11 are simply repeated for convenience. An important effect of undulation, as discussed in the previous section, is to delay separation at peaks and to anticipate it at valleys. This effect is clearly seen in Figure 11, especially at the bottom row of plots (case A11). In the latter, one can see that the shear layers behind a peak section are closer to one another as they bracket the recirculation region (near wake). They also originate from the cylinder at a greater azimuthal coordinate (separation angle). In contrast, for the valley section, the shear layers detach from the cylinder much earlier, becoming farther away from one another along the recirculation zone. These will visually look like strong "ejection-like" events subsequently to be discussed later. These observations are in line with studies on traditional wavy cylinders, although flow detachment, even when anticipated at valleys, is typically not as strong as the ones found here.

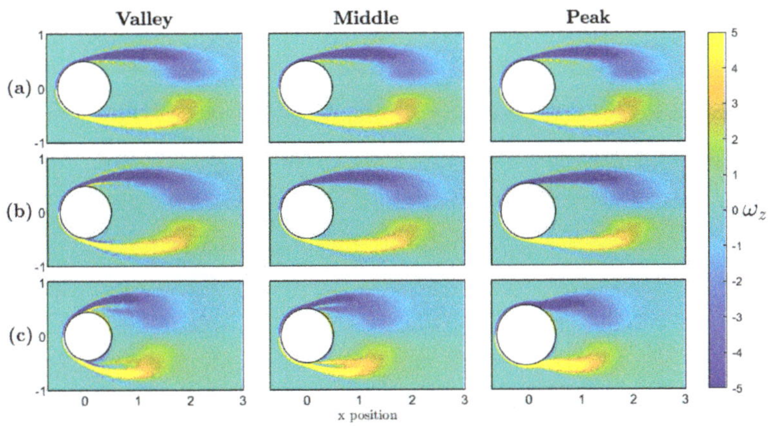

Figure 11. Two–dimensional contours of mean spanwise vorticity ω_z for the three relevant sections (see the header on top) of the simulated cylinders: baseline (**a**), A03 (**b**), and A11 (**c**). The baseline model plots are repeated on the top row.

Another effect observed in Figure 11 is that leading–edge undulation causes the recirculation region at the near wake to reduce its length in the streamwise direction. This effect, nevertheless, is in contrast to studies on traditional wavy cylinders, where the recirculation length typically becomes larger in comparison to that of the straight cylinder. In the literature, an increase in recirculation zone length is normally associated with a reduction in turbulent kinetic energy at the near wake. Here, we found that this trend is followed in the sense that the observed reduction in recirculation zone length leads to an increase in turbulent kinetic energy (TKE), as shown in Figure 12. This figure shows a carpet-like view in the $y = 0$ plane for the three cylinders, whose shapes (cross-sections) appear in white color.

For each cylinder, Figure 12 shows a relatively small TKE variation in the spanwise direction (note each plot repeats itself periodically along the span, following the undulation wavelengths), but it is clear that maximum TKE is reached behind valleys. However, when comparing the cylinders, it becomes clear that leading-edge undulation not only increases overall TKE levels in the near wake, but also brings the high-TKE zone closer to the cylinder which is consistent with the reduction in recirculation length. Turbulence along the wake can also be assessed through the energy spectrum of the velocity components, as given in Figure 13, which also shows the pressure spectrum (all these are span-averaged).

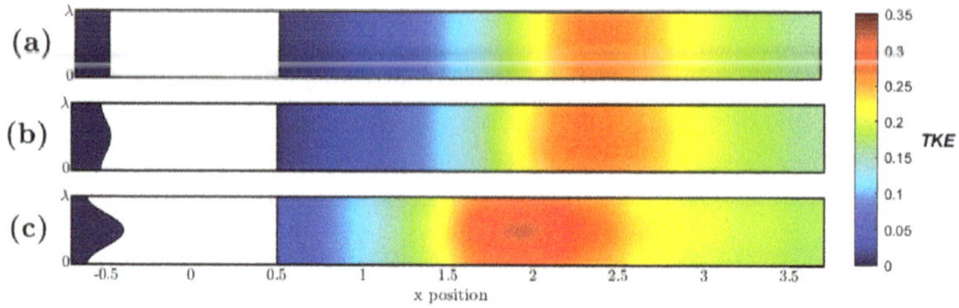

Figure 12. Carpet–like view of mean turbulent kinetic energy in the $y = 0$ plane: baseline (**a**), A03 (**b**), and A11 (**c**). As a spanwise periodic phenomenon, the spatial average referring to the repetition of wavelengths is calculated and exhibited here as a single wavelength.

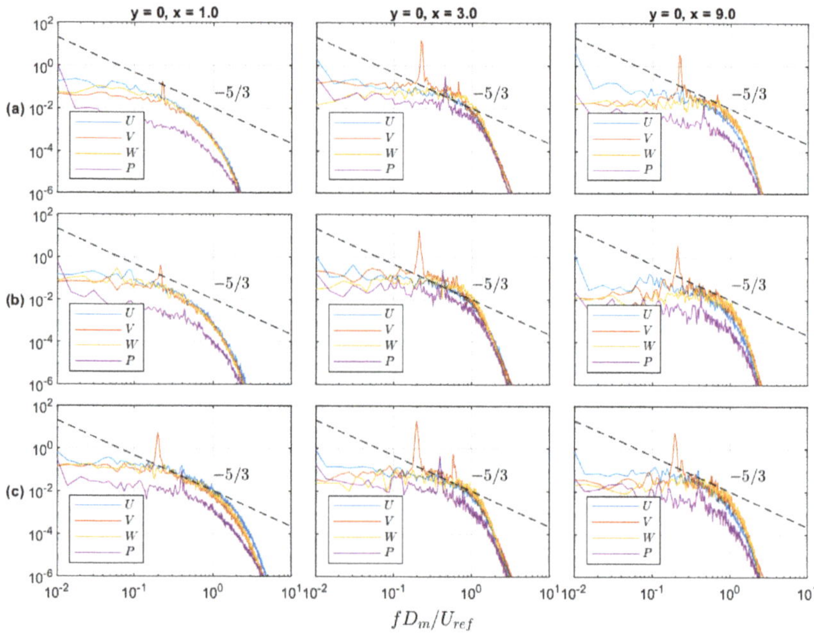

Figure 13. Energy spectrum of velocity components and of pressure at $y = 0$ and different x positions (see top header) for the three simulated cylinders: baseline (**a**), A03 (**b**), and A11 (**c**).

The positions along the wake where the spectra in Figure 13 have been measured can be correlated with the coordinates shown in Figures 11 and 12. Kolmogorov's $-5/3$ slope is included in the plots of Figure 13 for reference and its agreement with the measured spectra over the inertial range of the turbulence cascade further corroborates the suitability of the numerical methodology (including the mesh) employed. The rise in TKE intensity on the near wake due to undulation can be noted by comparing the spectra in the left column of Figure 13, as the curves move upwards from plots (a) to (c). Moreover, note that the peak in the y velocity component (V in the plots) increases especially for case A11 in the near wake, consistent with the rise in Cl_{RMS} discussed in the previous section, see Table 5. This also seems to be related to an anticipation of wake turbulence levels due to undulation, which can be noted by comparing the spectra of case A11 at $x = 1$ with that of the baseline case at $x = 3$. In this comparison, note that not only the peak in V is anticipated, but also the peak in pressure. This indicates that the turbulent features of the wake are anticipated, which is in line with the recirculation length being reduced. As the more intense turbulent activity is then brought closer to the cylinder, it is not that surprising that stronger oscillations will affect the cylinder, which helps explain at least partially the rise in Cl_{RMS}. In summary, the strong oscillations in y-velocity that would otherwise take place further downstream are now happening at the near wake, affecting the vortex shedding itself in a way that increases oscillations in Cl. Lastly, it is worth noting that the spectra measured at the far wake ($x = 9$) do not change significantly among cylinders, as the turbulent wakes seem to relax toward a canonical turbulent wake state.

Finally, we turn to the analysis of streamwise vorticity, which will reveal the formation of strong (streamwise) vortex pairs associated with how separation is anticipated along valleys and delayed along peaks. In order to track the evolution of said vorticity, we first consider the frontal part of the cylinders by using the azimuthal coordinate θ as a parameter, while looking at the vorticity component ω_t that is tangent to the (circular) cross-sections of the cylinder, as show in Figure 14 (left schematic). Hence, in Figure 15, ω_t will be shown in the r–θ plane that contains the original (baseline) axis of the cylinders, for increasing values of θ. After $\theta = 90$ degrees, i.e., for the rear part of the cylinders, the streamwise vorticity component ω_x will be considered instead, as this will more naturally allow for the tracking of the (streamwise) vortex pairs that, at this point, are no longer attached to the cylinders' surface. Thus, in Figure 16, streamwise vorticity is shown over the y–z plane, with the x coordinate being used as the parameter.

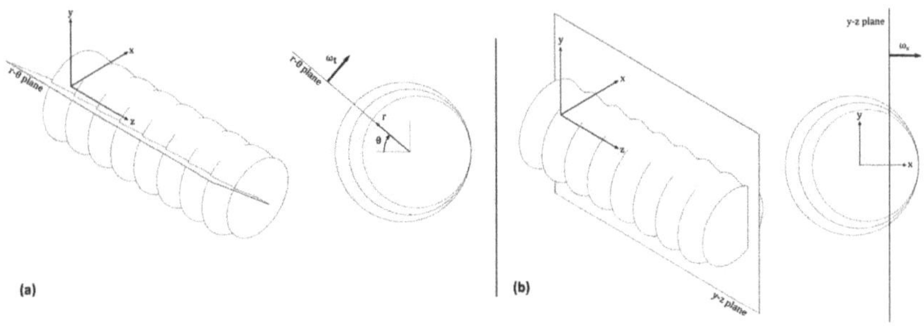

Figure 14. Planes and coordinates chosen to best follow the evolution of vorticity along the surface and after separation: tangent component ω_t for the cylinders' frontal part (**a**) and streamwise component ω_x for the cylinders' rear part (**b**).

In Figure 15, ω_t is shown for cases A03 (top) and A11 (bottom) for increasing values of θ. Note that ω_t would be zero for the baseline cylinder (at least for the mean flow) since it represents a distributed vorticity field associated with the boundary layer flow that exists along the spanwise direction (cross-flow) due to surface undulation. For $\theta < 20$ degrees, the plots basically indicate that, with respect to the spanwise direction, the boundary layer evolves from the peaks to the valleys, as expected near the leading-edge region. As θ increases, the direction of flow becomes reversed, as can be seen by how the colors have changed by $\theta = 80$ degrees. This reversal is likely associated with flow separation, which first occurs at valleys around $\theta = 80$ degrees, recall Figure 9. More specifically, the near-surface flow between peaks seems to be accommodating beforehand to bifurcate upon encountering the "ejections-like" by which separation at valleys occurs. These flow structures can be seen in Figure 17 and are surprisingly strong for case A11. Lastly, it is also worth noting in Figure 15 that the vorticity layer near the surface is thicker for case A11, which is likely caused by a stronger cross-flow due to larger undulation amplitude.

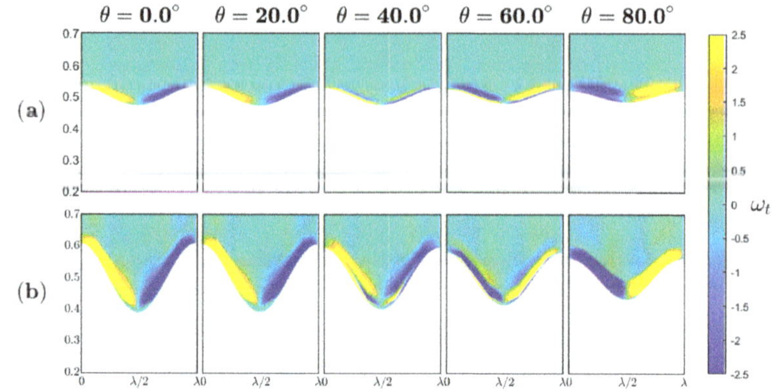

Figure 15. Contours of mean tangent vorticity ω_t over the r–θ plane for the two wavy models, A03 (a) and A11 (b), for increasing values of θ. Refer to Figure 14 (left schematic) for the planes and coordinates adopted. As a spanwise periodic phenomenon, the spatial average referring to the repetition of wavelengths is calculated and exhibited here as a single wavelength.

In Figure 16, it becomes clear that each side of the vorticity layer will, upon flow separation at valleys, detach from the surface and, once free, will roll-up upon itself and become a coherent streamwise vortex. Therefore, the two sides of the vorticity layer between peaks will become a pair of counter-rotating vortices. Globally, the array of undulation wavelengths along the span generates a corresponding array of streamwise vortex pairs. Clearly, from Figure 16, case A11 features much stronger vortices than case A03, as expected from the overall amount of streamwise vorticity held in the respective vorticity layers prior to separation (recall comment at the end of the previous paragraph). The right-most pair of plots in Figure 16 correspond to the plane $x = 0.5$, which is the one containing the trailing edge of the cylinders. At this point, the vortices of case A03 are already vanishing, whereas those of case A11 are still strong and will surely extend further into the near wake region. This will be made more evident in Figure 18, in which streamlines are employed to highlight the coherent pair of streamwise vortices. In fact, as per Figure 18, the vortices of case A03 are not strong enough to produce a cyclical streamline swirl, indicating that only case A11 effectively produces coherent vortices.

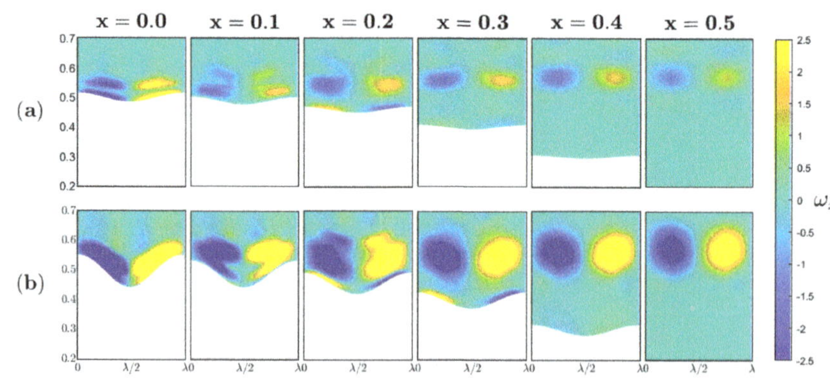

Figure 16. Contours of mean streamwise vorticity ω_x over the y–z plane for the wavy models, A03 (a) and A11 (b), for increasing values of x. Refer to Figure 14 (right schematic) for the planes and coordinates adopted. As a spanwise periodic phenomenon, the spatial average referring to the repetition of wavelengths is calculated and exhibited here as a single wavelength.

In Figure 17, iso-surfaces of mean streamwise velocity at a near zero value ($U \leq 10^{-3}$) are employed to highlight how the flow separation takes place for each cylinder. One can clearly see how separation first occurs at valleys by means of separated vortical shear layers of fluid moving away from the surface. These can be connected to the streamwise vortices by correlating Figure 17 to Figure 16. The vortex pair between two peaks rotates in such a way as to "lift" fluid from the surface at valleys and to "land" fluid on the surface at peaks. The former of these effects explains why such structures are stronger when vortices are stronger, whereas the latter explains why separation at peaks is further delayed when vortices are stronger, recall Figure 9.

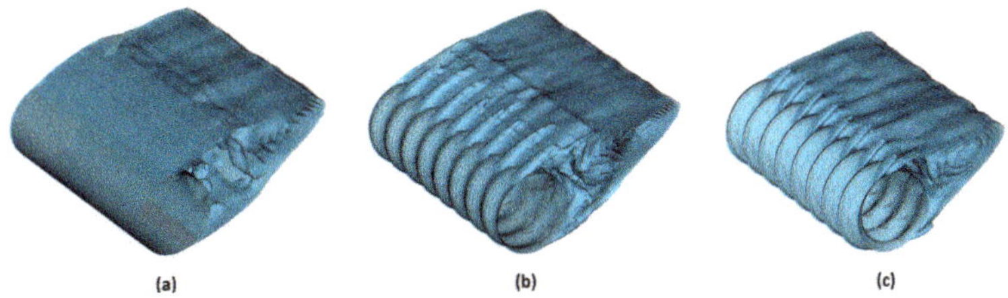

Figure 17. Iso−surfaces of mean streamwise velocity at a near zero value ($U \leq 10^{-3}$), showing how the flow separation at valleys occurs by means of "ejections−like" events. The three simulated cylinders are shown: baseline (**a**), A03 (**b**) and A11 (**c**).

In Figure 18, streamlines based on the mean flow are used to highlight the formation of the coherent vortices. They clearly show that only case A11 produces vortices that are effectively cyclical. The left-most figure of case A11 in Figure 18 shows how the vortex pair adjacent to a peak pushes outer fluid toward the surface. Since this happens prior to separation at peaks, the overall velocity increases in this region, and, therefore, pressure is reduced (Bernoulli equation). This is consistent with the C_p curve along peaks already shown (see Figure 10 for $80 < \theta < 120$ degrees). Hence, when separation occurs at peaks, the pressure at the surface is significantly reduced. This, combined with the fact that the earlier separation at valleys also happened at a location of reduced pressure ($\theta_s \approx 80$ degrees), ends up allowing for reduced pressure on the separation zone and near wake. In summary, the mechanism proposed to explain the increased drag of case A11 due to low base pressure is that strong coherent vortices increase the overall velocity just before separation at peaks, leading to reduced pressure on the separation zone.

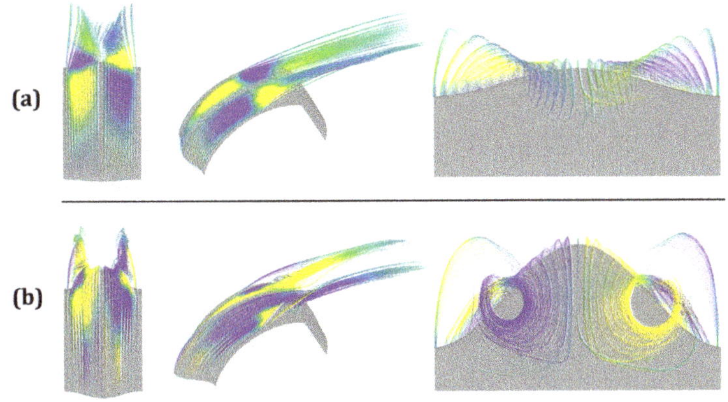

Figure 18. Mean flow-based streamlines showing the formation of coherent streamwise vortices for the wavy cylinders: A03 (**a**) and A11 (**b**). Only case A11 produces effectively complete cyclical vortices. The colors indicate local streamwise vorticity.

4. Conclusions

We presented a numerical study on the flow past cylinders with leading-edge waviness This type of waviness resembles the tubercles found on the flippers of humpback whales and has been often studied in the context of wavy wings. To the authors' knowledge, this was the first study to consider this type of waviness for cylinders, since traditional studies on wavy cylinders typically apply undulation both to leading and trailing edges.

Some of the results obtained here were similar to those of traditional wavy cylinders, such as anticipated separation at valleys (sections with smaller diameters) and delayed at peaks (sections with larger diameters). Still, most flow changes affected by leading-edge undulation were found to be in contrast to those observed in traditional wavy cylinders An explanation for these unexpected effects was then proposed based on a detailed flow field analysis. Specifically, the formation of counter−rotating streamwise coherent vortices was correlated to the delay of separation along peaks. Those vortices seem to push outer fluid toward the surface prior to separation at peaks, increasing overall velocity and reducing the pressure right before separation, especially for case A11. This defined a significantly lower base pressure (up to about 36%) and caused the observed rise in drag (up to about 28%). This decrease in base pressure likely draws the near wake features closer to the cylinder, for example, by reducing the recirculation length. Since the turbulent structures of the near wake become more "compressed" closer to the cylinder—compared to the straight cylinder, the turbulent kinetic energy rises on the near wake and the larger oscillations in y-velocity of the vortex formation zone impart larger oscillations in the lift As all these combined effects are far from simple, subsequent studies could further clarify the underlying dynamics.

It seems that the most relevant aspect in the reduction of base pressure is the strength of the coherent streamwise vortices. Their strength is nevertheless related to undulation (wavelength and amplitude), as well as to the Reynolds number. It is worth mentioning that an experimental study currently being conducted on similar geometries, see [13], but at much larger Reynolds numbers, showed a reduction in drag instead, which confirms its Reynolds number dependence.

Although the results did not replicate the feats of the wavy airfoil or even the typical wavy cylinders, this study was able to offer a good phenomenological description that may be useful in understanding and complementing these other studies. Furthermore, a practical application for it beyond flow control mechanisms, by taking advantage of the observed mixing layer effect promoted by the streamwise vortices, would be in circular tubes of heat exchangers. Additionally, more studies are being conducted to explore the possibilities of this geometric modification to different parameters combinations.

Author Contributions: Conceptualization, P.H.F., T.B.d.A. and R.C.M.; methodology, P.H.F. and R.C.M.; software, P.H.F., E.O.C. and L.D.F.; validation, P.H.F. and R.C.M.; formal analysis, P.H.F., T.B.d.A., E.O.C., L.D.F. and R.C.M.; investigation, P.H.F., T.B.d.A. and R.C.M.; data curation, P.H.F.; writing—original draft preparation, P.H.F. and R.C.M.; writing—review and editing, P.H.F.; visualization, P.H.F., E.O.C. and L.D.F.; supervision, R.C.M. and T.B.d.A.; funding acquisition, R.C.M. All authors have read and agreed to the published version of the manuscript.

Funding: This research was funded by FAPESP grant number 2020/10910-8. The APC was funded by CNPq (process number 141783/2019-2).

Data Availability Statement: Not applicable.

Acknowledgments: The authors would like to thank André F. C. da Silva and Mohsen Lahooti for their support regarding the efficient use of Nektar++ in the context of parallel computations. In addition, the first author acknowledges support from CNPq for his Ph.D. scholarship (process number 141783/2019-2).

Conflicts of Interest: The authors declare no conflict of interest.

References

1. Roshko, A. Perspectives on bluff body aerodynamics. *J. Wind Eng. Ind. Aerodyn.* **1993**, *49*, 79–100. [CrossRef]
2. Williamson, C.H.K. Three-dimensional wake transition. *J. Fluid Mech.* **1996**, *328*, 345–407. [CrossRef]
3. Zdravkovich, M.M. *Flow Around Circular Cylinders—Fundamentals*; Oxford University Press: Oxford, UK, 1997; Volume 1.
4. Bearman, P.W. Vortex Shedding from Oscillating Bluff Bodies. *Annu. Rev. Fluid Mech.* **1984**, *16*, 195–222. [CrossRef]
5. Wang, R.; Xin, D.; Ou, J. Experimental investigation on suppressing circular cylinder VIV by a flow control method based on passive vortex generators. *J. Wind Eng. Ind. Aerodyn.* **2019**, *187*, 36–47. [CrossRef]
6. Fish, F.E.; Battle, J.M. Hydrodynamic design of the humpback whale flipper. *J. Morphol.* **1995**, *225*, 51–60. [CrossRef]
7. Miklosovic, D.S.; Murray, M.M.; Howle, L.E.; Fish, F.E. Leading-edge tubercles delay stall on humpback whale (Megaptera novaeangliae) flippers. *Phys. Fluids* **2004**, *16*, L39–L42. [CrossRef]
8. Johari, H.; Henoch, C.; Custodio, D.; Levshin, A. Effects of leading-edge protuberances on airfoil performance. *AIAA J.* **2007**, *45*, 2634–2642. [CrossRef]
9. Hansen, K.; Kelso, R.; Dally, B. The effect of leading edge tubercle geometry on the performance of different airfoils. In Proceedings of the 7th World Conference on Experimental Heat Transfer, Fluid Mechanics and Thermodynamics, Krakow, Poland, 28 June–3 July 2009.
10. de Paula, A.A.; Meneghini, J.; Kleine, V.G.; Girardi, R.D. The wavy leading edge performance for a very thick airfoil. In Proceedings of the 55th AIAA Aerospace Sciences Meeting, Grapevine, TX, USA, 9–13 January 2017.
11. Rocha, F.A.; de Paula, A.A.; de Sousa, M.; Cavalieri, A.V.; Kleine, V.G. Lift enhancement by wavy leading edges at Reynolds numbers between 700,000 and 3,000,000. In Proceedings of the 36th AIAA Applied Aerodynamics Conference, Atlanta, GA, USA, 25–29 June 2018.
12. Seyhan, M.; Sarioglu, M.; Akansu, Y.E. Influence of Leading-Edge Tubercle with Amplitude Modulation on NACA 0015 Airfoil. *AIAA J.* **2021**, *59*, 3965–3978. [CrossRef]
13. Ferreira, P.H.; Moura, R.C.; de Araújo, T.B. Experimental Investigation of a Wavy Leading Edge Cylinder. In Proceedings of the AIAA Aviation 2022 Forum, Chicago, IL, USA, 27 June–1 July 2022.
14. Ahmed, A.; Bays-Muchmore, B. Transverse flow over a wavy cylinder. *Phys. Fluids A Fluid Dyn.* **1992**, *4*, 1959–1967. [CrossRef]
15. Zhang, W.; Daichin; Lee, S.J. PIV measurements of the near-wake behind a sinusoidal cylinder. *Exp. Fluids* **2005**, *38*, 824–832. [CrossRef]
16. Lam, K.; Lin, Y.F. Large eddy simulation of flow around wavy cylinders at a subcritical Reynolds number. *Int. J. Heat Fluid Flow* **2008**, *29*, 1071–1088. [CrossRef]
17. Lin, Z.; Yu-feng, L. Force reduction of flow around a sinusoidal wavy cylinder. *J. Hydrodyn. Ser. B* **2009**, *21*, 57–96.
18. Lam, K.; Lin, Y.F.; Zou, L.; Liu, Y. Investigation of turbulent flow past a yawed wavy cylinder. *J. Fluids Struct.* **2010**, *26*, 1078–1097. [CrossRef]
19. Zhao, R.; Xu, J.; Yan, C.; Yu, J. Scale-adaptive simulation of flow past wavy cylinders at a subcritical Reynolds number. *Acta Mech. Sin.* **2011**, *27*, 660–667. [CrossRef]
20. Bai, H.L.; Zang, B.; New, T.H. The near wake of a sinusoidal wavy cylinder with a large spanwise wavelength using time-resolved particle image velocimetry. *Exp. Fluids* **2019**, *60*, 15. [CrossRef]
21. Karthik, K.; Jeyakumar, S.; Sebastin, J.S. Optimization of wavy cylinder for aerodynamic drag and aeroacoustic sound reduction using computational fluid dynamics analysis. *Proc. Inst. Mech. Eng. Part C J. Mech. Eng. Sci.* **2021**, *235*, 1–13. [CrossRef]
22. Zhang, Z.; Tu, J.; Zhang, K.; Yang, H.; Han, Z.; Zhou, D.; Xu, J.; Zhang, M. Vortex characteristics and flow-induced forces of the wavy cylinder at a subcritical Reynolds number. *Ocean Eng.* **2021**, *222*, 108593. [CrossRef]
23. Moura, R.C. On the Use of Spectral Element Methods for Under-Resolved Simulations of Transitional and Turbulent Flows. Ph.D. Thesis, Imperial College London, London, UK, 2017.
24. Moura, R.C.; Mengaldo, G.; Peiró, J.; Sherwin, S.J. On the eddy-resolving capability of high-order discontinuous Galerkin approaches to implicit LES/under-resolved DNS of Euler turbulence. *J. Comput. Phys.* **2017**, *330*, 615–623. [CrossRef]
25. Moura, R.C.; Sherwin, S.J.; Peiró, J. Linear dispersion-diffusion analysis and its application to under-resolved turbulence simulations using discontinuous Galerkin spectral/*hp* methods. *J. Comput. Phys.* **2015**, *298*, 695–710. [CrossRef]
26. Moura, R.C.; Sherwin, S.J.; Peiró, J. Modified equation analysis for the discontinuous Galerkin formulation. In *Spectral and High Order Methods for Partial Differential Equations—ICOSAHOM 2014*; Springer: Cham, Switzerland, 2015; pp. 375–383.
27. Moura, R.C.; Sherwin, S.J.; Peiró, J. Eigensolution analysis of spectral/*hp* continuous Galerkin approximations to advection-diffusion problems: Insights into spectral vanishing viscosity. *J. Comput. Phys.* **2016**, *307*, 401–422. [CrossRef]
28. Moura, R.C.; Mengaldo, G.; Peiró, J.; Sherwin, S.J. An LES setting for DG-based implicit LES with insights on dissipation and robustness. In Proceedings of the 11th International Conference on Spectral and High Order Methods (under review), Rio de Janeiro, Brazil, 27 June–1 July 2016.
29. Mengaldo, G.; Moura, R.C.; Giralda, B.; Peiró, J.; Sherwin, S.J. Spatial eigensolution analysis of discontinuous Galerkin schemes with practical insights for under-resolved computations and implicit LES. *Comput. Fluids* **2017**, *169*, 349–364. [CrossRef]
30. Moura, R.C.; Peiró, J.; Sherwin, S.J. On the accuracy and robustness of implicit LES/under-resolved DNS approaches based on spectral element methods. In Proceedings of the 10th International Symposium on Turbulence and Shear Flow Phenomena, Chicago, IL, USA, 7–9 July 2017.

31. Mengaldo, G.; De Grazia, D.; Moura, R.C.; Sherwin, S.J. Spatial eigensolution analysis of energy-stable flux reconstruction schemes and influence of the numerical flux on accuracy and robustness. *J. Comput. Phys.* **2018**, *358*, 1–20. [CrossRef]
32. Fernandez, P.; Moura, R.C.; Mengaldo, G.; Peraire, J. Non-modal analysis of spectral element methods: Towards accurate and robust large eddy simulations. *Comput. Method. Appl. M.* **2019**, *346*, 43–62. [CrossRef]
33. Moura, R.C.; Peiró, J.; Sherwin, S.J. Implicit LES approaches via discontinuous Galerkin methods at very large Reynolds. In *Direct and Large-Eddy Simulation XI*; Springer: Cham, Switzerland, 2019; pp. 53–59.
34. Moura, R.C.; Peiró, J.; Sherwin, S.J. Under-resolved DNS of non-trivial turbulent boundary layers via spectral/*hp* CG schemes. In *Direct and Large Eddy Simulation XII*; Springer: Cham, Switzerland, 2020; pp. 389–395.
35. Moura, R.C.; Fernandez, P.; Mengaldo, G.; Sherwin, S.J. Viscous diffusion effects in the eigenanalysis of (hybridisable) DG methods. In *Spectral and High Order Methods for Partial Differential Equations ICOSAHOM 2018*; Springer: Cham, Switzerland, 2020; pp. 371–382.
36. Moura, R.C.; Fernandes, L.D.; Silva, A.F.C.; Mengaldo, G.; Sherwin, S.J. Spectral/hp element methods' linear mechanism of (apparent) energy transfer in Fourier space: Insights into dispersion analysis for implicit LES. *J. Comput. Phys.* **2022**, *471*, 111613. [CrossRef]
37. Martins, Y.W.G.; Malatesta, V.; Moura, R.C. Numerical study of the flow past an inclined flat plate via spectral/hp element methods with novel stabilization. In Proceedings of the 26th International Congress of Mechanical Engineering, Virtual Event, Brazil, 22–26 November 2021.
38. Moura, R.C.; Cassinelli, A.; da Silva, A.F.C.; Burman, E.; Sherwin, S.J. Gradient jump penalty stabilisation of spectral/hp element discretisation for under-resolved turbulence simulations. *Comput. Methods Appl. Mech. Eng.* **2022**, *388*, 114200. [CrossRef]
39. Moura, R.C.; Aman, M.; Peiró, J.; Sherwin, S.J. Spatial eigenanalysis of spectral/hp continuous Galerkin schemes and their stabilisation via DG-mimicking spectral vanishing viscosity: Application to high Reynolds number flows. *J. Comput. Phys.* **2019**, *406*, 109112. [CrossRef]
40. Cantwell, C.D.; Moxey, D.; Comerford, A.; Bolis, A.; Rocco, G.; Mengaldo, G.; de Grazia, D.; Yakovlev, S.; Lombard, J.; Ekelschot, D.; et al. Nektar++: An open-source spectral/hp element framework. *Comput. Phys. Commun.* **2015**, *192*, 205–219. [CrossRef]
41. Moxey, D.; Cantwell, C.D.; Bao, Y.; Cassinelli, A.; Castiglioni, G.; Chun, S.; Juda, E.; Kazemi, E.; Lackhove, K.; Marcon, J.; et al. Nektar++: Enhancing the capability and application of high-fidelity spectral/hp element methods. *Comput. Phys. Commun.* **2020**, *249*, 1–18. [CrossRef]
42. Lombard, J.W.; Moxey, D.; Sherwin, S.J.; Hoessler, J.F.A.; Dhandapani, S.; Taylor, M.J. Implicit large-eddy simulation of a wingtip vortex. *AIAA J.* **2016**, *54*, 506–518. [CrossRef]
43. Serson, D.; Meneghini, J.R.; Sherwin, S.J. Direct numerical simulations of the flow around wings with spanwise waviness. *J. Fluid Mech.* **2017**, *826*, 714–731. [CrossRef]
44. Wang, R.; Wu, F.; Xu, H.; Sherwin, S.J. Implicit large-eddy simulations of turbulent flow in a channel via spectral/hp element methods. *Phys. Fluids* **2021**, *33*, 035130. [CrossRef]
45. Mengaldo, G.; Moxey, D.; Turner, M.; Moura, R.C.; Jassim, A.; Taylor, M.; Peiro, J.; Sherwin, S.J. Industry-relevant implicit large-eddy simulation of a high-performance road car via spectral/hp element methods. *SIAM Rev.* **2021**, *63*, 723–755. [CrossRef]
46. Karniadakis, G.; Sherwin, S.J. *Spectral/hp Element Methods for Computational Fluid Dynamics*; Oxford University Press: Oxford, UK, 2005.
47. Raza, S.A.; Irawan, Y.H.; Chern, M. Effect of boundary conditions and domain size on the turbulent flow characteristics over a circular cylinder. *J. Chin. Inst. Eng.* **2021**, *44*, 659–672. [CrossRef]
48. Ali, H.; Khan, N.B.; Jameel, M.; Khan, A.; Sajid, M.; Munir, A.; Ahmed, A.E.; Alharbi, K.A.M.; Galal, A.M. Numerical investigation of the effect of spanwise length and mesh density on flow around cylinder at Re = 3900 using LES model. *PLoS ONE* **2022**, *17*, e0266065. [CrossRef] [PubMed]
49. Geuzaine, C.; Remacle, J.F. Gmsh: A three-dimensional finite element mesh generator with built-in pre- and post-processing facilities. *Int. J. Numer. Methods Eng.* **2009**, *79*, 1309–1331. [CrossRef]
50. Beaudan, P.; Moin, P. *Numerical Experiments on the Flow Past a Circular Cylinder at Sub-Critical Reynolds Number*; Technical Report No. TF-62; Thermosciences Division, Department of Mechanical Engineering, Stanford University: Stanford, CA, USA, 1994.
51. Kravchenko, A.G.; Moin, P. Numerical studies of flow over a circular cylinder at Re = 3900. *Phys. Fluids* **2000**, *12*, 403–417. [CrossRef]
52. Norberg, C. *Effects of Reynolds Number and a Low-Intensity Freestream Turbulence on the Flow around a Circular Cylinder*; Technical Report Publication 87/2; Department of Applied Thermodynamics and Fluid Mechanics, Chalmers University of Technology: Gothenburg, Sweden, 1987.
53. Ong, L.; Wallace, J.M. The velocity field of the turbulent very near wake of a circular cylinder. *Exp. Fluids* **1996**, *20*, 441–453. [CrossRef]
54. Lourenco, L.M.; Shih, C. Characteristics of the Plane Turbulent Near Wake of a Circular Cylinder. A Particle Image Velocimetry Study. *Engineering* **1993**.
55. Tremblay, F. Direct and Large-Eddy Simulation of Flow around a Circular Cylinder at Subcritical Reynolds Numbers. Ph.D. Thesis, Technische Universität München (TUM), Munich, Germany, 2002.
56. Norberg, C. Fluctuating lift on a circular cylinder: Review and new measurements. *J. Fluids Struct.* **2003**, *17*, 57–96. [CrossRef]

energies

MDPI

Article

Numerical Simulation of Flow and Heat Transfer of a Discontinuous Single Started Helically Ribbed Pipe

Simon Kügele [1,*], Gino Omar Mathlouthi [2], Peter Renze [1] and Thomas Grützner [3]

1 Institute of Energy and Drive Technology, Ulm University of Applied Sciences, Albert-Einstein-Allee 53, 89081 Ulm, Germany
2 Institute of Design and CA-Technology, Ulm University of Applied Sciences, Prittwitzstraße 10, 89075 Ulm, Germany
3 Laboratory of Thermal Process Engineering, Institute of Chemical Engineering, Ulm University, 89081 Ulm, Germany
* Correspondence: simon.kuegele@thu.de

Abstract: In the present study, the turbulent flow field and the heat transfer in a single started helically ribbed pipe with a discontinuous rib are investigated. A large-eddy simulation (LES) technique is applied in a pipe section with cyclic boundary conditions. The aim of this study is to explain and further analyze the findings from the heat transfer measurements at such complex structures with the help of detailed flow simulations. The simulation results are validated with measurements at a Reynolds number of Re = 21,100 and a Prandtl number of Pr = 7 with water as fluid. The comparison clearly shows that the current method delivers accurate results concerning average flow field, turbulence quantities and local heat transfer. The results demonstrate that the applied method is capable of correctly simulating flows with heat transfer in complex three-dimensional structures. The overall heat transfer performance of the helically ribbed pipe with a discontinuous rib is compared to a smooth pipe and a continuous rib configuration. The impact of the interruption of the rib structure on pressure drop and heat transfer are analyzed in detail.

Keywords: large eddy simulation; pipe flow; heat transfer; ribbed tube

Citation: Kügele, S.; Mathlouthi, G.O.; Renze, P.; Grützner, T. Numerical Simulation of Flow and Heat Transfer of a Discontinuous Single Started Helically Ribbed Pipe. *Energies* **2022**, *15*, 7096. https://doi.org/10.3390/en15197096

Academic Editors: Artur Bartosik and Dariusz Asendrych

Received: 30 August 2022
Accepted: 22 September 2022
Published: 27 September 2022

Publisher's Note: MDPI stays neutral with regard to jurisdictional claims in published maps and institutional affiliations.

Copyright: © 2022 by the authors. Licensee MDPI, Basel, Switzerland. This article is an open access article distributed under the terms and conditions of the Creative Commons Attribution (CC BY) license (https://creativecommons.org/licenses/by/4.0/).

1. Introduction

In order to improve the heat transfer in technical devices, pipes with a rough surface are used. In addition, there are many other possibilities to passively improve the heat transfer, for example various internals or the use of nanofluids, an overview of these methods can be found in the paper of Ajarostaghi et al. [1]. The roughness ensures that the boundary layer of the flow is disturbed, transport processes normal to the wall are enhanced, and, additionally, the heat transfer surface is increased. Due to larger frictional and pressure forces, the pressure loss is increased [2]. Most published studies of internally ribbed pipes are based on experimental methods. Due to the increasing computing power, it is now possible to perform detailed flow simulations. This enables an improved virtual product development to reduce the costs of expensive experiments and speed up the development.

In the past, numerous experimental studies have been performed on internally ribbed pipes. With the aim of improving heat transfer and pressure loss, helically ribbed pipes were investigated by Webb et al. [2], Gee and Webb [3], Withers [4,5], Han et al. [6], and Nakayama et al. [7]. These authors have established correlations for the heat transfer and the friction coefficient based on their experiments, which are valid for various physical and geometrical parameters. By applying measured data to a linear model, Ravigururajan and Bergles [8] developed general correlations for pressure loss and heat transfer in single-phase turbulent flow using the research data of Webb [9], Gee and Webb [3], Withers [4,5], and Kumar and Judd [10]. In addition, to develop correlations for heat transfer and friction in pipes with internal helical ribs, Zdaniuk et al. [11] used an artificial neural network approach based on experimental data.

For the investigation of internally structured pipes, the importance of simulations increases due to the ever increasing computing resources and the resulting reduced computing times. In the past, several studies have been performed in this area, and a selection of them are briefly described in the following.

A comparison between the experimental data of Ravigururajan and Bergles [8] with the results of Reynolds-averaged Navier–Stokes (RANS) simulations for a single started helically ribbed pipe can be found in Hossainpour and Hassanzadeh [12]. They show that it is possible to provide reasonably good results in the range of the Reynolds number from 25,000 to 80,000. In this study, no comparison of the local values such as the local heat transfer has been made. A comparison between RANS and a large-eddy simulation (LES) was performed by Vijapurapu and Cui [13]. Cauwenberge et al. [14] demonstrate that RANS simulations are unable to predict certain secondary flow phenomena that have an effect on heat transport and pressure loss. In the work of Wang [15], it has been shown by means of an LES simulation that the performance of helically corrugated pipes is superior to the transverse corrugated pipe. A multiple-started helically ribbed pipe is investigated in Akermann et al. [16] using an LES method, where the simulation setup is validated with experimental data for the Nusselt number and pressure drop for Reynolds numbers of 8000 and 16,000 and Prandtl numbers of 5, 7 and 9. Based on the previously mentioned experimental data from Mayo et al. [17], two studies have been carried out to validate a simulation with an LES model with the simpler continuous geometry. The first study was done by Cauwenberg et al. [18] and the second by Campet et al. [19]. Both studies show a good agreement between the simulation and measurement for the mean values of velocity and temperature. The authors were able to perform a successful validation for a continuous ribbed pipe geometry using an LES. The literature referenced above only deals with continuous structures at internal pipe walls. Measurements of three-dimensional structures were performed at the von Karman Institute, where Mayo et al. [17] determined experimentally the heat transfer and flow properties in a pipe with a helically structured rib. Further experiments for the identical pipe and a comparison with a modified version were performed in the publications of Virgillio et al. [20,21]. The flow and heat transfer between the ribs are investigated in these measurements and used for the comparison in the present study.

The aim of this study is to explain and further analyze the findings from these heat transfer measurements at such complex structures with the help of detailed flow simulations. Three-dimensional wall structures lead to very complex flow phenomena that have been rarely investigated so far. The experiments from the literature can usually only provide global values such as pressure loss and heat transfer, and local values can only be determined with complex measurement methods. With the current simulation method, it should be possible in the future to view the global results in connection with the local physical processes without complex measurement methods.

That is why in the present work an LES of a discontinuous single-started helically ribbed pipe is performed and validated with experimental data for the velocity and heat transfer between the ribs. The results demonstrate that it is possible to reproduce the measurement data even for complex geometries. The difficulty here is especially in the interruptions of the helix, so the local results are more difficult to compare than with a continuous helix. The aim of this study is to speed up the development process of new complex geometries and the associated correlations for the calculation of heat transfer and pressure loss.

2. Physical Modelling

The governing equations are given in the following section. The physical description of a turbulent flow with heat transport is based on the governing equation for mass, momentum and energy. The formulation used here is based on Anderson [22]. The continuity equation for a transient flow of a compressible fluid can be described as follows:

$$\frac{\partial \rho}{\partial t} + \nabla \cdot (\rho \mathbf{u}) = 0 \tag{1}$$

with the nabla-operator $\nabla = \left(\frac{\partial}{\partial x}, \frac{\partial}{\partial y}, \frac{\partial}{\partial z}\right)$, the density ρ, the velocity field $\mathbf{u}$ and time t.

The conservation of momentum is given by

$$\frac{\partial \rho \mathbf{u}}{\partial t} + \nabla \cdot (\rho \mathbf{u}\mathbf{u}) = -\nabla P + \rho \mathbf{g} + \nabla \cdot \left(2\mu_{eff}\mathbf{S}(\mathbf{u})\right) - \nabla\left(\frac{2}{3}\mu_{eff}(\nabla \cdot \mathbf{u})\right) \tag{2}$$

where P is the static pressure field and $\mathbf{g}$ is the gravitational acceleration. The effective viscosity μ_{eff} is the sum of molecular and turbulent viscosity. The rate of strain (deformation) tensor $\mathbf{S}(\mathbf{u})$ is defined as $\mathbf{S}(\mathbf{u}) = \frac{1}{2}(\nabla \mathbf{u} + (\nabla \cdot \mathbf{u})^T)$.
The conservation of energy in the fluid is defined in terms of the specific enthalpy h as

$$\frac{\partial(\rho h)}{\partial t} + \nabla \cdot (\rho \mathbf{u} h) + \frac{\partial(\rho k)}{\partial t} + \nabla \cdot (\rho \mathbf{u} k) - \frac{\partial P}{\partial t} = \nabla \cdot \left(\alpha_{eff}\nabla h\right) + \rho \mathbf{u} \cdot \mathbf{g} \tag{3}$$

where $k = \frac{|\mathbf{u}|^2}{2}$ is the specific turbulent kinetic energy. The effective thermal diffusivity α_{eff} is defined as the sum of laminar and turbulent thermal diffusivity

$$\alpha_{eff} = \frac{\rho \nu_t}{Pr_t} + \frac{\mu}{Pr}, \tag{4}$$

where Pr is the Prandtl number, Pr_t is the turbulent Prandtl number and ν_t is the turbulent kinematic viscosity.

To predict the effects of turbulence, the turbulent transport parameters require a turbulence model. In this work, the LES method is used to take turbulence into account. A RANS is clearly not suitable for the analysis of near wall turbulent transport phenomena. Since the use of a DNS significantly exceeds the computational effort of the LES, such a method is applied. Here, the large eddies, which contain most of the turbulent energy, are resolved by the conservation equations and only the small eddies are modeled. A filter function with a characteristic filter width of $\Delta = (\Delta_x \Delta_\omega \Delta_r)^{(1/3)}$ is applied to the conservation equations. This filter function splits up any field variable ϕ in a resolved $\hat{\phi}$ and non-resolved (subgrid) part ϕ' [23]. Following the Boussinesq approximation, the viscosity is replaced by an effective viscosity, which is the sum of molecular viscosity and the viscosity of the subgrid scales (eddy-viscosity), $\nu_{\mathrm{eff}} = \nu + \nu_{\mathrm{SGS}}$. The subgrid scale viscosity can then be modeled as $\nu_{\mathrm{SGS}} = C_{\mathrm{k}}\Delta\sqrt{k_{\mathrm{SGS}}}$, where $C_{\mathrm{k}} = 0.07$ is a model constant and k_{SGS} is the kinetic energy of the subgrid scale. The 0-equation WALE [24] (wall-adapting local eddy-viscosity) model calculates the kinetic energy of the subgrid scale using the following equation:

$$k_{\mathrm{SGS}} = \left(\frac{C_{\mathrm{w}}^2\Delta}{C_{\mathrm{k}}}\right)^2 \frac{\left(S_{\mathrm{ij}}^{\mathrm{d}}S_{\mathrm{ij}}^{\mathrm{d}}\right)^3}{\left(\left(\bar{S}_{\mathrm{ij}}\bar{S}_{\mathrm{ij}}\right)^{5/2} + \left(S_{\mathrm{ij}}^{\mathrm{d}}S_{\mathrm{ij}}^{\mathrm{d}}\right)^{5/4}\right)^2}. \tag{5}$$

where S_{ij} is the strain rate tensor of the resolved scale, $C_{\mathrm{w}} = 0.325$ and $C_{\mathrm{k}} = 0.094$ are model constants. This model takes into account the rotation of the flow field, so an additional damping function for ν_{SGS} in the near wall region is not necessary. Most of the turbulent energy is in the large eddies and these are difficult to model with a turbulence model because of their individual structure. Thus, the modelling part is very small compared to RANS.

OpenFOAM® (Open Source Field Operation and Manipulation) in the version of v1812 with the *buoyantPimpleFoam* solver was chosen for this study.

3. Geometry and Reference Cases

The validation reference case is based on experimental data published in the literature. The experiments were conducted at the "TU - Heat and Mass Transfer Laboratory" at the von Karman Institute in Sint Genesius Rode, Belgium. The flow data are taken from Virgilio et al. [20] and the heat transfer data from Virgilio et al. [21]. For the velocity measurements, a low-speed water tunnel was used. The measurement setup is configured in a way that the pressure at the inlet can be assumed to be constant; for a full optical access, the pipe is made of acrylic glass. Two different helical turbulators are measured, defined as RIB-1 and RIB-2, illustrated in Figure 1. The RIB-1 is the continuous helicoidal turbulator studied by Mayo et al. [17]. These are made of acrylonitrile butadiene styrene using a 3D printer. The diameter D of the pipe examined here is 150 mm; in the experiment, the pipe length is 15 × D. The rib pitch p corresponds to 63 mm, rib height e is 5.4 mm, angle between the axial and helix-wise directions is 80°, and the rib width is 10.8 mm. RIB-2 is different in that the height of the semi-circular rib changes along the helical pattern direction, but the maximum height remains the same for both ribs. After an angular displacement of 11°, this height becomes zero in both directions of the helix. This results in six obstacles on one pitch of the helix, with a total of nine pitches. For RIB-1 and RIB-2, Reynolds numbers of 24,400 and 21,100 have been investigated in the simulation. The stereoscopicv particle image velocimetry (S-PIV) images are taken between the seventh and eighth pitch. Data are only available in the area between the ribs. Liquid Crystals Thermography (LCT) is used to measure the stationary heat transfer. To reduce the angular influence on the measurement inaccuracy, a lately developed calibration technique for the narrow-band LCT was successfully applied. Since the simulations are computationally expensive and RIB-1 has already been examined in detail by Campet et al. [19] and Cauwenberg et al. [18] with an LES, in this article, RIB-2 has been investigated with the use of LES.

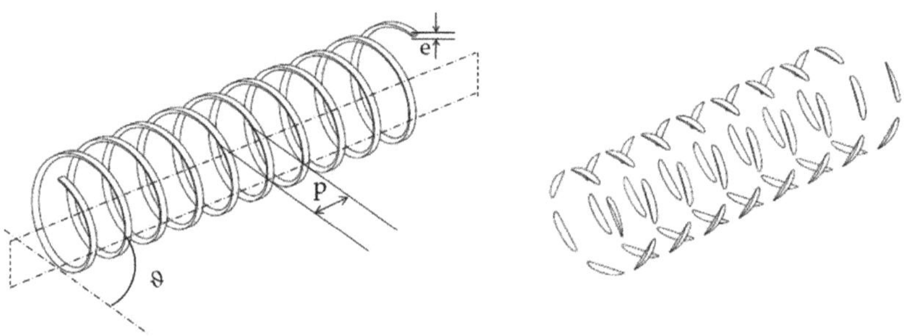

Figure 1. Schematics of the continuous helicoidal turbulator RIB-1 (**left**) and RIB-2 (**right**) from Virgillio et al. [20].

4. Numerical Methodology

The accurate calculation of heat transfer in turbulent pipe flow requires a fully developed turbulent flow inside the pipe. Limitations of computational resources did not allow simulations of a long pipe to ensure a fully developed flow. By simulating a small section of a pipe with cyclic boundary conditions at the inlet and outlet, a fully developed turbulent flow could be achieved, if a sufficient number of flow cycles were carried out. Table 1 lists the boundary conditions used, and the schematic illustration is shown in Figure 2. For the calculation, it is necessary to insert a source term for momentum and the energy equation due to the cyclic boundary conditions. This setup is also applied in the studies by

Campet et al. [19], Kügele et al. [25] and Akermann et al. [16]. Second-order methods are used for all discretisations. A Courant number of 0.3 is used for the computation, and, after a quasi-steady state of the flow has set in, the results are averaged over seven cycles as in the study by Campet et al. [19].

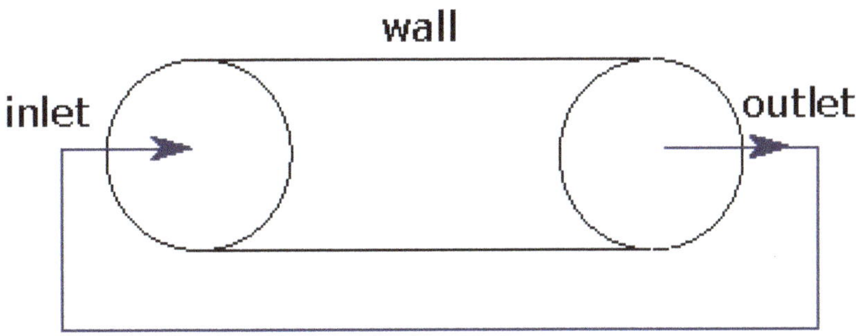

Figure 2. Schematic representation of the boundary conditions used.

Table 1. Definition of the boundary conditions used.

	U	p	T
wall	u = 0	$\partial p/\partial n = 0$	$\partial T/\partial n = -10{,}000$
inlet	cyclic	cyclic	cyclic
outlet	cyclic	cyclic	cyclic

The length of the calculation area is 4 × D. For the cyclical boundary conditions, it must be ensured that inlet and outlet are exactly identical. The mesh without the boundary layers is created with cfMesh; snappyHexMesh is then used to generate the final mesh. The maximum cell size within the geometry is 0.002 mm. Near the wall, the grid resolution is reduced to about 0.0005 mm to be able to better represent the structure of the ribs and to keep the transition between layers and cells small. The number of cells adds up to 23.33 million, with 67% of the cells forming the layers. The total number of layers used is ten, the smallest layer has a height of 0.03116 mm, and the growth rate of the layers is set to 1.2. The influence of the refinement on the wall and the addition of layers can be seen on the inlet and on the ribs in Figure 3. With these techniques, a y^+ value of less than 0.2 can be achieved. The values for x^+ and z^+ correspond to 7, which is sufficient for LES as shown in Akermann et al. [16].

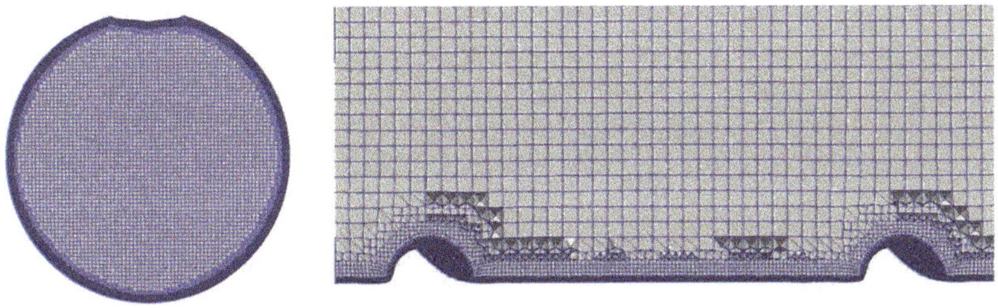

Figure 3. Display of the grid with layers at the inlet (**left**) and along the center (**right**) of the rib in RIB-2 geometry.

To ensure a sufficient grid resolution for the LES, the LSR (Length Scale Resolution) parameter is investigated. It corresponds to the ratio between the actually resolved energy level and the corresponding lower limit of the inertial subrange and is defined as:

$$LSR = \frac{\Delta}{60 \cdot \eta_{kol}} \tag{6}$$

where Δ corresponds to the grid width. The Kolmogorov length scale is calculated with:

$$\eta_{kol} = (\frac{\nu^3}{\epsilon})^{1/4} \tag{7}$$

$$\epsilon = \frac{k^{3/2}}{D/6} \tag{8}$$

and the turbulent kinetic energy k is determined using the simulation results. If the value is equal to 1, the turbulent scales are resolved up to the dissipation range. Therefore, a link between resolved energy levels and the local filter sizes can be created [26]. The evaluation in Figure 4 shows that the value is below 1 in all ranges. This ensures that the grid resolution is high enough.

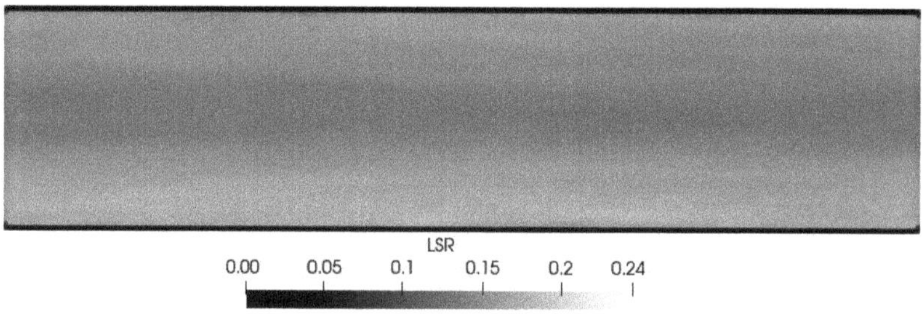

Figure 4. Display of the grid with layers at the inlet of the RIB-2 geometry.

5. Flow Field Analysis

A Reynolds number of 21,100 and a Prandtl number of 7 has been chosen for the simulation, the operating fluid is water and the thermophysical properties are constant To validate the simulation results, the flow field is analyzed in a range between 40 and 100 percent of the inner radius to focus on the near wall area. First, the velocity profiles calculated by the simulations are compared with the experimental data. All velocities are averaged with U_0, where for the variance U_0^2 is applied. The results in Figure 5 show that the simulation can reflect the measured values both in the area near the wall and towards the centre. For the first two velocity profiles, the values of simulation and measurement near the wall falls below zero, which is an indication of the backflow being correctly reflected by the model. The maximum velocity is 1.4 times the bulk velocity.

The investigation of the azimuthal velocity in Figure 6 also shows a good agreement with the measured values. The deviation of simulation and measurement is less than 10%. Especially at the first two measurement lines after the rib, the deviation between simulation and measurement is relatively high. Here, because of the flow recirculation, the averaging in simulation and measurement is very sensitive to pertubations and therefore susceptible to errors. Due to the detachment of the flow and the resulting backflow after the rib, the maximum velocity at the first measurement lines is lower compared to the measurement lines directly before the rib. Thus, the maximum value of the first curve is 0.069 compared to 0.143 at the last curve, which is an increase of 48%. Furthermore, in simulation and experiments, the azimuthal velocity above the ribs approaches zero. From this, it can be concluded that the described effects only occur close to the wall.

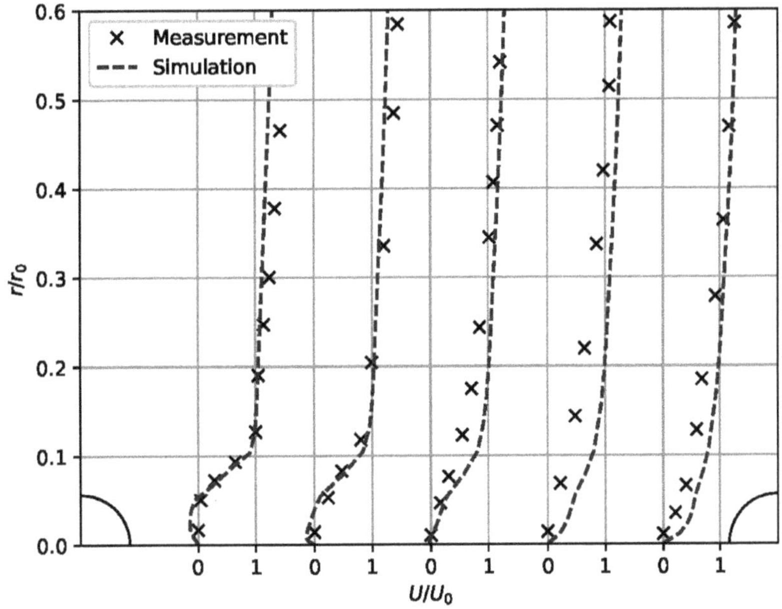

Figure 5. Comparison between measurement and simulation with the velocity in the flow direction.

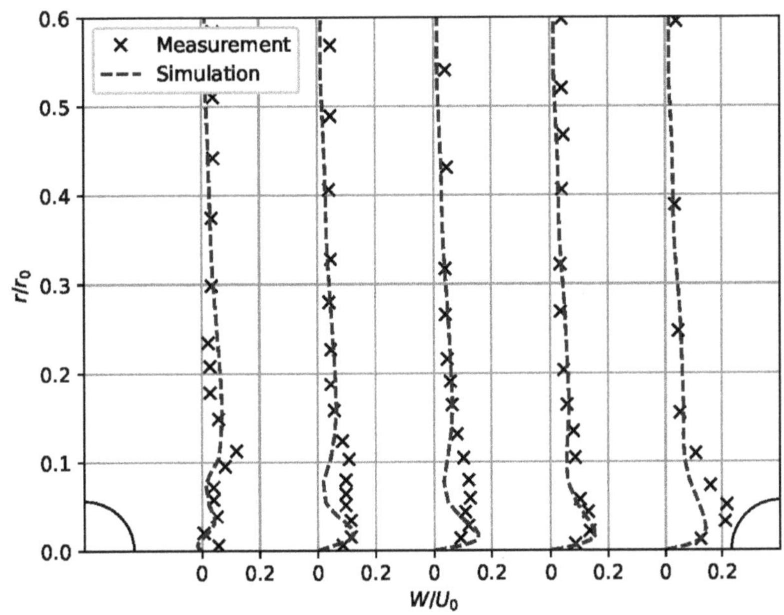

Figure 6. Comparison between measurement and simulation with the velocity in the circumferential direction.

The variance <*uu*> in the modelled and measured flow shows a good agreement as shown in Figure 7. The greatest deviation is seen in the first two curves. Here the variance is significantly lower than in the experimental data. The maximum value is 43% smaller at the first curve and 13% smaller at the second curve. It is assumed that the complex structures, caused by the detachment of the flow, are responsible for this difference. Nevertheless, the measurements can be replicated with the simulation. The comparison in Figure 8 for <*uv*> also shows a good agreement in this case. Overall, the simulation is able to reproduce the measurements for both the velocity profiles and the turbulent fluctuations well. Therefore, the simulation model is able to reproduce the turbulent flow in a ribbed pipe and is suitable for further investigations of the flow.

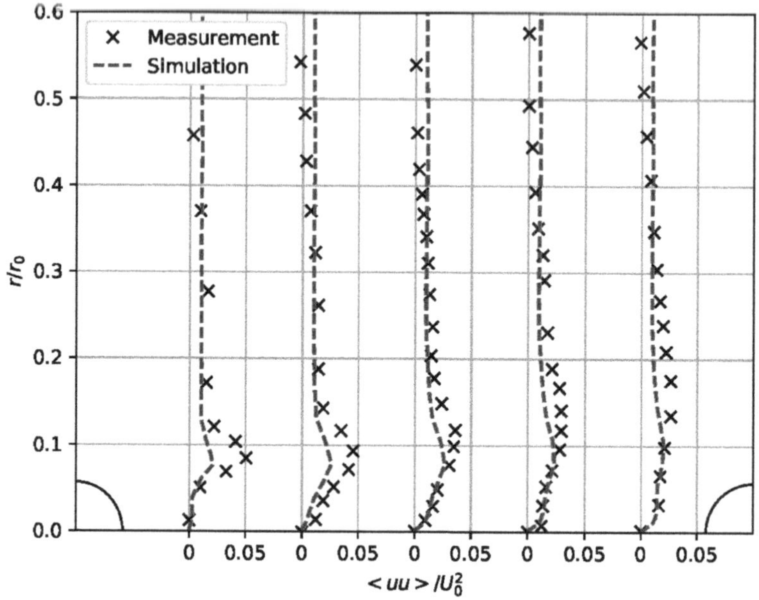

Figure 7. Comparison between measurement and simulation with the velocity variations <*uu*>.

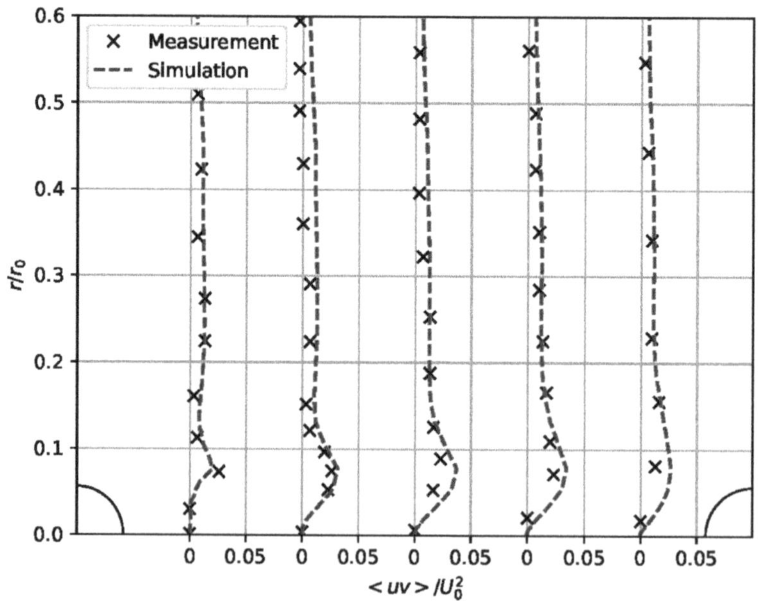

Figure 8. Comparison between measurement and simulation with the velocity variations <*uv*>.

6. Heat Transfer Analysis

For heat transport, the turbulent kinetic viscosity plays a major role, as can be seen from Equation (4). Especially near the wall, the heat transfer is dominated by the thermal diffusivity, while it can be neglected in the main flow. In the main flow, the thermal diffusivity has a negligible contribution compared to the turbulent transport. In order to be able to neglect the turbulent Prandtl number in the modeling approach, the ratio of molecular to turbulent viscosity is examined in Figure 9. From the results, it can be seen that the turbulent kinetic viscosity ν_t near the wall area is approaching zero, since the resolution of the mesh is very high here. This relationship can be explained by the following equation which is used for the calculation of ν_t:

$$\nu_t = (C_S \Delta)^2 \mathbf{OP}. \tag{9}$$

where Δ is the cell width, C_S is a model constant and **OP** is the LES model operator. The strong increase at $y/r \approx 0.05$ can be explained by the quadratic relationship of the cell width. Since the cells in this range become larger, the value of ν_t increases as can be concluded using Equation (9).

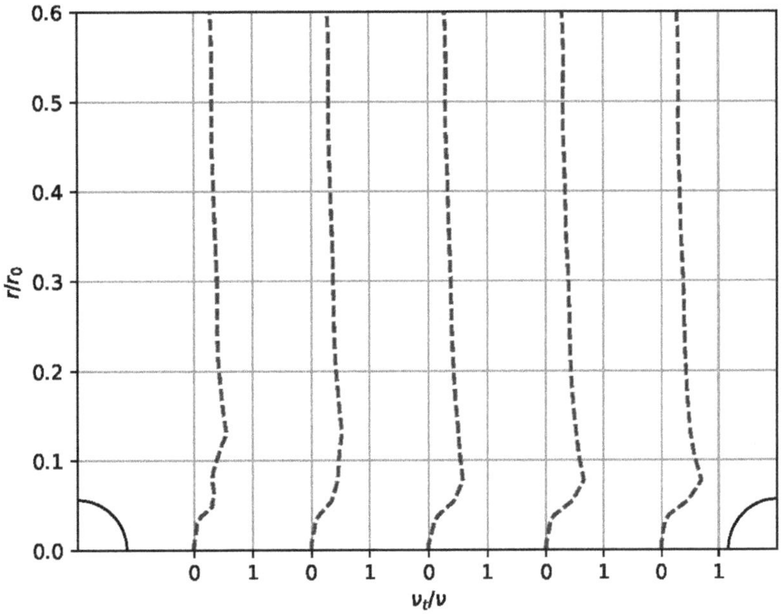

Figure 9. Representation of the turbulent viscosity in relation to the molecular viscosity.

In Figure 10, the measured Nusselt number is compared with the results of the simulation. The relationship between the Nusselt number of the examined structure and a smooth pipe is displayed. This ration is defined as EF with:

$$EF = Nu/Nu_S, \tag{10}$$

where Nu_S is calculated with the Dittus–Boelter correlation [27]:

$$Nu_S = 0.023 \cdot Re^{0.8} \cdot Pr^{0.4} \tag{11}$$

Measurements can only be made for the area between the ribs, and a detailed explanation can be found in Virgillio et al. [21]. The grey area corresponds to the tolerance range of the measurement errors. For a better overview, this representation is used instead of error bars. The comparison shows a very good agreement between the simulation and the measurement, except for the area directly before and after the rib. Investigations by Campet et al. [19] were able to demonstrate that, analogous to the results of this research, the deviation between simulation and experiment are significantly higher at these points. It can therefore be assumed that a measurement of the correct values for the heat transfer at these points is critical.

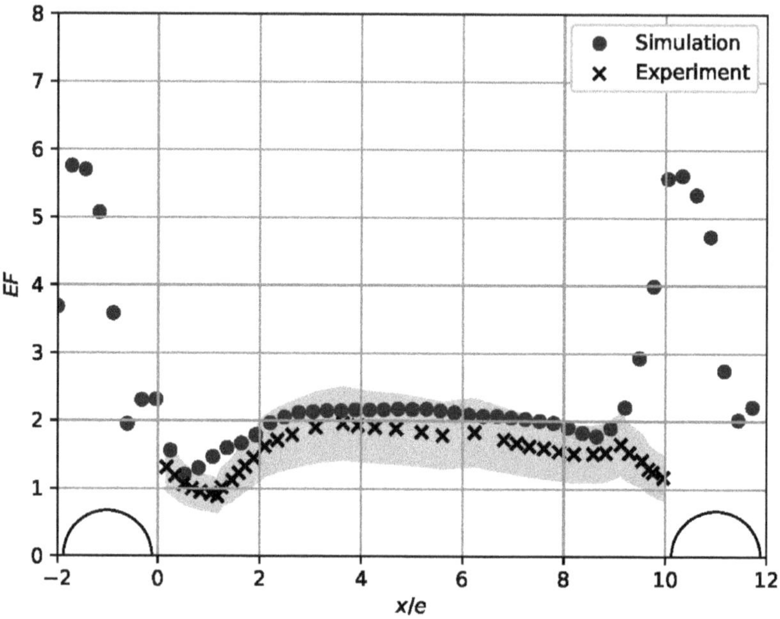

Figure 10. Comparison between measurement and simulation of the Nusselt number in relation to the smooth pipe between the ribs; the grey area shows the maximum and minimum values of the measurement.

As previously proven in Mayo et al. [17], there is a local maximum for the Nusselt number between two ribs along the axial flow direction in the pipe, from the point where the flow is reattached. This can be determined by evaluating the wall shear stress τ_x in the flow direction. In Figure 11 , τ_x is displayed in the range $x/e = 2$ and 8, and the reattachment point is located at the point where the value of τ_x changes its sign. It can thus be determined at $x/e \approx 5.05$. The local maximum of EF between the ribs, shown in Figure 12, is located in the same area. This is caused by the cold fluid transported from the bulk flow in wall direction, thus resulting in a higher temperature difference between wall and fluid. This increase is linear in relation to the Nusselt number. After the reattachment of the flow at $x/e \approx 5.05$, the Nusselt number becomes smaller again.

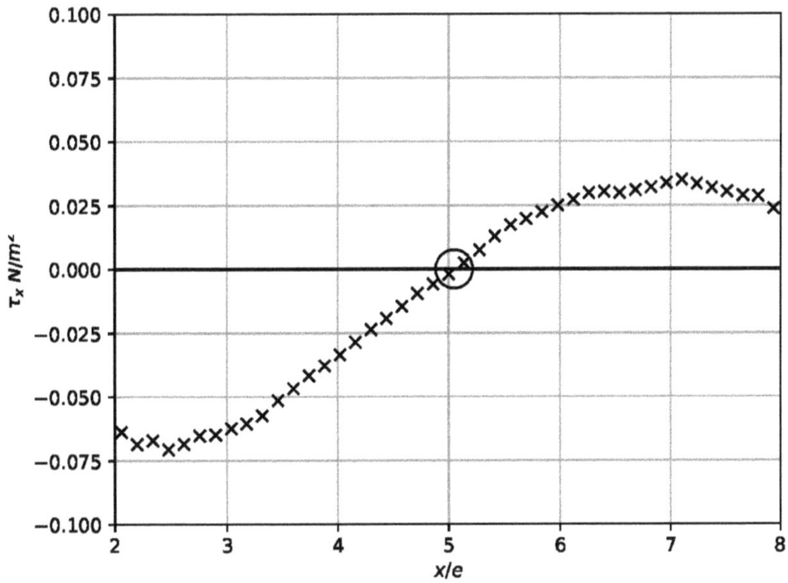

Figure 11. Representation of τ_x to determine the reattachment point of the flow after detachment at the rib.

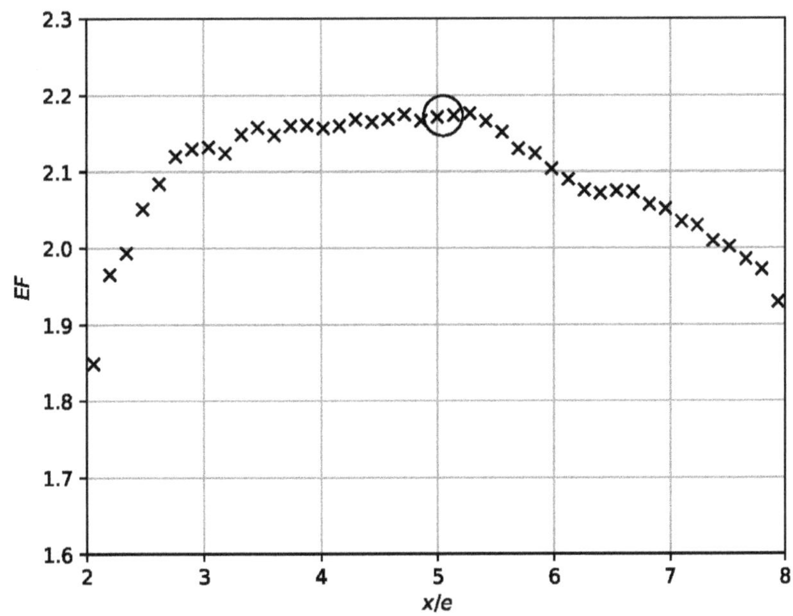

Figure 12. Display of the Nusselt number in the range from $x/e = 2$ to $x/e = 8$ to show the influence of the reattachment point of the flow on the Nusselt number.

For RIB-1, a reattachment point of the flow at $x/e \approx 4.25$ has been determined in Virgilio et al. [20]. The author points out that the reattachment point of the flow for RIB-2 is located further upstream than with RIB-1, which can be confirmed by the results shown here. The authors state that the interruptions of the helix prevent the flow from a blockage in comparison to the geometry in RIB-1. Using Figure 13, it can be illustrated with the streamlines that the fluid can also flow around the interrupted rib.

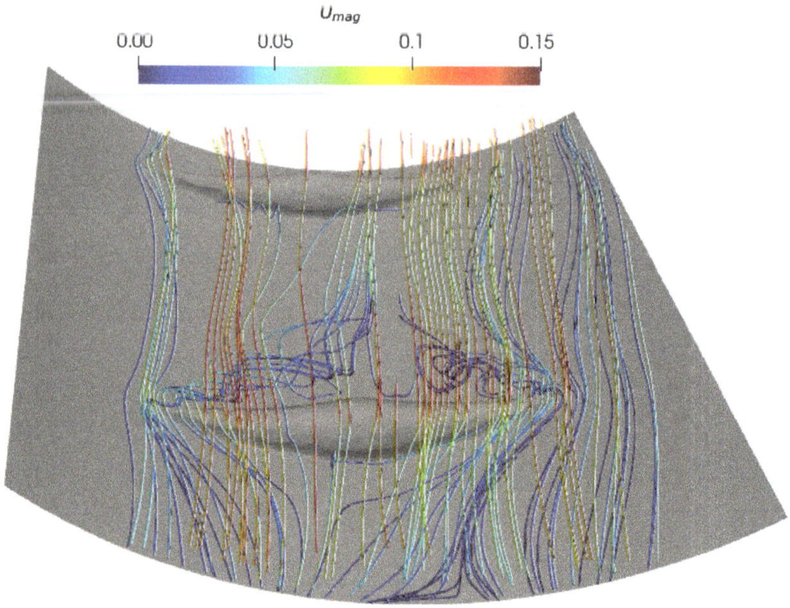

Figure 13. Streamlines of the magnitude velocity, over an interrupted rib.

The analysis of EF in Figure 14 demonstrates that an increase in efficiency compared to the smooth pipe is achieved almost everywhere, with the exception of the region highlighted in orange. In contrast to the RIB-1 with a continuous corrugation, the interruption and reduction at the sides results in reduced global heat transfer. Because the area in front of the centre of the rib before the flow separates has the highest heat transfer, and this area becomes smaller on both sides with this kind of geometry. The global value of EF for the measurement is 1.44 ± 0.41 for a Reynolds number of 21,160. In the simulation, the average value is 1.75 for a Reynolds number of 21,100 and is therefore still within the tolerance range of the measurement. Compared to the simulation of the RIB-1 geometry at a Reynolds number of 20,000, the value for EF decreases by 32%, showing that the interruption of the ribs reduces the heat transfer.

To compare the performance of the structured pipe with a smooth pipe, the performance evaluation factor (PEC) is used, defined as:

$$PEC = \frac{Nu/Nu_S}{(f/f_S)^{1/3}} \tag{12}$$

Nu, f, Nu_S and f_S are the Nusselt number and the friction factor of the investigated geometry and the ones of a smooth pipe (index S). To calculate the friction factor, the correlation of Gnielinski [28] is applied. If the value of the PEC factor is higher than 1, the improvement of the heat transfer is greater than the increased costs due to the pressure loss. A PEC factor of 1.53 is determined for the pipe examined here. Thus, it can be shown that the additional heat transfer is higher than the additional loss due to the additionally required pump power.

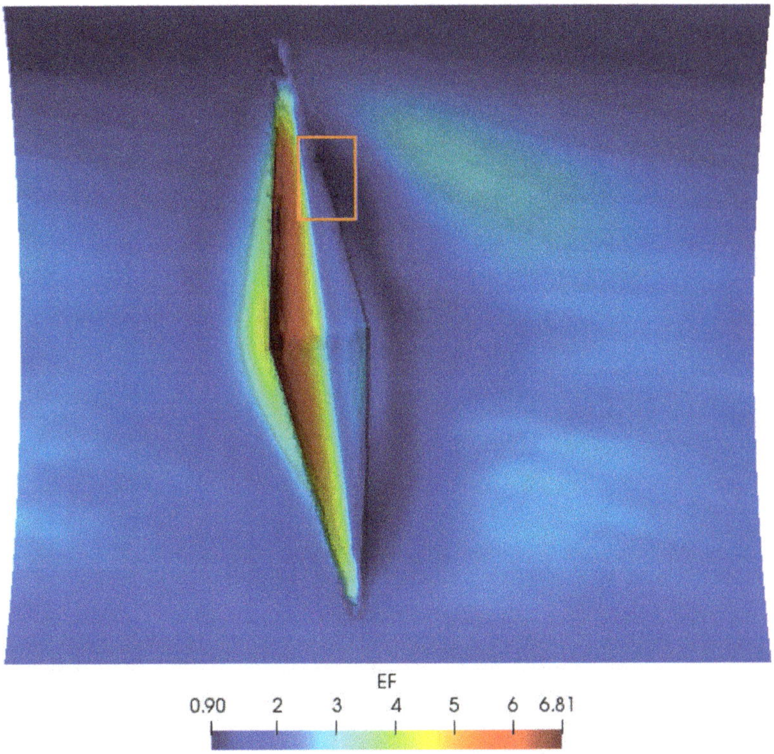

Figure 14. Representation of the ratio of Nusselt number to Nusselt number smooth pipe over a complete rib.

7. Pressure Loss

The experiments do not provide local pressure data; therefore, the results are compared with the values of the LES for RIB-1 from Campet et al. [19]. Here, the local friction coefficient C_f and the pressure coefficient P_{Norm} from Equations (13) and (14) are compared. A derivation of the two coefficients is provided by Campet et al. [19]. C_F has been validated in the above-mentioned paper using measurement data between the ribs and is defined as follows:

$$C_f = \frac{\tau_x}{0.5 \cdot U_0^2 \cdot \rho} \tag{13}$$

with the bulk velocity U_0 and the wall shear stress τ_x in the flow direction. The pressure coefficient P_{norm} is defined with:

$$P_{norm} = \frac{P - P_{ref}}{0.5 \cdot U_0^2 \cdot \rho} \tag{14}$$

where P_{ref} corresponds to the pressure on the wall at $x/e = 0$, so P_{norm} becomes zero at this point. Figure 15 shows the comparison between the two rib structures for C_f in the centre of the interrupted rib. It can be seen that the change of sign of C_f takes place where the reattachment point of the flow is located. This is caused by the velocity gradient being directly included in the calculation of τ_x. Furthermore, there is a backflow at RIB-2 shortly before the rib shown in Figure 16, which causes τ_x to obtain a negative value. This swirling is not visible at RIB-1, as can be concluded from the results of Campet et al. [19].

In addition to that, the peak in the simulation of RIB-2 is clearly above the values of RIB-1 and the maximum of the first mentioned geometry is 0.0686 compared to 0.0185, which corresponds to an increase of 370%. It is assumed that the velocity gradient on the rib due to the backflow in front of the rib results in a strong increase in local velocity.

The global C_f value used to calculate the pressure drop in the Darcy–Weißbach equation has a value of 0.0167 for RIB-2 and 0.0304 for RIB-1. The pressure drop in the geometry examined here is therefore lower by a factor of 1.82, since the influence of C_f is directly linear in the equation for the global pressure loss.

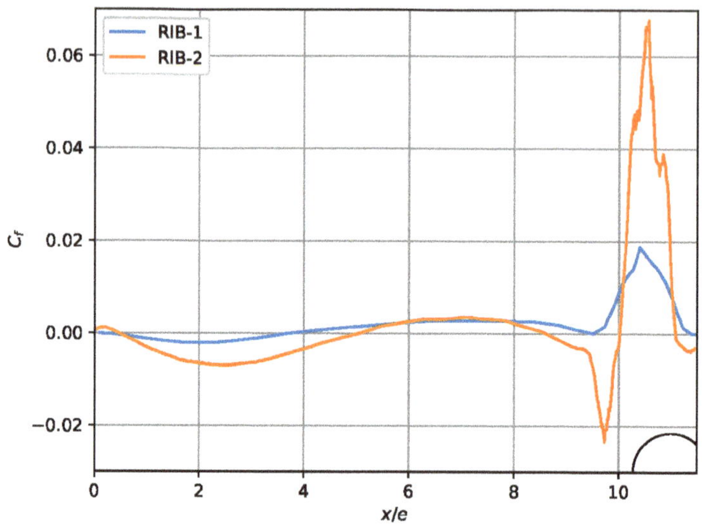

Figure 15. Comparison of the values of C_f between RIB-1 with values from Campet et al. [19] and RIB-2.

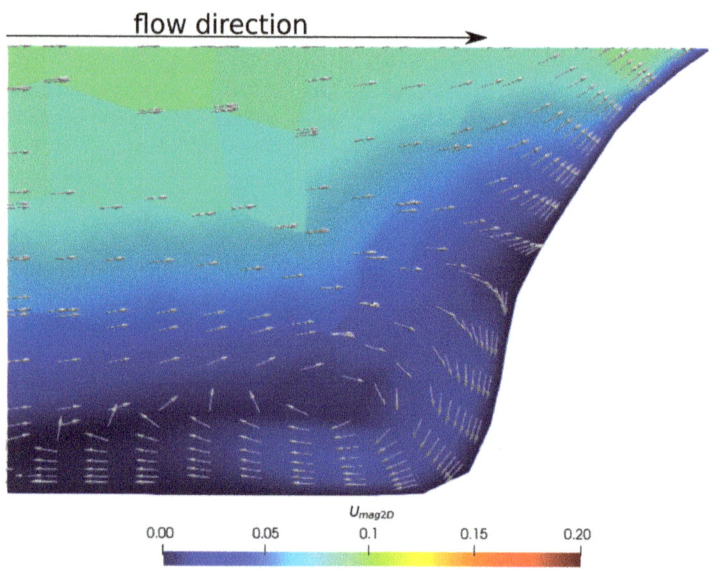

Figure 16. Two-dimensional velocity combined with vector arrows in front of the rib to show the backflow that takes place there.

Figure 17 compares the pressure coefficients over the centre of the interrupted rib. The maximum value in the range at x/e = 10, shows that the pressure coefficient is almost identical. The pressure coefficient of RIB-1 is almost always above RIB-2 in the positive range and below RIB-2 in the negative range. While the flow has to pass over continuous rib structure at RIB-1, it is partly redirected through the gaps in the interrupted rib at RIB-2, which explains the differences in the pressure coefficient plot.

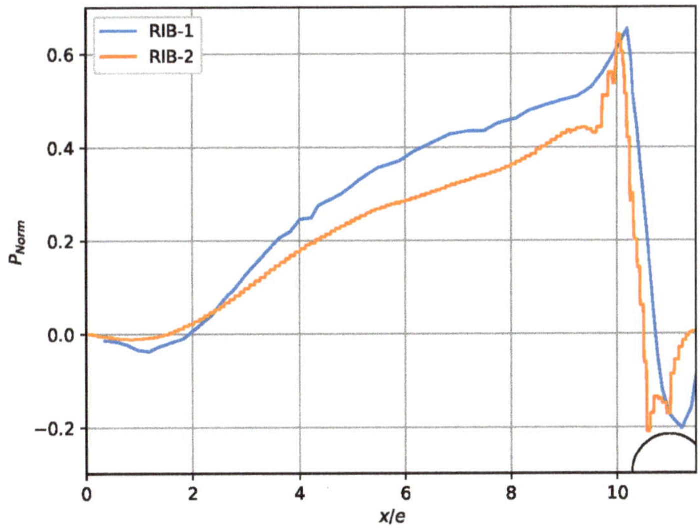

Figure 17. Comparison of the values of P_{Norm} between RIB-1 with values from Campet et al. [19] and RIB-2.

8. Conclusions

In this article, a numerical flow simulation is successfully validated by measurements. A y^+ value of less than 0.2 is maintained in order to be able to represent the thermal boundary layer as well as the flow boundary layer at a Prandtl number of 7. The local velocity profiles and fluctuations of the velocity between the ribs show a good agreement with the measurements. This proves that the method is capable of reproducing the flow correctly. Using the Nusselt number validation, it can be demonstrated that a maximum Nusselt number is reached between the ribs until the flow reattaches to the wall. This effect is described by Mayo [17] using the RIB-1 geometry and was again observed in this study. Overall heat transfer is reduced compared to the continuous rib, which is proven by both experiment and simulation. The analysis of the wall shear stress showed that the detachment of the flow is shifted further backwards by the interruption, in contrast to the continuous rib. As was previously the case for the flow field, a successful validation is performed for the heat transfer. An analysis of the pressure drop was realized with the help of the results of Campet et al. [19], who performed an LES of the RIB-1 geometry; here, the results could be compared with each other. It could be shown that the C_f value for RIB-2 is higher on the rib than for RIB-1, due to backflow being formed in front of the rib. The pressure coefficient for RIB-2 is lower in comparison to RIB-1, due to the reduction through the rib geometry being interrupted. The simulation was therefore able to reflect the results of the measurement for both the flow and the heat transfer well. Through the results, it can be shown that it is possible to simulate complex internally structured pipes, even without a continuous structure. Thus, further structures with increased complexity can be simulated and evaluated.

Author Contributions: Conceptualization, S.K.; methodology, S.K. and G.O.M.; software, S.K. and G.O.M.; validation, S.K. and P.R.; formal analysis, S.K.; investigation, S.K.; resources, S.K.; data curation, S.K.; writing—original draft preparation, S.K.; writing—review and editing, S.K. and P.R. and T.G.; visualization, S.K.; supervision, P.R. and T.G.; project administration, P.R.; funding acquisition, P.R. All authors have read and agreed to the published version of the manuscript.

Funding: This research was funded by the Federal Ministry of Education and Research of Germany (BMBF) Grant No. 13FH119PX6.

Institutional Review Board Statement: Not applicable.

Informed Consent Statement: Not applicable.

Data Availability Statement: Not applicable.

Acknowledgments: The authors would like to gratefully acknowledge support by the Steinbuch Center for Computing (SCC) at the Karlsruhe Institute of Technology (KIT) for providing computational resources. Most of the simulations have been performed on the bwUniCluster and the forHLR II cluster, where access has been granted to the authors. The article processing charge was funded by the Baden-Württemberg Ministry of Science, Research and the Arts and the University of Applied Sciences Ulm in the funding programme Open Access Publishing.

Conflicts of Interest: The authors declare no conflict of interest.

Abbreviations

The following abbreviations are used in this manuscript:

LES	large-eddy simulation	
RANS	Reynolds-Averaged Navier–Stokes	
OpenFOAM	Open Source Field Operation and Manipulation	
WALE	wall adapting local eddy-viscosity	
S-PIV	stereoscopic particle image velocimetry	
Nu	Nusselt number	
Pr	Prandtl number	
Nomenclature		
t	time	s
$\mathbf{u}$	velocity vector	m/s
U_0	Bulk velocity	m/s
g	gravitational acceleration	m/s^2
k	turbulent kinetic energy	m^2/s^2
P	pressure	Pa
h	enthalpy	J
T	Temperature	K
p	rib pitch	mm
e	rib height	mm
C_f	friction coefficient	-
P_{Norm}	Pressure coefficient	-
Greek symbols		
ρ	density	kg/m^3
μ	dynamic viscosity	$Pa \cdot s$
ν	kinematic viscosity	m^2/s
α	thermal diffusivity	m^2/s
η_{kol}	Kolmogorov length scale	m
τ_x	wall shear stress	m/s^2

References

1. Mousavi Ajarostaghi, S.S.; Zaboli, M.; Javadi, H.; Badenes, B.; Urchueguia, J.F. A Review of Recent Passive Heat Transfer Enhancement Methods. *Energies* **2022**, *15*, 986. [CrossRef]
2. Webb, R.L.; Narayanamurthy, R.; Thors, P. Heat Transfer and Friction Characteristics of Internal Helical-Rib Roughness. *J. Heat Transf.* **1999**, *122*, 134–142. [CrossRef]

3. Gee, D.L.; Webb, R.L. Forced convection heat transfer in helically rib-roughened tubes. *Int. J. Heat Mass Transf.* **1980**, *23*, 1127–1136. [CrossRef]
4. Withers, J.G. Tube-Side Heat Transfer and Pressure Drop for Tubes Having Helical Internal Ridging with Turbulent/Transitional Flow of Single-Phase Fluid. Part 1. Single-Helix Ridging. *Heat Transf. Eng.* **1980**, *2*, 48–58. [CrossRef]
5. Withers, J.G. Tube-Side Heat Transfer and Pressure Drop for Tubes Having Helical Internal Ridging with Turbulent/Transitional Flow of Single-Phase Fluid. Part 2. Multiple-Helix Ridging. *Heat Transf. Eng.* **1980**, *2*, 43–50. [CrossRef]
6. Han, J.; Glicksman, L.; Rohsenow, W. An investigation of heat transfer and friction for rib-roughened surfaces. *Int. J. Heat Mass Transf.* **1978**, *21*, 1143–1156. [CrossRef]
7. Nakayama, M.; Takahashi, K.; Daikoku, T. Spiral ribbing to enhance single-phase heat transfer inside tubes. *Proc. ASME-JSME Therm. Eng. Jt. Conf.* **1983**, *1*, 365–372.
8. Ravigururajan, T.S.; Bergles, A.E. Development and verification of general correlations for pressure drop and heat transfer in single-phase turbulent flow in enhanced tubes. *Exp. Therm. Fluid Sci.* **1996**, *13*, 55–70. [CrossRef]
9. Webb, R.L.; Eckert, E.R.G.; Goldstein, R.J. Heat transfer and friction in tubes with repeated-rib roughness. *Int. J. Heat Mass Transf.* **1971**, *14*, 601–617. [CrossRef]
10. Kumar, P.; Judd, R.L. Heat transfer with coiled wire turbulence promoters. *Can. J. Chem. Eng.* **1970**, *48*, 378–383. [CrossRef]
11. Zdaniuk, G.J.; Chamra, L.M.; Mago, P.J. Experimental determination of heat transfer and friction in helically-finned tubes. *Exp. Therm. Fluid Sci.* **2008**, *32*, 761–775. [CrossRef]
12. Hossainpour, S.; Hassanzadeh, R. Numerical investigation of tub side heat transfer and pressure drop in helically corrugated tubes. *Int. J. Energy Environ. Eng.* **2011**, *2*, 2008–916365.
13. Vijiapurapu, S.; Cui, J. Performance of turbulence models for flows through rough pipes. *Appl. Math. Model.* **2010**, *34*, 1458–1466. [CrossRef]
14. Cauwenberge, D.J.V.; Schietekat, C.M.; Floré, J.; Geem, K.M.V.; Marin, G.B. CFD-based design of 3D pyrolysis reactors: RANS vs. LES. *Chem. Eng. J.* **2015**, *282*, 66–76. [CrossRef]
15. Wang, W.; Zhang, Y.; Liu, J.; Li, B.; Sundén, B. Large eddy simulation of turbulent flow and heat transfer in outward transverse and helically corrugated tubes. *Numer. Heat Transf. Part A Appl.* **2019**, *75*, 456–468. [CrossRef]
16. Akermann, K.; Renze, P.; Dietl, J.; Schröder, W. Large-Eddy Simulation of turbulent heat transfer in a multiple-started helically rib-roughened pipe. *Int. J. Heat Mass Transf.* **2020**, *154*, 119667. [CrossRef]
17. Mayo, I.; Cernat, B.C.; Virgilio, M.; Pappa, A.; Arts, T. Experimental Investigation of the Flow and Heat Transfer in a Helically Corrugated Cooling Channel. *J. Heat Transf.* **2018**, *140*, 071702. [CrossRef]
18. Van Cauwenberge, D.J.; Dedeyne, J.N.; Van Geem, K.M.; Marin, G.B.; Floré, J. Numerical and experimental evaluation of heat transfer in helically corrugated tubes. *AIChE J.* **2018**, *64*, 1702–1713. [CrossRef]
19. Campet, R.; Zhu, M.; Riber, E.; Cuenot, B.; Nemri, M. Large Eddy Simulation of a single-started helically ribbed tube with heat transfer. *Int. J. Heat Mass Transf.* **2019**, *132*, 961–969. [CrossRef]
20. Virgilio, M.; Mayo, I.; Dedeyne, J.; Geem, K.V.; Marin, G.; Arts, T. Effects of 2-D and 3-D helical inserts on the turbulent flow in pipes. *Exp. Therm. Fluid Sci.* **2020**, *110*, 109923. [CrossRef]
21. Virgilio, M.; Mayo, I.; Dedeyne, J.; Geem, K.V.; Marin, G.; Arts, T. Influence of obstacles on the wall heat transfer for 2D and 3D helically ribbed pipes. *Int. J. Heat Mass Transf.* **2020**, *148*, 119087. [CrossRef]
22. Anderson, J. *Computational Fluid Dynamics*; Computational Fluid Dynamics: The Basics with Applications; McGraw-Hill Education: New York, NY, USA, 1995.
23. Ferziger, J.; Peric, M. *Numerische Strömungsmechanik*; Springer: Berlin/Heidelberg, Germany, 2008.
24. Nicoud, F.; Ducros, F. Subgrid-Scale Stress Modelling Based on the Square of the Velocity Gradient Tensor. *Flow Turbul. Combust.* **1999**, *62*, 183–200. [CrossRef]
25. Kügele, S.; Renze, P.; Dietl, J.; Grützner, T. Investigation of heat transfer and pressure drop for a multiple-started ribbed pipe using large-eddy simulation. *AIChE J.* **2022**, e17808. [CrossRef]
26. Pope, S.; Eccles, P. *Turbulent Flows*; Cambridge University Press: Cambridge, UK, 2000.
27. Bergman, T.; Lavine, A.; Incropera, F. *Fundamentals of Heat and Mass Transfer*, 7th ed.; John Wiley & Sons, Incorporated: Hoboken, NJ, USA, 2011.
28. Gnielinski, V. New equations for heat and mass transfer in turbulent pipe and channel flow. *Int. Chem. Eng.* **1976**, *16*, 359–368.

energies

Article

Strongly Heated Turbulent Flow in a Channel with Pin Fins

Chien-Shing Lee [1], Tom I. -P. Shih [1,*], Kenneth Mark Bryden [2], Richard P. Dalton [3] and Richard A. Dennis [3]

1 School of Aeronautics and Astronautics, Purdue University, West Lafayette, IN 47907, USA
2 Ames Laboratory, U.S. Department of Energy, Ames, IA 50010, USA
3 National Energy Technology Laboratory, U.S. Department of Energy, Morgantown, WV 26507, USA
* Correspondence: tomshih@purdue.edu

Abstract: Large-eddy simulations (LES) were performed to study the turbulent flow in a channel of height H with a staggered array of pin fins with diameter D = H/2 as a function of heating loads that are relevant to the cooling of turbine blades and vanes. The following three heating loads were investigated—wall-to-coolant temperatures of T_w/T_c = 1.01, 2.0, and 4.0—where the Reynolds number at the channel inlet was 10,000 and the back pressure at the channel outlet was 1 bar. For the LES, two different subgrid-scale models—the dynamic kinetic energy model (DKEM) and the wall-adapting local eddy-viscosity model (WALE)—were examined and compared. This study was validated by comparing with data from direct numerical simulation and experimental measurements. The results obtained show high heating loads to create wall jets next to all heated surfaces that significantly alter the structure of the turbulent flow. Results generated on effects of heat loads on the mean and fluctuating components of velocity and temperature, turbulent kinetic energy, the anisotropy of the Reynolds stresses, and velocity-temperature correlations can be used to improve existing RANS models.

Keywords: LES; internal cooling; pin fin; heat loads; turbulence

Citation: Lee, C.-S.; Shih, T.I.-P.; Bryden, K.M.; Dalton, R.P.; Dennis, R.A. Strongly Heated Turbulent Flow in a Channel with Pin Fins. *Energies* **2023**, *16*, 1215. https://doi.org/10.3390/en16031215

Academic Editors: Artur Bartosik and Dariusz Asendrych

Received: 30 December 2022
Revised: 18 January 2023
Accepted: 19 January 2023
Published: 22 January 2023

Copyright: © 2023 by the authors. Licensee MDPI, Basel, Switzerland. This article is an open access article distributed under the terms and conditions of the Creative Commons Attribution (CC BY) license (https://creativecommons.org/licenses/by/4.0/).

1. Introduction

The thermal efficiency of gas turbines can be increased by increasing the temperature of the gas entering the turbine component. In advanced gas turbines, the gas temperature entering the turbine can far exceed the melting point of the turbine's material. Thus, the turbine component must be cooled to maintain its mechanical strength for reliable operation. One part that is especially difficult to cool is the trailing-edge regions of the turbine's blades and vanes. In those regions, embedded cooling passages with pin fins have been found to be effective [1–4].

Since the physical processes that take place in channels with pin fins are quite complicated, many investigators have performed experimental and computational studies to understand how design and operating parameters affect heat transfer and pressure drop (see e.g., [1–22] and references cited there). Parameters studied include pin-fin shape (circular, square, diamond), pin-fin hydraulic diameter to channel height, arrangement of pin fins (staggered and inline), pin-to-pin streamwise and spanwise spacings, and the Reynolds number of the flow. Of the experimental studies, most were based on time-averaged measurements. Only Ames et al. [9–11] and Ostanek and Thole [16] examined time-resolved measurements; these studies showed the flow in channels with a staggered array of pin fins at Reynolds numbers from 3000 to 30,000 to be highly unsteady due to vortex shedding and unsteady separation.

Most computational studies were also based on steady analyses by using steady RANS (i.e., Reynolds-averaged continuity, Navier-Stokes, and energy equations) closed by one or more of the following models: k-ω, Shear-Stress Transport (SST), Explicit Algebraic Reynolds Stress (EARS), realizable k-ε, Renormalization Group (RNG) k-ε, and

v^2-f [15,19]. Steady RANS was found to predict trends with reasonable accuracy. However, quantitatively, relative errors could be 10–30% or more.

Relatively few investigators performed time-accurate simulations of the unsteady flow and heat transfer in channels with pin fins. Delibra et al. [12–14] used unsteady RANS (URANS) with the elliptic-relaxation eddy-viscosity (ζ-f) turbulence model. By capturing the large-scale wake shedding structures, the relative error was reduced to less than 10%. Paniagua et al. [18] used large-eddy simulation (LES) and a hybrid LES-RANS method with LES away from walls and URANS next to walls. Their study showed that the hybrid LES-RANS was able to capture the dominant large-scale eddies and mean flow quantities with reasonable accuracy. Their study also assessed three LES subgrid-scale (SGS) models–WALE, QR, and VMS and found the WALE model to predict the mean velocity distribution the best and the QR model to predict the mean pressure coefficient the best.

The aforementioned studies only investigated low heat loads where the wall-to-coolant temperature was near unity. However, that ratio can be as high as two or more in cooling passages in turbine vanes and blades. Shih et al. [20] and Lee et al. [21–23] studied the effects of heating load on flow, heat transfer, and pressure loss by using steady RANS, URANS, LES, and hybrid LES. Shih et al. [20] showed scaling formulas that accounted for heating developed for passages with smooth walls to yield large errors when applied to passages with pin fins. They developed a method to scale experimental data obtained at low heat loads in laboratory conditions to high heat loads in gas turbines. Lee et al. [21–23] showed that high heating load significantly increased the length of the entrance region and reduced the Nusselt number (Nu) in the entrance and post-entrance regions. Lee et al. [22] used LES to examine the turbulence statistics created by high heating loads. Lee et al. [23] showed RANS to underpredict Nu because it did not account for vortex shedding and URANS to first underpredict and then overpredict Nu because the RANS models were unable to predict at which rows shedding occurred and then overpredicted its effects on heat transfer. Thus, better RANS models are needed for this class of flows with unsteady separation under high heat loads.

Though the study by Lee et al. [21–23] provided considerable understanding on how heat load affects flow and heat transfer, the details of the turbulent flow field, such as Reynolds stresses and velocity-temperature correlations, were not provided and analyzed. Since such information could assist the development of better RANS models to study this class of flows as well as enable improved designs of pin-fin arrays under high heat loads, the objective of this study is to use LES to study the statistics of the turbulent flow structure created by high heat loads.

The remainder of this paper is organized as follows. The problem studied is first described. Then, the problem formulation, the numerical method of solution, and the verification and validation are presented. This is followed by results obtained on the effects of heat load and the turbulent statistics.

2. Problem Description

A schematic of the problem studied is shown in Figure 1, where all dimensions are given in terms of D = 2.54 cm, the diameter of the pin fin. In this figure, the cooling channel is bound by two flat plates with length 130D and height H = 2D. This channel is made up of three sections – a test section of length 12.5D and two smooth sections connected to the test section. The test section has five rows of pin fins that are arranged in a staggered fashion. The center of the first pin fin is located at x = 1.25D, and the spacing between centers of pin fins is S_x = 2.5D in the streamwise (x) and S_z = 2.5D in the spanwise directions (z). All solid surfaces of the test section, including the surfaces of the pin fins, are isothermal. For the two smooth sections, where there are no pin fins, all solid surfaces are adiabatic. The smooth section of length 115D is attached to the inlet of the test section to ensure that flow entering the test-section has a "fully developed" velocity profile in the sense that it is no longer affected by the entrance region. The smooth section of length 2.5D is attached to the exit of the test section to ensure no reversed flow at the outflow boundary.

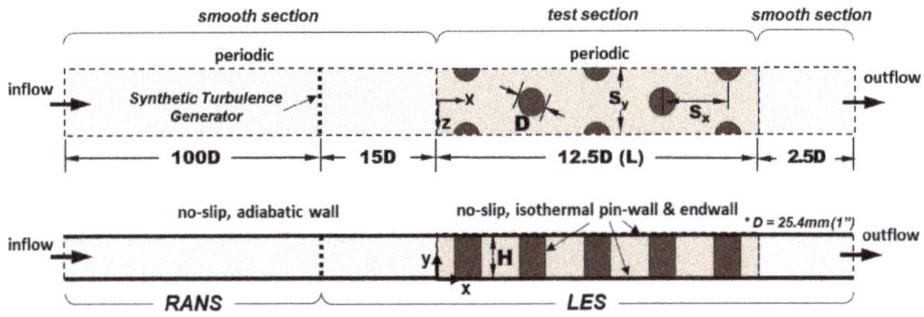

Figure 1. Schematic of cooling channel studied.

For this cooling channel, the cooling flow that enters at x = −115D is air and has a uniform temperature of T_c = 300 K and a mass flow rate of $\dot{m}_c$ = 0.014 kg/s. The pressure at the channel's outlet located at x = 15D is maintained at P_b = 1 bar. The Reynolds number (Re_D) is 10,000, which is based on the pin-fin diameter D, the averaged momentum of $4\dot{m}_c/\left(\pi D^2\right)$, and the dynamic viscosity evaluated at T_c. As noted, all solid surfaces of the test section are isothermal, and the following surface or wall temperatures were investigated: T_w = 303 K, 600 K, and 1200 K, representing negligible, moderate, and strong heat loads, respectively. Table 1 summarizes all of the cases studied.

Table 1. Summary of simulations performed.

Case	T_w (K)	T_w/T_c	SGS Model
NH-DKEM	303	1.01 (negligible heating, NH)	
MH-DKEM	600	2.0 (moderate heating, MH)	DKEM
SH-DKEM	1200	4.0 (strong heating, SH)	
NH-WALE	303	1.01 (NH)	
MH-WALE	600	2.0 (MH)	WALE
SH-WALE	1200	4.0 (SH)	

T_c = 300 K for all cases.

The configuration just described matches the experimental study of Ames and Dvorak [11] with detailed measurements of the mean turbulent flow field. Their data was used to validate this computational study. Here, it is also noted that this configuration differs from the configurations studied by Lee and Shih [23] by having a channel height of 2D instead of D so that the pin fins have height-to-diameter ratio of two instead of one.

3. Governing Equations and Solution Procedure

Since the cooling air that flowed through the channel with pin fins was subjected to high heat loads, the density and temperature of the air were expected to change considerably along the channel. Thus, though the Mach number of the flow throughout the channel was low, the compressible formulation with temperature-dependent properties was needed. In this study, steady RANS was used from x = −115D to −15D, and LES was used at x = −5D and downwards. The density-weighted Reynolds-averaged and spatially-filtered continuity, momentum, and energy equations for RANS and for LES can be written as [24]

$$\frac{\partial U}{\partial t} + \nabla \cdot (F_I - F_V) = 0 \quad (1)$$

where $U, F_I = (F_{I,1}, F_{I,2}, F_{I,3})$, and $F_V = (F_{V,1}, F_{V,2}, F_{V,3})$ are given by

$$U = \begin{bmatrix} \overline{\rho} \\ \overline{\rho}\widetilde{u}_1 \\ \overline{\rho}\widetilde{u}_2 \\ \overline{\rho}\widetilde{u}_3 \\ \overline{\rho}\widetilde{e} \end{bmatrix}$$

$$F_{I,1} = \begin{bmatrix} \overline{\rho}\widetilde{u}_1 \\ \overline{\rho}\widetilde{u}_1\widetilde{u}_1 + \overline{p} \\ \overline{\rho}\widetilde{u}_1\widetilde{u}_2 \\ \overline{\rho}\widetilde{u}_1\widetilde{u}_3 \\ (\overline{\rho}\widetilde{e} + \overline{p})\widetilde{u}_1 \end{bmatrix} F_{I,2} = \begin{bmatrix} \overline{\rho}\widetilde{u}_2 \\ \overline{\rho}\widetilde{u}_1\widetilde{u}_2 \\ \overline{\rho}\widetilde{u}_2\widetilde{u}_2 + \overline{p} \\ \overline{\rho}\widetilde{u}_2\widetilde{u}_3 \\ (\overline{\rho}\widetilde{e} + \overline{p})\widetilde{u}_2 \end{bmatrix} F_{I,3} = \begin{bmatrix} \overline{\rho}\widetilde{u}_3 \\ \overline{\rho}\widetilde{u}_1\widetilde{u}_3 \\ \overline{\rho}\widetilde{u}_2\widetilde{u}_3 \\ \overline{\rho}\widetilde{u}_2\widetilde{u}_3 + \overline{p} \\ (\overline{\rho}\widetilde{e} + \overline{p})\widetilde{u}_3 \end{bmatrix}$$

$$F_{V,1} = \begin{bmatrix} 0 \\ \sigma_{11} \\ \sigma_{12} \\ \sigma_{13} \\ \sigma_{1k}\widetilde{u}_k - \widetilde{q}_1 \end{bmatrix} F_{V,2} = \begin{bmatrix} 0 \\ \sigma_{12} \\ \sigma_{22} \\ \sigma_{23} \\ \sigma_{2k}\widetilde{u}_k - \widetilde{q}_2 \end{bmatrix} F_{V,3} = \begin{bmatrix} 0 \\ \sigma_{13} \\ \sigma_{23} \\ \sigma_{33} \\ \sigma_{jk}\widetilde{u}_k - \widetilde{q}_3 \end{bmatrix}$$

$$\sigma_{ij} = 2\mu S_{ij}^* + \tau_{ij},\ S_{ij}^* = S_{ij} - \frac{1}{3}\frac{\partial \widetilde{u}_k}{\partial x_k}\delta_{ij},\ S_{ij} = \frac{1}{2}\left(\frac{\partial \widetilde{u}_i}{\partial x_j} + \frac{\partial \widetilde{u}_j}{\partial x_i}\right),\ K = \frac{1}{2}\widetilde{u_k' u_k'}$$

$$\widetilde{q}_j = -\left(k + C_p\frac{\mu_t}{\Pr_t}\right)\frac{\partial \widetilde{T}}{\partial x_j}$$

In the above equations, $\overline{p}$ is the Reynolds-averaged or spatially-filtered pressure and is connected to density and temperature through the thermally perfect equation of state The turbulent stresses in RANS and LES both invoke the Boussinesq concept, namely,

$$\tau_{ij,RANS} = 2\mu_{t,RANS}S_{ij}^* - \frac{2}{3}\overline{\rho}k\delta_{ij}$$

$$\tau_{ij,LES} = 2\mu_{t,LES}S_{ij}^* - \frac{2}{3}\overline{\rho}k\delta_{ij}$$

For RANS, $\mu_{t,RANS}$ is modelled by the shear-stress transport (SST) model [25]. For LES, two different subgrid-scale models were used for $\mu_{t,LES}$: the dynamic kinetic energy model (DKEM) [26] and the wall-adapting local eddy-viscosity model (WALE) [27] Two subgrid models were examined because of the complexity of the unsteady shedding about pin fins, the horseshoe vortices about the bases of pin fin, and the interactions among them in the near-wall region under high heat-load conditions. In this study, the temperature-dependence of the constant pressure specific heat, dynamic viscosity, and thermal conductivity were accounted for. The turbulent thermal conductivity was modelled by connecting it to the turbulent viscosity through the turbulent Prandtl number, $\Pr_t$, which was set to 0.85 for both RANS and LES.

The boundary conditions imposed are as follows: At the inflow boundary (x = −115D), uniform mean temperature and mass flow rate were imposed for the RANS. RANS was used from x = −115D to −15D to obtain a solution that was no longer affected by the entrance region (referred to as "fully" developed flow if the flow was incompressible). LES was used for x > −15D. To get the turbulent fluctuations started and self-sustaining for LES, the synthetic turbulence generator (STG) method of Shur et al. [28], a Fourier based method, was applied at x = −15D. The input into STG was the RANS solution obtained at x= −15D. 15D was the distance between where LES started and where the test section started. That distance, 15D, was obtained by numerical experiments to ensure that the correct turbulent structures were produced before the cooling flow entered the test section, as will be explained in the section on verification and validation. The boundary condition imposed at the outflow boundary (x = 15D) was constant static pressure at P_b. At periodic boundaries (x = −1.25D and +1.25D), periodic conditions were imposed.

Solutions to the governing equations were obtained by using version 19.2 of the ANSYS Fluent code [29]. For both RANS and LES, the finite-volume method with the SIMPLE algorithm was used. For RANS, the fluxes at cell faces were interpolated by using the second-order upwind scheme, and the Poisson equation for pressure was computed by using a second-order scheme. For LES, the time derivatives were approximated by a second-order accurate in time bounded implicit scheme. The fluxes at the cell faces were interpolated by using a bounded central difference scheme, and the pressure equation was computed by using a second-order central scheme.

For RANS, only steady-state solutions were of interest. Iterations were continued until all residuals plateaued. At convergence, the scaled residuals were always less than 10^{-5} for the continuity equation and the three momentum equations, less than 10^{-7} for the energy equation and less than 10^{-5} for the turbulent transport equations. For LES, time-accurate solutions were of interest. The number of iterations needed to get a converged solution at each time step ranged from 15 to 20 once initial transients washed out. The time-step size used was obtained via a time-step-size sensitivity study described in the next section.

Although the flow was highly unsteady, the flow field was statistically stationary. Thus, all results presented on the flow field and the turbulence statistics were obtained by time-averaging the time-resolved solutions. The time-averaging started once the turbulent flow became statistically stationary and continued until the time-averaged values no longer changed. The maximum number of flow-through time needed to achieve this was nine. One cycle time was defined as L/V, where L = 30D was the length of the LES domain and V was the mean flow speed at the inlet of the LES domain. Since the length of the test section with pin fins was 12.5D, nine flow-through time for the entire LES domain was equivalent to 21.6 flow-through time for the test section. Also, since the flow speed increased along the duct because of heating, the maximum number of flow-through time for the test section based on actual flow speed was higher than 21.6.

4. Verification and Validation

To verify this study, a grid-and time-step-size sensitivity study was performed. On validation, it was conducted in two parts. Data from direct numerical simulation (DNS) was used to validate the turbulent flow predicted in the smooth part of the channel upstream of the test section, and available experimental data was used to validate the flow predicted in the test section with pin fins. The details of these studies are given below.

4.1. LES of Flow in Channel without Pin Fins

To ensure that the synthetic turbulence inflow BC imposed at x = −15D could provide the correct turbulent flow to enter into the test section with pin fins at x = 0, three grids were examined along with time-step sizes that would yield stable and accurate solutions. The three grids used, denoted as coarse, baseline, and fine, are shown in Figure 2 and summarized in Table 2. Note that h-refinement was used when finer resolution was needed next to the wall to maintain a consistent aspect ratio as the grid is refined. The coarse grid in the LES region from x = −100D to x = 0 consisted of 1.64 million cells with maximum/minimum non-dimensional cell spacings of 16/16 in the streamwise direction, 13.9/0.9 in the wall-normal direction, and 16/16 in the spanwise direction. The baseline grid consisted of 2.79 million cells with max./min. non-dimensional cell spacings of 16/8, 13.7/0.5, and 16/8 in the streamwise, wall-normal, and spanwise directions, respectively. The fine grid consists of 5.08 million cells with max./min. cell spacings of 16/4, 13.5/0.3, and 16.4/0.3 in the streamwise, wall-normal direction, and spanwise directions, respectively. Here, it is noted that the region upstream of the LES region (x = −115D to −15D), where RANS with the SST model was used to provide the "fully-developed" mean flow at x = −15D for the LES region, had a grid with 2.5 M cells. This grid provided grid-independent solutions for the RANS region, where the first cell next to the channel walls had y+ less than unity, and there were five cells within y+ of five.

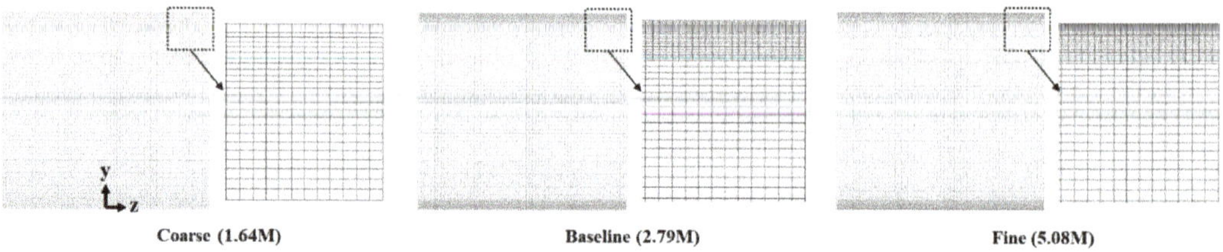

Figure 2. Grids used for LES of flow in the test section without pin fins.

Table 2. Summary of grid used for channel without pin fins.

Grid		Coarse	Baseline	Fine
Total Number of Cells (million)		1.64	2.79	5.08
Cell Spacing (max./min.)	Δx^+	≈16/16	≈16/8	≈16/4
	Δy^+	13.9/0.9	13.7/0.5	13.5/0.3
	Δz^+	≈16/16	≈16/8	≈16/4
y^+ of Cell next to Wall		0.9	0.5	0.3

Figure 3 shows the mean flow profiles normalized by the friction velocity at $Re_\tau = 365$ obtained by using the coarse, baseline, and fine grids along with results from DNS [30] and experimental measurements [31]. From this figure, the solution from the coarse grid can be seen to overpredict the DNS and experimental data in the log-law region with y+ between 10 and 100, indicating inadequate resolution in the near-wall region. Solutions from both the baseline and fine grids were in good agreement with the log-law equation and the DNS and experimental data.

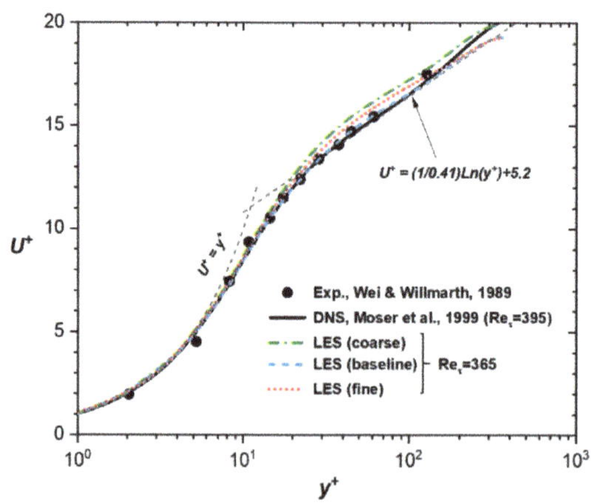

Figure 3. Mean flow profiles normalized by friction velocity at $Re_\tau = 365$.

Figure 4 shows the Reynolds stress and the streamwise and spanwise velocity fluctuations normalized by the friction velocity. The solution from the coarse grid can be seen to overpredict the DNS data, whereas solutions from the baseline and fine grids match well.

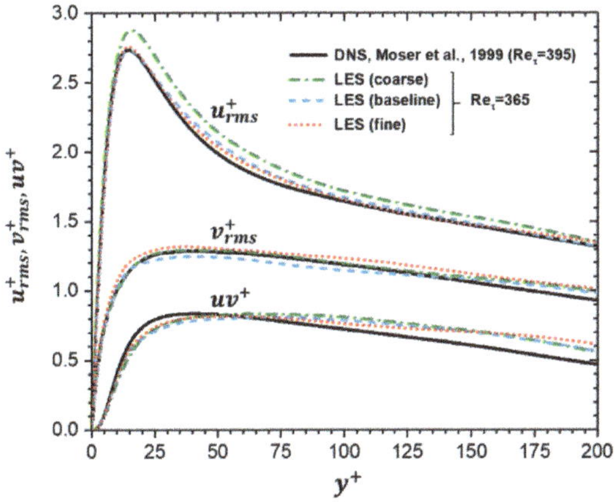

Figure 4. Mean flow fluctuations normalized by friction velocity at Re_τ = 365.

Figure 5 shows the power spectrum of turbulent kinetic energy at (x, y) = (56.25H, 0.5D) in the "fully-developed" region of the channel obtained by the baseline grid. The spectrum in the inertial subrange follows the Kolmogorov's −5/3 energy decay slope, indicating that most of the energetic large-scale eddies were resolved by the grid. According to these comparisons, the baseline grid had sufficient near-wall resolution to resolve the turbulent boundary layer and required less computation costs than the fine grid. Thus, the grid resolution based on the baseline grid was used for all LES simulations in the region between x = −15D and x = 0.

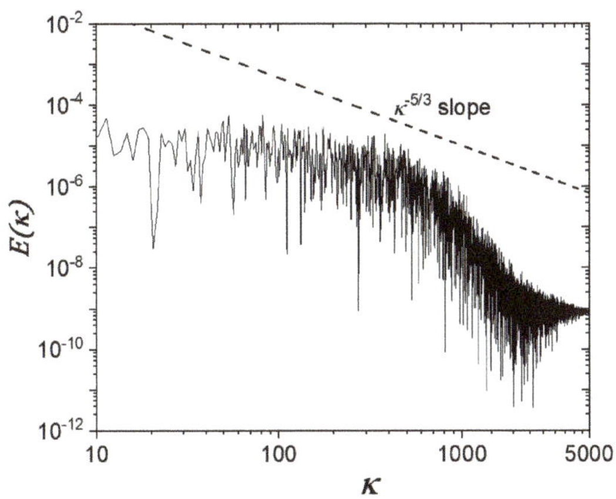

Figure 5. Energy spectrum at (x, y) = (56.25H, 0.5D) obtained by the baseline grid.

4.2. LES of Flow in Channel Flow with Pin Fins

Figure 6 shows a close-up view of the grid used in the test section with pin fins. As shown in the figure, a wrap-around grid was used about each pin fin. Also, grid points were clustered to all solid surfaces. The grid spacings used were guided by the "baseline grid" described in the previous section to ensure that the turbulent structures about all solid surfaces were adequately resolved, including those about pin fins. Table 3 summarizes the grid spacings and time-step sizes used for all the LES cases performed. Note that the grid spacings and time-step sizes had to be reduced to get the required spatial and temporal accuracy when the heating load was increased.

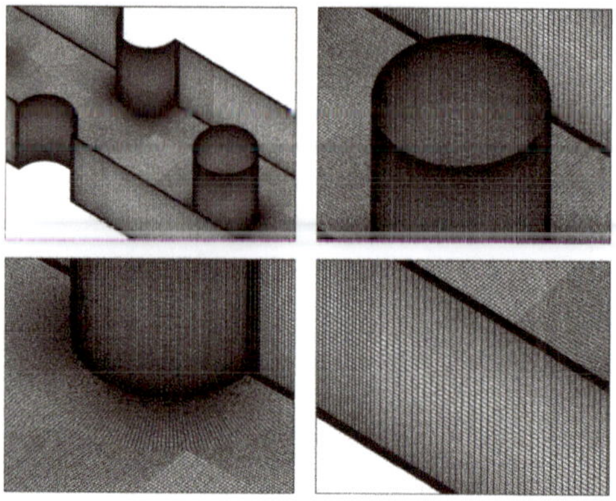

Figure 6. Grid used in the test section with pin fins.

Table 3. Summary of grid and time-step size used for channel with pin fins.

Total Number of Cells: 11.5 Million		T_w/T_c = 1.01 (Negligible Heating)	T_w/T_c = 2.0 (Moderate Heating)	T_w/T_c = 4.0 (Strong Heating)
	Δx^+	11.4	10.2	8.4
Cell Spacing (max.)	Δy^+	16.1/0.5	14.3/0.4	10.3/0.3
	Δz^+	7.6/0.5	6.8/0.4	5.1/0.3
Wall y^+ (endwall/pin-wall)		0.9/0.5	0.8/0.4	0.5/0.3
Time-step Size, Δt^+		0.1378	0.0841	0.0111

Figure 7 shows the power spectra of turbulent kinetic energy at (x/D, y/D, z/D) = (5, 1, 0) and (10, 1, 0) behind Pin 2 and Pin 4 at different heating loads. In the figure, the energy spectra can be seen to follow the Kolmogorov's -5/3 energy decay slope in the inertial subrange before their cut-off wave numbers.

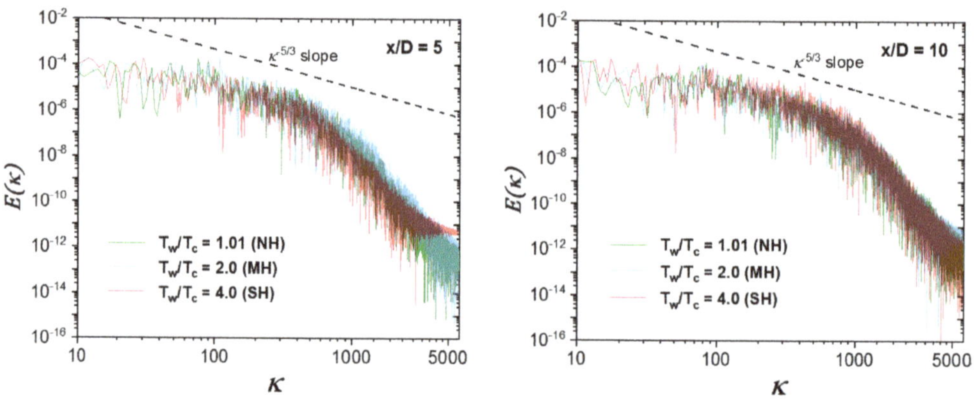

Figure 7. Energy Spectra obtained with different heating loads at x/D = 5 and 10.

To further assess the grid resolutions, the index of resolution quality for LES, $LESIQ_v$, introduced by Celik et al. [32], given by the equation below was computed:

$$LESIQ_v = \frac{1}{1 + 0.05\left(\frac{(\mu+\mu_{SGS})}{\mu}\right)^{0.53}} \tag{2}$$

Figure 8 shows the $LESIQ_v$ distributions from the LES solutions obtained under three levels of heating. Since the minimum value of $LESIQ_v$ is 0.94, which is higher than 0.8, it shows that the grids used were able to resolve a significant portion of turbulent kinetic energy.

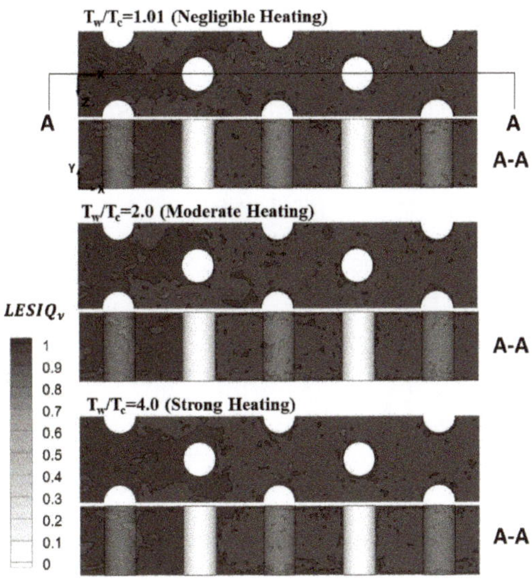

Figure 8. Celik's index of resolution quality for the LES solutions generated.

To validate the LES solutions, the results generated were compared with the experimental data of Ames & Dvorak [9–11]. Figure 9 shows a comparison between computed and measured turbulent boundary layer profiles normal to the pin-fin wall and normal to the endwall. From that figure, LES can be seen to predict the mean flow velocity well with maximum relative difference less than 2%. Note that the profile normal to the pin-wall at Row#1 does show noticeable overprediction in the outer region with about 8% relative difference, but matches well to the near-wall region.

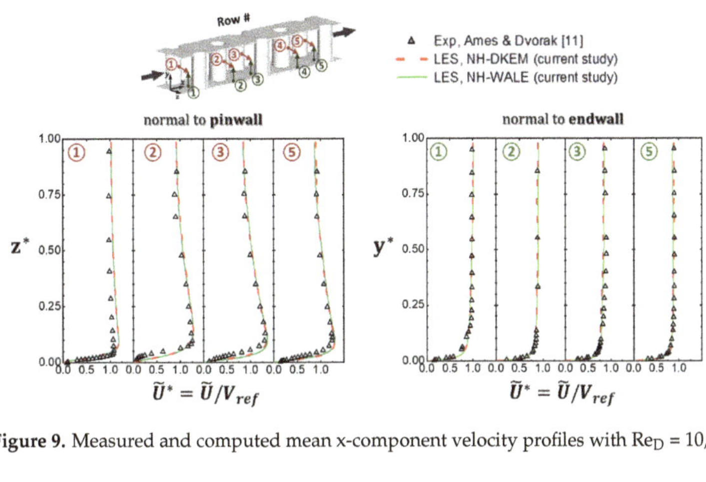

Figure 9. Measured and computed mean x-component velocity profiles with Re_D = 10,000.

4.3. DKEM vs. WALE Models

DKEM sub-grid scale model was used in the grid-sensitivity and validation study described in the previous section. In this section, DKEM is compared with the wall adapting local eddy-viscosity sub-grid scale model (WALE). Figure 10 shows profiles of the mean velocity, streamwise velocity fluctuation, and velocity-temperature correlations obtained by using the DKEM and WALE SGS models. From this figure, it can be seen that both DKEM and WALE models predict similar results in all of these quantities. Also, the results obtained match the experimental data well, which gives confidence to the results obtained by LES with DKEM.

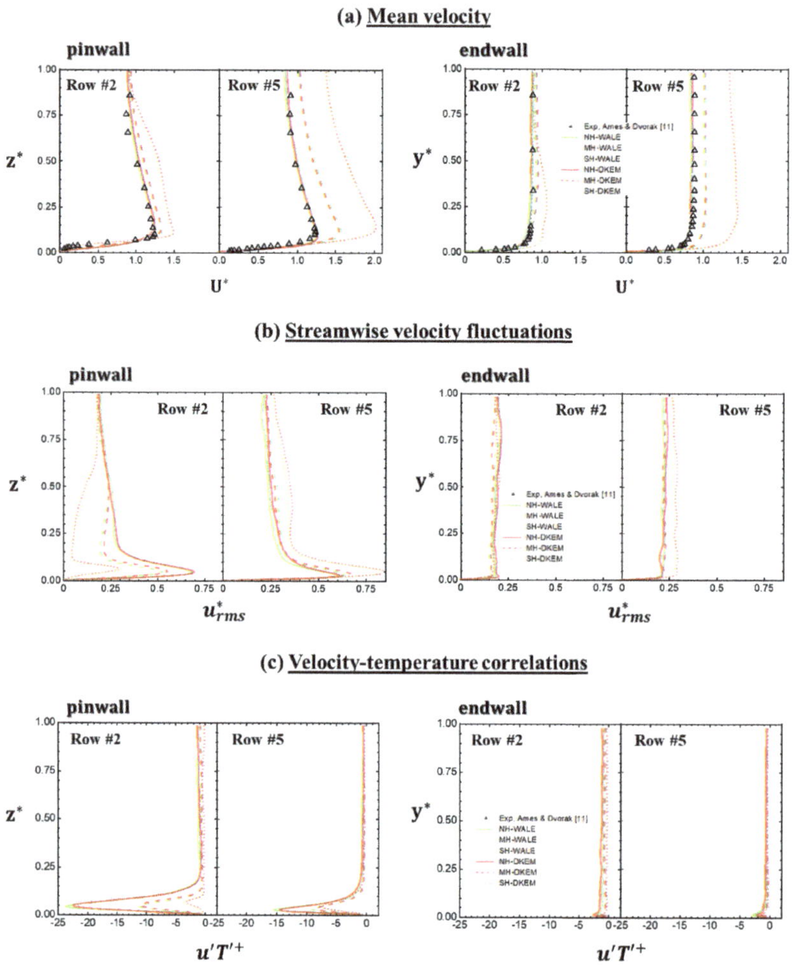

Figure 10. (**a**) Mean velocity, (**b**) streamwise velocity fluctuations, and (**c**) velocity-temperature correlations obtained by DKEM and WALE SGS models. Open circle is experimental data obtained under negligible heating.

5. Results

As noted in the introduction, the objective of this study was to examine the details of the turbulent flow created by increasing heat load from $T_W/T_c = 1.01$ to 4.0. In this section, the results are presented in the following order. First, the structure of turbulent flow induced by heat loads is described. Next, results for the mean velocity and temperature profiles and their root-mean-square of fluctuation are described. This is followed by results obtained for the turbulent kinetic energy, Reynolds stresses and measures of their anisotropy, and velocity-temperature correlations. The results obtained are plotted in two midplanes (an x-z plane at y = H/2 and an x-y plane at z = 0) and along several coordinate lines emanating from the endwall and from pin fins with row numbers 1 to 5 designating x/D = 1.25, 3.75, 6.25, 9.75, and 11.25 locations, respectively.

5.1. Flow Structures

Figure 11 shows the turbulent structures in the cooling channel studied by using iso-surfaces of the velocity gradient tensor based on the λ_2-criterion [33] colored by the magnitude of the mean velocity. The turbulent structures shown in this figure are dominated by three flow mechanisms and their interactions, and they are as follows: (1) horseshoe vortices about each pin fin; (2) jet impingement on the leading edge of each pin fin; and (3) the recirculating flows in the wake of each pin fin due to flow separation. Of these, the recirculating flows in the wake were found to be the most affected by heat load. At all heat loads, recirculating flows about pin fins in the first row did not shed. At low heat loads, those recirculating flows shed for pin fins in the second and higher rows. However, at high heat loads, those recirculating flows did not shed until the third row of pin fins. This is because when the heat load was high, the temperature of the air in the recirculating flows about second-row pin fins was significantly higher than the temperature of the air that flowed around them. With higher temperature, density is lower. As a result, even though the magnitudes of the velocity of the recirculating flows and the flow around them were similar where they interacted, the momentum of the recirculating flows was significantly less and so could not shed. When recirculating flows did shed, the vortical structures created were large. At low heat loads, the large vortical structures broke up into clusters of smaller-scale structures by the third and fourth pin fins. However, at high heat loads, this breakup did not occur until the fifth-row pin fins. High heat loads delayed breakup because they reduced density, which increased velocity magnitude and acceleration, and acceleration reduced turbulence. Thus, high heat loads reduced mixing and heat transfer by delaying the shedding and breakup of the large-scale vortical structures. Additional details of the flow field can be found in Lee & Shih [23].

Figure 12 shows the mean streamwise velocity contours and profiles (u^*) as a function of heat load. From this figure, it can be seen that the magnitude of the streamwise velocity increased as the heat load increased. This is because increasing the heat load increased the cooling flow's bulk temperature, which reduced its density and hence increased its velocity. Though the velocity magnitude of the flow increased with heat load, the velocity gradient next to the wall decreased as heat load increases. This is because heating causes expansion. This expansion also created jet-like flows about each pin fin, and the strength of the jet increased with heat load.

Figure 13 shows the mean temperature distributions (T^*) along the cooling channel as a function of heat load. From this figure, higher heating loads can be seen to lead to not just higher temperature in the flow but also higher temperature gradients in the thermal boundary layers. In addition, because of the expansion of the cooling air next to the walls, the thicknesses of the thermal boundary layer increased with increases in heat load.

Figure 14 shows the velocity fluctuations—streamwise, wall normal, and spanwise—at three heat loads. From Figure 14a, it can be seen that the streamwise fluctuating velocity (u^*_{rms}) about the pin fin decreased as the heat load increased for the first two rows of pin fins. This is because the flow there was highly accelerated by the expansion of the cooling air created by heating, and acceleration dampens turbulence. However, once shedding started about the pin fins in the third row, streamwise fluctuating velocity about the pin fin increased as the heat load increased. In the region midway between pin fins, the effect of heat loads was less because most of the acceleration created by the heating occurred near the pin fins. Figure 14b shows that heating of the endwall increased velocity fluctuation normal (v^*_{rms}) to the wall as the heat load increased. This was caused by the expansion of the air and jet-like flow created by the heating. However, like streamwise velocity fluctuation, normal velocity fluctuation about pin fins in the second row decreased as the heat load increased, but increased with the heat load once shedding started. Figure 14c shows that the spanwise velocity fluctuations (w^*_{rms}) had the same trend as the normal velocity fluctuations (v^*_{rms}).

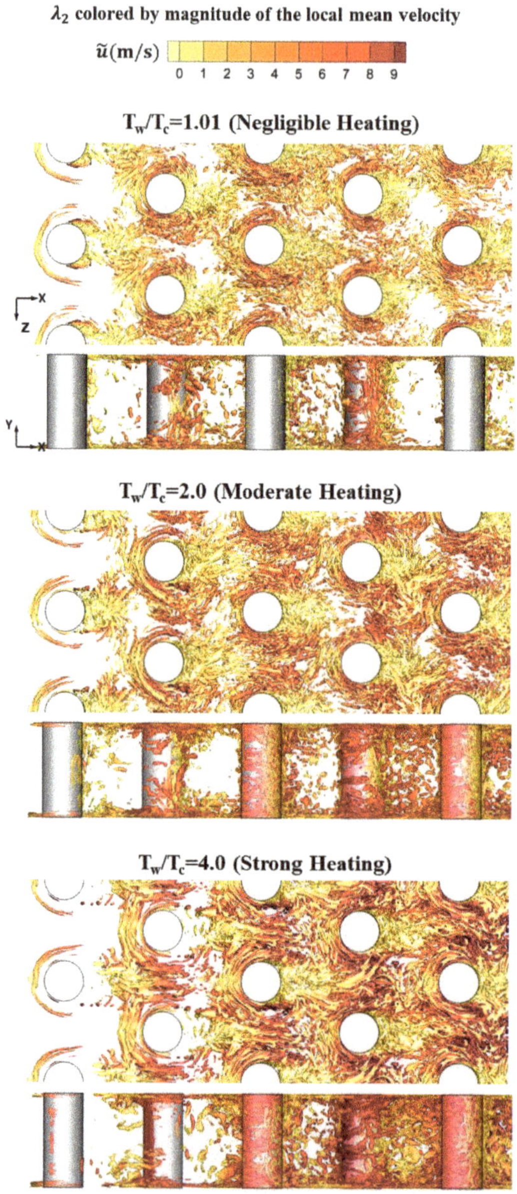

Figure 11. λ_2-criterion colored by magnitude of mean velocity.

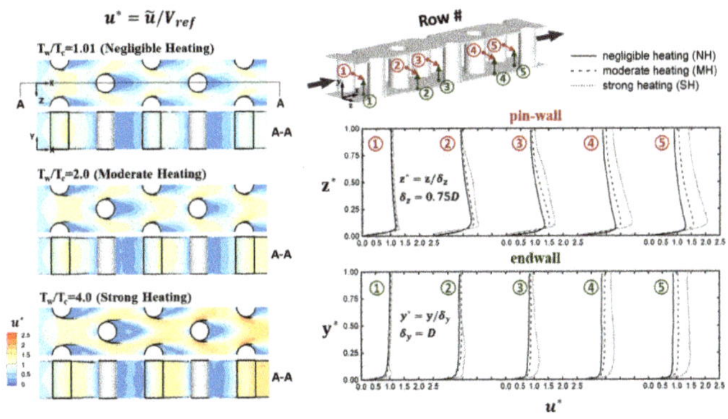

Figure 12. Mean streamwise velocity contours and profiles for T_w/T_c = 1.01 (NH), 2.0 (MH), and 4.0 (SH).

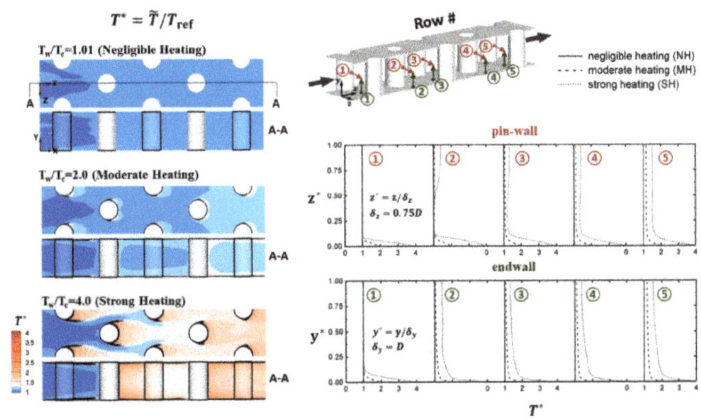

Figure 13. Mean temperature contours and profiles for T_w/T_c = 1.01 (NH), 2.0 (MH), and 4.0 (SH).

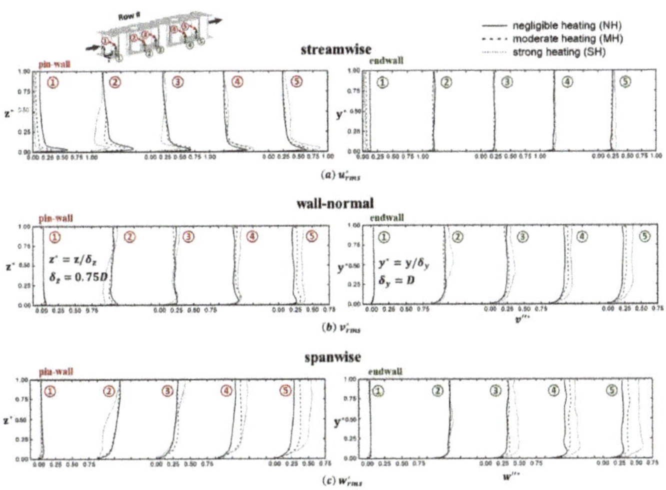

Figure 14. Root-mean square (rms) of the (**a**) streamwise, (**b**) wall normal, and (**c**) spanwise fluctuating velocities as a function of heating load.

Figure 15 shows the variance of the temperature fluctuation (T''^*). With T_w/T_c = 1.01 (NH), there was no temperature variance as expected. As T_w/T_c increased, the fluctuating temperature got higher, particularly in the near-wall region, where the mean temperature gradient was also large. For the moderate and strong heating cases, the peak temperature fluctuations next to the endwall increased steadily along the channel. After pin fins in the fourth row, the mean temperature and the normalized temperature fluctuation became periodic from row to row.

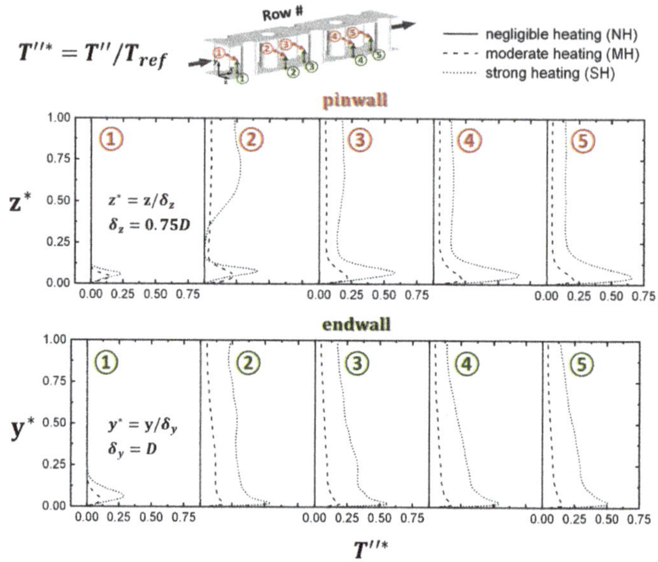

Figure 15. Temperature variance profiles for NH, MH, and SH wall heating.

5.2. Turbulent Statistics

Figure 16 shows the "total" turbulent kinetic energy normalized by the reference velocity (K^*) as a function of heat load. The "total" K is the sum of the turbulent kinetic energy resolved by LES and the "unresolved" turbulent kinetic energy modelled by the SGS. The turbulent kinetic energy was generated in regions of high mean velocity gradients–although there was a delay between the gradients and the production. Regions of high gradients were next to walls created by vortex shedding, horseshoe vortices, and the shear layer where different flow structures interacted. With the heat load at T_w/T_c = 1.01 (which is negligible heating), K was essentially periodic in the streamwise direction with the periodicity repeating from one row of pin fins to the next. With the heat load increased to T_w/T_c = 2.0, K increased along the channel, and the periodicity from one row of pin fins to the next no longer existed. With heat load further increased to T_w/T_c = 4.0, K in the first two rows decreased significantly. This is because once the heat load reaches a critical value, shedding about pin fins is suppressed. With T_w/T_c = 4.0, shedding did not start until pin fins in the third row. Once shedding did start, K with T_w/T_c = 4.0 was significantly higher than the K at lower heat loads. As noted, once there was heat load with shedding, K increased from row to row because the heat load increased the velocity along the channel. This steady increase in K along the channel explains why Lee and Shih [23] observed the row-averaged Nusselt number to be nearly constant though the Reynolds number was steadily decreasing along the channel.

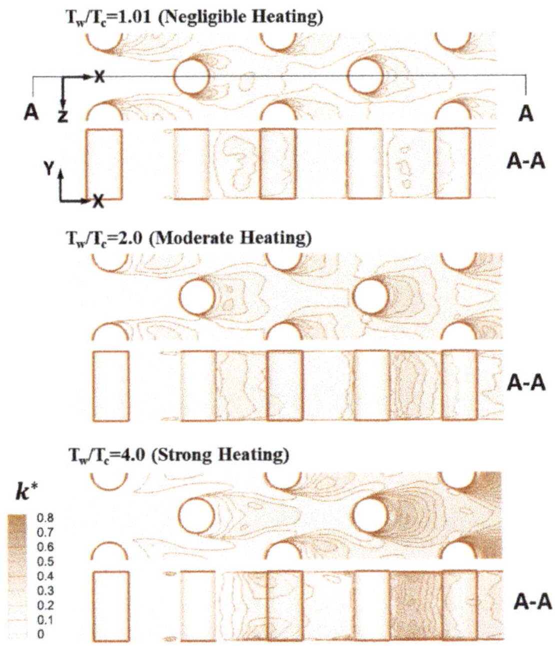

Figure 16. Total turbulent kinetic energy distribution for various wall heating levels.

Figure 17 shows the Reynolds stresses normalized by the turbulent kinetic energy $(-\widetilde{u_i'' u_j''}^*)$. From this figure, it can be seen that only $-\widetilde{u'' w''}^*$ was significantly affected by the heat load. However, this is only true at the few coordinate lines where data are shown. Figure 18 shows the distribution of the anisotropy flatness parameter at two planes for the three heating loads. The anisotropy flatness parameter, A, proposed by Lumley and Newman [34] is as follows:

$$A = 1 - \frac{9}{8}(A_2 - A_3) \tag{3}$$

$$A_2 = a_{ij}a_{ji}$$

$$A_3 = a_{ij}a_{jk}a_{ki}$$

$$\text{where } a_{ij} = \frac{\overline{u_i' u_j'}}{k} - \frac{2}{3}\delta_{ij}$$

where A = 1 implies isotropic turbulence, and A = 0 implies maximum anisotropy. From Figure 18, it can be seen that anisotropy in the Reynolds stresses was highest in the separated regions about the pin fins due to shedding of the recirculating flows and the horseshoe vortices. With T_W/T_c = 1.01, the anisotropy was highest next to the pin fins. Increasing heat load from T_W/T_c = 1.01 to 2.0 increased anisotropy throughout the flow field. However, further increase to T_W/T_c = 4.0 decreased anisotropy in the first three rows. Downstream of the third row, anisotropy did increases with the increase in heat load from T_W/T_c = 2.0 to 4.0.

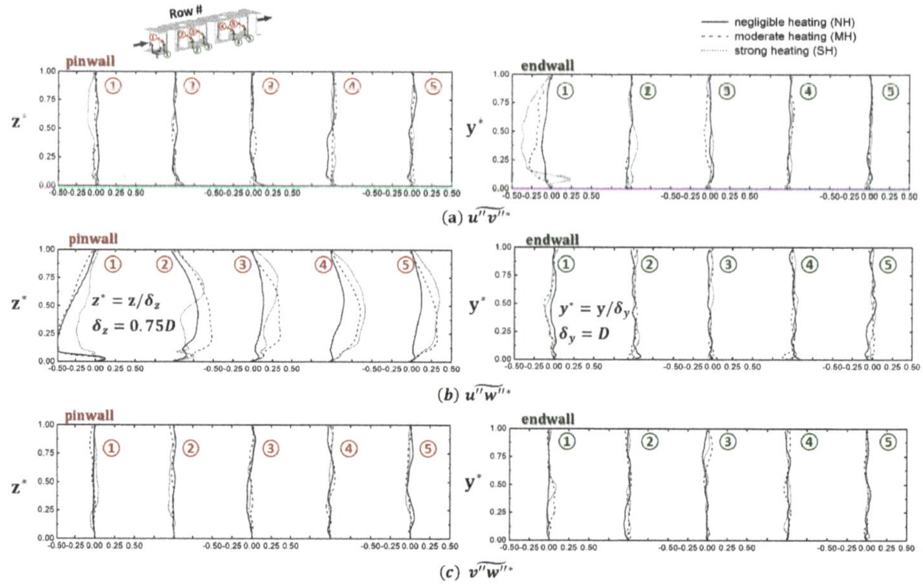

Figure 17. Reynolds stress profiles of (a) $\widetilde{u''v''}^*$, (b) $\widetilde{u''w''}^*$, (c) $\widetilde{v''w''}^*$ components for various wall heating

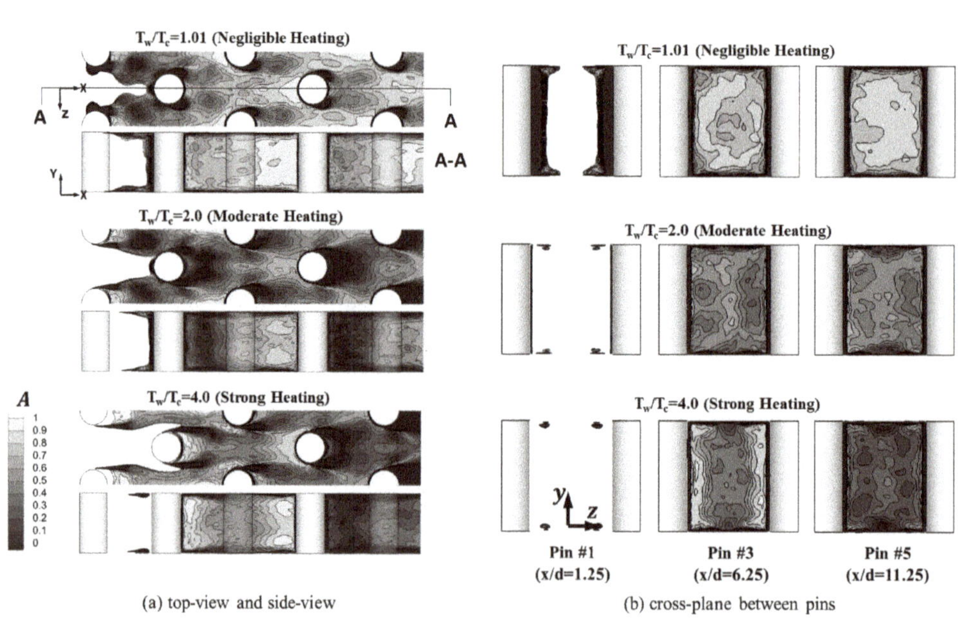

Figure 18. The distribution of the anisotropy flatness parameter, A.

Figure 19 shows the velocity-temperature correlations, which represent turbulent heat flux components, normalized by the product of the reference velocity and temperature ($-\widetilde{u''T''}^+$, $-\widetilde{v''T''}^+$, and $-\widetilde{w''T''}^+$). This figure shows the velocity-temperature correlations to be highest next to walls where shedding, horseshoe vortices, and shear-layer interactions took place. Of the correlations, $-\widetilde{u''T''}^+$ was the highest, and it is higher about pin-fin walls because of shedding and about endwalls because of horseshoe vortices, where streamwise fluctuating components played a dominant role. $-\widetilde{v''T''}^{+}$ was higher about the endwall because of the horse vortices, which had dominant fluctuations normal to the wall. $-\widetilde{w''T''}^{+}$ was higher about pin fins because of shedding, which had dominant fluctuations in the spanwise direction. Figure 20 shows the distribution of $-\widetilde{w''T''}^{*}$ in two midplanes, where the effects of shedding about pin fins can be clearly seen.

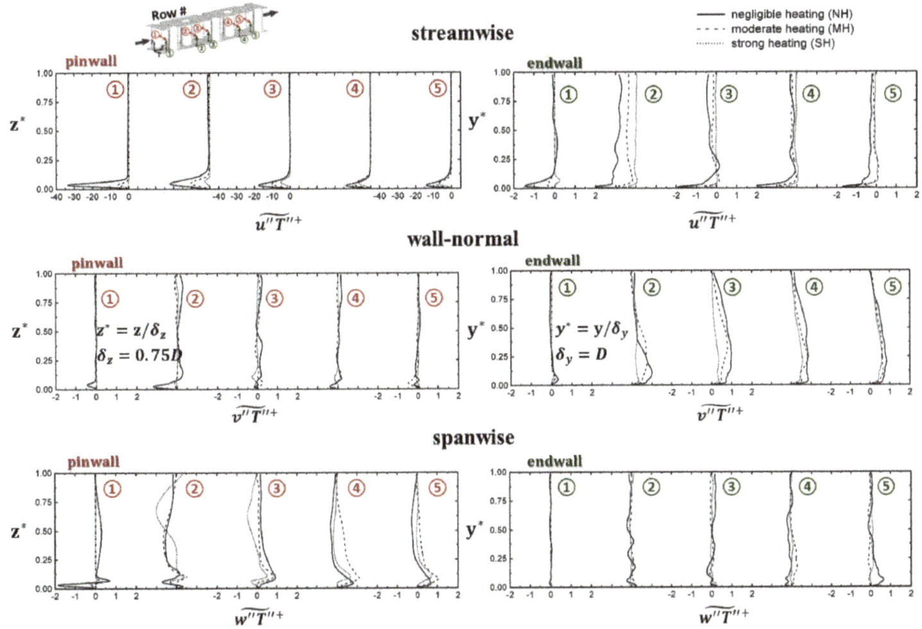

Figure 19. Resolved turbulent heat flux distributions for various heat loads.

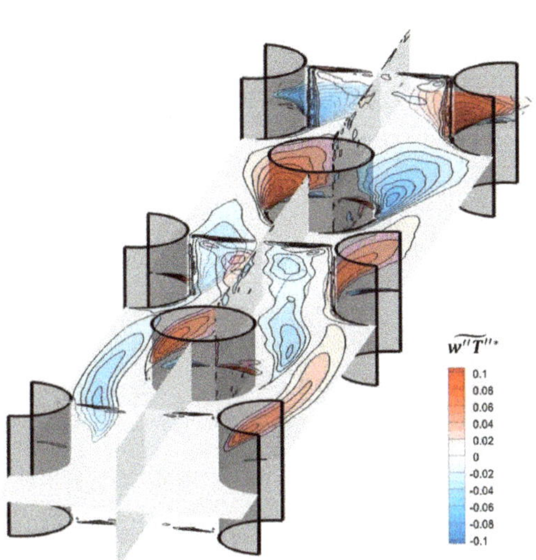

Figure 20. Resolved $-\widetilde{w''T''}^{*}$ in two midplanes ($T_w/T_c = 4.0$).

6. Conclusions

Results obtained from large-eddy simulations of the unsteady flow and heat transfer in a channel with a staggered array of pin fins show the heating load to significantly affect the nature of the flow and the statistics of the turbulence. Though the turbulent structures were dominated by horseshoe vortices at the bases of each pin fin, jet impingement on the leading edge of each pin fin, and vortex shedding about each pin fin, only vortex shedding was the most affected by heat load. Basically, once the heat load reached a critical value, it could suppress vortex shedding about pin fins and breakup of large vortical structures into clusters of smaller-scale structures, which significantly affect turbulence statistics. Velocity fluctuations in the streamwise direction were found to decrease as the heat load increased for the first two rows of pin fins because the flow there was highly accelerated by the expansion of the cooling air created by heating. However, once vortex shedding started, streamwise fluctuating velocity increased as the heat load increased. Increase heating of the endwall also increased velocity fluctuation normal to the wall. Velocity fluctuations in the normal and spanwise directions had similar trends. As expected, temperature fluctuations increased as heat load increased When there was negligible heat load, the turbulent kinetic energy was found to be essentially periodic in the streamwise direction. However, once heat load was increased, the turbulent kinetic energy increased along the channel, and periodicity broke down. When shedding about pin fins in the first few rows was suppressed by high heat loads, the turbulent kinetic energy was significantly reduced. On the Reynolds stresses, the anisotropy was highest next to pin fins if the heat load was negligible. Increasing heat load increased anisotropy throughout the flow field. However, further increase in heat load decreased anisotropy in the first three rows because shedding was suppressed. Of the velocity-temperature correlations, the streamwise-component correlation was the highest, and it was higher about pin fins because of shedding and about endwalls because of horseshoe vortices. Normal-component correlation was higher about endwalls because of horseshoe vortices, and spanwise-component correlations were higher about pin fins because of vortex shedding.

Author Contributions: C.-S.L.: Methodology, Validation, Analysis, Writing—original draft. T.I.-P.S.: Conceptualization, Analysis, Writing—original draft. K.M.B.: Writing—review—editing. R.P.D.: Supervision, Writing—review—editing. R.A.D.: Conceptualization, Writing—review & editing. All authors have read and agreed to the published version of the manuscript.

Funding: This research was supported by the US Department of Energy's Ames Laboratory and the National Energy Technology Laboratory. The Contract No. is DE-AC02-07CH11358, and the Agreement No. is 26110-AMES-CMI. The support is gratefully acknowledged.

Data Availability Statement: The datasets generated and supporting the findings of this article are obtainable from the corresponding author upon reasonable request.

Conflicts of Interest: The authors declare no conflict of interest.

Nomenclature

C_f	skin friction coefficient: $C_f = \tau_w/(0.5\ \rho_f V_i^2)$
C_p	pressure coefficient: $C_p = (P - P_o)/(0.5\ \rho V_{ref}^2)$
C_{cp}	constant pressure specific heat [J/(kg·K)]
D	diameter of pin fins [m]
$E(\kappa)$	energy spectrum [m^3/s^2]
H	height of pin fin and cooling passage [m]
k	thermal conductivity [W/(m·K)]
K	turbulent kinetic energy (TKE) [m^2/s^2]
K^*	normalized turbulent kinetic energy: $K^* = K/V_{ref}^2$
$\dot{m}_c$	mass flow rate of cooling flow [kg/s]
p	static pressure [Pa]
$\overline{p}$	mean pressure [Pa]

P_b	back pressure at the exit of cooling passage [Pa]
Pr	Prandtl number
q''_w	wall heat flux $[W/m^2]$
Re_D	Reynolds number: $Re_D = \rho U_{max} D/\mu$, $\mu(T_b)$
T	temperature [K]
T_b	bulk temperature of the cooling flow [K]
T_c	bulk temperature at cooling passage's inlet [K]
T_{ref}	reference flow temperature = T_c [K]
T_w	wall temperature [K]
T_τ	friction temperature: $T_\tau = q_w/(\rho C_{cp} U_\tau)$, [K]
T^*	normalized mean temperature: $T^* = /T_{ref}$
T''	rms of temperature fluctuation [K]
$T''*$	normalized rms of temperature fluctuation $T''* = /T_{ref}$
Δt^+	dimensionless time-step size: $\Delta t^+ = \Delta t(U_\tau^2/\nu)$
U_c	velocity magnitude at the cooling passage's inlet [m/s]
U_τ	friction velocity: $(\tau_w/\rho)^{0.5}$ [m/s]
u^*	normalized mean flow velocity: $u^* = \overline{u}/V_{ref}$
$\tilde{u}$	mean streamwise (x-component) velocity [m/s]
u_{rms}	rms streamwise velocity fluctuation [m/s]
U^+	normalized streamwise velocity : $U^+ = \tilde{u}/U_\tau$
u^*_{rms}	normalized rms streamwise velocity fluctuation: $u^*_{rms} = u_{rms}/V_{ref}$
$-\widetilde{u''_i u''_j}$	Reynolds stresses $[N/m^2]$
$-\widetilde{u''_i u''_j}^*$	Reynolds stresses normalized by turbulent kinetic energy
$\widetilde{u''_i T''}$	turbulent heat flux $[W/m^2]$
$\widetilde{u''_i T''}^+$	normalized turbulent heat flux: $\widetilde{u''_i T''}^+ = \widetilde{u''_i T''}/(U_\tau T_\tau)$
$\lvert V \rvert$	velocity magnitude [m/s]
$\overline{V}_{max}$	mean streamwise velocity through the minimum flow area [m/s]
V_{ref}	reference velocity magnitude = $\overline{V}_{max@x/D=1\cdot25}$, [m/s]
v_{rms}	rms wall-normal velocity fluctuation [m/s]
v^*_{rms}	normalized rms wall-normal velocity fluctuation: $v^*_{rms} = v_{rms}/V_{ref}$
w_{rms}	rms spanwise velocity fluctuation [m/s]
w^*_{rms}	normalized rms spanwise velocity fluctuation: $w^*_{rms} = w_{rms}/V_{ref}$
x	coordinate in the streamwise direction [m]
y	coordinate in the endwall-normal direction [m]
y	normal distance from wall [m]
y^+	normalized distance from wall: $y^+ = \rho U_\tau y/\mu$
y^*	normalized coordinate in the streamwise direction: $y^* = y/\delta_y$
z	coordinate in the spanwise direction [m]
z^*	normalized coordinate in the streamwise direction: $z^* = z/\delta_z$
Greek Symbols	
δ_y	half-channel height = D [m]
δ_z	half pin-pin spanwise spacing = 0.75D [m]
κ	wavenumber
ρ	density $[kg/m^3]$
μ	dynamic viscosity [Pa·s]
ν	kinematic viscosity $[m^2/s]$
Φ	power spectrum $[m^3/s^2]$

References

1. Han, J.C.; Dutta, S.; Ekkad, S.V. *Gas Turbine Heat Transfer and Cooling Technology*; Taylor & Francis: New York, NY, USA, 2000.
2. Goldstein, R. *Heat Transfer in Gas Turbine Systems*; Annals of the New York Academy of Sciences: Hoboken, NJ, USA, 2001; Volume 934.
3. Sundén, B.; Faghri, M. Heat Transfer in Gas Turbines. WIT Press: Southampton, UK, 2001.
4. Shih, T.I.-P.; Yang, V. Turbine Aerodynamics, Heat Transfer, Materials, and Mechanics. In *Progress in Astronautics and Aeronautics*; American Institute of Aeronautics and Astronautics: Reston, VA, USA, 2014; Volume 243.

5. Metzger, D.E.; Fan, C.S.; Shepard, W.B. Pressure Loss and Heat Transfer Through Multiple Rows of Short Pins. *Heat Transf.* **1982**, *3*, 137–142.
6. Chyu, M.K.; Hsing, Y.C.; Natarajan, V.V. Convective Heat Transfer of Cubic Fin Arrays in a Narrow Channel. *ASME J. Turbomach.* **1998**, *120*, 362–367. [CrossRef]
7. Chyu, M.K.; Hsing, Y.C.; Shih, T.I.-P.; Natarajan, V. Heat Transfer Contributions of Pins and Endwall in Pin-Fin Arrays: Effects of Thermal Boundary Condition, Modelling. *ASME J. Turbomach.* **1999**, *121*, 257–263. [CrossRef]
8. Hwang, J.-J.; Lai, D.-Y.; Tsia, Y.-P. Heat Transfer and Pressure Drop in Pin-Fin Trapezoidal Ducts. *ASME J. Turbomach.* **1999**, *121*, 264–271. [CrossRef]
9. Dvorak, L.A. Turbulent Augmentation of Heat Transfer off Pin and Endwall Surfaces in a Staggered Pin Fin Array. Master's Thesis, Mechanical Engineeing. University of North Dakota, Grand Forks, ND, USA, 2004.
10. Ames, F.E.; Dvorak, L.A.; Morrow, M.J. Turbulent Augmentation of Internal Convection Over Pins in Staggered-Pin Fin Arrays. *Trans. ASME J. Turbomach.* **2005**, *127*, 183–190. [CrossRef]
11. Ames, F.E.; Dvorak, L.A. Turbulent Transport in Pin Fin Arrays: Experimental Data and Predictions. *Trans. ASME J. Turbomach.* **2006**, *128*, 71–81. [CrossRef]
12. Delibra, G.; Borello, D.; Hanjalic, K.; Rispoli, F. URANS of Flow and Endwall Heat Transfer in a Pinned Passage Relevant to Gas-Turbine Blade Cooling. *Int. J. Heat Mass Transfer* **2009**, *30*, 549–560. [CrossRef]
13. Delibra, G.; Borello, D.; Hanjalic, K.; Rispoli, F. LES of flow and heat transfer in a channel with a staggered cylindrical pin matrix. In *Direct and Large-Eddy Simulation VII—Proceedings of 7th Inter. ERCOFTAC Workshop on Direct and Large-Eddy Simulation University of Trieste, Trieste, Italy, 8–10 September 2008*; Armenio, V., Geurts, B., Fröhlich, J., Eds.; ERCOFTAC Serie; Springer Heidelberg/Berlin, Germany, 2008; Volume 13.
14. Delibra, G.; Hanjalic, K.; Borello, D.; Rispoli, F. Vortex structures and heat transfer in a wall-bounded pin matrix: LES with a RANS wall-treatment. *Int. J. Heat Fluid Flow* **2010**, *31*, 740–753. [CrossRef]
15. Chi, X.; Shih, T.I.-P.; Bryden, K.M.; Siw, S.; Chyu, M.K.; Ames, R.; Dennis, R.A. Effects of Pin- Fin Height on Flow and Heat Transfer in a Rectangular Duct. In *Turbo Expo: Power for Land, Sea, and Air*; ASME: Vancouver, BC, Canada, 2011; pp. 1435–1445.
16. Ostanek, J.K.; Thole, K.A. Wake Development in Staggered Short Cylinder Arrays within a Channel. *Exp. Fluids* **2012**, *53*, 673–697. [CrossRef]
17. Pandit, J.; Thompson, M.; Ekkad, S.V.; Huxtable, S.T. Effects of Pin Fin to Channel Height Ratio and Pin Fin Geometry on Heat Transfer Performance for Flow in Rectangular Channels. *Int. J. Heat Mass Transf.* **2014**, *77*, 359–368. [CrossRef]
18. Paniagua, L.; Lehmkuhl, O.; Oliet, C.; Pérez-Segarra, C.D. Large Eddy Simulations (LES) on the Flow and Heat Transfer in a Wall-Bounded Pin Matrix. *Numer. Heat Transf. Part B Fundam.* **2014**, *65*, 103–128. [CrossRef]
19. Li, W.; Ren, J.; Hongde, J.; Luan, Y.; Ligrani, P. Assessment of Six Turbulence Models for Modeling and Predicting Narrow Passage Flows, Part 2: Pin Fin Arrays. *Numer. Heat Transf. Part A* **2016**, *69*, 445–463. [CrossRef]
20. Shih, T.I.-P.; Lee, C.-S.; Bryden, K.M. *Scaling Heat-Transfer Coefficients Measured Under Laboratory Conditions to Engine Conditions*; ASME: Vancouver, BC, Canada, 2017; p. 64039.
21. Lee, C.-S.; Shih, T.I.-P.; Bryden, K.M.; Dennis, R.A. *Effects of High Heating Loads on Unsteady Flow and Heat Transfer in a Cooling Passage with a Staggered Array of Pin Fins*; ASME: Vancouver, BC, Canada, 2019; p. 90567.
22. Lee, C.-S.; Bryden, K.M.; Shih, T.I.-P. *Strongly Heated Turbulent Boundary Layer in a Channel with Staggered Pin Fins*; ASME: Vancouver, BC, Canada, 2019; p. 4180.
23. Lee, C.-S.; Shih, T.I.-P. Effects of Heat Loads on Flow and Heat Transfer in the Entrance Region of a Cooling Duct with a Staggered Array of Pin Fins. *Int. J. Heat Mass Transf.* **2021**, *175*, 121302. [CrossRef]
24. Zhang, W.; Shih, T.I.-P. Hybrid LES Method with Adaptive Downstream Anisotropic Eddy Viscosity Model. *Int. J. Numer. Methods Fluids* **2022**, *94*, 1–20. [CrossRef]
25. Smirnov, P.E.; Menter, F.R. Sensitization of the SST Turbulence Model to Rotation and Curvature by Applying the Spalart-Shur Correction Term. *ASME J. Turbomach.* **2009**, *131*, 041010. [CrossRef]
26. Kim, S.-E. Large Eddy Simulation Using Unstructured Based Finite-Volume Solver; AIAA 2004-2548, 34th AIAA Fluid Dynamics Conference and Exhibit. Portland, OR, USA, 28 June–1 July 2004.
27. Nicoud, F.; Ducros, F. Subgrid-Scale Stress Modelling Based on the Square of the Velocity Gradient Tensor. *Flow Turbul. Combust.* **1999**, *62*, 183–200. [CrossRef]
28. Shur, M.L.; Spalart, P.R.; Strelets, M.K.; Travin, A.K. Synthetic Turbulence Generators for RANS-LES Interfaces in Zonal Simulations of Aerodynamic and Aeroacoustic Problems. *Flow Turbul. Combust.* **2014**, *93*, 63–92. [CrossRef]
29. ANSYS FLUENT Computational Fluid Dynamic Code, Ver.19.2. Available online: https://www.ansys.com/en-in/products/fluids/ansys-fluent (accessed on 1 January 2023).
30. Moser, D.; Kim, J.; Mansour, N.N. Direct numerical simulation of turbulent channel flow up to Reτ=590. *Phys. Fluids* **1999**, *11*, 943–945. [CrossRef]
31. Wei, T.; Willmarth, W.W. Reynolds-number effects on the structure of a turbulent channel flow. *J. Fluid Mech.* **1989**, *204*, 57–95. [CrossRef]
32. Celik, I.; Cehreli, Z.N.; Yavuz, I. Index of Resolution Quality for Large Eddy Simulations. *ASME J. Fluids Eng.* **2005**, *127*, 949–958. [CrossRef]

33. Jeong, J.; Hussain, F. On the Identification of a Vortex. *J. Fluid Mech.* **1995**, *285*, 69–94. [CrossRef]
34. Lumley, J.L.; Newman, G.R. The return to isotropy of homogeneous turbulence. *J. Fluid Mech.* **1977**, *82*, 161–178. [CrossRef]

Disclaimer/Publisher's Note: The statements, opinions and data contained in all publications are solely those of the individual author(s) and contributor(s) and not of MDPI and/or the editor(s). MDPI and/or the editor(s) disclaim responsibility for any injury to people or property resulting from any ideas, methods, instructions or products referred to in the content.

energies

MDPI

Article

Identification of Transient Steam Temperature at the Inlet of the Pipeline Based on the Measured Steam Temperature at the Pipeline Outlet

Karol Kaczmarski

Department of Thermal Processes Air Protection and Waste Utilization, Cracow University of Technology, 31155 Cracow, Poland; karol.kaczmarski@pk.edu.pl

Abstract: A solution to the inverse heat transfer problem (IHP) occurring in steam pipelines is presented in the paper. The transient steam temperature at the pipeline inlet was determined from the steam temperature measured at the pipeline outlet. Temporary changes of steam temperature at the turbine inlet are set by the turbine manufacturer and result from the conditions of safe starting of the turbine and maintaining high durability of its components. The boiler start-up should be carried out so that the time-temperature changes at the boiler outlet equal the time-temperature changes determined using the inverse problem. In this paper, the inverse problem of heat transfer in the pipeline was solved by the finite volume method using data smoothing, future times steps, and Tikhonov regularization that stabilized the solution of the inverse problem. The determined transient steam temperature at the pipeline inlet was compared with the measured temperatures. The steam temperature at the inlet to the pipeline, which is the solution to the inverse problem, agrees very well with the measured temperature, as the absolute value of the relative difference ε_T between measured and calculated temperature is between 0.045% and 0.3%, and the root mean square error RMSE is within the range of 0.038 K to 0.322 K.

Keywords: inverse heat conduction problem; numerical modelling; steam pipeline

Citation: Kaczmarski, K. Identification of Transient Steam Temperature at the Inlet of the Pipeline Based on the Measured Steam Temperature at the Pipeline Outlet. *Energies* **2022**, *15*, 5804. https://doi.org/10.3390/en15165804

Academic Editors: Artur Bartosik and Dariusz Asendrych

Received: 21 July 2022
Accepted: 8 August 2022
Published: 10 August 2022

Publisher's Note: MDPI stays neutral with regard to jurisdictional claims in published maps and institutional affiliations.

Copyright: © 2022 by the author. Licensee MDPI, Basel, Switzerland. This article is an open access article distributed under the terms and conditions of the Creative Commons Attribution (CC BY) license (https://creativecommons.org/licenses/by/4.0/).

1. Introduction

In inverse problems, the boundary conditions are identified, or the fluid physical properties are determined based on the measured responses of the system.

Typically, inverse heat conduction problems (IHCP) are solved, and much less attention is paid to inverse convective heat transfer issues in the literature.

A method for solving non-linear inverse heat conduction using the space marching method is presented in [1]. The internal surface temperature of the cylindrical element and the heat flux was determined from the measured wall temperature using a temperature sensor near the inner surface of the element. The resulting temperature distribution across the component wall was used to calculate the thermal stresses at the inner surface. The excellent accuracy of the method presented in [1] was achieved by eliminating random disturbances of measured wall temperature and its first-order derivative using moving digital filters.

Inverse problems are often solved for unsteady heat conduction [2,3]. The paper [2] presents the solution to the IHCP using the Trefftz method. The authors presented two methods for solving IHCPs. The conjugate gradient method with the Tikhonov regularization method was used to stabilize the inverse solution when measured temperatures were perturbed with random errors. They showed that the regularization results in a shorter computation time, while methods using iterations do not always lead to convergent solutions. The paper [2] shows that Trefftz numerical functions can be used to solve non-linear IHCP.

An algorithm is presented in [3] for the solution of an IHCP: for determining the steady-state distribution of the heat transfer coefficient (HTC) on one surface of a slab wall from the known temperature distribution in a plane inside the slab. It was assumed that the thermal boundary conditions on the other wall surfaces were known.

The algorithm is based on the finite volume discretization of the slab and on the formulation and the subsequent inversion of square matrices linking the wall surface temperature and heat flux to that of measured temperature at the inner plane [3].

The paper [4] presents a sequential gradient-based method for non-linear one-dimensional heat conduction. A quasi-Newton update strategy was used to determine the transient HTC on the solid surface. Compared to the traditional sequential conjugate gradient method, the proposed method gave more accurate, reliable, and stable results.

There are considerable stability problems with the solutions of the IHCPs, which are very sensitive to random temperature measurement errors. The paper [5] developed a hybrid algorithm for selecting regularization parameters which give low error variances for estimated parameters. Consequently, the algorithm can reduce the total error and provide better stability for the IHCP solution.

Determining the unsteady temperature of flowing fluid at high pressure from thermometer indications is also an inverse problem [6,7]. Accurate fluid temperature measurement is critical since correct identification of thermal stresses depends heavily on it.

A new method for determining the fluid temperature based on measuring the temperature of the pipeline wall near its inner surface and the readings of a thermometer is presented in [6]. The thermometer for measuring the temperature of a flowing fluid has the form of a solid cylinder in the axis of which the temperature is measured using a thin thermocouple. The fluid temperature determined from the pipeline wall temperature measurement and the temperature determined from the thermometer readings should be equal. From this temperature equality condition, the correlation to Nusselt's number was determined on the outer surface of a thermometer transversely swept by steam or another fluid.

Jaremkiewicz et al. [7] present a method for measuring the unsteady steam temperature based on a new fast response thermometer. The proposed thermometer can be adapted to a wide range of temperatures and steam pressures by optimum design and suitable materials for the thermometer housing. This paper [7] demonstrates the effect of temperature measurement accuracy on the values of stresses calculated in pressure components.

The permissible time variations of fluid temperature and pressure in pipelines and other pressure components can be determined using the procedure outlined in [8], which is based on the European Standard [9] for calculating allowable heating and cooling rates for thick-walled components. The fluid temperature was determined from the solution of the first-order ordinary differential equation for time, considering that the allowable rate of fluid temperature change is a linear function of fluid pressure.

An essential issue of the flexibility of thermal power blocks is to optimize pipeline heating and cooling so that the sum of circumferential thermal and pressure stress at the edge of the pipeline opening is less than the allowable stress. The optimum time variations of the fluid temperature are determined using the solution to the IHP [10]. The heating time of the cylindrical element from the initial temperature to the given final temperature is about 40% shorter than the heating time determined using the European Standard [9].

The interesting IHP was solved in [11]. The rotational speed of the fan forcing air through the heat exchanger was sequentially determined so that the calculated water temperature at the heat exchanger outlet was equal to the set temperature. The golden section method was used to solve the non-linear IHP. Three methods were used to stabilize the determined fan speed: the Tikhonov regularization method, Beck's future time steps and smoothing of the measured outlet water temperature using a moving averaging filter.

IHPs also occur in identifying leakages in various types of aboveground and underground pipelines. Leakages are usually identified by measuring the flow and thermal parameters of the flowing fluid or medium around the pipeline.

A review and comparative study of computer-based methods for pipeline leak detection were carried out in [12].

A method for leak detection in buried pipelines based on measurements of the temperature and moisture of the soil was developed in [13]. A CFD (Computational Fluid Dynamics) modelling was used to validate the proposed method.

During start-up, shut-down, and load changes, thick-walled steam boiler elements operating under high pressure and high-temperature conditions cause high thermal stresses. In thick walls, significant temperature differences occur during transient operation, accompanied by the formation of high thermal stresses. The largest, in value, stresses and deformations usually occur at the edges of openings. Circumferential stresses at the hole's edge of varying signs, compressive during heating and tensile during cooling, causing fatigue cracking. The stresses lead to low-cycle cracking, failure, and accelerated degradation of the block components [14]. For this reason, thick-walled boiler elements limit the maximum heating and cooling rates during the start-up or shut-down of the boiler.

The heating of the steam pipeline connecting the boiler to the turbine is essential for the start-up of the boiler and turbine [15]. The pipeline design, internal and external diameters, length, and material for a steam boiler type are different and depend on the operating parameters of the steam. The working fluid, i.e., superheated steam fed to the turbine, must have appropriate parameters (temperature and pressure). Moreover, the large wall pipeline thickness and the length pipeline influence the fluid temperature drop. Significant changes in the working fluid parameters substantially affect the lifetime of turbine components. It is not only the turbine rotor at risk but also the turbine casing as thick-walled components.

The literature on modelling thermal-flow phenomena in pipelines is scarce despite its very high practical relevance.

It is difficult to find information in the literature on modelling steam pipelines' transient operation. The steam and pipeline wall temperatures at the given boundary and initial conditions were determined numerically in [16]. The direct heat transfer problem was solved using the finite volume method.

Flow and thermal phenomena in superheater tubes are much more frequently analyzed [17–21]. Due to the small wall thickness of superheater tubes, little attention is paid to the temperature distribution over the wall thickness. The wall thickness of the pipeline is much greater. There are often holes in the walls of the pipelines with a high concentration of stresses at their edges. A high concentration of thermal stresses also occurs in Y- or T-shaped tees in the pipelines. For this reason, pipelines' flow-thermal and strength analysis is highly important for their safe long-term operation.

Two types of models analyze heat transfer in pipelines and tube heat exchangers. The first is a model with distributed parameters, in which the system of partial differential equations is solved to determine the fluid and wall parameters [21,22]. The second model is a model with concentrated parameters, described by the system of ordinary differential equations [23]. The solution to the IHCP can be used to determine the temperature distribution in the wall of pressure thick-walled elements [24–26]. Solving inverse problems is very sensitive to random measurement errors [24,25]. Therefore, the measured time changes in the temperature are approximated by an appropriate function, or digital filters are used to eliminate random errors from the input data [18,27].

This paper developed a new numerical model of the steam pipeline. The pipeline connects the boiler to the turbine. The steam turbine works at specific input parameters of superheated steam. To ensure its safe and trouble-free start-up operation, the changes in time of the input steam parameters cannot be rapid, and the steam temperature cannot differ more from the rotor and turbine casing temperature compared to allowed values. Too rapid a steam temperature change may cause high stresses in the pipeline and the turbine's structural components. The steam temperature at the outlet of the pipeline connecting the boiler and the turbine, i.e., before the turbine, depends on the steam temperature at the steam pipeline inlet.

In this work, the new IHP was solved to determine the time variation of temperature at the pipeline inlet with a known temperature at the turbine inlet. The time variation of the temperature at the turbine inlet is due to the conditions of safe turbine start-up or the steam temperature before the turbine is known from measurements. To the author's knowledge, the IHP solved in this paper has not yet been analyzed in the available literature.

2. Numerical Model

A scheme of the steam pipeline connecting the boiler with the turbine in a 120 MW unit is shown in Figure 1. Superheated steam from the last superheater stage flows into the outlet chambers, connected from both sides of the boiler with steam pipelines ($2\ r_{in}$ = 245 mm, s_w = 30 mm). Next, the pipelines are connected through a T-pipe to the main steam pipeline ($2\ r_{in}$ = 324 mm, s_w = 40 mm).

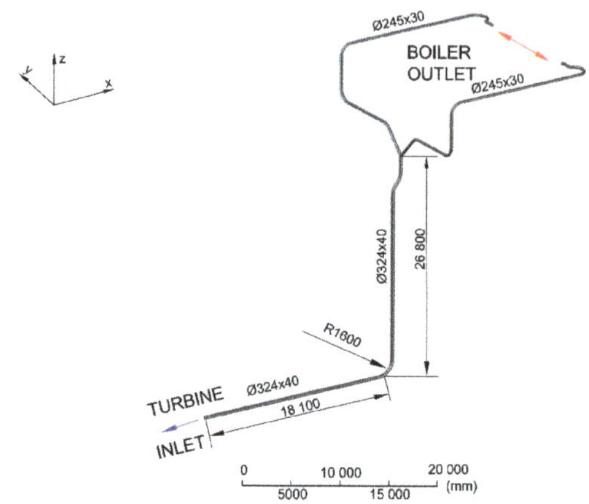

Figure 1. The pipeline connecting the boiler and turbine.

First, the analyzed domain consisting of steam pipeline and steam was divided into control volumes (Figure 2). In the radial direction, the pipeline is divided into n finite volumes, and in the longitudinal direction into m finite volumes. Only half of the pipeline was analyzed due to symmetry.

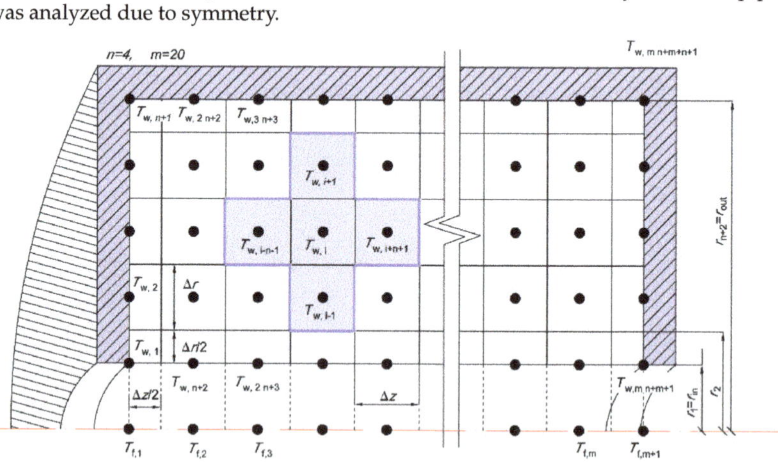

Figure 2. Division of the computational domain into finite volumes.

The equation of the transient heat transfer for the pipeline wall has the form

$$\rho_w \, c_{p,w} \frac{\partial T_w}{\partial t} = \nabla \cdot [k_w(T_w) \, \nabla T_w] \tag{1}$$

The heat balance Equation (1) in a cylindrical coordinate system is as follows

$$\rho_w \, c_{pw} \frac{\partial T_w}{\partial t} = \frac{1}{r} \frac{\partial}{\partial r} \left[r \, k_w(T_w) \frac{\partial T_w}{\partial r} \right] + \frac{\partial}{\partial z} \left[k_w(T_w) \frac{\partial T_w}{\partial z} \right] \tag{2}$$

Heat balance equations were formed for each node, including nodes in control volumes near the boundary with the steam. For example, the equation for a node i for a control volume located in the wall area is

$$\begin{aligned} \frac{dT_{w,i}}{dt} = \frac{a\left(T_{w,i}\right)}{k_w\left(T_{w,i}\right)} \Bigg[& \frac{r_j}{\Delta r} \frac{k_w\left(T_{w,i}\right) + k_w\left(T_{w,(i-1)}\right)}{r_{j+1}^2 - r_j^2} \left(T_{w,(i-1)} - T_{w,i}\right) + \frac{r_{j+1}}{\Delta r} \frac{k_w\left(T_{w,(i+1)}\right) + k_w\left(T_{w,i}\right)}{r_{j+1}^2 - r_j^2} \left(T_{w,(i+1)} - T_{w,i}\right) \\ & + \frac{k_w\left(T_{w,(i+n+1)}\right) + k_w\left(T_{w,i}\right)}{2(\Delta z)^2} \left(T_{w,(i+n+1)} - T_{w,i}\right) \\ & + \frac{k_w\left(T_{w,(i-n-1)}\right) + k_w\left(T_{w,i}\right)}{2(\Delta z)^2} \left(T_{w,(i-n-1)} - T_{w,i}\right) \Bigg] \end{aligned} \tag{3}$$

where:

$$\Delta r = \frac{r_{out} - r_{in}}{n}, \quad \Delta z = \frac{L}{m} \tag{4}$$

The symbol n denotes the number of the control volumes in radial directions in the pipeline wall. Over the length of the pipeline, we have m finite volumes of length Δz. Similarly, the heat balance equations for the steam region were formed. For example, after transformation, the heat balance equation for a node i takes the form

$$\frac{dT_{f,i+1}}{dt} = -\frac{\dot{m}}{\rho_f\left(T_{f,i}\right) A} \frac{T_{f,i+1} - T_{f,i}}{\Delta z} - \frac{h_{in}(T_{cz,i}) \, U_{in}}{A \, \rho_f\left(T_{f,i}\right) c_{p,f}\left(T_{f,i}\right)} \left[\frac{T_{f,i+1} + T_{f,i}}{2} - \frac{T_{w,i(n+1)+1} + T_{w,(i-1)\cdot(n+1)+1}}{2} \right] \tag{5}$$

After formulating the energy conservation equation for all control volumes, a system of ordinary differential equations was obtained. The Runge-Kutta method of the fourth order was used to solve the formed system of ordinary differential equations.

The following boundary and initial conditions were assumed

$$T_f \Big|_{t=0} = T_0 \tag{6}$$

$$T_w|_{t=0} = T_{w,0} \tag{7}$$

$$T_f \Big|_{z=0} = f(t) \tag{8}$$

$$k_w \frac{\partial T_w}{\partial r} \Big|_{r=r_{in}} = h_{in} \left(T_w|_{r=r_{in}} - T_f \right) \tag{9}$$

$$k_w \frac{\partial T_w}{\partial r} \Big|_{r=r_{out}} = 0 \tag{10}$$

$$k_w \frac{\partial T_w}{\partial z} \Big|_{z=0} = 0 \tag{11}$$

$$k_w \frac{\partial T_w}{\partial z} \Big|_{z=L} = 0 \tag{12}$$

The heat transfer coefficient h_{in} (Figure 2) on the inside surface of the pipeline was determined from the following formula

$$h_{in} = \frac{Nu\, k_f}{d_{in}} \tag{13}$$

The correlation proposed by Taler [28] was used to determine Nusselt numbers Nu in Equation (13)

$$Nu = Nu_{m,q}(Re = 2300) + \frac{\left(\frac{\xi}{8}\right)(Re-2300)Pr^{1.008}}{1.08+12.4\left(\frac{\xi}{8}\right)^{\left(\frac{1}{2}\right)}\left(Pr^{\left(\frac{2}{3}\right)}-1\right)}\left[1+\left(\frac{d_{in}}{L}\right)^{\left(\frac{2}{3}\right)}\right] \tag{14}$$
$$2300 < Re < 10^6, \;\; 0.1 < Pr < 1000$$

The friction factor ξ in Equation (14) is given by the correlation of Taler [29]

$$\xi = [1.2776\log(Re) - 0.406]^{-2.246} \tag{15}$$

The symbol $Nu_{m,q}(Re = 2300)$ designates the Nusselt number at $Re = 2300$ for laminar flow at the tube with constant wall heat flux [30]. At the beginning of the transitional flow i.e., the end of the laminar flow for Reynolds number $Re = 2300$, the second term on the right-hand side of Equation (14) is equal to zero.

3. Inverse Problems in Heating up the Pipeline Connecting the Boiler with the Turbine during the Start-Up of the Unit

The paper presents a numerical method to determine the steam temperature as a function of time at the pipeline inlet $T_f(t)\,|_{z=0m}$, at which the steam temperature at the pipeline outlet $T_f(t)\,|_{z=45m}$ (turbine inlet) is known $f(t)$ from the measurement. The problem formulated in this way is an inverse transient heat transfer problem. The inverse problem is much more difficult to solve than the direct one, as random errors influence the stability and accuracy of the determination of the inlet steam temperature in the measured transient steam temperature at the outlet of the pipeline.

Random measurement errors influence the solution of IHP significantly. Therefore, the measured temperature variation was approximated by a local polynomial of third-degree with respect to time. Similarly, the accuracy of approximation of the time derivatives from measured steam temperature $f(t)$ is essential. In numerical methods, for example, moving digital filters or so-called future steps are used to reduce the impact of random measurement errors.

The following assumptions were adopted in the solution of the IHP:

- the steam temperature $g(t)$ at the outlet of the pipeline is known from the measurements,
- the external surface of the pipeline as well as the cross-section of the pipeline wall at the inlet and outlet of the pipeline, are perfectly thermally insulated,
- the transient temperature field in the pipeline wall is two-dimensional, while in the steam region, it is one-dimensional, i.e., the temperature of the steam flowing through the pipeline varies only along the steam flow path,
- the pipeline inlet steam temperature is determined from the measured time variations of the pipeline outlet temperature, i.e., from the solution of IHP.

Random measurement errors from the measured temperature and its first-order derivative after time are partly eliminated using moving digital filters. The Beck future time steps [24] in solving IHCP are also applied. Future steps are an effective tool for increasing the stability of solutions to inverse problems and making it possible to determine the time variation of the inlet temperature with a smaller time step. Beck's concept of future time steps is extended in this paper. The time step of the solution of the inverse problem is several to a dozen times larger than the integration step of the system of differential equations present in the solution of the direct problem. The single time step in solving the inverse problem is divided into smaller time steps, the size of which is derived from the stability condition for the solution of the direct problem.

The temperature $T_f(t)\,|_{z=0} = T_{f1}$ of the steam at the pipeline inlet was determined sequentially (Figure 2).

The steam inlet temperature in the time interval $t_{M-1} < t < t_{M+F}$ (Figure 3) was determined using the least squares method with F future time steps. The following squared differences between the calculated steam temperature $T^{calc}_{f,m+1}(t)$, and measured fluid temperatures $T^{meas}_{f,m+1}(t)$ over the time interval $[t_{M-1}, t_{M+F}]$ must be minimum (Figure 3).

$$S\left[T^{calc}_{f,1}(t_M)\right] = \int\limits_{t_{M-1}}^{t_{M+F}} \left[T^{calc}_{f,m+1}(t) - T^{meas}_{f,m+1}(t)\right]^2 dt + w_r\left(\left.\frac{dT^{calc}_{f,1}(t)}{dt}\right|_{t=t_M}\right)^2 \rightarrow min \quad (16)$$

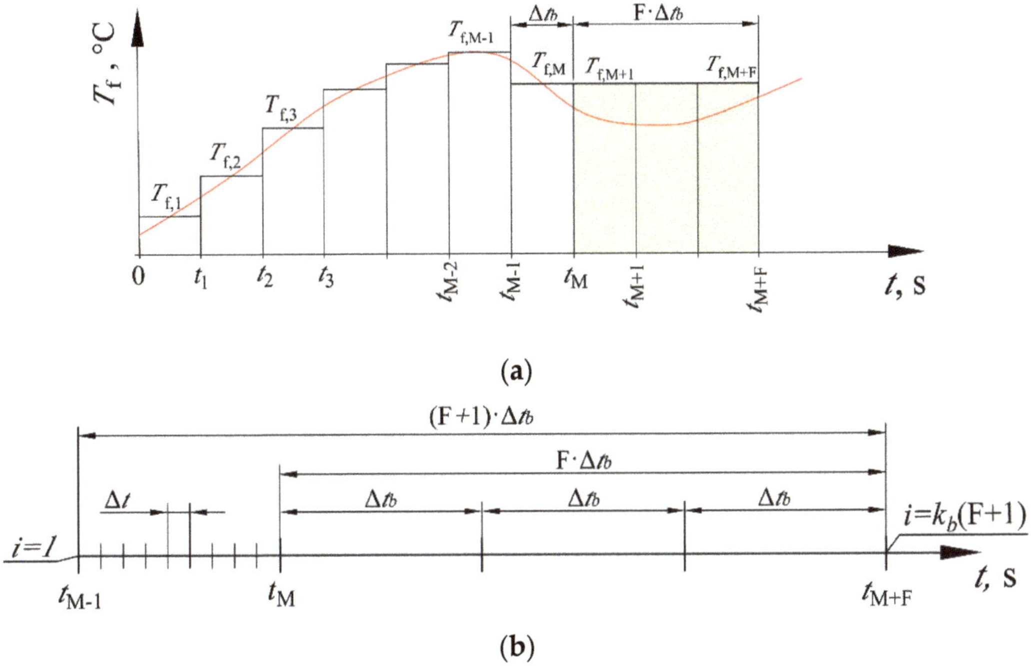

Figure 3. Approximation the time changes of the steam at the pipe inlet by stepwise curve (**a**), and the concept of future time steps, $F = 3$ (**b**).

The second term on the right-hand side of Equation (16) represents the regularization term of order one in the Tikhonov regularization [31] that is used in this paper. The inlet steam temperature $T_{f,1}(t_M)$ was determined with a basic time step equal to $\Delta t_b = t_M - t_{M-1} = k_b\,\Delta t$. The time step Δt_b is a multiple of the step Δt used to solve a direct heat transfer problem using the finite volume method. At the time $t = t_{M-1}$, the steam inlet temperature $T_{f,1}(t_{M-1})$ is known, while the steam temperature $T_{f,1}(t_M)$ at the time $t = t_M$ was sought. The time step Δt_b should be chosen so that the change of steam temperature at the pipeline inlet $T_{f,1}$ at time $t = t_{M-1}$ caused the change of steam temperature at the pipeline outlet at time $t = t_M$.

Step Δt_b is k_b times larger than step Δt used in determining the temperatures of the pipeline wall and steam from Equations (3) and (5), respectively. The step Δt must not be too large for the solution of the system of Equations (3)–(5) to be stable.

To ensure the stability of determining the wall and steam temperature should be the Fourier stability condition for the wall, and the Courant-Friedrichs-Lewy condition for the steam should be satisfied. The allowable time step Δt results from the Courant-Friedrichs-Lewy condition [32].

$$\frac{w_{z,i}\Delta t}{\Delta z} \leq 1, \qquad i = 1, \quad , m+1 \tag{17}$$

The steam velocity w_i at the i-th finite volume inlet is calculated from the following equation

$$w_{z,i} = \frac{4\,\dot{m}}{\pi\,\rho_i\,d_{in}^2} \tag{18}$$

where the symbol ρ_i denotes the steam density at the i-th finite volume inlet.

The steam temperature at the pipeline inlet was determined with the time step Δt_b based on the preset steam temperature at the pipeline outlet. The steam temperature at the pipeline outlet was calculated with a time step Δt using the mathematical model of the pipeline developed. The steam temperature $T_{f,1}^{calc}(t_M)$ was determined by minimizing the sum given by Equation (16).

$$S\left[T_{f,1}^{calc}(t_M)\right] = \sum_{i=1}^{k_b(F+1)} \left[T_{f,m+1}^{calc}(t_i) - T_{f,m+1}^{meas}(t_i)\right]^2 + w_r\left(\frac{T_{f,1}^{calc}(t_{M+F}) - T_{f,1}^{calc}(t_{M-1})}{t_{M+F} - t_{M-1}}\right)^2 \tag{19}$$

where $w_r = w_r'/\Delta t$, $t_i = t_{M-1} + i\Delta t$, $i = 1, \ldots, k_2(F+1)$.

Equation (19) is the discrete form of Equation (16), calculating the integral in Equation (16) using the rectangular method.

For too small values of basic step Δt_b, it is not possible to determine the steam temperature variation $T_{f,1}$ at the pipeline inlet in the time interval $t_{M-1} \leq t \leq t_M$ based on the set or measured outlet steam temperature variation $T_{f,m+1}^{meas}(t)$. For a too small time step Δt_b, there are instabilities in the determined pipeline inlet temperature.

For the solution stabilization, the future time interval $t_M \leq t \leq t_{M+F}$ is used (Figure 3).

By increasing the analyzed time interval from $t_{M-1} \leq t \leq t_M$ to $t_{M-1} \leq t \leq t_{M+F}$, there is a measurable change in steam temperature at the pipeline outlet $T_{f,m+1}^{calc}(t)$ as a result of the temperature change of the fluid at the pipeline inlet $T_{f,1}^{calc}(t)$ during the time t_{M-1}. After determination of the fluid temperature $T_{f,1}(t_{M+F})$ at time point t_{M+F}, it is assumed that this temperature value occurs only in the basic range $t_{M-1} \leq t \leq t_M$. The analysis at the next time step $[t_M, t_{M+1}]$ is repeated with the time t_M as the starting point and not the time point t_{M+F}.

The temperature of the fluid $T_{f,1}$ (t_M) was determined by the golden-section search method [32], for which the sum $S\left[T_{f,1}^{calc}(t_M)\right]$ given by Equation (19) attained minimum.

Figure 4 shows the block diagram of the program for the sequential steam temperature at the pipeline inlet $T_{f,1}$ using the golden section method.

First, the limits of the interval $[T_{f,MIN}, T_{f,MAX}]$ are set in which the desired inlet temperature $T_{f,1}^{calc}(t)$ lies, for which the S sum is defined by Equation (19) reaches a minimum. The steam temperature calculated values at points X_L and X_R takes values in the range $T_{f,MIN} < X_L < X_R < T_{f,MAX}$. The value of the factor k is $k \approx 0.61803398$. The golden division factor k is a constant factor that reduces the interval value at each iteration until the condition $(T_{f,MAX} - T_{f,MIN}) \leq \epsilon$ is satisfied.

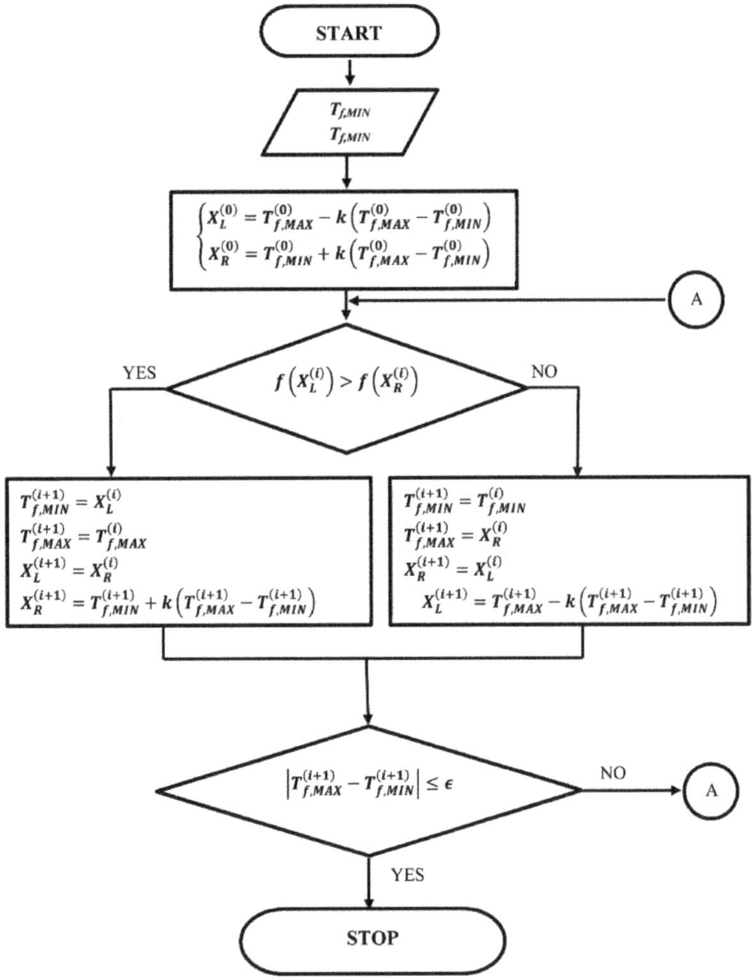

Figure 4. Block diagram of a program to determine the fluid temperature using the golden division method.

4. Thermal Stresses

The temperature distribution throughout the pipeline wall, obtained by solving the IHCP, is used to determine the thermal stresses in the whole surface and considered on boundary surfaces.

The radial, longitudinal and circumferential thermal stresses are determined, assuming that the pipeline ends can move freely. The thermal stress equations are given by [33].

$$\sigma_r = \frac{E\beta_T}{2(1-\nu)}\left(1 - \frac{r_{in}^2}{r^2}\right)\left[\overline{T}_w(t) - \overline{T}_w(r,t)\right] \tag{20}$$

$$\sigma_\varphi = \frac{E\beta_T}{2(1-\nu)}\left[\left(1 - \frac{r_{in}^2}{r^2}\right)\overline{T}_w(t) + \left(1 - \frac{r_{in}^2}{r^2}\right)\overline{T}_w(r,t) - 2T_w(r,t)\right] \tag{21}$$

$$\sigma_z = \frac{E\beta_T}{1-\nu}\left[\overline{T}_w(t) - T_w(r,t)\right] \tag{22}$$

Equations (20)–(22) giving the thermal stress components consider the radial temperature distribution in the pipeline, as the temperature drop in the pipeline wall in the direction of steam flow is minor.

In Equations (20)–(22), the symbols $\overline{T}_w(t)$ and $\overline{T}_w(r,t)$ denote the mean wall temperature that is given by the following formulas.

$$\overline{T}_w(t) = \frac{2}{r_{out}^2 - r_{in}^2} \int_{r_{in}}^{r_{out}} rT_w dr \approx \frac{2\Delta r}{r_{out}^2 - r_{in}^2} \left[r_2 \frac{T_{w_1} + T_{w_2}}{2} + r_{n+1} \frac{T_{w_n} + T_{w_{n+1}}}{2} + \sum_{i=2}^{n-1} \frac{r_i + r_{i+1}}{2} T_{w_i} \right] \tag{23}$$

$$\overline{T}_w(r,t) = \overline{T}_w(r_i,t) = \frac{2}{r^2 - r_{in}^2} \int_{r_{in}}^{r} rT_w dr \approx \frac{2\Delta r}{r_i^2 - r_{in}^2} \left[r_2 \frac{T_{w_1} + T_{w_2}}{2} + \sum_{j=2}^{i} \frac{r_j + r_{j+1}}{2} T_{w_j} \right] \tag{24}$$

where the symbols r_2, r_i, $i = 3 \ldots n+1$, and r_j, $j = 1 \ldots n+2$ denote the radiuses (Figure 2).

The radial stresses σ_r is equal to zero on the inner and outer surface of the tube $(\sigma|_{r=r_{in}} = \sigma|_{r=r_{out}} = 0)$. The circumferential σ_φ and the axial σ_z thermal stresses on these surfaces are equal $\sigma_\varphi|_{r=r_{in}} = \sigma_z|_{r=r_{in}}$, $\sigma_\varphi|_{r=r_{out}} = \sigma_z|_{r=r_{out}}$).

5. Result

The calculations were carried out for the steam pipeline with the following data r_{out} = 0.162 m, r_{in} = 0.122 m, s_w = 0.04 m, and L = 45 m. The steam tube is made of steel 14MoV63. The pipeline wall was divided into 100 control volumes. The number of control volumes in the radial direction is $n + 1 = 5$ and $m + 1 = 21$ in the axial direction (Figure 2). The pipeline wall and steam temperatures were calculated for each node lying in the center of the control volumes.

Data from the direct heat conduction problem solution were treated as "exact measurement data". Both exact and measurement data were used to carry out many simulations. Figure 5 shows the temperature variation $T_{f,m+1}^{meas}(t)$, pressure p, and steam mass flow rate $\dot{m}$ as functions of time t used in the first calculation test.

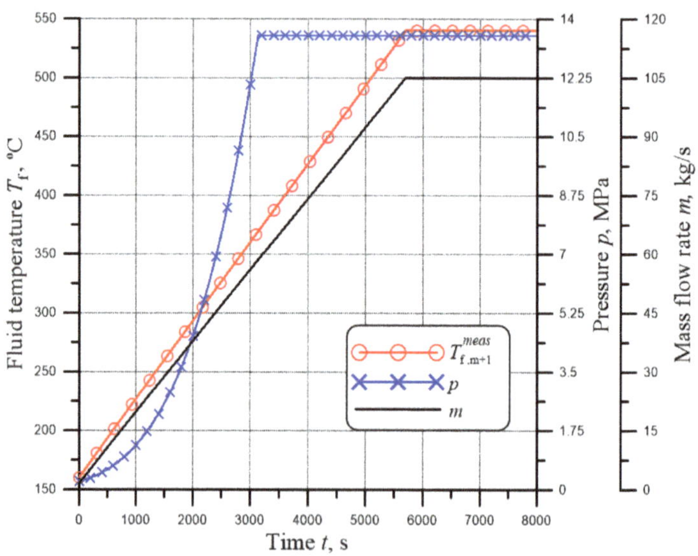

Figure 5. The temperature variation $T_{f,m+1}^{meas}(t)$, pressure p and steam mass flow rate $\dot{m}$ as functions of time t at the pipeline outlet z = 45 m.

The temperature $T_{f,1}^{calc}(t)$ at the pipeline inlet obtained from the IHP solution for "exact measurement data" is shown in Figure 6. Figure 6b compares the steam temperature determined by $T_{f,1}^{calc}(t)$ with the expected steam temperature $T_{f,1}^{meas}(t)$.

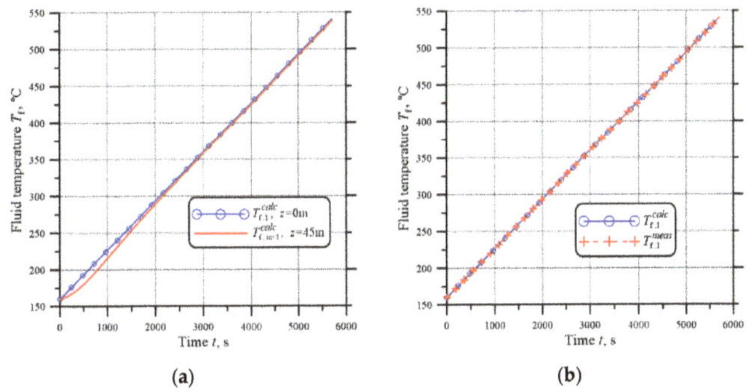

Figure 6. The steam temperature variation as a function of time (**a**) at the inlet of the pipeline $T_{f,1}^{calc}(t)$ and at the turbine inlet $T_{f,m+1}^{calc}(t)$, (**b**) comparison of the temperature $T_{f,1}^{calc}(t)$ obtained by solving IHP with measured temperature $T_{f,1}^{meas}(t)$.

To assess the accuracy of solving IHCP, a relative difference between the inlet steam temperature $T_{f,1}^{calc}(t)$ obtained from the IHP solution and measured temperature $T_{f,1}^{meas}(t)$ was calculated as follows.

$$\varepsilon_T = \left| \frac{T_{f,1}^{meas}(t) - T_{f,1}^{calc}(t)}{T_{f,1}^{meas}(t)} \right| \times 100\% \tag{25}$$

Root-Mean Square Error (*RMSE*) was calculated as follows.

$$\text{RMSE} = \sqrt{\sum_{j=1}^{N_{meas}} \frac{\left[T_{f,1}^{meas}(t_j) - T_{f,1}^{calc}(t_j)\right]^2}{N_{meas}}} \tag{26}$$

where the symbol N_{meas} stands for the number of measurement points.

The relative difference ε_T as a function of time is depicted in Figure 7. The relative error does not exceed 0.016 %, while the Root-Mean Square Error (RMSE) is 0.087 K.

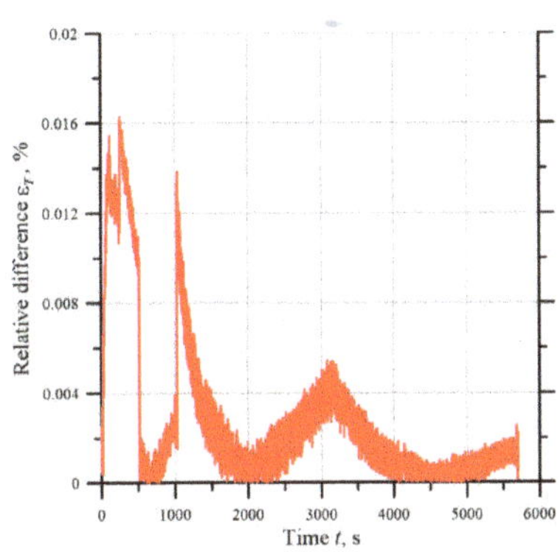

Figure 7. Relative difference between the inlet steam temperature $T_{f,1}^{calc}(t)$ obtained from the IHP solution and "exact" measured temperature $T_{f,1}^{meas}(t)$.

The analysis of the results shown in Figures 5 and 6 shows excellent conformity of the calculated temperature $T_{f,1}^{calc}(t)$ and measured temperature $T_{f,1}^{meas}(t)$. The IHP was solved using data from a power plant to verify the model's effectiveness and accuracy. Based on the measured steam temperature at the end of the pipeline $T_{f,m+1}^{meas}(t)$, the steam temperature $T_{f,1}^{calc}(t)$ was estimated using the developed method.

Figure 8 illustrates the steam temperature variation $T_{f,m+1}^{meas}(t)$, pressure p, and mass flow rate $\dot{m}$ as a function of time t obtained from the measurement.

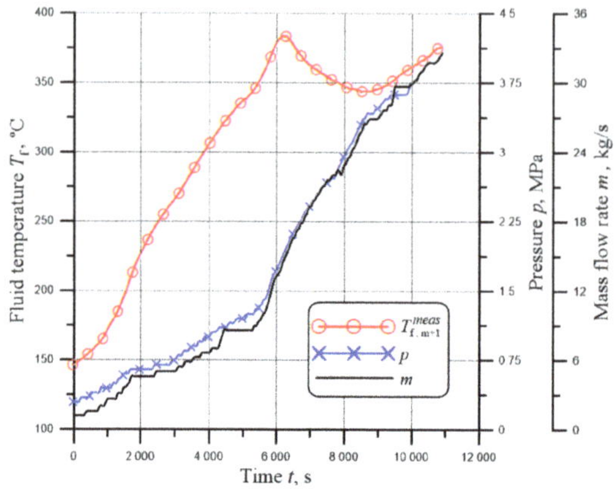

Figure 8. The steam temperature variation $T_{f,m+1}^{meas}(t)$, pressure p and steam mass flow rate $\dot{m}$ as a function of time t on the pipeline outlet.

Then the inverse problem was solved based on actual measurement data. Figure 8 depicts the measured steam temperature $T_{f,m+1}^{meas}(t)$, pressure p, and mass flow rate $\dot{m}$ of steam measured at the end of the pipeline. The IHP was solved to determine $T_{f,1}^{calc}(t)$. The basic time step Δt_b was equal to $\Delta t_b = k_b \cdot \Delta t = 30 \times 0.04 = 1.2$ s with the time step $\Delta t = 0.04$ s used for the solution of the direct problem. The number of future time intervals and the regularization factor were $F = 2$, $w_r = 0.01$, respectively.

Figure 9a shows the calculated fluid temperature at the inlet $T_{f,1}^{calc}(t)$ and outlet $T_{f,m+1}^{calc}(t)$ of the pipeline. A comparison of the calculated steam temperature $T_{f,1}^{calc}(t)$ and the measured steam temperature $T_{f,1}^{meas}(t)$ is shown in Figure 9b.

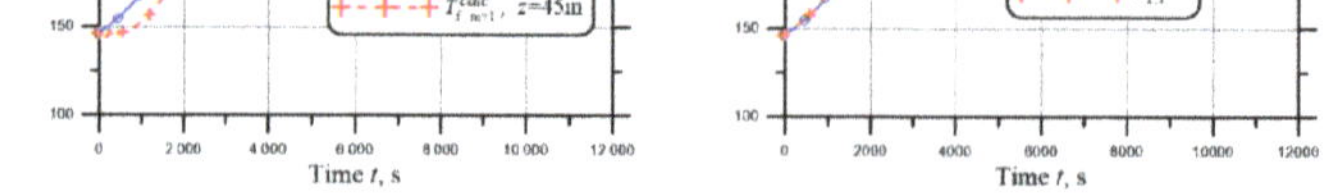

(a) (b)

Figure 9. The steam temperature variation as a function of time (**a**) on the inlet pipeline $T_{f,1}^{calc}(t)$ and on the turbine inlet $T_{f,m+1}^{calc}(t)$, (**b**) comparison of the temperatures determined $T_{f,1}^{calc}(t)$ and the expected $T_{f1,m}$.

For the determined steam temperature T_f at the pipeline inlet, the relative error was determined using Equation (25); the change in time of this error is depicted in Figure 10.

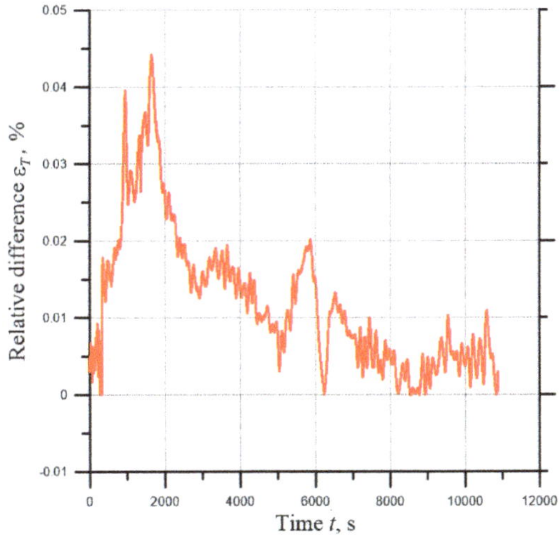

Figure 10. The relative difference between the inlet steam temperature $T_{f,1}^{calc}(t)$ obtained from the IHP solution and the actual measured temperature $T_{f,1}^{meas}(t)$.

The analysis of the results shown in Figure 9b reveals that the steam temperature at the pipeline inlet $T_{f,1}^{calc}(t)$ determined from the IHCP solution, differs slightly from the steam temperature obtained from the $T_{f,1}^{calc}(t)$ measurements. The RMSE = 0.038 K.

Figure 11 shows the time variations in temperature, pressure, and steam mass flow rate obtained from the measurements used in the next calculation test. The calculations were performed for a basic time step equal to $\Delta t_b = k_b \cdot \Delta t = 1.2$ s ($\Delta t = 0.04$ s), where the number of basic time steps was $k_b = 30$. The number of future time intervals $F = 2$ and the regularization factor $w_r = 0.01$ were assumed.

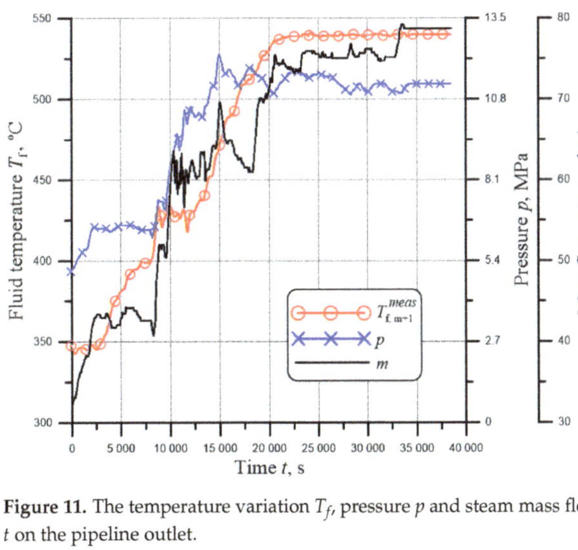

Figure 11. The temperature variation T_f, pressure p and steam mass flow rate $\dot{m}$ as a function of time t on the pipeline outlet.

Figure 12a presents the time variations in the outlet steam temperature $T_{f,m+1}^{calc}(t)$ of the pipeline and the variations in the inlet steam temperature $T_{f,1}^{calc}(t)$ determined by solving the IHP. A comparison between the IHP solution $T_{f,1}^{calc}(t)$ and the measured steam temperature $T_{f,1}^{meas}(t)$ is shown in Figure 12b.

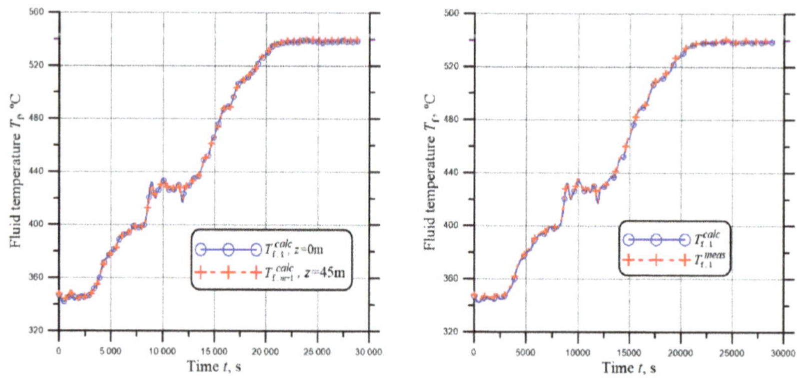

(a) (b)

Figure 12. The steam temperature variation as a function of time (**a**) at the pipeline inlet $T_{f,1}^{calc}(t)$ and at the turbine inlet $T_{f,m+1}^{calc}(t)$, (**b**) comparison of the temperatures $T_{f,1}^{calc}(t)$ by solving the IHP and the measured temperature $T_{f,1}^{meas}(t)$.

Analyzing the results depicted in Figure 12, it can be seen that there is very good agreement between the calculated steam temperature $T_{f,1}^{calc}(t)$ and the measured temperature $T_{f,1}^{meas}(t)$. The average value of the RMSE is 0.322 K for this case.

The relative difference ε_T between the fluid temperature obtained from the inverse solution and the measured temperature at the pipeline inlet is shown in Figure 13. The maximum value of the relative difference is ε_T = 0.296%.

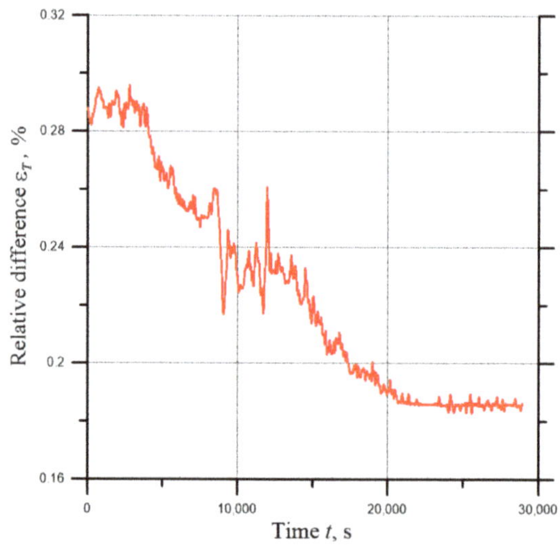

Figure 13. The relative difference between the inlet steam temperature $T_{f,1}^{calc}(t)$ obtained from the IHP solution and the actual measured temperature $T_{f,1}^{meas}(t)$.

The analysis of the results (Figures 11 and 12) shows that the differences between the steam temperature obtained from the inverse solution and the fluid temperature assumed in the direct solution are small. The small value of RMSE indicates a very good agreement between the inverse solution and measured steam temperature.

6. Conclusions

A steam turbine operates with prescribed parameters of the superheated steam at the turbine inlet. To ensure a safe and trouble-free operation of the turbine, the time changes of the inlet steam parameters must not be abrupt and do not differ from permissible values. Too rapid steam temperature changes can cause high stress in the pipeline and the turbine construction elements. To ensure trouble-free operation, the variation of steam temperature at the turbine inlet is necessary.

The paper developed the procedure for solving IHP to determine the time change of pipeline inlet temperature directly behind the boiler based on the prescribed measured steam temperatures before the steam turbine inlet.

The procedure proposed in this paper can be used to control the operation of the boiler so that, at its outlet, the steam temperature is equal to the steam temperature determined from the inverse solution. In this way, changes in the steam temperature upstream of the turbine will correspond to the requirements of the turbine user or changes due to conditions for safe turbine operation.

In summary, it can be stated that the developed mathematical model can be used to simulate transient temperature changes in pressure pipelines used in power plants. It allows us to determine the temperature distribution and stresses occurring in the pipeline wall and to calculate the allowable medium temperature change rate in time so that the allowable stress values are not exceeded.

The permissible temperature time changes at the pipeline outlet can also be determined from the condition of not exceeding thermal stresses, e.g., in components of the quick-closing valve before the turbine or the turbine rotor. The procedure for determining the allowable steam temperature variations downstream of the boiler will be identical to that presented in the paper.

The developed computational algorithm is efficient, allowing quick results of transient processes that occur in the pipeline during its operation.

Future studies will determine the allowable time variation of steam temperature at the turbine inlet due to not exceeding the allowable thermal stresses in the turbine quick-closing valve and the rotor and turbine body. These will be the basis for determining the corresponding changes in boiler outlet steam temperature and, subsequently, for the complete automation of the steam power plant start-up.

Funding: This research received no external funding.

Institutional Review Board Statement: Not applicable.

Informed Consent Statement: Not applicable.

Conflicts of Interest: The author declares no conflict of interest.

Nomenclature

a	thermal diffusivity $a = k/(c_p\,\rho)$, m^2/s
A	cross-section area, m^2
CFD	Computational Fluid Dynamics,
$c_{p,f}$	specific heat capacity at constant pressure, J/(kg·K)
$c_{p,w}$	specific heat capacity of the heat tube material, J/(kg·K)
d_{in}	inner diameter, m
E	modulus of elasticity (Young's modulus), Pa
f	fluid temperature at the inlet of the pipeline, °C

F	future time steps, s
g	fluid temperature at the outlet of the pipeline, °C
h	heat transfer coefficient, W/(m^2·K)
HTC	Heat Transfer Coefficient,
IHCP	Inverse Heat Conduction Problems,
IHP	Inverse Heat Problem,
k	golden division factor $k \approx 0.61803398$,
k_b	number of basic time steps,
k_w	thermal conductivity, W/(m·K)
L	length of the pipeline, m
m	number of nodes in the longitudinal direction,
$\dot{m}$	steam mass flow rate, kg/s
n	number of nodes in a radial direction,
Nu	Nusselt numbers,
$Nu_{m,q}$	mean Nusselt number for laminar tube flow,
p	absolute pressure, Pa
Pr	Prandtl number,
r	radius, m
Re	Reynolds number,
RMSE	Root-Mean Square Error, K
r_{in}	inner radius, m
r_{out}	outer radius, m
s_w	wall thickness, m
t	time, s
T_0	initial temperature, °C
T_f	fluid temperature, °C
T_w	wall temperature, °C
$\overline{T}(t)$	mean temperature on the wall thickness, °C or K
$\overline{T}(r,t)$	mean temperature of the wall between r_{in} and r, °C or K
U_{in}	overall heat transfer coefficient, W/(m^2·K)
w_r	regularization factor,
$w_{z,i}$	steam velocity at the inlet of the i-th finite volume, m/s
Greek symbols	
β_T	linear thermal expansion coefficient, 1/K
Δr	radial step $\Delta r = (r_{out} - r_{in})/n$, m
Δt	time step, s
Δt_b	basic time step, s
Δz	axial step $\Delta z = L/m$, m
ε_T	relative difference, %
ν	Poisson's ration,
ξ	friction factor,
ρ	density, kg/m^3
σ_r	radial stresses component, Pa
σ_φ	longitudinal stresses component, Pa
σ_z	circumferential stresses component, Pa
∇	gradient operation (nabla),
Subscripts	
0	initial value,
f	fluid (steam),
i	node number,
in	inner,
MIN	minimum,
MAX	maximum,
out	outer,
w	wall,
Superscripts	
$calc$	calculated,
$meas$	measured,

References

1. Taler, J. A new space marching method for solving inverse heat conduction problems. *Forsch. Lngenieurwesen* **1999**, *64*, 296–306. [CrossRef]
2. Frąckowiak, A.; Wróblewska, A.; Ciałkowski, M. Trefftz numerical functions for solving inverse heat conduction problems. *Int. J. Therm. Sci.* **2022**, *177*, 107566. [CrossRef]
3. Ciofalo, M. Solution of an inverse heat conduction problem with third-type boundary conditions. *Int. J. Therm. Sci.* **2022**, *175*, 107466. [CrossRef]
4. Tourn, B.A.; Hostos, J.C.A.; Fachinotti, V.D. A modified sequential gradient-based method for the inverse estimation of transient heat transfer coefficients in non-linear one-dimensional heat conduction problems. *Int. Commun. Heat Mass Transf.* **2021**, *127*, 105488. [CrossRef]
5. Ahn, C.; Park, C.; Park, D.; Kim, J. Optimal hybrid parameter selection for stable sequential solution of inverse heat conduction problem. *Int. J. Heat Mass Transf.* **2022**, *183*, 122076. [CrossRef]
6. Jaremkiewicz, M.; Taler, J. Measurement of Transient Fluid Temperature in a Pipeline. *Heat Transf. Eng.* **2018**, *39*, 1227–1234. [CrossRef]
7. Jaremkiewicz, M.; Taler, D.; Dzierwa, P.; Taler, J. Determination of Transient Fluid Temperature and Thermal Stresses in Pressure Thick-Walled Elements Using a New Design Thermometer. *Energies* **2019**, *12*, 222. [CrossRef]
8. Taler, J.; Taler, D.; Kaczmarski, K.; Dzierwa, P.; Trojan, M.; Jaremkiewicz, M. Allowable Rates of Fluid Temperature Variations and Thermal Stress Monitoring in Pressure Elements of Supercritical Boilers. *Heat Transf. Eng.* **2019**, *40*, 1430–1441. [CrossRef]
9. *European Standard EN 12952-3*; Water-Tube Boilers and Auxiliary Installations. Part 3: Design and Calculation for Pressure Parts. European Committee for Standardization: Brussels, Belgium, 2011.
10. Taler, D.; Dzierwa, P.; Kaczmarski, K.; Taler, J. Increase the flexibility of steam boilers by optimization of critical pressure component heating. *Energy* **2022**, *250*, 123855. [CrossRef]
11. Taler, D. Direct and Inverse Heat Transfer Problems in Dynamics of Plate Fin and Tube Heat Exchangers. In *Heat Transfer—Mathematical Modelling, Numerical Methods and Information Technology*; InTech: Rijeka, Croatia, 2011; pp. 77–100.
12. Sekhavati, J.; Hashemabadi, S.H.; Soroush, M. Computational methods for pipeline leakage detection and localization: A review and comparative study. *J. Loss Prev. Process Ind.* **2022**, *77*, 104771. [CrossRef]
13. Wang, X.; Liu, L.; Duan, R.; Liu, Y.; Wei, Z.; Yang, X.; Liu, X.; Li, Z. A method for leak detection in buried pipelines based on soil heat and moisture. *Int. Commun. Heat Mass Transf.* **2022**, *135*, 106123. [CrossRef]
14. Rusin, A.; Bieniek, M. Maintenance planning of power plant elements based on avoided risk value. *Energy* **2017**, *134*, 672–680. [CrossRef]
15. Taler, J.; Taler, D.; Kaczmarski, K.; Dzierwa, P.; Trojan, M.; Sobota, T. Monitoring of thermal stresses in pressure components based on the wall temperature measurement. *Energy* **2018**, *160*, 500–519. [CrossRef]
16. Taler, D.; Kaczmarski, K. Mathematical modelling of the transient response of pipeline. *J. Therm. Sci.* **2016**, *25*, 549–557. [CrossRef]
17. Chaibakhsh, A.; Ghaffari, A.; Moosavian, S.A.A. A simulated model for a once-through boiler by parameter adjustment based on genetic algorithms. *Simul. Model. Pract. Theory* **2007**, *15*, 1029–1051. [CrossRef]
18. Taler, D.; Trojan, M.; Dzierwa, P.; Kaczmarski, K.; Taler, J. Numerical simulation of convective superheaters in steam boilers. *Int. J. Therm. Sci.* **2018**, *129*, 320–333. [CrossRef]
19. Taler, J.; Węglowski, B.; Pilarczyk, M. Monitoring of thermal stresses in pressure components using inverse heat conduction methods. *Int. J. Numer. Methods Heat Fluid Flow* **2017**, *27*, 740–756. [CrossRef]
20. Trojan, M.; Taler, D. Thermal simulation of superheaters taking into account the processes occurring on the side of the steam and flue gas. *Fuel* **2015**, *150*, 75–87. [CrossRef]
21. Serov, E.P.; Korolkov, B.P. *Dynamics of Steam Generators*; Energoizdat: Moscow, Russia, 1981.
22. Wang, H.; Wang, H.; Zhu, T.; Deng, W. A novel model for steam transportation considering drainage loss in pipeline networks. *Appl. Energy* **2017**, *188*, 178–189. [CrossRef]
23. Roffel, B.; Betlem, B. *Process Dynamics and Control, Modeling for Control and Prediction*; John Wiley & Sons Ltd.: Chichester, UK, 2006.
24. Beck, J.V.; Blackwell, B.; Haji-Sheikh, A. Comparison of some inverse heat conduction methods using experimental data. *Int. J. Heat Mass Transf.* **1996**, *39*, 3649–3657. [CrossRef]
25. Blackwell, B.; Beck, J.V. A technique for uncertainty analysis for inverse heat conduction problems. *Int. J. Heat Mass Transf.* **2010**, *53*, 753–759. [CrossRef]
26. Mohebbi, F.; Sellier, M.; Rabczuk, T. Estimation of linearly temperature-dependent thermal conductivity using an inverse analysis. *Int. J. Therm. Sci.* **2017**, *117*, 68–76. [CrossRef]
27. Jarny, Y.; Maillet, D. *Lecture 10: Linear Inverse Heat Conduction Problems–Two Basic Examples*; Metti 5 Spring School: Roscoff, France, 2011; pp. 1–18.
28. Taler, D. A new heat transfer correlation for transition and turbulent fluid flow in tubes. *Int. J. Therm. Sci.* **2016**, *108*, 108–122. [CrossRef]
29. Taler, D. Determining velocity and friction factor for turbulent flow in smooth tubes. *Int. J. Therm. Sci.* **2016**, *105*, 109–122. [CrossRef]
30. Gnielinski, V. Heat Transfer in Pipe Flow. In *VDI Heat Atlas*, 2nd ed.; VDI-Gesellschaft Verfahrenstechnik und Chemieingenieurwesen, Ed.; Springer: Berlin/Heidelberg, Germany, 2010; pp. 693–699.

31. Tikhonov, A.N.; Arsenin, V.Y. *Solution of Ill-Posed Problems*; V.H. Winston & Sons: New York, NY, USA, 1977.
32. Dahlquist, G.; Björck, Å. *Numerical Methods*; Dover Publications: Mineola, NY, USA, 2012.
33. Hetnarski, R.B.; Noda, N.; Tanigawa, Y. *Thermal Stresses*, 2nd ed.; Taylor & Francis Inc.: Boca Raton, FL, USA, 2002.

energies

MDPI

Article

Cooling Modelling of an Electrically Heated Ceramic Heat Accumulator

Dawid Taler [1], Jan Taler [2,*], Tomasz Sobota [1] and Jarosław Tokarczyk [1]

1 Department of Thermal Processes, Air Protection, and Waste Utilization, Faculty of Environmental Engineering and Energy, Cracow University of Technology, ul. Warszawska 24, 31-155 Cracow, Poland
2 Department of Energy, Faculty of Environmental Engineering and Energy, Cracow University of Technology, Al. Jana Pawła II 37, 31-864 Cracow, Poland
* Correspondence: jan.taler@pk.edu.pl; Tel.: +48-12-628-3554

Abstract: This paper presents a simple novel mathematical model of a heat accumulator with an arranged packing in the form of ceramic cylinders. The accumulator analysed in the paper can be heated with inexpensive electricity overnight or excess electricity from wind farms. It can be used as a heat source in a hydronic heating system or for domestic hot water. The differential equations describing the transient temperature of the accumulator packing and flowing air were solved using the explicit Euler and Crank–Nicolson methods. The accuracy of both methods was assessed using exact analytical solutions and the superposition method for a uniform initial temperature and accounted for time changes in inlet air temperature. A numerical simulation of the accumulator cooled by flowing air was carried out. The correlation for the air-side Nusselt number was determined using the method of least squares based on experimental data. The calculated exit air temperature was compared with the measured data. The accumulator can operate as a heat source with dynamic discharge. The developed mathematical model of the accumulator can be used in a system to adjust the fan rotational speed so that the air temperature in the room is equal to the preset temperature.

Keywords: heat accumulation; ceramic filling; electrical resistance heating of the storage pack; numerical model of storage-pack cooling; experimental study

Citation: Taler, D.; Taler, J.; Sobota, T.; Tokarczyk, J. Cooling Modelling of an Electrically Heated Ceramic Heat Accumulator. *Energies* **2022**, *15*, 6085. https://doi.org/10.3390/en15166085

Academic Editors: Artur Bartosik and Dariusz Asendrych

Received: 26 July 2022
Accepted: 19 August 2022
Published: 22 August 2022

Publisher's Note: MDPI stays neutral with regard to jurisdictional claims in published maps and institutional affiliations.

Copyright: © 2022 by the authors. Licensee MDPI, Basel, Switzerland. This article is an open access article distributed under the terms and conditions of the Creative Commons Attribution (CC BY) license (https://creativecommons.org/licenses/by/4.0/).

1. Introduction

In recent years, there has been increasing interest in both heat and electricity storage. This is mainly due to the time-varying power generated by renewable energy sources such as wind and photovoltaic farms. Ceramic-filled heat accumulators or electrically heated hot water storage tanks can utilise surplus electrical energy from photovoltaics and wind farms, which will improve the controllability of the entire energy system.

Water, solid materials, and phase-change materials (PCMs) are most commonly used as thermal storage materials [1]. Large water heat storage tanks with hot and cold water stratification are widely used in urban central heat supply systems. The paper [2] presents CFD modelling of a water heat accumulator with water temperature stratification used in a large municipal thermal power plant comprising coal-fired grate boilers and natural gas-fired internal combustion engines. Pressurised hot water storage tanks are also used to improve the flexibility of thermal power units. One way of managing excess electricity at night; i.e., during times of reduced electricity demand, is through the accumulation of hot water in pressurised storage tanks. The heat flow rate supplied to the power plant with the fuel is partly used during the night period to heat up the hot water in the accumulators, resulting in a reduction in the electrical power generated. The boiler in this situation can operate above the technical minimum without burning heavy fuel oil in the chamber. In the case of boilers fired with pulverised coal, when the boiler load is below the technical minimum, pulverised coal and fuel oil are burnt in the boiler simultaneously to avoid the pulverised coal burners being extinguished. Thanks to these tanks, it is possible to increase the maximum load of the power unit by a few percent and decrease the value of the minimum acceptable load of the power unit by dozens of percent [3]. A concept of a nuclear plant with improved flexibility was described in [4]. Water tanks and borehole thermal energy-storage systems (BTES) can be used for thermal energy storage for district heating systems utilising waste heat from data centres [5]. The application of water-to-water heat pumps in a solar system for space heating and domestic hot water preparation in a cold climate was presented by Pinamonti et al. [6]. A seasonal water storage tank for a single-family house was used to store the excess heat. The long-term storage tank was completely buried under the ground near the house. Electric water heaters affected the operation of the entire energy system. The flexibility potential of electric water heaters was strongly influenced by the daily hot water demand [7]. Hot water storage tanks can be used to capture the excess electricity generated by photovoltaic cells [8]. A system consisting of photovoltaic cells and an electric water heater can essentially meet a household's electricity and heat requirements for heating a building [9]. A novel hot water storage unit with low thermal inertia was proposed for floor heating and domestic hot water preparation using solar collectors. The proposed solution ensured stable system operation even after sunset [10].

The use of a proportional water flow controller in a hot water preparation system with a hot water reservoir was presented in [11]. The effect of thermometer inertia on temperature control quality in an electrically heated hot water reservoir was investigated in [12]. The correct operation of the temperature control in the hot water storage tank was strongly influenced by random temperature measurement errors. The accidental errors in the thermometer indications could result in the faulty operation of the PID controller, since the water temperature in the tank could rise steadily despite exceeding the set temperature [13].

Solid-fill accumulators are often used for heating buildings and domestic hot water preparation. The filling of the accumulator can be either structured or random. A ceramic filling resistant to high temperatures in the order of 800 °C is used in the designed battery. As a result, a large amount of heat can be stored in the lightweight heat storage unit, which is used to heat the building. The use of fireproof ceramic cylinders in the proposed accumulator is an inexpensive and safe solution. The accumulator filling can be made of other materials; e.g., magnetite, as it has a high specific gravity and high heat capacity. In the general case, high-density, high-heat-capacity, high-temperature-resistant materials are the materials from which the filling can be made. Due to the high temperature of the accumulator packing when charged, materials such as concrete or stones are unsuitable due to their low resistance to thermal stress cracking. The ceramic elements are arranged in an orderly manner so that the porosity of the filling can be reduced and the pressure loss when air flows through the accumulator is reduced. This results in a reduction in the energy consumed by a fan pushing air through the accumulator.

A numerical model of a channel heat accumulator heated by hot air was presented in [14]. Computer simulations showed that the number of finite elements along the duct length had to be large to ensure satisfactory accuracy of the air temperature calculations at a low air velocity.

The standard approach in heat accumulator modelling, in which the average temperature of the flowing gas along the length of one cell is approximated by the arithmetic mean of the inlet and outlet temperatures of the cell, can lead to substantial errors, especially for small mass flow rates of the gas [14]. It is necessary to divide the accumulator model into a considerable number of cells; e.g., into 50 or more, to achieve a numerical solution of the system of differential equations for an accumulator that does not change with the increasing number of cells and agrees well with the experiment. This requirement applies to the finite-difference method, including both the explicit Euler method and the Crank–Nicolson method. Furthermore, in contrast to heat conduction issues, the stability and accuracy of modelling heat accumulators with different numerical methods have not been sufficiently investigated in the literature.

A hybrid finite volume method–finite element method was used to simulate the operation of the accumulator with an organised structure [15]. The results of the calculations carried out in [15] were compared with the measurements. A heat storage unit made of steel or concrete cylindrical elements was modelled using the COMSOL Multiphysics 4.3a software [16]. The operation of a honeycomb ceramic heat accumulator was simulated using a one-dimensional model in the object-oriented modeling language Modelica [17]. A simple transient floor heating model was presented in [18]. The solution of the one-dimensional transient heat conduction was solved using a Laplace transform. The storage capacity of the floor and the energy transferred from the floor to the heated space were estimated. A resistance-heated energy storage unit made of firebricks was studied in [19]. The thermal performance of the industrial heat accumulator was studied numerically using the finite volume method in conjunction with the Crank–Nicolson method. The paper [20] presented the concept of a graphite heat accumulator. The heat was accumulated in a graphite block in which steel water tubes were installed. The advantage of such an accumulator over this type of concrete accumulator was the higher operating temperature of the graphite block and lower thermal stresses in the graphite block due to the high value of the graphite thermal conductivity of about 88 W/(m·K). A combined heat and compressed air energy-storage system using a packed bed accumulator and an electrical heater was proposed in [21]. The thermal oil was resistively heated to a high temperature, stored in a reservoir, and then directed to an interstage air superheater located between the air turbine's high- and low-pressure parts. An air-based high-temperature heat and power storage system that cogenerated heat and electricity with high efficiency was presented in [22]. During the charging process, the packed bed of stones was heated up to 685 °C. The effect of the thermal conductivity of the solid particles that made up the filling of a heat accumulator was analysed in [23].

In the last few decades, there has been a lot of research on the application of phase-change materials (PCMs) for heat storage in technical systems. Numerical modelling of the combined sensible–latent heat-storage unit with a periodic flow used in concentrated solar power plants was carried out in [24]. Solid (GF/Sn-Bi composite) and PCM (GF/$MgCl_2$ composite) were used in numerical simulations of the proposed solution. A significant problem in the broader use of PCM-based heat storage is the high thermal inertia. Usually, a small area close to the components to be heated or cooled; e.g., pipes or flat surfaces, is used. A large part of the tank does not undergo a phase change in either the heating or cooling phase. The appropriate tank structure and fin arrangement [25] or metal foam inside the storage tank [26] can improve the heat performance of thermal storage systems. A heat accumulator filled entirely with rocks and an accumulator filled in the upper part with PCM capsules and in the lower part with rocks was simulated numerically and studied experimentally in [27]. The performance of the heat accumulator during charging and discharging processes, as well as its cyclic operation, were investigated. Sodium nitrate ($NaNO_3$), which has a melting point of 306 °C, was used as the PCM [28]. The PCM was placed between two steel plates. To improve the thermal performance of the multi-slab phase-change thermal energy storage block, 0.14 mm-thick fins were used inside the PCM. The effect of the position of the PCM layer on the temperature distribution in the wall and on heat loss to the surroundings was investigated in [29]. The PCM layer was placed under a layer of plaster near the interior of the building or on the exterior surface of the brick wall and covered with a layer of extruded polystyrene. The variability of electricity production from wind farms causes major disruption to electricity and heat networks. Distributed electric heating storage units were used to capture excess electricity generated at wind farms [30]. To improve solar energy utilisation and the stability of solar heating systems, an air-type solar collector with heat accumulation was designed [31]. Phase-change material was placed inside the solar vacuum tube to reduce the impact of solar radiation fluctuations on indoor heating.

This paper presents a mathematical model for the cooling of a ceramic electric accumulator with an ordered structure. Conventional heat accumulators are inconvenient to use. The disadvantages of traditional electric heat accumulators for space heating are the high casing temperature and the significant decrease in the heat flow rate given off by the accumulator with time. Dynamic discharge heat storage units that are placed inside rooms have the disadvantages of fan noise and dust deposits on the room walls. The high-temperature accumulator analysed in this paper can be used in replacement of a conventional solid-fuel boiler or oil- or gas-fired boilers that are located in the boiler room. By locating the accumulator and the finned air–water heat exchanger in a separate room e.g., the previous boiler room, they are not troublesome for the room occupants in the heated building. The central heating system does not have to be replaced. The coupling element between the heat accumulator and the central heating system is a finned air–water heat exchanger. The accumulator filling is heated by low-cost electricity at low loads in the power system or excess electricity from photovoltaic cells. The term "low-cost electricity" refers to the price of 1 kWh of electricity at night, when electricity demand is low, which is much lower than the price of 1 kWh during peak load times, which typically occur in the evening.

The new idea of replacing traditional coal-, oil-, or gas-fired central heating boilers with heat storage units is presented in this paper. Accumulators heated using inexpensive electricity will help to reduce the danger of carbon monoxide poisoning or explosions present in traditional heating systems. In addition, carbon dioxide emissions and low-particulate-matter emissions will be reduced compared to traditional home heating systems.

Solid-fill heat accumulators are typically used for direct space heating. The electrically heated storage unit analysed in this paper can be used as a heat source in a hydronic central heating system for a building. In this way, solid-fuel-, oil-, or gas-fired boilers can be eliminated from existing building heating systems. The electrically heated accumulators can be used to capture excess electricity from wind farms. In summer, the heat stored in the accumulator can be used to heat domestic hot water. An example of using a heat accumulator to heat a building is shown in Figure 1.

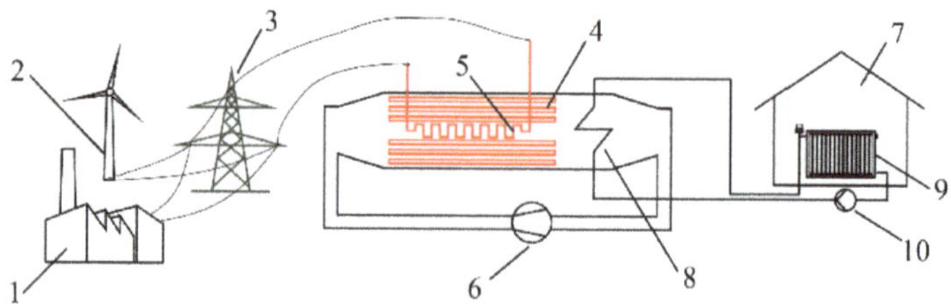

Figure 1. Use a heat accumulator heated with low-cost electricity to heat a building or provide domestic hot water. 1—thermal power plant, 2—wind farm, 3—power network, 4—solid heat accumulator, 5—electric resistance heater, 6—fan, 7—building, 8—finned air–water heat exchanger, 9—hydronic central heating system, 10—circulation pump.

During the period of overproduction of electricity in the system, which usually occurs during the night, the heat storage unit is heated using electricity. During the night period, the demand for electricity is low. Thermal power plants operate with the minimum allowable load, as it is not worthwhile to shut them down due to the very high restart costs. In addition, the wind velocity during the night period is usually higher than during the day, resulting in increased output from wind farms. Due to the excess of electricity during the night period when electricity demand is low, the electricity supply is greater than the electricity demand. This results in an extreme reduction in the electricity price per kWh at night. One way to utilise this excess electricity is through heat accumulators with a fixed packing, which can be heated to a high temperature; e.g., up to 600 °C. When electricity prices rise due to increased demand during the daytime period, the heat accumulator is not supplied with electricity. During the day, the accumulator is discharged using the air flowing through it. The accumulator can be located in a heated room, or it can replace a conventional fossil fuel-fired boiler situated in the boiler room. The latter solution is very beneficial for historic city centres, where there are many old buildings without access to a central heating network. A favourable alternative to replacing a masonry heater (tiled stove) or other heat sources is a designed accumulator that does not emit harmful substances. By heating the accumulator using inexpensive electricity in times of low demand, the heating costs are competitive with other techniques to heat buildings. A central heating system in which an accumulator has replaced a conventional gas-fired boiler is shown in Figure 1. The accumulator, which is heated up during the night, is cooled down by flowing air that is then cooled down in a finned air–water heat exchanger. This heat exchanger works as a central heating boiler. The proposed solution is simple and inexpensive.

This paper developed a new simple numerical model of the heat storage unit. A new method for determining the correlation for the heat transfer coefficient from the packing surface to the flowing air was proposed using the least-squares method. The accuracy of the numerical accumulator models was assessed by comparing the air and packing temperatures at the accumulator outlet with those obtained using an analytical exact solution that was obtained using the superposition method. A numerical mathematical model of the accumulator based on the explicit Euler method and the Crank–Nicolson method was verified by exact solutions for a step change in air temperature at the inlet to the accumulator and a ramp change in inlet air temperature. The verification of numerical solutions is of great importance in the case of heat accumulators. If the number of finite volumes is too low, the accuracy of the numerical solutions can frequently be unsatisfactory [14]. Compared to heat conduction, the number of finite volumes along the length of the accumulator needs to be greater or several times higher to achieve satisfactory accuracy of the accumulator modelling. The computer calculation time using the developed model, especially using the Crank–Nicolson method, was very short. This made it possible to apply the developed numerical model in an automatic air temperature control system. In the predictive air temperature control, the proposed mathematical model of the heat storage unit could be used to determine the fan rotational speed so that the air temperature at its outlet was equal to the set temperature. In this way, despite the discharging of the accumulator during the day; i.e., a decrease in the packing temperature, it was possible to maintain the set temperature in the heated room by changing the airflow velocity.

2. Design of the Heat Accumulator

A laboratory test stand was built to test an electric–water heating system for a building. The air heated in the accumulator gave up heat to the water flowing in the finned heat exchanger (car radiator), which in turn was the heat source for the central heating system.

Figure 2a shows a hybrid electric–water building heating system using a dynamic discharge heat accumulator, while a photograph of the heating system test rig is shown in Figure 2b. The central element of the heating system was a ceramic dynamic discharge heat accumulator with an ordered packing that was heated by low-cost electricity. The following systems can be specified in the test rig:

- An electrical sub-system for packing heating using resistance heaters; the filling of the accumulator was composed of ceramic and steel elements;
- A water circuit comprising a plate-finned and tube heat exchanger (PFTHE), central heating radiators, a water circulation pump, and pipes through which the water flowed;
- An air-circulation sub-system consisting of the heat accumulator, a radial forced draft fan, and the PFTHE;
- A power and control unit;
- Measuring equipment with a data-acquisition system.

The hybrid electric–water heating system for a building analysed in the paper is presented in Figure 2. The Heat accumulator ① (Figure 2a) as a heat source in a central heating system was heated by resistance heaters during periods of low electricity demand; i.e., at night or during midday hours when electricity was inexpensive. This type of heating system is particularly cost-effective in countries where electricity is inexpensive; e.g., in countries with a high proportion of nuclear, hydroelectric, or wind power stations in the energy system.

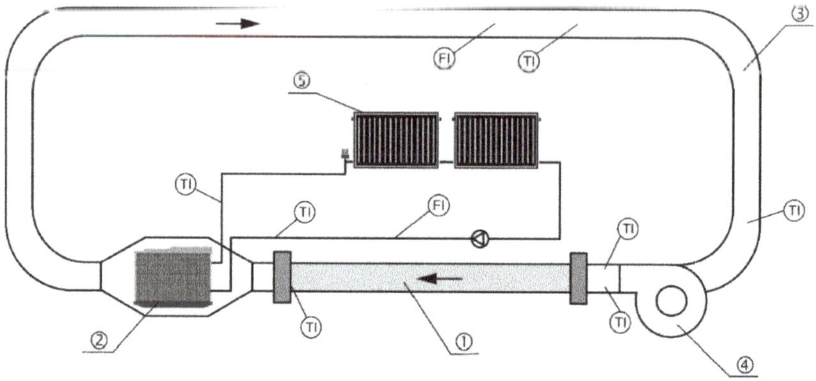

(**a**)

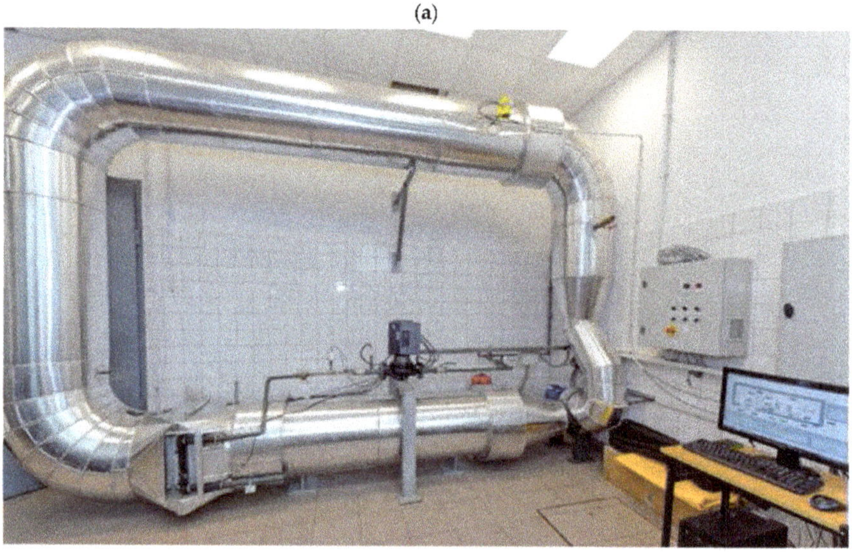

(**b**)

Figure 2. Test rig for a hybrid electric–water heating system for a building: (**a**) flow system of the test stand; (**b**) general view of the test stand. ①, Ceramic bed heat accumulator; ②, air–water heater (plate-finned and tube heat exchanger); ③, reversal duct; ④, forced draft fan; ⑤, water-heating radiators; FI, flow rate meter; TI, pre-calibrated thermocouples.

In the heat storage unit ①, a dynamic heat discharge was used. On the outer surface, the accumulator was well thermally insulated. Heat was extracted from the high-temperature filling by flowing air that, after heating, was cooled in the PFTHE ②, which was composed of oval tubes with continuous fins. The capacity of the PFTHE was about 20 kW when the air velocity before entering the heat exchanger was approximately 3 m/s. The area of continuous fins in the heat exchanger was several times larger than the area of the plain tubes, so the PFTHE dimensions and its cost were small. The water temperature in the central heating system supply had to be maintained with a controller to ensure a constant temperature in the room. A centrifugal pump forced the water flow in the central heating system. The water temperature at the PFTHE outlet could be changed by varying the rotational speed of the fan ④ or by changing the number of pump impeller revolutions. The centrifugal fan ④ forced the air through the PFTHE.

The PFTHE ② was located at the outlet of the heat accumulator in the left lower part of the test stand. The fan ④ forced air through the accumulator. A water flow meter was situated in the centre of the stand above the accumulator.

The water circuit in the hybrid heating system consisted of a PFTHE, a central heating system with plate radiators, and a water circulation pump. The PFTHE was an automotive radiator used in spark-ignition engines with a capacity of 1600 cm^3. The air heated in the accumulator was used to heat the water in the PFTHE, which performed the function of a classic boiler in a central heating system. The purpose of plate radiators in the installation was to dissipate the heat absorbed by the air as it flowed through the accumulator. The panel radiators were located in an adjacent room to avoid heating the air where the test rig was located. The water temperature in the central heating radiators decreased by the same number of degrees. This ensured a stable internal temperature during the tests.

The hot water at the outlet of the lamella heat exchanger was the feed water for the central heating system. The water temperature in the central heating radiators dropped by the same amount as it heated up in the PFTHE. This ensured that the entire hybrid system operated under steady-state conditions. The water flow in the central heating system was regulated by changing the rotational speed of the circulation pump via a current frequency converter. A centrifugal fan with a drive motor of 1.5 kW forced the air through the heat accumulator. The maximum volumetric capacity of the fan was 660 m^3/h and the maximum air pressure was 4.2 kPa.

The heat accumulator with dynamic discharge was the main component in the hybrid heating system. The principle of the heat accumulator operation was based on the use of clean energy such as electricity, including electricity from renewable energy sources.

The heat storage unit consisted of two basic components: the outer casing and the core (packing). The outer casing was a steel tube made of heat-resistant steel. The cylindrical shape of the casing provided advantages that are not present in traditional accumulation heaters. The air flow through the circular cross-section of the battery was more uniform compared to the rectangular cross-section of the housing.

The filling of the accumulator consisted of cylindrical ceramic elements heated during the night using inexpensive electric energy. A view of the accumulator packing is illustrated in Figure 3a, and the dimensions of the outer shell and regenerator components are shown in Figure 3b.

Eight steel tubes (*4*) with an external diameter of 101.6 mm and a wall thickness of 3.6 mm were inside the outer casing (*3*) (Figure 3). Three electric heaters (*2*) were placed in each tube. Seven rows of cylinders (*1*) formed by cylinders of an equal diameter and height of 30 mm were inside each of the steel tubes along their entire lengths. The ceramic cylinders were made of heat-resistant corundum concrete. The packing of the accumulator then had an arranged structure. The air flowed longitudinally around the ceramic elements inside the steel tubes. An additional air stream flowed between the steel tubes on the shell side. Air enters the inter-tube space through 14 holes (*6*) that were 20 mm in diameter (Figure 3b). The front view of the accumulator illustrating the distribution of the ceramic elements in the steel tubes with visible connections for resistance heaters is presented in Figure 4c.

During the day, air flowed through the heat accumulator, which was pre-heated by the accumulator packing. Internal energy stored in the heated packing could be used directly for space heating or via an air–water heat exchanger to heat water in the central heating system.

The density and specific heat of the ceramic cylinders were high. As a result, the heat capacity of the accumulator also was high, and the heat stored in the accumulator was large, which in turn made it possible to heat a house with a large space. The thermal conductivity of cylinders was also significant. The thermal stresses in the cylindrical ceramic elements were therefore low, avoiding scratches and cracks.

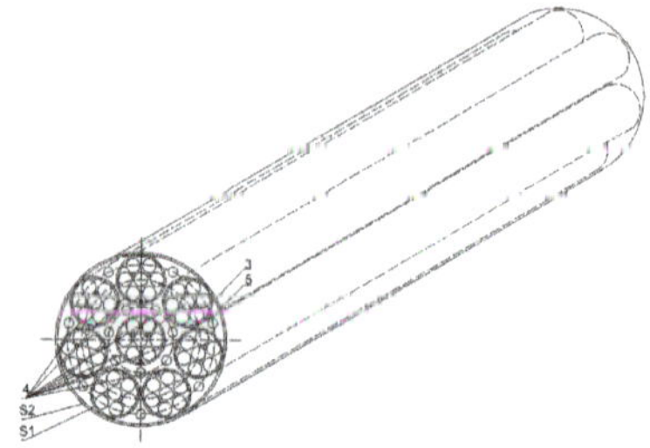

(a)

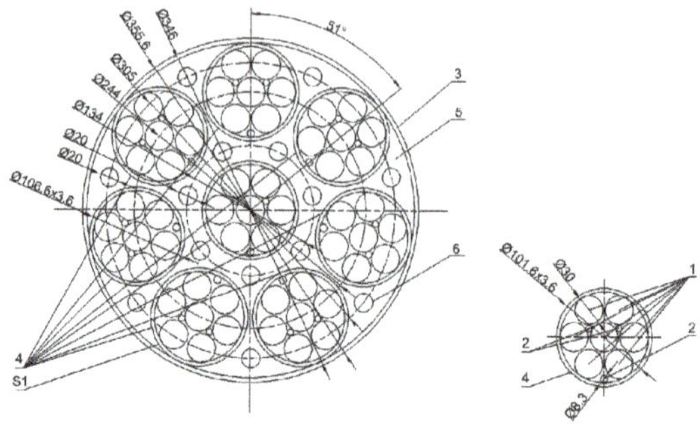

(b)

Figure 3. Ceramic heat accumulator: (**a**) view of the accumulator packing; (**b**) outer shell and regenerator components. 1—ceramic cylinders, 2—resistance electric heaters, 3—accumulator outer casing, 4—steel tubes (inside of which there are seven rows of ceramic cylinders), 5—a bottom with openings through which air flowed into the inter-tube space, 6—the air intake opening to the inter-tube space.

(a)

(b)

(c)

Figure 4. Heat accumulator with dynamic discharge: (**a**) view of cylindrical filling elements; (**b**) external shell of the accumulator with visible steel tubes, inside of which the ceramic cylinders are located; (**c**) front view of the accumulator illustrating the distribution of the ceramic filling elements in the steel tubes with visible connections for resistance heaters.

3. Mathematical Model of a Heat Accumulator

A mathematical model of a heat accumulator was developed using the following assumptions:

- The change in temperature of the packing over time was described by a first-order ordinary differential equation;
- The air temperature was a function of time and the axial x-coordinate, and it did not change in the radial direction;
- The external surface of the accumulator was perfectly thermally insulated;
- The ceramic and steel parts of the filling and the accumulator shell were at the same temperature at a given cross-section and time.

The first assumption was commonly made in the literature on thermal modelling of heat accumulators [1]. The modelled heat storage unit met the requirements to consider the filling as an object with lumped thermal capacity. The filling element consisted of cylindrical elements and steel pipes. A body can be treated as an object of lumped heat capacity when the Biot number (Bi) for a given element is less than 0.1. A more detailed explanation of when a body can be treated as an object with a concentrated heat capacity can be found in the books by Kreith [32] and Taler and Duda [33]. In the case of the heat accumulator analysed in this paper, the maximum value of the Biot number ($\mathrm{Bi}_w = h_g d_c/(2k_w)$) was less than 0.01, where d_c is the outer diameter of the ceramic cylinder, h_g is the heat transfer coefficient on the outer cylinder surface, and k_w is the ceramic cylinder thermal conductivity. In addition, for steel tubes with a wall thickness s_w equal to 3.6 mm located inside the accumulator, the Biot number (Bi), while taking into account that the tube was cooled on the external and internal surface, was $\mathrm{Bi}_w = h_g s_w/(2k_m)$, where k_m is the thermal conductivity of the alloyed steel of which the tubes were made. The maximum Biot number (Bi_w) shall not exceed 0.0005. Longitudinal heat conduction in exchanger walls is of secondary importance and is usually neglected [32]. This is due to the small temperature gradient of the air along its flow path. An additional factor that hinders heat conduction in cylindrical elements in the axial direction is the contact resistance at the interface between adjacent cylindrical elements. There were 67 ceramic cylinders arranged in a line along the length of the accumulator.

Figure 5 shows a circular heat storage unit, where D_w is the internal diameter of the accumulator and L_r is its length. The ordered accumulator packing consists of ceramic cylinders of diameter d_c and height H_c (Figure 5).

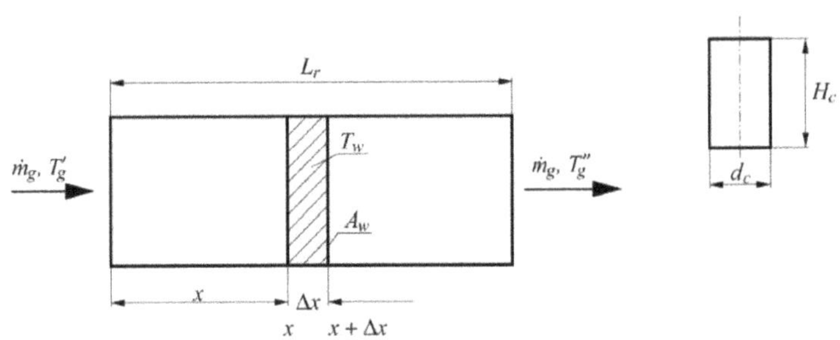

Figure 5. Heat accumulator with a finite volume of a Δx thickness.

The energy conservation equation for air (gas) for a control volume of thickness Δx (Figure 5) has the following form:

$$\dot{m}_g \, c_{pg}\big|_0^{T_g} T_g\big|_x + hA_{pack}\frac{\Delta x}{L_r}\left(T_w\big|_{x+\frac{\Delta x}{2}} - T_g\big|_{x+\frac{\Delta x}{2}}\right) = \dot{m}_g \, c_{pg}\big|_0^{T_g} T_g\big|_{x+\Delta x} + p_r A_w \Delta x \rho_g c_{pg} \left.\frac{\partial T_g}{\partial t}\right|_{x+\frac{\Delta x}{2}} \quad (1)$$

where $\dot{m}_g$—air mass flow rate, kg/s; $A_w = \pi D_w^2/4$—cross-section area of the packing, m^2; D_w—inner diameter of the accumulator casing, m; $c_{pg}|_0^{T_g}$—the mean specific heat of air at constant pressure in the temperature range from 0 °C to T_g, J/(kg·K); T_g—air temperature, °C; T_w—packing temperature, °C; h—heat transfer coefficient, W/(m^2·K); A_{pack}—the surface area of ceramic and steel elements where heat exchange with air takes place, m^2; $p_r = V_g/V_c$—porosity; V_g—air volume in the accumulator, m^3; V_c—the total capacity of the accumulator, m^3; Δx—the thickness of the control area, m; ρ_g—air density, kg/m^3; L_r—accumulator length, m; $c_{pg}(T_g)$—specific heat of air at constant pressure at temperature T_g, J/(kg·K); and t—time, s.

Introducing the mean specific heat in the control volume over the temperature interval from $T_g|_x$ to $T_g|_{x+\Delta x}$ is defined as:

$$c_{pg}(T) = \frac{c_{pg}|_0^{T_g} T_g|_{x+\Delta x} - c_{pg}|_0^{T_g} T_g|_x}{T_g|_{x+\Delta x} - T_g|_x} \tag{2}$$

which gives:

$$\dot{m}_g c_{pg}(T)\left(T_g|_{x+\Delta x} - T_g|_x\right)\frac{1}{\Delta x} + p_r A_w \rho_g c_{pg}\frac{\partial T_g}{\partial t} = hA_{pack}\frac{1}{L_r}\left(T_w|_{x+\frac{\Delta x}{2}} - T_g|_{x+\frac{\Delta x}{2}}\right) \tag{3}$$

where $c_{pg}(T)$ is the specific heat of the air at constant pressure at temperature T.

When $\Delta x \to 0$, then Equation (3) takes the form:

$$\dot{m}_g c_{pg}(T) L_r \frac{\partial T_g}{\partial x} + p_r A_w L_r \rho_g c_{pg}(T)\frac{\partial T_g}{\partial t} = hA_{pack}(T_w - T_g) \tag{4}$$

By introducing a dimensionless coordinate $x^+ = x/L_r$, Equation (4) can be written as:

$$\frac{1}{N_g}\frac{\partial T_g}{\partial x^+} + \tau_g\frac{\partial T_g}{\partial t} = T_w - T_g \tag{5}$$

The number of units N_g and the time constant τ_g are defined by the following expressions:

$$N_g = \frac{hA_{pack}}{\dot{m}_g c_{pg}} \quad \tau_g = \frac{m_g c_{pg}}{hA_{pack}} \tag{6}$$

where $m_g = p_r A_w L_r \rho_g(\overline{T}_g)$—mass of air in the accumulator, kg; p_r—packing porosity; $\rho_g(\overline{T}_g) = \rho_{gn}(\overline{T}_g + 273.15)/273.15$; ρ_{gn}—air density in standard temperature and pressure conditions, kg/m^3; $\rho_g(\overline{T}_g)$—air density at mean air temperature $\overline{T}_g$, kg/m^3; and $\dot{m}_g$—mass flow rate of the air flowing through the accumulator, kg/s.

Next, the differential equation for the packing was derived. The conservation energy equation for the packing is:

$$hA_{pack}\frac{\Delta x}{L_r}\left(T_w|_{x+\frac{\Delta x}{2}} - T_g|_{x+\frac{\Delta x}{2}}\right) + (m_m c_m + m_c c_c)\frac{\Delta x}{L_r}\frac{\partial T_w}{\partial t} = 0 \tag{7}$$

where T_w—packing temperature, °C; m_m—the mass of the steel structural components inside the accumulator including the mass of the accumulator casing, kg; c_m—stainless steel specific heat, J/(kg·K); m_c—the mass of the ceramic packing elements, kg; and c_c—specific heat of packing elements, J/(kg·K).

Transformation of Equation (7) to dimensionless form gives:

$$\tau_w\frac{\partial T_w}{\partial t} = -(T_w - T_g) \tag{8}$$

where τ_w is given by the following relation:

$$\tau_w = \frac{m_m c_m + m_c c_c}{h A_{pack}} \tag{9}$$

Equations (5) and (8) describe the changes in air and packing temperature as a function of time.

The boundary condition for air and the initial temperatures for air and packing have the following forms:

$$T_g\big|_{x^+=0} = T_g'(t) \tag{10}$$

$$T_g\big|_{t=0} = T_0 \tag{11}$$

$$T_w\big|_{t=0} = T_0 \tag{12}$$

where $T_g'(t)$—the air temperature at the accumulator inlet, °C; and T_0—initial packing and air temperature, °C.

4. Modelling the Operation of a Heat Accumulator Using Exact Analytical Methods

Solutions of Equations (5) and (8) with the boundary condition (10) and initial conditions (11) and (12) can only be found for specific boundary and initial conditions when using exact analytical methods. In the general case, when the air temperature at the inlet to the accumulator is a function of time, it is not easy to find an exact analytical solution. In this paper, solutions to two problems are presented. In the first problem, the initial temperature of the packing and the air was uniformly equal to $T_{g,0}$. The air temperature at the inlet to the accumulator decreased stepwise by the value ΔT_g. In the second problem, the initial temperature of the packing and air was also equal to T_0. The air temperature at the inlet to the accumulator increased first at a constant rate v_T and then equalled the set nominal temperature T_{gnom}. Exact analytical solutions were used to assess the accuracy of the numerical solutions obtained using the finite-difference method.

4.1. Exact Analytical Solution for a Step Change in the Air Temperature at the Heat Accumulator Inlet

The temperature of the air flowing through the heat accumulator was a function of time and position. The operation of the heat accumulator was modelled when the air temperature at the inlet to the accumulator was stepped down (Figure 6).

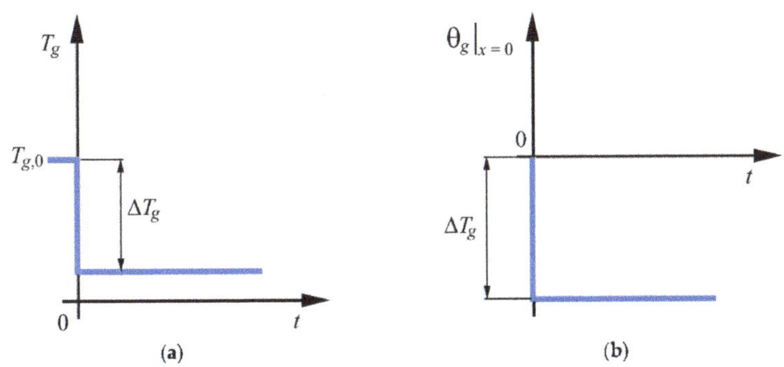

Figure 6. Changes in air temperature at the accumulator inlet: **(a)** air temperature changes T_g; **(b)** changes in temperature difference $\theta_g\big|_{x=0} = T_g\big|_{x=0} - T_0$.

The differential equations describing the air and packing temperatures as a function of time and position were derived under the following simplifying assumptions:

- The physical properties of the air and the ceramic packing were constant;
- The packing temperature was a function of time only;

- The air temperature was a function of time t and the x-coordinate, and in a given cross-section, the air temperature is uniform;
- The external surface of the accumulator was wholly insulated, and heat transfer took place only between the packing and the air flowing through the accumulator.

Changes in the air temperature inside the accumulator were described by the conservation of energy Equation (5) and the packing temperature by Equation (8).

The system of Equations (5) and (8) was solved with the following boundary condition

$$T_g|_{x=0} = T_{g,in} \tag{13}$$

The initial temperature of the air and the accumulator packing was constant and equal to T_0; i.e., the initial condition for air and wall has the following form:

$$T_g|_{t=0} = T_{g,0} \tag{14}$$

$$T_w|_{t=0} = T_{g,0} \tag{15}$$

By introducing temperature differences:

$$\theta_g = T_g - T_{g,0} \tag{16}$$

$$\theta_w = T_w - T_{g,0} \tag{17}$$

Equation (5) becomes:

$$\tau_g \frac{\partial \theta_g}{\partial t} + \frac{1}{N_g} \frac{\partial \theta_g}{\partial x^+} = -(\theta_g - \theta_w) \tag{18}$$

The conservation of energy Equation (8) for accumulator packing can be written in the form

$$\tau_w \frac{\partial \theta_w}{\partial t} = \theta_g - \theta_w \tag{19}$$

The boundary condition (13) and the initial conditions (14) and (15) can be written in the following manner (Figure 6):

$$\theta_g|_{x=0} = -\Delta T_g \tag{20}$$

$$\theta_g|_{t=0} = 0 \tag{21}$$

$$\theta_w|_{t=0} = 0 \tag{22}$$

The analytical solution of the system of Equations (18) and (19) for a stepwise increase in the medium temperature by $-\Delta T_g$ and initial conditions (21) and (22) has the form [31]:

$$\frac{\theta_g}{\Delta T_g} = -U(\xi,\eta)\exp[-(\xi+\eta)],\ t \geq t_{pr} \tag{23}$$

$$\frac{\theta_w}{\Delta T_g} = -\left[U(\xi,\eta) - I_0\left(2\sqrt{\xi\eta}\right)\right]\exp[-(\xi+\eta)],\ t \geq t_{pr} \tag{24}$$

where:

$$\xi = \frac{xN_g}{L_x} \quad \eta = \frac{t-t_{pr}}{\tau_w} \quad t_{pr} = x^+ N_g \tau_g \tag{25}$$

$$U = U(\xi,\eta) = e^{\xi+\eta} - \sum_{n=1}^{\infty}\left(\frac{\xi}{\eta}\right)^{\frac{n}{2}} I_n\left(2\sqrt{\xi\eta}\right) \tag{26}$$

In Equations (23), (24) and (26), $I_0(x)$—modified function of order zero $I_n(x)$—modified function of n-th order [34], and t_{pr}—the time of the particle passage from the inlet (x = 0) to the point of x coordinate determined by the formula $t_{pr} = x/w_g$.

4.2. Exact Analytical Solution for Changing the Air Temperature at the Inlet to the Heat Accumulator in the Form of a Ramp

Modelling of the heat accumulator was carried out using the explicit finite-difference method and the Crank–Nicolson method to test their suitability for determining the air and filling temperature distribution with time-dependent air temperature at the inlet to the accumulator. For this reason, the heat transfer in the accumulator was modelled under the particularly challenging boundary condition of a step change in temperature. Exact analytical solutions made it possible to assess the accuracy of the heat storage unit modelling using numerical methods for different time-dependent boundary conditions, as so far, there are no methods to assess the accuracy of numerical methods. There was no guarantee that both numerical methods used in the paper would give correct results.

The time–space temperature distribution during the flow of hot medium through the accumulator was determined. The initial temperature of the air and packing was constant and was equal to T_0. At the beginning of the heating process, the air temperature increased in steps by ΔT_g (Figure 7). The air temperature then rose at a constant rate v_T for the time t_{cn} until the temperature of the medium reached T_{nom}. For time $t_{cn} \leq t$, the air temperature was constant and equal to T_{nom}.

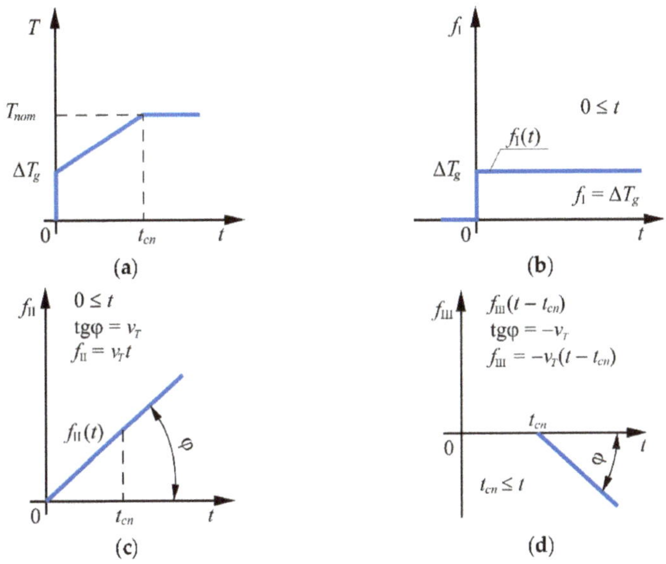

Figure 7. Changes in gas temperature over time at the accumulator inlet (**a**), which were decomposed into three components according to the superposition method: $f_I(t)$ (**b**), $f_{II}(t)$ (**c**), and $f_{III}(t - t_{cn})$ (**d**).

The initial air and packing temperatures were assumed to be zero to simplify considerations. This assumption did not reduce the generality of the solutions obtained because the initial temperature $T_{g,0}$ had to be added to the packing and air temperatures obtained at the zero initial temperature.

The temperature distribution in the air and the pipe wall at the boundary condition shown in Figure 7a was determined using the superposition method [35]. According to this method, the air and packing temperatures were defined as the following:

$$T_g(x,t) = T_1(x,t) + T_2(x,t),\ 0 \leq t \leq t_{cn} \tag{27}$$

$$T_w(x,t) = T_{w1}(x,t) + T_{w2}(x,t),\ 0 \leq t \leq t_{cn} \tag{28}$$

$$T_g(x,t) = T_1(x,t) + T_2(x,t) + T_3(x,t - t_{cn}),\ t_{cn} \leq t \tag{29}$$

$$T_w(x,t) = T_{w1}(x,t) + T_{w2}(x,t) + T_{w3}(x, t - t_{cn}),\ t_{cn} \le t \tag{30}$$

The boundary conditions for solutions T_1, T_2, and T_3; i.e., the change in air temperature over the time at the accumulator inlet, are shown in Figure 7b–d, respectively.

4.2.1. Solution $T_1(x, t)$ for Air Temperature

The following differential equation describes the air temperature variation in the accumulator:

$$\tau_g \frac{\partial T_1}{\partial t} + \frac{1}{N_g}\frac{\partial T_1}{\partial x^+} = -(T_1 - T_w) \tag{31}$$

where T_1 is the excess temperature over the ambient temperature T_0. The following differential equation describes the variation of the packing temperature over the time:

$$\tau_w \frac{\partial T_{w1}}{\partial t} = T_1 - T_{w1} \tag{32}$$

The time changes of the air inlet temperature are described by the boundary condition (Figure 7b):

$$T_1|_{x=0} = \Delta T_g,\ 0 < t \tag{33}$$

The initial conditions have the following form:

$$T_1 = 0 \tag{34}$$

$$T_{w1} = 0 \tag{35}$$

The solution of the initial-boundary problem (31)–(33) is defined as:

$$\frac{T_1 - T_{g,0}}{\Delta T_g} = U(\xi,\eta)\exp[-(\xi+\eta)],\ t \ge t_{pr} \tag{36}$$

$$\frac{T_{w1} - T_{g,0}}{\Delta T_g} = \left[U(\xi,\eta) - I_o\left(2\sqrt{\xi\eta}\right)\right]\exp[-(\xi+\eta)],\ t \ge t_{pr} \tag{37}$$

4.2.2. Solution $T_2(x, t)$ for Air Temperature

The differential equations describing the changes in air temperature and packing have the following forms:

$$\tau_g \frac{\partial T_2}{\partial t} + \frac{1}{N_g}\frac{\partial T_2}{\partial x^+} = -(T_2 - T_w) \tag{38}$$

$$\tau_w \frac{\partial T_{w2}}{\partial t} = T_2 - T_{w2} \tag{39}$$

The air temperature T_2 at the inlet to the accumulator increases linearly with time and the initial air and packing temperatures are equal to zero. The boundary condition and initial conditions have the form:

$$T_2|_{x=0} = f_I(t),\ 0 \le t \le t_{cn} \tag{40}$$

$$T_2|_{t=0} = 0 \tag{41}$$

$$T_{w2}|_{x=0} = 0 \tag{42}$$

where $f_I(t) = v_T\, t$. The symbol v_T denotes the rate of change of the air temperature at the inlet to the accumulator (Figure 7c).

The solution of the initial problem (38)–(42) has the following form:

$$T_2 = v_T \tau_w \left\{ e^{-(\xi+\eta)}\left[(\eta-\xi)U + \xi I_0\left(2\sqrt{\xi\eta}\right) + \sqrt{\xi\eta} I_1\left(2\sqrt{\xi\eta}\right)\right]\right\},\ 0 \le t \le t_{cn}, \tag{43}$$

$$T_{w2} = T_2 - v_T \tau_w e^{-(\xi+\eta)} \left[U - I_0\left(2\sqrt{\xi\eta}\right) \right],\ 0 \le t \le t_{cn}, \tag{44}$$

where the function U is defined as follows:

$$U = U(\xi, \eta) = e^{\xi+\eta} - \sum_{n=1}^{\infty} \left(\frac{\xi}{\eta}\right)^{\frac{n}{2}} I_n\left(2\sqrt{\xi\eta}\right) \tag{45}$$

where $I_0(x)$, $I_1(x)$, and $I_n(x)$ denote modified Bessel functions of the zero, first, and n-th order, respectively.

4.2.3. Solution T_3(x, t) for Air Temperature

The differential equations describing the changes in air temperature and packing have the following forms:

$$\tau_g \frac{\partial T_3}{\partial t} + \frac{1}{N_g} \frac{\partial T_3}{\partial x^+} = -(T_3 - T_w) \tag{46}$$

$$\tau_w \frac{\partial T_{w3}}{\partial t} = T_3 - T_{w3} \tag{47}$$

The air temperature T_3 at the inlet to the accumulator increases linearly with time and the initial air and packing temperatures are equal to zero. The boundary condition and initial conditions take the form:

$$T_3|_{x=0} = f_{\rm II}(t - t_{cn}) = -v_T(t - t_{cn}),\ t \ge t_{cn} \tag{48}$$

$$T_3|_{t=0} = 0 \tag{49}$$

$$T_{w3}|_{t=0} = 0 \tag{50}$$

where $f_{II}(t) = v_T(t - t_{cn})$, $t_{cn} \le t$. The symbol v_T is the rate of change of the air temperature at the inlet to the accumulator (Figure 7d).

$$T_3 = -v_T \tau_w \left\{ e^{-(\xi+\eta_1)} \left[(\eta_1 - \xi)U + \xi I_0\left(2\sqrt{\xi\eta_1}\right) + \sqrt{\xi\eta_1} I_1\left(2\sqrt{\xi\eta_1}\right) \right] \right\},\ t \ge t_{cn} \tag{51}$$

$$T_{w3} = T_3 + v_T \tau_w e^{-(\xi+\eta_1)} \left[U - I_0\left(2\sqrt{\xi\eta_1}\right) \right],\ t \ge t_{cn} \tag{52}$$

where $\xi = \frac{xN_g}{L_x}$, $\eta = \frac{t-t_{pr}}{\tau_w}$, $\eta_1 = \frac{t-t_{cn}-t_{pr}}{\tau_w}$, and $t_{pr} = x^+ N_g \tau_g$.

The number of heat transfer units N_g, the time constant τ_g, and the wall time constant τ_w are defined by Equations (6) and (9), respectively.

5. Numerical Modelling of a Heat Accumulator

The differential equations describing the air and packing temperature changes were solved using the finite-difference method. Two finite difference schemes were used: the explicit Euler method and the Crank–Nicolson method.

The accuracy of the explicit Euler method is of the first order [36], and that of the Crank–Nicolson method [36] is the second order; i.e., the latter method is more accurate with the same number of nodes in the difference mesh. The advantage of these methods was that they could consider that the thermophysical properties of the packing and air were temperature dependent. The air temperature at the inlet to the accumulator could change over time. With the numerical methods, the accumulator air and packing temperature distribution could be determined using any temporal variation of the air temperature at its inlet. Unlike the exact analytical method, the finite difference procedure did not require the boundary condition to be a step or linear change in the air temperature.

Cooling of the accumulator filling, during which heat was extracted by the flowing air and then transferred to the finned air–water heat exchanger, was modelled using two methods: the explicit Euler method and the implicit Crank–Nicolson method. To assess the accuracy of these methods, an exact analytical solution was also found for selected temporal variations in the air temperature at the accumulator inlet. The mutual comparison of the two different numerical methods and their comparison with the exact analytical solution was necessary because, as shown in [14], the accuracy of the solutions obtained by the finite-difference (finite volume) method depends very strongly on the number of divisions of the accumulator length into calculation cells. The air temperature over the length of one cell was approximated using the arithmetic mean of the inlet and outlet temperatures for the cell. However, it turned out that the temperature distribution inside a given cell was exponential, and at low air velocities, the air temperature dropped quickly over the short length from the air inlet to the cell. In this case, the arithmetic mean did not give the true average temperature over the length of a single cell. Computational tests of air heating in an accumulator [14] have shown that when the length of the accumulator was divided into 10 control volumes, inaccurate—one might say absurd—results were obtained. For example, at an accumulator filling temperature of 300 °C, the air temperature at the heat accumulator outlet reached 500 °C, which was obviously not possible. For this reason, the assessment of the accuracy of the two numerical methods used was given more attention.

5.1. Modelling a Heat Accumulator Using an Explicit Finite-Difference Method

The system of Equations (5) and (8) with boundary condition (10) and initial conditions (11) and (12) was solved using the explicit finite-difference method. An accumulator of length L_r was divided into N finite volumes of length Δx (Figure 8). By introducing a dimensionless coordinate, the dimensionless length of the finite volume is given by the formula:

$$\Delta x^{+} = \frac{1}{N}, \tag{53}$$

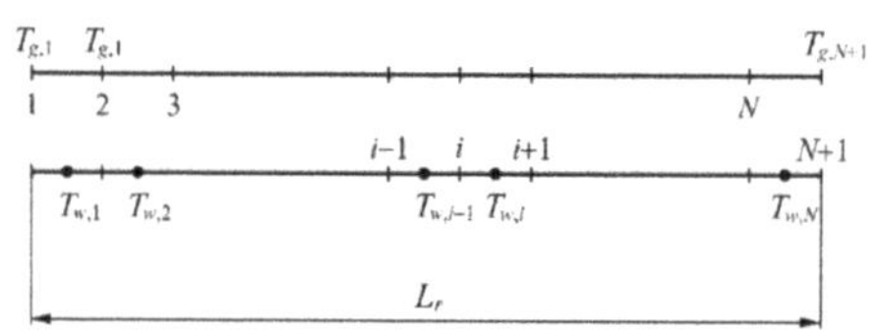

Figure 8. Finite difference grid used in air (T_g) and packing (T_w) temperature calculations.

The derivative of the air temperature T_g after the x-coordinate in Equation (5) was approximated by the forward difference quotient at the old time step n, and the derivative after time by the forward difference quotient for node $i + 1$, which was located at the outlet of the control region contained between nodes i and $i + 1$. The remaining terms in Equation (5) were calculated at the old time step $t_n = n\Delta t$, $n = 0, 1, \ldots$. The approximate form of Equation (5); i.e., the difference equation, is as follows:

$$\tau_{g,i}^{n} \frac{T_{g,i+1}^{n+1} - T_{g,i+1}^{n}}{\Delta t} + \frac{1}{N_{g,i}^{n}} \frac{T_{g,i+1}^{n} - T_{g,i}^{n}}{\Delta x^{+}} = T_{w,i}^{n} - \frac{T_{g,i}^{n} + T_{g,i+1}^{n}}{2} \tag{54}$$

The finite-difference grid is defined as follows:

$$x_i = (i-1)\Delta x,\ \Delta x = L_r/N,\ x_i^{+} = x_i/L_r\ i = 1,\ldots,(N+1)\ t_n = n\Delta t,\ n = 0,1,\ldots \tag{55}$$

The differential Equation (8) was approximated in a similar way:

$$\tau_{w,i}^{n} \frac{T_{w,i}^{n+1} - T_{w,i}^{n}}{\Delta t} = -\left(T_{w,i}^{n} - \frac{T_{g,i}^{n} + T_{g,i+1}^{n}}{2} \right) i = 1,\ldots,(N+1)\ n = 0,1,\ldots \tag{56}$$

Solving Equation (54) for $T_{g,i+1}^{n+1}$ gives:

$$T_{g,i+1}^{n+1} = T_{g,i+1}^{n} + \frac{\Delta t}{\tau_{g,i}^{n}}\left[\left(T_{w,i}^{n} - \frac{T_{g,i}^{n} + T_{g,i+1}^{n}}{2}\right) - \frac{1}{N_{g,i}^{n}}\frac{T_{g,i+1}^{n} - T_{g,i}^{n}}{\Delta x^{+}}\right], i = 1,\ldots,N\ n = 0,\ldots \tag{57}$$

The wall temperature $T_{w,i}^{n+1}$ was determined using Equation (56):

$$T_{w,i}^{n+1} = T_{w,i}^{n} - \frac{\Delta t}{\tau_{w,i}^{n}}\left(T_{w,i}^{n} - \frac{T_{g,i}^{n} + T_{g,i+1}^{n}}{2}\right), i = 1,\ldots,N\ n = 0,1,\ldots \tag{58}$$

The system of difference Equations (57) and (58) was solved with the boundary condition (10), which takes the form:

$$T_{g,1} = \left(T_g'\right)^{n} \tag{59}$$

The initial conditions for gas and packing are as follows:

$$T_{g,i}^{0} = T_0\ \ i = 1,\ldots,N+1 \tag{60}$$

$$T_{w,i}^{0} = T_0\ \ i = 1,\ldots,N \tag{61}$$

From Equations (57) and (58), the air and packing temperatures were determined as a function of location and time while taking into account the boundary condition (59) and the initial conditions (60) and (61). The time step Δt should satisfy the Courant–Friedrichs–Lewy condition [36]:

$$\frac{\Delta t}{N_g \tau_g \Delta x^{+}} \leq 1 \tag{62}$$

Due to the small value of the product $\rho_g c_{pg}$, the heat accumulated in the air is minimal. Therefore, air temperature changes over time occur very quickly. Disregarding the heat accumulation in the air; i.e., assuming $\tau_g = 0$, a simplified form of the differential Equation (5) is as follows:

$$\frac{1}{N_g}\frac{\partial T_g}{\partial x^{+}} = T_w - T_g \tag{63}$$

Replacing the derivatives in Equation (63) with the difference quotient yields:

$$\frac{1}{N_{g,i}^{n}}\frac{T_{g,i+1}^{n+1} - T_{g,i}^{n+1}}{\Delta x^{+}} = T_{w,i}^{n+1} - \frac{T_{g,i}^{n+1} + T_{g,i+1}^{n+1}}{2} \tag{64}$$

The solution of Equation (64) for $T_{g,i+1}^{n+1}$ has the following form:

$$T_{g,i+1}^{n+1} = \frac{1}{1 + \frac{\Delta x^{+}}{2N_{g,i}^{n}}}\left[\frac{\Delta x^{+}}{N_{g,i}^{n}}T_{w,i}^{n+1} + T_{g,i}^{n+1}\left(1 - \frac{\Delta x^{+}}{2N_{g,i}^{n}}\right)\right],\ i = 1,\ldots,N\ n = 0,1,\ldots \tag{65}$$

The packing temperature is calculated using the relationship (58) with boundary condition (59) and initial conditions (60) and (61).

Assuming zero air heat capacity, the computer simulation time is greatly reduced, as there is no Courant–Friedrichs–Lewy stability condition on time-step length Δt (62). At low air velocities, the air temperature rise per control volume is large, and the approximation of the average temperature by the arithmetic mean of the inlet and outlet temperatures, as in Equation (54) or Equation (64), is insufficiently accurate [14]. The accumulator length must be divided into a large number of control volumes; e.g., assume $N = 50$ or more to ensure high calculation accuracy at low air velocities [14].

5.2. Modelling a Heat Accumulator Using the Crank–Nicolson Method

In the Crank–Nicolson method [36], the derivative after time in the differential equations for air and packing was approximated by the forward difference quotient. The arithmetic mean of the values approximated the remainder of the equation at the beginning and end of a given time step. Equation (5) can be written in the following form:

$$\tau_g \frac{\partial T_g}{\partial t} = -\frac{1}{N_g} \frac{\partial T_g}{\partial x^+} + (T_w - T_g) \tag{66}$$

Equation (66) was approximated by the following difference scheme using the Crank-Nicolson method:

$$\begin{aligned} &\frac{1}{2}\left(\tau_{g,i}^{n} + \tau_{g,i}^{n+1}\right) \frac{T_{g,i+1}^{n+1} - T_{g,i+1}^{n}}{\Delta t} = \frac{1}{2}\left\{\left[-\frac{1}{N_{g,i}^{n}} \frac{T_{g,i+1}^{n} - T_{g,i}^{n}}{\Delta x^+} + \left(T_{w,i}^{n} - \frac{T_{g,i}^{n} + T_{g,i+1}^{n}}{2}\right)\right] + \right. \\ &\left. + \left[-\frac{1}{N_{g,i}^{n+1}} \frac{T_{g,i+1}^{n+1} - T_{g,i}^{n+1}}{\Delta x^+} + \left(T_{w,i}^{n+1} - \frac{T_{g,i}^{n+1} + T_{g,i+1}^{n+1}}{2}\right)\right]\right\} \; i = 1, \ldots, N,\; n = 0, 1, \ldots \end{aligned} \tag{67}$$

Equation (8) for the accumulator packing was transformed in a similar manner:

$$\begin{aligned} &\frac{1}{2}\left(\tau_{w,i}^{n} + \tau_{w,i}^{n+1}\right) \frac{T_{w,i}^{n+1} - T_{w,i}^{n}}{\Delta t} = -\frac{1}{2}\left[\left(T_{w,i}^{n} - \frac{T_{g,i}^{n} + T_{g,i+1}^{n}}{2}\right) + \left(T_{w,i}^{n+1} - \frac{T_{g,i}^{n+1} + T_{g,i+1}^{n+1}}{2}\right)\right] \\ &i = 1, \ldots, N,\; n = 0, 1, \ldots \end{aligned} \tag{68}$$

The system of Equations (67) and (68) was solved for $T_{g,i+1}^{n+1}$ and $T_{w,i}^{n+1}$. To solve the system of Equations (67) and (68), the Thomas method [37], also known as the tri-diagonal matrix algorithm, was used. The Thomas algorithm is a particular case of the Gauss elimination method that does not require inversion of the coefficient matrix. The unknowns in the system of equations were determined using simple analytical formulas. Therefore, the time needed for numerical calculation was very short. The set of Equations (67) and (68) can also be solved with the iterative method of Gauss–Seidel. The temperatures $T_{g,i+1}^{n}$ and $T_{w,i}^{n}$ were taken as initial values in the iterative process. Due to the insignificant differences between $T_{g,i+1}^{n+1}$ and $T_{g,i+1}^{n}$ and between $T_{w,i}^{n+1}$ and $T_{w,i}^{n}$ the number of iterative steps required to obtain the solution was small.

5.3. Computational Tests

Two cases of a change in inlet air temperature were considered. In the first case, the air temperature decreased stepwise from an initial temperature T_0 = 600 °C to an ambient temperature of 20 °C (Figure 9). In the second case, the initial air and packing temperature increased stepwise from initial temperature T_0 = 0 °C to 100 °C, then rose at a constant rate of 0.25 K/s and then remained constant after reaching 600 °C (Figure 10).

The results shown in Figures 9 and 10 were obtained for the following data calculated for the accumulator discussed in Section 2: N_g = 1.275, τ_w = 1013.627 s, τ_g = 0.3577 s, h_g = 11.55 W/(m^2·K), N = 20, and Δt = 0.005. Figure 10 shows the results of the calculations obtained for the same data except for the number of nodes, which was (N + 1) = 65.

The analysis of the results shown in Figure 9 demonstrated that even with a small number of nodes along the length of the regenerator, the agreement between the air and packing temperatures at the outlet of the regenerator calculated using the exact analytical method and the explicit finite-difference method was very good.

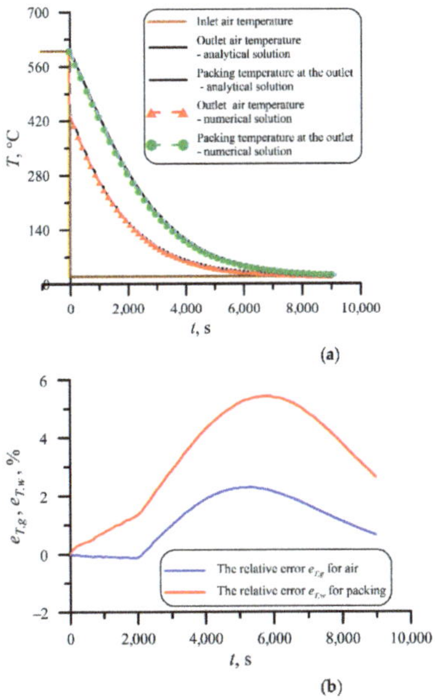

Figure 9. Air and packing temperature variations at the accumulator outlet determined for a step change in the air temperature from 600 °C to 20 °C of the air temperature at the accumulator inlet (**a**); and the relative difference between the temperature determined using the exact analytical formula and the finite-difference method (**b**).

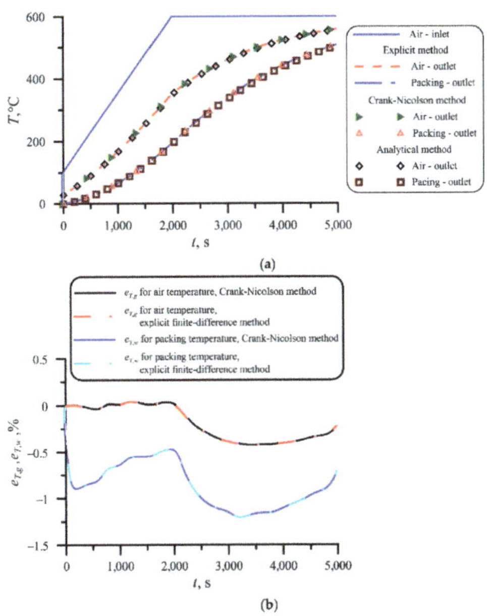

Figure 10. Air and packing temperature changes at the outlet of the accumulator; at the beginning, the air temperature increased in steps from 0 to 100 °C, then increased at a constant rate of 0.25 K/s, then remained stable after reaching 600 °C (**a**); and the relative difference between temperature determined using the exact analytical formula, the explicit finite-difference method, and the Crank–Nicolson method (**b**).

The relative values of the difference between the air temperature determined by the exact analytical formula and the finite-difference method were calculated using the following formula:

$$e_{T,g} = 100\frac{T_g^{exact}\Big|_{x=L_r} - T_g^{num}\Big|_{x=L_r}}{T_g^{exact}\Big|_{x=L_r}}, \ \% \quad (69)$$

The relative difference was calculated analogously for the packing temperatures:

$$e_{T,w} = 100\frac{T_w^{exact}\big|_{x=L_r} - T_w^{num}\big|_{x=L_r}}{T_w^{exact}\big|_{x=L_r}}, \ \% \quad (70)$$

where T_g^{exact} and T_w^{exact}—air and packing temperatures calculated using an exact analytical method, respectively; and T_g^{num} and T_w^{num}—air and packing temperatures calculated using a numerical method, respectively.

The maximum value of $e_{T,g}$ = 2.3% occurred at time t = 5400 s and the maximum value of $e_{T,w}$ = 5.43% at time t = 5700 s (Figure 9b).

The relative differences between the exact analytical solution and the numerical solution were smaller for the second case analysed due to the larger number of nodes equal to (N + 1) = 65 and the use of the Crank–Nicolson method to solve the system of equations, which had a second order of accuracy; i.e., it was more accurate than the explicit finite-difference method, the order of accuracy of which was one.

The absolute maximum relative difference values were $e_{T,g}$ = 0.425% and $e_{T,w}$ = 1.20% for the explicit finite-difference method and $e_{T,g}$ = 0.424% and $e_{T,w}$ = 1.20% for the Crank-Nicolson method (Figure 10b).

From a comparison of the results shown in Figures 9b and 10b, it can be seen that the accuracy of calculations using the explicit finite-difference method increased as the number N of finite volumes increased. When increasing N from 20 to 64, the relative error $e_{T,g}$ for air decreased by 5.4 times and for packing $e_{T,w}$ by 4.5 times. When dividing the pipe into N = 64 finite volumes (Figure 10b), the Crank–Nicolson method gave similar results to the explicit finite-difference method due to a large number of finite volumes. The computational tests carried out demonstrated that both the explicit finite-difference method and the Crank–Nicolson method could be used to simulate the operation of the accumulator under analysis. The accuracy of such a simulation was significantly affected by the number of finite volumes, especially at low air velocity. In order to achieve sufficient calculation accuracy, it is recommended that $N \geq 50$.

6. Experimental Verification of the Developed Numerical Model of Accumulator Discharge

The packing of the heat accumulator under test consisted of cylindrical ceramic elements and pipes, as well as other steel elements. For modelling, the following thermophysical properties and dimensions of the accumulator construction elements were assumed: Prandtl number for air, Pr_g = 0.7; air thermal conductivity, k_g = 0.236 W/(m·K); kinematic viscosity of the air, $\nu_g = 16.35 \cdot 10^{-6}$ m^2/s; air density at 50 °C, ρ_g = 1.09 kg/m^3; air specific heat at constant pressure, $c_{p,g}$ = 1000.0 J/(kg·K); air velocity before the packing, w_g = 2.0 m/s; stainless steel thermal conductivity, k_m = 14.7 W/(m·K); stainless steel density, ρ_m = 7800 kg/m^3; steel specific heat, c_m = 519 J/(kg·K); total mass of steel elements, m_m = 252.8 kg; heat accumulator casing cross-sectional area, A_r = 2.17 m^2; length of heat accumulator, L_r = 2 m; thermal conductivity of ceramic cylinder, k_w = 6.4 W/(m·K); density of ceramic cylinder, ρ_w = 2700 kg/m^3; specific heat of ceramic cylinder, c_w = 519 J/(kg·K); total mass of ceramic packing, m_w = 252.8 kg; and ceramic packing outer surface area, A_w = 22.57 m^2.

Air temperature measurements were carried out using K-type sheathed thermocouple sensors with a grounded junction. All thermocouples were pre-calibrated using a Pt100 platinum resistance sensor. The thermocouples were selected so that when measuring air temperatures below 100 °C, their readings would be the same with an accuracy of ±0.1 K. Uncertainty of temperature measurement in the air temperature range of −20 °C to +100 °C was assumed as for a resistance temperature sensor Pt100; i.e., ±0.35 K

Air velocity was measured using a digital vane anemometer with a maximum resolution of 0.01 m/s. The uncertainty of the sensor was ±0.5% of the upper limit of the measuring range or ±1.5% of the measured value.

Due to the lack of dependence in calculating the heat transfer coefficient between the air and the elements of the heat accumulator with such a complex packing structure, a procedure for determining the mean heat transfer coefficient was developed.

The accumulator we built and tested was characterised by a complex flow and heat system inside the packing. Part of the air stream flowed inside the tubes in which the cylindrical filling elements were placed, and the other part flowed outside between the steel tubes. The air-side heat transfer coefficient h_a was approximated using the following formula:

$$h_g = a + b\overline{w}_g^c \tag{71}$$

where a, b, and c are constants to be determined.

The heat transfer coefficient h_g in Formula (71) is expressed in W/(m^2·K) and the mean air velocity through the accumulator $\overline{w}_g^c$ in m/s.

The mean velocity of air flow through the whole cross-section of the accumulator is calculated as:

$$\overline{w}_g = \frac{\dot{m}_g}{\overline{\rho}_g A_g} \tag{72}$$

The free area A_g of the cross-section of the accumulator through which air flows is given by:

$$A_g = \varepsilon \pi R_w^2 \tag{73}$$

where ε is the porosity of the entire accumulator packing and R_w is the internal diameter of the accumulator casing.

All constants in Equation (71) were selected so that the measured and calculated air temperature at the outlet from the accumulator agreed as well as possible. Constants a, b, and c were determined from the condition:

$$S = S(a,b,c) = \sum_{i=1}^{n_m}\left[T_{g,out}^{meas}(t_i) - T_{g,out}^{calc}(t_i)\right]^2 \rightarrow \min \tag{74}$$

The air temperature $T_{g,out}^{calc}$ at the outlet from the accumulator for selected time points t_i, $i = 1, \ldots, n_m$ was calculated using the developed mathematical model.

The constants a, b, and c in correlation (71) were determined from condition (74). The constants a^*, b^*, and c^* at which the function S reached its minimum were determined using the Nelder–Mead method [38]. A library program for optimisation using the Nelder–Mead method can also be found in the IMSL library [39]. Calculations were performed with the use of MATLAB [40] using the FMINSEARCH program. In this paper, the number of measurement points was 1956. The air temperature at the battery outlet was measured with a step equal to 5 s. The total air temperature measurement time was 9775 s.

Calculations of the heat accumulator under test were carried out for the following input data:

- The porosity of the accumulator packing, ε = 0.4025;
- The initial temperature of the packing and the air inside the accumulator, T_0 = 199 °C;
- Heat delivered to the accumulator packing during the charging period (heat accumulated in the packing), ΔQ_{tot} = 13.21 kWh.

The time changes in air velocity ($w_{g,inlet}$) and temperature at the accumulator inlet ($T_{g,inlet}$) are shown in Figure 11.

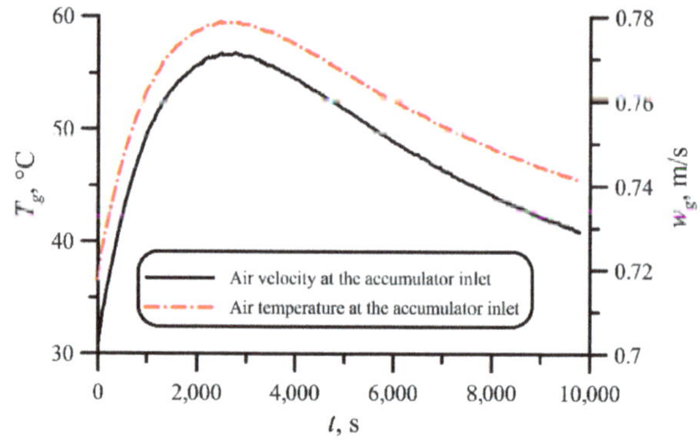

Figure 11. Time changes of air velocity and temperature at the accumulator inlet.

The following values of parameters appearing in Equation (71) were estimated using FMINSEARCH [39]:

$$a = 1.2192,\ b = 8.42 \cdot 10^{-5},\ c = 13.1104 \tag{75}$$

The time changes in the heat transfer coefficient calculated using Formula (71) and the coefficients determined using Equation (74) are shown in Figure 12.

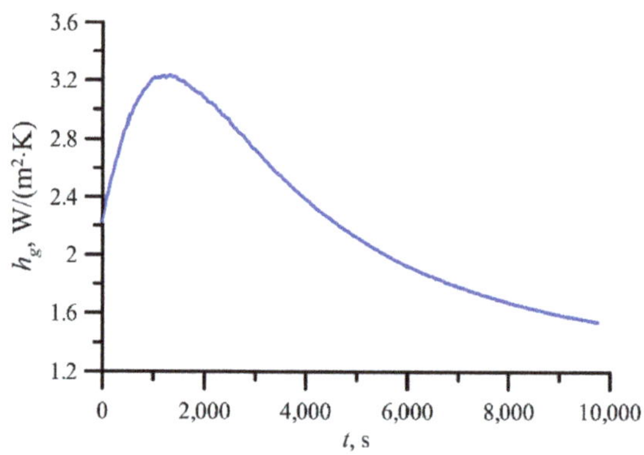

Figure 12. Changes in the heat transfer coefficient h_g on the surface of the heat accumulator packing elements as a function of time.

When analyzing the results shown in Figure 12, it can be seen that the average heat penetration coefficient on all air-flown fill elements was low. This was due to the low air flow velocity inside the accumulator, as the free cross-section through which the air flowed was large.

When analyzing the results presented in Figure 13, it can be seen that at the beginning of accumulator cooling, the differences between the measured and calculated air temperature at the accumulator outlet were more significant. The maximum value of the relative difference e_T between the measured $\left(T_g''\right)^{meas}$ and the calculated $\left(T_g''\right)^{calc}$ of the accumulator outlet air temperature, which was calculated using the relationship $e_T = \left[\left(T_g''\right)^{meas} - \left(T_g''\right)^{calc}\right] / \left(T_g''\right)^{meas} \cdot 100\%$, did not exceed 25%. For times greater than about 700 s, the relative difference e_T between the calculated and measured values of air temperature decreased to values in the order of a few percent. This was mainly due to the approximate heat transfer model in the ceramic cylinders and steel filling elements. The mathematical model of the accumulator assumed that these elements were modelled as objects with concentrated heat capacity. This meant that the temperature difference inside the ceramic cylinders and the walls of the steel tubes was not taken into account. In reality, the temperature of the elements changed with time and inside the elements as well.

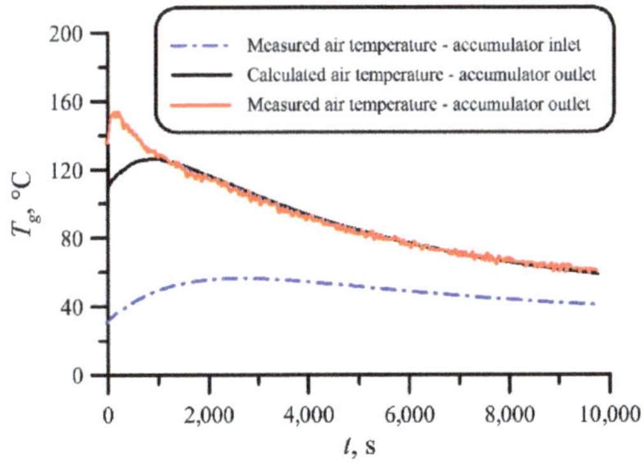

Figure 13. Measured air temperature at the accumulator inlet and comparison of the measured and calculated air temperature at the accumulator outlet as a function of time.

7. Conclusions

Based on the calculations and measurements, the following conclusions were reached:

The investigated heat accumulator was heated at night during the period of low electricity demand when the cost of 1 kWh is low. The packing of the accumulator was heated to a temperature of about 600 °C so that the mass of the accumulator could be smaller at a given heat demand.

The developed mathematical models of the accumulator—an analytical model, a numerical model based on the explicit finite-difference method, and the implicit Crank–Nicolson method—gave similar results.

The comparison between the calculated and measured air temperatures at the accumulator outlet was satisfactory except for the initial air heating period, when the differences between the numerical solution and the measured results were more significant.

The differences in the calculated and measured values of the air temperatures at the accumulator outlet in the initial phase of the cooling of the accumulator packing were partly due to the assumption that the temperature of the filling elements was constant in the entire volume and changed only in time (model with lumped heat capacity). However, at the beginning of the cooling of the filling elements, the temperature difference inside ceramic cylinders or steel tubes could be significant. The second reason for the differences between the calculated and measured temperatures at the air outlet from the accumulator was the complex structure of the accumulator filling, which, for example, made it difficult to determine the velocity distribution in the cross-section of the packing, as well as to find appropriate correlations for calculating the heat transfer coefficient. Due to the small length of the accumulator, the heat transfer between the packing elements and the flowing air was strongly influenced by the inlet section with a developing flow, where the heat transfer coefficient was much higher than for the developed air flow.

The developed numerical model of the accumulator can be used to calculate the appropriate sizing so that it can be used to heat a building during a period of high electricity prices. The model can also be used to calculate the changes in the air temperature at the outlet of the accumulator during the entire period of its discharge (heating of a building). The elaborated dynamic mathematical model of the accumulator can also be used in an automatic temperature control system for a heated room (building) based on the mathematical model. The designed heat accumulator can replace a conventional fossil-fuel-fired boiler in a boiler room or operate as an independent unit in a heated room. In both cases, the mathematical model of the accumulator can be used in the system for automatic air temperature control in the heated space.

Author Contributions: Methodology, formal analysis, writing—original draft, supervision, D.T.; conceptualisation, investigation, software, validation, T.S.; conceptualisation, methodology, software, validation, supervision, writing—review and editing, J.T. (Jan Taler); investigation, data curation, J.T. (Jarosław Tokarczyk). All authors have read and agreed to the published version of the manuscript.

Funding: This research received no external funding.

Institutional Review Board Statement: Not applicable.

Informed Consent Statement: Not applicable.

Conflicts of Interest: The authors declare no conflict of interest.

Nomenclature

a, b, c	Constants in Equation (71)
A	Cross-section area, m^2
Bi	Biot number, Bi = hd/(2k)
c_p	Specific heat at constant pressure, J/(kg·K)
D	Internal diameter, m
d	Outer diameter, m
e_T	Relative error, %
FI	Flow indicator
h	Heat transfer coefficient, W/(m^2·K)
H	Height, m
I_0	Modified Bessel function of zero order
I_1	Modified Bessel function of first order
I_n	Modified function of n-th order
k	Thermal conductivity, W/(m·K)
L_r	Accumulator length, m
m	Mass, kg
$\dot{m}$	Mass flow rate, kg/s
N	Number of control volumes
N_g	Number of heat transfer units $N_g = hA_{pack}/(\dot{m}_g c_{pg})$
PFTHE	Plate-finned and tube heat exchanger
p_r	Packing porosity
Pr	Prandtl number, Pr = $c_p\mu/k$
S	Sum of the least squares
T	Temperature, °C
t	Time, s
TI	Temperature indicator
V	Volume, m^3
v_T	Rate of change of the air temperature at the inlet to the accumulator, K/s
w	Velocity, m/s
$\overline{w}$	Mean air velocity, m/s
x	Axial coordinate, m
x^+	Dimensionless axial coordinate, $x^+ = x/L_r$
Greek symbols	
ν	Kinematic viscosity, m^2/s
ΔQ_{tot}	Heat delivered to the accumulator packing during the charging period, J
Δt	Time step, s
Δx	Thickness of the control area, m
μ	Dynamic viscosity, Pa·s
θ	Temperature difference, °C
ρ	Density, kg/m^3
τ	Time constant, s

Subscripts

0	Initial condition
c	Ceramic cylinder
g	Air
i, j, n	Iterative step
in	Inlet
m	Metal (steel)
pack	Package
w	Heat accumulator

Superscripts

calc	Calculated
meas	Measured
′	At the accumulator inlet
″	At the accumulator outlet

References

1. Li, P.-W.; Chan, C.L. *Thermal Energy Storage Analyses and Designs*; Academic Press Elsevier: London, UK, 2017.
2. Dzierwa, P.; Taler, J.; Peret, P.; Taler, D.; Trojan, M. Transient CFD simulation of charging hot water tank. *Energy* **2022**, *239*, 122241. [CrossRef]
3. Trojan, M.; Taler, D.; Dzierwa, P.; Taler, J.; Kaczmarski, K.; Wrona, J. The use of pressure hot water storage tanks to improve the energy flexibility of the steam power unit. *Energy* **2019**, *173*, 926–936. [CrossRef]
4. Rämä, M.; Leurent, M.; Devezeaux de Lavergne, J.-L. Flexible nuclear co-generation plant combined with district heating and a large-scale heat storage. *Energy* **2020**, *193*, 116728. [CrossRef]
5. Li, H.; Hou, J.; Hong, T.; Ding, Y.; Nord, N. Energy, economic, and environmental analysis of integration of thermal energy storage into district heating systems using waste heat from data centres. *Energy* **2021**, *219*, 119582. [CrossRef]
6. Pinamonti, M.; Beausoleil-Morrison, I.; Prada, A.; Baggio, P. Water-to-water heat pump integration in a solar seasonal storage system for space heating and domestic hot water production of a single-family house in a cold climate. *Sol. Energy* **2021**, *213*, 300–311. [CrossRef]
7. Lakshmanan, V.; Sæle, H.; Degefa, Z. Electric water heater flexibility potential and activation impact in system operator perspective-Norwegian scenario case study. *Energy* **2021**, *236*, 121490. [CrossRef]
8. Yildiz, B.; Bilbao, J.I.; Roberts, M.; Heslop, S.; Dore, J.; Bruce, A.; MacGill, I.; Egan, R.J.; Sproul, A.B. Analysis of electricity consumption and thermal storage of domestic electric water heating systems to utilize excess PV generation. *Energy* **2021**, *235*, 121325. [CrossRef]
9. Behzadi, A.; Arabkoohsar, A. Feasibility study of a smart building energy system comprising solar PV/T panels and a heat storage unit. *Energy* **2020**, *210*, 118528. [CrossRef]
10. Fan, Y.; Zhao, X.; Lia, J.; Cheng, Y.; Badiei, A.; Zhou, J.; Yu, M.; Li, G.; Du, Z.; Ji, J.; et al. Operational performance of a novel fast-responsive heat storage/exchanging unit (HSEU) for solar heating systems. *Renew. Energy* **2020**, *151*, 137–151. [CrossRef]
11. Araújo, A.; Pereira, V. Solar thermal modeling for rapid estimation of auxiliary energy requirements in domestic hot water production: Proportional flow rate control. *Energy* **2017**, *138*, 668–681. [CrossRef]
12. Taler, D.; Sobota, T.; Jaremkiewicz, M.; Taler, J. Influence of the thermometer inertia on the quality of temperature control in a hot liquid tank heated with electric energy. *Energies* **2020**, *13*, 4039. [CrossRef]
13. Taler, D.; Sobota, T.; Jaremkiewicz, M.; Taler, J. Control of the temperature in the hot liquid tank by using a digital PID controller considering the random errors of the thermometer indications. *Energy* **2021**, *239*, 122771. [CrossRef]
14. Taler, D.; Dzierwa, P.; Trojan, M.; Sacharczuk, J.; Kaczmarski, K.; Taler, J. Numerical modeling of transient heat transfer in heat storage unit with channel structure. *Appl. Therm. Eng.* **2019**, *149*, 841–853. [CrossRef]
15. Sacharczuk, J.; Taler, D. Numerical and experimental study on the thermal performance of the concrete accumulator for solar heating systems. *Energy* **2019**, *170*, 967–977. [CrossRef]
16. Rao, C.R.C.; Niyas, H.; Muthukumar, P. Performance tests on lab–scale sensible heat storage prototypes. *Appl. Therm. Eng.* **2018**, *129*, 953–967. [CrossRef]
17. Li, Q.; Bai, F.; Yang, B.; Wang, Y.; Xua, L.; Chang, Z.; Wang, Z.; El Hefni, B.; Yang, Z.; Kubo, S.; et al. Dynamic simulations of a honeycomb ceramic thermal energy storage in a solar thermal power plant using air as the heat transfer fluid. *Appl. Therm. Eng.* **2018**, *129*, 636–645. [CrossRef]
18. Guo, J.; Jiang, Y. A semi-analytical model for evaluating the thermal storage capacity and heat use efficiency of flexible thermal storage heating floor. *Appl. Therm. Eng.* **2021**, *198*, 117448. [CrossRef]
19. Stack, D.C.; Curtis, D.; Forsberg, C. Performance of firebrick resistance-heated energy storage for industrial heat applications and round-trip electricity storage. *Appl. Energ.* **2019**, *242*, 782–796. [CrossRef]
20. Guan, C.; Lu, H.; Zhang, L.; Yu, Z. Regulation of the output temperature in a novel water heating system using solid graphite as sensible heat thermal energy storage medium: Effects of water tank. *Energ. Rep.* **2020**, *6*, 160–171. [CrossRef]

21. Wang, P.; Zhao, P.; Xu, W.; Wang, J.; Dai, Y. Performance analysis of a combined heat and compressed air energy storage system with packed bed unit and electrical heater. *Appl. Therm. Eng.* **2019**, *162*, 114321. [CrossRef]
22. Hussam, W.K.; Rahbarib, H.R.; Arabkoohsar, A. Off-design operation analysis of air-based high-temperature heat and power storage. *Energy* **2020**, *196*, 117149. [CrossRef]
23. Zheng, B.; Xu, J.; Sun, P.; Zhang, K.; Han, X.; Wang, Z.; Wang, Q.; Liu, Y. Effects of solid particle thermal conductivity on heat storage performance of heat storage bed. *Sustain. Energy Technol. Assess.* **2021**, *43*, 100983. [CrossRef]
24. Zuo, H.; Zhou, Y.; Wu, M.; Zeng, K.; Chang, Z.; Chen, S.; Lu, W.; Flamant, G. Development and numerical investigation of parallel combined sensible-latent heat storage unit with intermittent flow for concentrated solar power plants. *Renew. Energ.* **2021**, *175*, 29–43. [CrossRef]
25. Mao, Q.; Li, Y.; Li, G.; Badiei, A. Study on the influence of tank structure and fin configuration on heat transfer performance of phase change thermal storage system. *Energy* **2021**, *235*, 121382. [CrossRef]
26. Huang, S.; Lu, J.; Li, Y. Numerical study on the influence of inclination angle on the melting behaviour of metal foam-PCM latent heat storage units. *Energy* **2021**, *239*, 122489. [CrossRef]
27. Liao, Z.; Zhao, G.; Xu, C.; Yang, C.; Jin, Y.; Ju, X.; Du, X. Efficiency analyses of high temperature thermal energy storage systems of rocks only and rock-PCM capsule combination. *Sol. Energ.* **2018**, *162*, 153–164. [CrossRef]
28. Zhou, D.; Wu, S.; Wu, Z.; Yu, X. Thermal performance analysis of multi-slab phase change thermal energy storage unit with heat transfer enhancement approaches. *Renew. Energ.* **2021**, *172*, 46–56. [CrossRef]
29. Pirasaci, T. Investigation of phase state and heat storage form of the phase change material (PCM) layer integrated into the exterior walls of the residential-apartment during heating season. *Energy* **2020**, *207*, 118176. [CrossRef]
30. Wang, H.; Yang, J.; Chen, Z.; Li, G.; Liang, J.; Ma, Y.; Dong, H.; Ji, H.; Feng, J. Optimal dispatch based on prediction of distributed electric heating storages in combined electricity and heat networks. *Appl. Energ.* **2020**, *267*, 114879. [CrossRef]
31. Serov, E.P.; Korolkov, B.P. *Dynamics of Steam Generators*; Energoizdat: Moscow, Russia, 1981.
32. Kreith, F.; Manglik, R.M.; Bohn, M. *Principles of Heat Transfer*, 7th ed.; Cengage Learning: Stamford, CT, USA, 2011.
33. Taler, J.; Duda, P. *Solving Direct and Inverse Heat Conduction Problems*; Springer: Berlin/Heidelberg, Germany, 2006.
34. Thomson, W.J. *Atlas for Computing Mathematical Function*; Wiley-Interscience: New York, NY, USA, 1997.
35. Taler, J.; Taler, D. Superposition Method for Time-Dependent Thermal Boundary Conditions. In *Encyclopedia of Thermal Stresses*; Hetnarski, R., Ed.; Springer: Dordrecht, The Netherlands; Berlin/Heidelberg, Germany, 2014; Volume VIII, pp. 4718–4729.
36. Press, W.H.; Teukolsky, S.A.; Vetterling, W.T.; Flannery, B.P. *Numerical Recipes. The Art of Scientific Computing*, 3rd ed.; Cambridge University Press: New York, NY, USA, 2007.
37. Ford, W. *Numerical Linear Algebra with Applications: Using MATLAB*; Academic Press Elsevier: London, UK, 2015.
38. Lagarias, J.C.; Reeds, J.A.; Wright, M.H.; Wright, P.E. Convergence properties of the Nelder-Mead Simplex Method in low dimensions. *SIAM J. Optim.* **1998**, *9*, 112–147. [CrossRef]
39. Fortran, IMSL Thread Safe. *IMSL Fortran Library User's Guide Math/Library*; Visual Numerics, Inc.: San Ramon, CA, USA, 2003; Volume 2.
40. *MATLAB R2020*; The MathWorks: Natick, MA, USA, 2020.

energies

MDPI

Article

An Asymptotic Energy Equation for Modelling Thermo Fluid Dynamics in the Optical Fibre Drawing Process

Giovanni Luzi [1,*], Seunghyeon Lee [2], Bernhard Gatternig [3] and Antonio Delgado [1]

1 LSTME Busan, Busan 46742, Korea
2 Department of Energy and Machine Engineering, Gyeongsang National University, Tongyeong 53064, Korea
3 Department of Environmental Engineering, Hochschule Weihenstephan-Triesdorf, 91746 Weidenbach, Germany
* Correspondence: giovanni.luzi@lstme.org

Abstract: Microstructured optical fibres (MOFs) are fibres that contain an array of air holes that runs through the whole fibre length. The hole pattern of these fibres can be customized to manufacture optical devices for different applications ranging from high-power energy transmission equipment to telecommunications and optical sensors. During the drawing process, the size of the preform is greatly scaled down and the original hole pattern result might be modified, potentially leading to unwanted optical effects. Because only a few parameters can be controlled during the fabrication process, mathematical models that can accurately describe the fibre drawing process are highly desirable, being powerful predictive tools that are significantly cheaper than costly experiments. In this manuscript, we derive a new asymptotic energy equation for the drawing process of a single annular capillary and couple it with existing asymptotic mass, momentum, and evolution equations. The whole asymptotic model only exploits the small aspect ratio of a capillary and relies on neither a fitting procedure nor on any empirical adjustable parameters. The numerical results of the simplified model are in good accordance with experimental data available in the literature both without inner pressurization and when internal pressure is applied. Although valid only for annular capillaries, the present model can provide important insights towards understanding the MOF manufacturing process and improving less detailed approaches for more complicated geometries.

Keywords: asymptotic analysis; optical fibre drawing; creeping flow; energy equation; MOFs

Citation: Luzi, G.; Lee, S.; Gatternig, B.; Delgado, A. An Asymptotic Energy Equation for Modelling Thermo Fluid Dynamics in the Optical Fibre Drawing Process. *Energies* **2022**, *15*, 7922. https://doi.org/10.3390/en15217922

Academic Editors: Artur Bartosik and Dariusz Asendrych

Received: 20 September 2022
Accepted: 20 October 2022
Published: 25 October 2022

Publisher's Note: MDPI stays neutral with regard to jurisdictional claims in published maps and institutional affiliations.

Copyright: © 2022 by the authors. Licensee MDPI, Basel, Switzerland. This article is an open access article distributed under the terms and conditions of the Creative Commons Attribution (CC BY) license (https://creativecommons.org/licenses/by/4.0/).

1. Introduction

Microstructured optical fibres (MOFs) or Photonic Crystal Fibres (PCFs) are a new kind of optical fibres, appearing for the first time approximately thirty years ago [1]. This new type of fibres possesses an array of air holes arranged in a specific pattern that spans the whole fibre length. Light guidance within PCFs relies either on the index guiding or on the photonic bandgap (PBG) mechanism. If the central air capillary is removed from the structure, the electromagnetic waves are guided by a modified total internal reflection mechanism. Conversely, if the central air capillary is replaced with another one of a different size, the PBG mechanism is realized [2]. The network of air holes can be suitably designed to allow for the guidance of selected modes. This can be achieved with specific ratios between the diameter of an air capillary and the crystal lattice constant. Many advantages of this new type of fibres are represented by the high degree of flexibility and many possibilities they offer. PCFs find a large number of applications, ranging from high-power and energy transmission [3], fibre lasers [4] and amplifiers [5], Kerr-related non-linear effects [6], Brillouin scattering [7], telecommunications [1], and optical sensors [8–12], among others.

The manufacturing process of glass optical fibres presupposes two steps. First, a fibre-preform is manufactured, and afterward it is drawn inside a high-tech furnace incorporated in a tower set-up. Fibre preforms are built by stacking silica capillary tubes and solid rods. This allows for quick, low-cost, and flexible manufacturing of preforms. After stacking, capillaries and rods are seized together by thin wires and eventually fused in an intermediate drawing process in which the structures do not achieve the required final dimensions, but are instead drawn into an intermediate preform-cane. A large number of preform-canes are usually manufactured, as they can be utilized for the development and optimization of different PCFs structures. During the drawing process, holes might experience distortion and positions and sizes might be altered. This occurs because the drawing process takes place at elevated temperatures and surface tension may lead to the collapse of the internal holes that form the air lattice. Therefore, internal pressurization is commonly applied to prevent hole collapse. An alternative to glass MOFs is represented by microstructured polymer optical fibres (MPOFs) [13]. They have several advantages compared to glass MOFs. For instance, the processing temperature of polymers is much lower than that of the glass, and the polymerization processes are easier to control. This entails the utilization of different techniques to produce polymer preforms of arbitrary cross section arrangements, such as extrusion, polymer coating, polymerization in a mold, and injection molding [14]. Another advantage of MPOFs is that they can be drawn over a wide temperature range without significant changes in the fibre structure, unlike glass MOFs for which temperature variations of just a few percentage points can induce significant variations in the fibre microstructure. In addition, the high temperatures involved in the fabrication of glass MOFs hinder the possibility of modifying the optical properties using dopants, as phase separation may occur [14]. Conversely, MPOFs can be easily doped with atomic species, molecular components, dispersed molecules, and phases. Moreover, the lower temperatures involved in the fabrication process of MPOFs reduce the chances of hole collapse and allow low cost production in large volumes, as both the material and the production process are cheaper [15]. Independently of the material used to fabricate MOFs, accurate control of fibre structures concerning hole dimensions and position is essential for the manufacturing of PCFs with specific properties. The inclusion or elimination of interstitial holes can have dramatic consequences on the final fibre properties. The key element in the drawing process of PCFs is the ability to maintain the highly regular structure down to the final dimensions.

Mathematical models and numerical simulations that can describe the fibre drawing process are highly desirable, as they allow for understanding and quantifying the transport phenomena and main physical quantities involved in the process. Furthermore, they represent more valid predictive and process control tools compared to expensive experiments. To this end, many theoretical and numerical studies have been carried out over the past fifty years. Early studies mainly focused on one-dimensional drawing models of solid fibres. Peak and Runk [16] derived a simplified model consisting of axial momentum and energy equations with a simplified radiation model to predict the neck shape and the temperature distribution of a silica rod during the drawing process. Myers [17] extended Glicksman's model [18] by introducing a radiative heat transfer model that considers the slope of preform surfaces, the spectral variation of the glass properties, and the dependencies of the emissivity on the fibre diameter. Fitt et al. [19] utilized asymptotic techniques to derive a model that describes the drawing process of a capillary and examined it on selected asymptotic limits to isolate and quantify the effects of main physical parameters on the drawing process. Luzi et al. [20] numerically solved the asymptotic model of Fitt et al. [19] by assuming Gaussian distributions of the temperature profile inside the furnace. The numerical results are in good accordance with the experimental ones, both for the case of an unpressurized capillary and for the case when the inner pressurization is applied. In a series of contributions, Voyce et al. [21–23] extended the previous work by Fitt et al. [19] by including the effect of preform rotation in their model. Rotation is particularly useful to control Polarization Mode Dispersion (PMD) and fibre birefringence effects, as well as to tune the capillary size. More recently, Taroni et al. [24] utilized asymptotic analysis to derive simplified momentum and energy equations to describe the drawing process of a solid fibre while considering the heat transport within the fibre via conduction, convection, and radiative heating.

Numerical investigations of the drawing process of fibres with more complicated cross-sectional structures have been initially carried out using finite element-based commercial software. In a two-series contribution, Xue et al. [25] first performed a scaling analysis of the governing equations to determine the importance of the main parameters involved in the drawing process. Afterwards, they simulated the transient drawing process of MOFs containing five holes, showing that the shape of the holes changes dramatically in the vicinity of the neck-down region. In the subsequent manuscript, Xue et al. [26] investigated the steady-state process, focusing on the effects of surface tension, viscosity, and stress redistribution within the fibre. Non-isothermal simulations revealed that the slope of the neck-down region is highly sensitive to the viscosity profile, and therefore to temperature gradients. In a different work, Xue et al. [27] scrutinized the mechanism of hole deformation for silica and polymer fibres. To this end, they simulated the drawing process of a five-hole and polarization-maintaining structure, focusing on hole deformation and hole expansion in terms of the capillary number, draw, and aspect ratio. Luzi et al. [28] modeled the drawing process of six-hole MOFs, obtaining very good agreement between numerical simulations and experiments as long as the applied inner pressurization was not too high. Even in that case, the shape of the deformed holes was in qualitatively good agreement with the experimental one. Nevertheless, solving the full three-dimensional problem proves numerically expensive, and significant computational resources are often needed.

To cope with this issue, Stokes et al. [29] presented a general mathematical framework to model the drawing process of optical fibres of general cross-sectional shape, with the only requirement that the fibre must be slender. Chen et al. [30] extended the work of Stokes et al. [29] by including channel pressurization. Buchak et al. [31] developed the generalized Elliptical Pore Model (EPM), a very efficient method for cases in which the fibre cross-section contains elliptical holes. The evolution of an inner hole is determined by the solution of a set of ordinary integrodifferential equations that determine the centroid position, orientation, area, and eccentricity along the drawing direction. In a different contribution, Buchak and Crowdy [32] employed spectral methods with conformal mapping to obtain a very accurate reconstruction of the cross-sectional shape. Chen et al. [33] utilized the numerical approach of Buchak and Crowdy [32] to model the drawing process of a six-hole MOF and compared the results with the experiments and the Finite Element Method (FEM)-based simulations by Luzi et al. [28]. The approach used by [33] allows for accurate computation of the hole-interface curvature and is in better agreement with experimental results compared with the FEM simulations, and is significantly more computationally efficient. However, in these contributions, the heat transfer between the furnace and the fibre is not modelled, and the drawing is assumed to be isothermal with an assumed fibre temperature profile. In a recent manuscript, Stokes et al. [34] derived an asymptotic energy equation for the full three-dimensional problem utilizing only asymptotic analysis based on the small fibre aspect ratio and coupled it with the generalized EPM of Buchak et al. [31]. Jasion et al. [35] proposed the MicroStructure Element Method (MSEM) for modeling the drawing process of MOFs with a high filling fraction and thin glass membranes, such as Hollow Core Photonic Crystal fibres (HC-PCFs). They used the model of Fitt et al. [19] to describe the evolution of the external jacket that surrounds the microstructured array of air-holes with a network of fluid struts linked through nodes where surface tension, viscous, and pressure force act. However, the energy equations utilized in these two contributions consider neither the heat transfer across the fibre cross-section nor the viscous diffusion.

Detailed numerical investigations concerning the conjugate heat transfer between the fibre and the furnace have been carried out by different researchers. Lee and Jaluria [36] simulated the conjugate heat transfer between the furnace and solid fibre assuming an axis-symmetric geometry and a given distribution of the fibre shape. Chodhury and Jaluria [37] investigated the effects of the fibre draw speed, inert gas velocity, furnace dimensions, and gas properties on the temperature distribution within a solid glass fibre and an oven. Yin and Jaluria [38] utilized the zonal method and the optically thick approximation to compute the radiative heat exchange between the furnace and solid glass fibre. Their numerical investigations reveal that the zonal method can predict the radiative flux with reasonable accuracy independently of the temperature distribution within the fibre, although the optically thick approximation can only predict a correct temperature distribution when the radial temperature distribution is small. In a subsequent contribution, Yin and Jaluria [39] utilized the zonal method and the optically thick approximation with a force balance to generate neck-down profiles of a solid fibre. On the one hand, the difference between the profiles generated using the zonal method and the optically thick approximation is not very large. On the other hand, other parameters have a significant influence on the thermal neck-down profiles, that is, the fibre drawing speed and the furnace wall temperature, while the purge gas velocity and gas type have only minor effects. Xue et al. [40] modeled the transient heat transfer through an eight-hole MOF and found that a MOF heats up faster than a solid one, as there is less material to heat. In addition, the inclusion of radiative heat transfer across the holes accelerates the heating in the whole fibre. However, numerical treatment of the full heat transfer problem requires significant computational efforts, even for the two-dimensional case of a solid fibre where axis symmetry can be exploited to simplify the problem.

In this work, we derive an asymptotic energy equation for the drawing process of annular capillaries by extending the work of Taroni et al. [24] and build a complete asymptotic fibre drawing model with the equations obtained by Fitt et al. [19]. In addition, we include the effects of viscous dissipation in our model. This simplified system of equations can be solved numerically quite readily compared to the full three-dimensional problem, and its predictions are in very good agreement with the experimental results by Luzi et al. [20]. The rest of this manuscript is organized as follows. In Section 2, we develop the theoretical formulation of the problem, concisely yet comprehensively providing the mass, momentum, and energy equations with the associated boundary conditions that govern the drawing process of an annular capillary. In Section 3, we derive the final asymptotic equations for the drawing process of a capillary, unifying the mass, momentum, and evolution equations of Fitt et al. [19] with a simplified energy equation. In Section 4, we compare the numerical outcomes of the asymptotic model with the experimental results of Luzi et al. [20], and in Section 5 we discuss our results, highlighting possible ways to improve the present model. Finally, in Appendix A, we derive an asymptotic energy equation for rotating capillaries and show how it can be coupled with the asymptotic model of Voyce et al. [21].

2. Problem Description

2.1. Governing Equations

A fibre preform is slowly inserted from the top of a furnace through an opening iris and is pulled from the bottom. During drawing, the size of the preform is significantly reduced, and it achieves the required final dimensions at the exit of the furnace. The process is schematically depicted in Figure 1. To develop a suitable mathematical model for describing the fibre drawing process, we begin our study with the full three-dimensional continuity, momentum, and energy equations written in cylindrical coordinates.

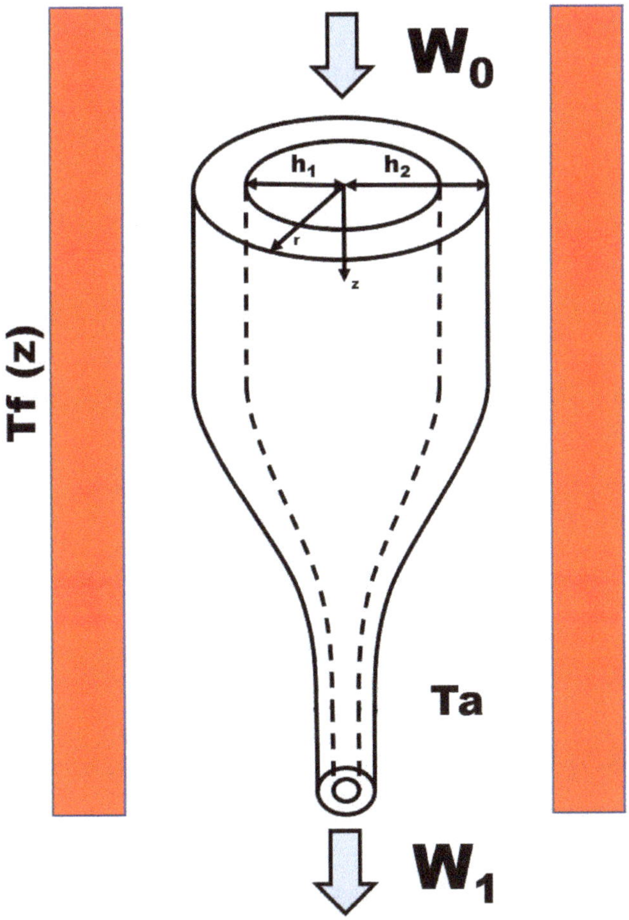

Figure 1. Schematic diagram of the drawing process of an optical glass capillary. The furnace temperature depends on the axial position z, and is indicated by T_f, while the furnace ambient temperature is indicated by T_a.

2.1.1. Mass and Momentum Equations

The continuity and momentum equations in cylindrical coordinates read [41]

$$\frac{\partial \rho}{\partial t} + \frac{1}{r}\frac{(\rho r u)}{\partial r} + \frac{1}{r}\frac{\partial(\rho v)}{\partial \phi} + \frac{\partial(\rho w)}{\partial z} = 0 \quad (1)$$

$$\rho\left(\frac{\partial u}{\partial t} + u\frac{\partial u}{\partial r} + \frac{v}{r}\frac{\partial u}{\partial \phi} - \frac{u^2}{r} + w\frac{\partial u}{\partial z}\right) = f_r - \frac{\partial p}{\partial r} + \frac{1}{r}\frac{\partial(r\tau_{rr})}{\partial r} + \frac{1}{r}\frac{\partial(r\tau_{r\phi})}{\partial \phi} + \frac{\partial(\tau_{rz})}{\partial z} - \frac{\tau_{\phi\phi}}{r} \quad (2)$$

$$\rho\left(\frac{\partial v}{\partial t} + u\frac{\partial v}{\partial r} + \frac{v}{r}\frac{\partial v}{\partial \phi} + \frac{uv}{r} + w\frac{\partial v}{\partial z}\right) = f_\phi - \frac{1}{r}\frac{\partial p}{\partial \phi} + \frac{1}{r^2}\frac{\partial(r^2\tau_{r\phi})}{\partial r} + \frac{1}{r}\frac{\partial(\tau_{\phi\phi})}{\partial \phi} + \frac{\partial(\tau_{\phi z})}{\partial z} \quad (3)$$

$$\rho\left(\frac{\partial w}{\partial t}+u\frac{\partial w}{\partial r}+\frac{v}{r}\frac{\partial w}{\partial \phi}+w\frac{\partial w}{\partial z}\right)=f_z-\frac{\partial p}{\partial z}+\frac{1}{r}\frac{\partial (r\tau_{rz})}{\partial r}+\frac{1}{r}\frac{\partial (r\tau_{\phi z})}{\partial \phi}+\frac{\partial (\tau_{zz})}{\partial z} \tag{4}$$

Herein, ρ, t, and p denote the density, the time, and the pressure, respectively, and r, z, and ϕ represents the radial, axial, and azimuthal coordinates. The fluid velocity is denoted by $\boldsymbol{q}=w\boldsymbol{e}_z+u\boldsymbol{e}_r+v\boldsymbol{e}_\phi$, and $\boldsymbol{e}_z$, $\boldsymbol{e}_r$, and $\boldsymbol{e}_\phi$ are the unit vectors in the axial, radial and azimuthal directions, respectively. u, v, and w are the radial, azimuthal, and axial components of the velocity field, and $\boldsymbol{f}=f_z\boldsymbol{e}_z+f_r\boldsymbol{e}_r+f_\phi\boldsymbol{e}_\phi=\rho g(\boldsymbol{e}_z+\boldsymbol{e}_r+\boldsymbol{e}_\phi)$ represents the body force. The components of the viscous stress tensor read

$$\tau_{rr}=\mu\left(2\frac{\partial u}{\partial r}-\frac{2}{3}(\boldsymbol{\nabla}\cdot\boldsymbol{q})\right)$$

$$\tau_{\phi\phi}=\mu\left(2\left(\frac{1}{r}\frac{\partial v}{\partial \phi}+\frac{u}{r}\right)-\frac{2}{3}(\boldsymbol{\nabla}\cdot\boldsymbol{q})\right)$$

$$\tau_{zz}=\mu\left(2\frac{\partial w}{\partial z}-\frac{2}{3}(\boldsymbol{\nabla}\cdot\boldsymbol{q})\right)$$

$$\tau_{r\phi}=\mu\left(r\frac{\partial}{\partial r}\left(\frac{v}{r}\right)+\frac{1}{r}\frac{\partial u}{\partial \phi}\right)$$

$$\tau_{rz}=\mu\left(\frac{\partial w}{\partial r}+\frac{\partial u}{\partial z}\right)$$

$$\tau_{\phi z}=\mu\left(\frac{\partial v}{\partial z}+\frac{1}{r}\frac{\partial w}{\partial \phi}\right)$$

The divergence of the velocity field may be written as

$$\boldsymbol{\nabla}\cdot\boldsymbol{q}=\frac{1}{r}\frac{\partial (ru)}{\partial r}+\frac{1}{r}\frac{\partial v}{\partial \phi}+\frac{\partial w}{\partial z} \tag{5}$$

The left-hand side of the momentum equations denotes the temporal and convective inertial acceleration, while the right-hand side incorporates the pressure gradient, divergence of the stress tensor, and sum of the body forces.

2.1.2. Two-Dimensional Mass and Momentum Equations

We assume the flow to be incompressible, and consider axis symmetry and a non-rotating capillary, say, $\rho = const$, $\frac{\partial}{\partial \phi}=0$, and $v=0$. Therefore, the mass conservation equation reduces to

$$\frac{1}{r}\frac{\partial (ru)}{\partial r}+\frac{\partial w}{\partial z}=0 \tag{6}$$

and many terms of the components of the stress tensor vanish:

$$\tau_{rr}=2\mu\left(\frac{\partial u}{\partial r}\right)$$

$$\tau_{\phi\phi}=2\mu\left(\frac{u}{r}\right)$$

$$\tau_{zz}=2\mu\left(\frac{\partial w}{\partial z}\right)$$

$$\tau_{r\phi}=0$$

$$\tau_{rz} = \mu\left(\frac{\partial w}{\partial r} + \frac{\partial u}{\partial z}\right)$$

$$\tau_{\phi z} = 0$$

Thereby, the previous momentum equations reduce to

$$\begin{aligned} \rho\left(\frac{\partial u}{\partial t} + u\frac{\partial u}{\partial r} + w\frac{\partial u}{\partial z}\right) &= f_r - \frac{\partial p}{\partial r} \\ +\frac{2}{r}\frac{\partial}{\partial r}\left(r\mu\left(\frac{\partial u}{\partial r}\right)\right) + \frac{\partial}{\partial z}\left(\mu\left(\frac{\partial w}{\partial r} + \frac{\partial u}{\partial z}\right)\right) &- 2\mu\frac{u}{r^2} \end{aligned} \tag{7}$$

$$\begin{aligned} \rho\left(\frac{\partial w}{\partial t} + u\frac{\partial w}{\partial r} + w\frac{\partial w}{\partial z}\right) = f_z - \frac{\partial p}{\partial z} + \frac{1}{r}\frac{\partial}{\partial r}\left(r\mu\left(\frac{\partial w}{\partial r} + \frac{\partial u}{\partial z}\right)\right) \\ +2\frac{\partial}{\partial z}\left(\mu\frac{\partial w}{\partial z}\right) \end{aligned} \tag{8}$$

because the momentum equation in the azimuthal direction vanishes identically. To close the problem, boundary and initial conditions must be specified. Concerning boundary conditions, we impose initial, kinematic, and dynamic boundary conditions. The kinematic boundary conditions require that the normal components of the velocity at each interface of the capillary be continuous across the interfaces. In addition, the tangential component of the velocity must be continuous at the interfaces. At each fibre surface, they read

$$\frac{\partial h_1}{\partial t} + w\frac{\partial h_1}{\partial z} = u \tag{9a}$$

$$\frac{\partial h_2}{\partial t} + w\frac{\partial h_2}{\partial z} = u \tag{9b}$$

The normal boundary conditions represent a balance of forces across the surfaces of the capillary. They may be written as

$$-\boldsymbol{n}_1^T \cdot \boldsymbol{\tau} \cdot \boldsymbol{n}_1 + \frac{\gamma}{h_1} = p_H \tag{10a}$$

$$-\boldsymbol{n}_2^T \cdot \boldsymbol{\tau} \cdot \boldsymbol{n}_2 - \frac{\gamma}{h_2} = p_a \tag{10b}$$

in the normal direction, and

$$\boldsymbol{t}_1^T \cdot \boldsymbol{\tau} \cdot \boldsymbol{n}_1 = 0 \tag{10c}$$

$$\boldsymbol{t}_2^T \cdot \boldsymbol{\tau} \cdot \boldsymbol{n}_2 = 0 \tag{10d}$$

in the tangential direction. In Equation (10), $\boldsymbol{\tau}$, $\boldsymbol{t}$, and $\boldsymbol{n}$ denote the stress tensor and the unit vectors in the tangential and normal directions, respectively. They read

$$\boldsymbol{\tau} = \begin{bmatrix} 2\mu\left(\frac{\partial w}{\partial z}\right) - p & \mu\left(\frac{\partial w}{\partial r} + \frac{\partial u}{\partial z}\right) \\ \mu\left(\frac{\partial w}{\partial r} + \frac{\partial u}{\partial z}\right) & 2\mu\left(\frac{\partial u}{\partial r}\right) - p \end{bmatrix} \tag{11a}$$

$$\boldsymbol{t}_i^T = \frac{(-1)^{i+1}}{\sqrt{1 + \left(\frac{\partial h_i}{\partial z}\right)^2}}\left(1\boldsymbol{e}_z + \frac{\partial h_i}{\partial z}\boldsymbol{e}_r\right) \tag{11b}$$

$$\boldsymbol{n}_i^T = \frac{(-1)^{i+1}}{\sqrt{1 + \left(\frac{\partial h_i}{\partial z}\right)^2}}\left(\frac{\partial h_i}{\partial z}\boldsymbol{e}_z - 1\boldsymbol{e}_r\right) \tag{11c}$$

Moreover, γ, p_H, and p_a are the surface tension, the hole, and the ambient pressure, respectively. To close the problem, we impose the velocity at the beginning and end of the drawing process as initial conditions:

$$w(t,r,z=0) = W_0, \qquad w(t,r,z=L) = W_1 \tag{11d}$$

2.1.3. Energy Equation

The three-dimensional energy equation in cylindrical coordinates assumes the form

$$\begin{aligned}\rho c_p\left(\frac{\partial T}{\partial t}+u\frac{\partial T}{\partial r}+\frac{v}{r}\frac{\partial T}{\partial \phi}+w\frac{\partial T}{\partial z}\right) &= \frac{1}{r}\frac{\partial}{\partial r}\left(kr\frac{\partial T}{\partial r}\right)\\ +\frac{1}{r}\frac{\partial}{\partial \phi}\left(\frac{k}{r}\frac{\partial T}{\partial \phi}\right)+\frac{\partial}{\partial z}\left(k\frac{\partial T}{\partial z}\right)&+\beta T\left(\frac{\partial p}{\partial t}+u\frac{\partial p}{\partial r}+\frac{v}{r}\frac{\partial p}{\partial \phi}+w\frac{\partial p}{\partial z}\right)+\Phi\end{aligned} \tag{12}$$

In Equation (12), k and c_p denote the thermal conductivity and the specific heat capacity. In addition, T, β, and Φ are the temperature, the coefficient of thermal expansion at constant pressure, and the viscous dissipation, respectively. The viscous dissipation Φ reads

$$\begin{aligned}\frac{\Phi}{\mu} = 2&\left[\left(\frac{\partial u}{\partial r}\right)^2+\left(\frac{1}{r}\frac{\partial v}{\partial \phi}+\frac{u}{r}\right)^2+\left(\frac{\partial w}{\partial z}\right)^2\right]+\left[\frac{1}{r}\frac{\partial w}{\partial \phi}+\frac{\partial v}{\partial z}\right]^2\\ &+\left[r\frac{\partial}{\partial r}\left(\frac{v}{r}\right)+\frac{1}{r}\frac{\partial u}{\partial \phi}\right]^2+\left[\frac{\partial u}{\partial z}+\frac{\partial w}{\partial r}\right]^2-\frac{2}{3}(\nabla\cdot\boldsymbol{q})\end{aligned} \tag{13}$$

Assuming axis symmetry and constant density, Equation (12) reduces to

$$\begin{aligned}\rho c_p\left(\frac{\partial T}{\partial t}+u\frac{\partial T}{\partial r}+w\frac{\partial T}{\partial z}\right) &= \frac{1}{r}\frac{\partial}{\partial r}\left(kr\frac{\partial T}{\partial r}\right)+\frac{\partial}{\partial z}\left(k\frac{\partial T}{\partial z}\right)\\ +2\mu\left[\left(\frac{\partial u}{\partial r}\right)^2+\left(\frac{u}{r}\right)^2+\left(\frac{\partial w}{\partial z}\right)^2\right]&+\mu\left[\frac{\partial u}{\partial z}+\frac{\partial w}{\partial r}\right]^2\end{aligned} \tag{14}$$

The left-hand side of the energy equation denotes the change of temperature with time and thermal convection. The right-hand side encompasses thermal conduction and viscous dissipation. Because glass is a transparent material, radiative heat transfer within the fibre is relevant, especially at high temperatures. Nevertheless, we avoid the full description of the problem in detail and opt for the common approximation in which the capillary is optically thick. Thereby, we assume that the capillary thickness is much greater than the absorption length scale. Following Taroni et al. [24], we utilize the Rosseland approximation and add a radiative contribution to the total thermal conductivity, say, $k(T) = k_c + k_r(T)$

$$k_r(T) = \frac{16 n_0^2 \sigma T^3}{3\chi} \tag{15}$$

where n_0 and χ are the refractive index and the absorption coefficient, respectively. Therefore, Equation (14) assumes the form

$$\begin{aligned}\rho c_p\left(\frac{\partial T}{\partial t}+u\frac{\partial T}{\partial r}+w\frac{\partial T}{\partial z}\right) &= \frac{1}{r}\frac{\partial}{\partial r}\left(k_c r\frac{\partial T}{\partial r}\right)\\ +\frac{1}{r}\frac{\partial}{\partial r}\left(\tilde{k}_r r\frac{\partial T^4}{\partial r}\right)+\frac{\partial}{\partial z}\left(k_c\frac{\partial T}{\partial z}\right)&+\frac{\partial}{\partial z}\left(\tilde{k}_r\frac{\partial T^4}{\partial z}\right)\\ +2\mu\left[\left(\frac{\partial u}{\partial r}\right)^2+\left(\frac{u}{r}\right)^2+\left(\frac{\partial w}{\partial z}\right)^2\right]&+\mu\left[\frac{\partial u}{\partial z}+\frac{\partial w}{\partial r}\right]^2\end{aligned} \tag{16}$$

where $\tilde{k}_r = \frac{k_r(T)}{4T^3}$. The thermal boundary conditions at the outer glass-air interface read

$$-k\frac{\partial T}{\partial r} = \sigma\epsilon_r\left(T^4 - T_f^4\right) + k_h(T - T_a) \quad (17)$$

where σ and ϵ_r are the Stefan-Boltzmann constant and the specific emissivity of the furnace. In addition, k_h, T_f, and T_a are the convective heat exchange coefficient, the furnace, and the ambient temperature, respectively. The boundary condition at the inner surface assumes the form

$$-k\frac{\partial T}{\partial r} = k_h(T - T_a) \quad (18)$$

Finally, we impose the temperature at the beginning of the drawing by numerically solving Equation (17) with the left-hand side set equal to zero and specified functional forms of the furnace and ambient temperature.

2.1.4. Non-Dimensionalization

We non-dimensionalize the previous equations by exploiting the slenderness of the geometry, that is, utilizing the ratio $\epsilon = \frac{h_{20}-h_{10}}{L} << 1$, where L is a "hot-zone" length and $h_{20} - h_{10}$ is the initial size of a capillary. We set

$$\begin{aligned} r &= \epsilon L\bar{r} & z &= L\bar{z} & u &= \epsilon W_1\bar{u} \\ w &= W_1\bar{w} & h_1 &= \epsilon L\bar{h}_1 & h_2 &= \epsilon L\bar{h}_2 \\ p &= \frac{\mu_s W_1}{\epsilon^2 L}\bar{p} & T &= T_s\bar{T} & t &= \frac{L}{W_1}\bar{t} \end{aligned}$$

where overbars indicate non-dimensional quantities, W_1 is a typical draw speed, and T_s and μ_s are typical glass softening temperature and viscosity at the glass softening temperature, respectively. The continuity equation becomes

$$\frac{1}{\bar{r}}\frac{\partial(\bar{r}\,\bar{u})}{\partial\bar{r}} + \frac{\partial\bar{w}}{\partial\bar{z}} = 0 \quad (19)$$

The momentum equation in the z direction reads

$$\begin{aligned} \epsilon^2 Re\left(\frac{\partial\bar{w}}{\partial\bar{t}} + \bar{u}\frac{\partial\bar{w}}{\partial\bar{r}} + \bar{w}\frac{\partial\bar{w}}{\partial\bar{z}}\right) &= \epsilon^2\frac{Re}{Fr^2} - \frac{\partial\bar{p}}{\partial\bar{z}} + \frac{1}{\bar{r}}\frac{\partial}{\partial\bar{r}}\left(\bar{r}\bar{\mu}\left(\frac{\partial\bar{w}}{\partial\bar{r}}\right)\right) \\ &+\epsilon^2\frac{1}{\bar{r}}\frac{\partial}{\partial\bar{r}}\left(\bar{r}\bar{\mu}\left(\frac{\partial\bar{u}}{\partial\bar{z}}\right)\right) + 2\epsilon^2\frac{\partial}{\partial\bar{z}}\left(\bar{\mu}\frac{\partial\bar{w}}{\partial\bar{z}}\right) \end{aligned} \quad (20)$$

while the one in the r direction becomes

$$\begin{aligned} \epsilon^4 Re\left(\frac{\partial\bar{u}}{\partial\bar{t}} + \bar{u}\frac{\partial\bar{u}}{\partial\bar{r}} + \bar{w}\frac{\partial\bar{u}}{\partial\bar{z}}\right) &= \epsilon^3\frac{Re}{Fr^2} - \frac{\partial\bar{p}}{\partial\bar{r}} \\ +2\frac{\epsilon^2}{\bar{r}}\frac{\partial}{\partial\bar{r}}\left(\bar{r}\bar{\mu}\left(\frac{\partial\bar{u}}{\partial\bar{r}}\right)\right) + \epsilon^2\frac{\partial}{\partial\bar{z}}\left(\bar{\mu}\left(\frac{\partial\bar{w}}{\partial\bar{r}}\right)\right) &+ \epsilon^4\frac{\partial}{\partial\bar{z}}\left(\bar{\mu}\left(\frac{\partial\bar{u}}{\partial\bar{z}}\right)\right) - 2\epsilon^2\bar{\mu}\left(\frac{\bar{u}}{\bar{r}^2}\right) \end{aligned} \quad (21)$$

where we have set $\mu = \mu_s\bar{\mu}$ and $\boldsymbol{f} = \rho\boldsymbol{g}$. Moreover,

$$Re = \frac{L\rho W_1}{\mu_s}$$

$$Fr = \frac{W_1^2}{\sqrt{Lg}}$$

where Re and Fr are the Reynolds and the Froude numbers, respectively. The energy equation assumes the form

$$\epsilon^2 Pe\left(\frac{\partial \overline{T}}{\partial \overline{t}}+\overline{u}\frac{\partial \overline{T}}{\partial \overline{r}}+\overline{w}\frac{\partial \overline{T}}{\partial \overline{z}}\right)=\frac{1}{\overline{r}}\frac{\partial}{\partial \overline{r}}\left(\overline{r}\frac{\partial \overline{T}}{\partial \overline{r}}\right)+\frac{1}{\overline{r}}\frac{\partial}{\partial \overline{r}}\left(\gamma_R\overline{r}\frac{\partial \overline{T}^4}{\partial \overline{r}}\right)+\epsilon^2\frac{\partial}{\partial \overline{z}}\left(\frac{\partial \overline{T}}{\partial \overline{z}}\right)$$
$$+\epsilon^2\frac{\partial}{\partial \overline{z}}\left(\gamma_R\frac{\partial \overline{T}^4}{\partial \overline{z}}\right)+2\epsilon^2 Br\overline{\mu}\left[\left(\frac{\partial \overline{u}}{\partial \overline{r}}\right)^2+\left(\frac{\overline{u}}{\overline{r}}\right)^2+\left(\frac{\partial \overline{w}}{\partial \overline{z}}\right)^2\right]+\epsilon^2 Br\overline{\mu}\left[\frac{\partial \epsilon \overline{u}}{\partial z}+\frac{\partial \overline{w}}{\partial \epsilon \overline{r}}\right]^2 \quad (22)$$

where

$$Pe=\frac{L\rho W_1 c_p}{k_c}$$
$$Br=\frac{\mu_s W_1^2}{T_s k_c}$$
$$\gamma_R=\frac{4n_0^2\sigma T_s^3}{3\chi k_c}$$

In the previous equation, Pe and Br are the Peclet and the Brinkman number, respectively. The former denotes the ratio between the convective transport of thermal energy to the fluid to the conduction of thermal energy within the fluid, and the latter is a ratio between the heat originated by mechanical dissipation to the heat transferred by conduction. Furthermore, γ_R is a parameter that indicates the strength of bulk diffusion [24]. The dimensionless kinematic boundary conditions for the inner and the outer surfaces read

$$\frac{\partial \overline{h}_1}{\partial \overline{t}}+\overline{w}\frac{\partial \overline{h}_1}{\partial \overline{z}}=\overline{u} \quad (23a)$$
$$\frac{\partial \overline{h}_2}{\partial \overline{t}}+\overline{w}\frac{\partial \overline{h}_2}{\partial \overline{z}}=\overline{u} \quad (23b)$$

while the dimensionless dynamic boundary conditions in the normal and tangential direction assume the form

$$-2\overline{\mu}\epsilon^2\frac{\partial \overline{w}}{\partial \overline{z}}\left(\frac{\partial \overline{h}_1}{\partial \overline{z}}\right)^2+2\overline{\mu}\frac{\partial \overline{h}_1}{\partial \overline{z}}\frac{\partial \overline{w}}{\partial \overline{r}}\bigg|_{\overline{r}=\overline{h}_1}+2\epsilon^2\overline{\mu}\frac{\partial \overline{h}_1}{\partial \overline{z}}\frac{\partial \overline{u}}{\partial \overline{z}}$$
$$-2\overline{\mu}\frac{\partial \overline{u}}{\partial \overline{r}}\bigg|_{\overline{r}=\overline{h}_1}+\frac{\overline{\gamma}}{\overline{h}_1}\left(1+\epsilon^2\left(\frac{\partial \overline{h}_1}{\partial \overline{z}}\right)^2\right) \quad (24a)$$
$$=\frac{1}{\epsilon^2}(\overline{p}_H-\overline{p})\left(1+\epsilon^2\left(\frac{\partial \overline{h}_1}{\partial \overline{z}}\right)^2\right)$$

$$-2\overline{\mu}\epsilon^2\frac{\partial \overline{w}}{\partial \overline{z}}\left(\frac{\partial \overline{h}_2}{\partial \overline{z}}\right)^2+2\overline{\mu}\frac{\partial \overline{h}_2}{\partial \overline{z}}\frac{\partial \overline{w}}{\partial \overline{r}}\bigg|_{\overline{r}=\overline{h}_2}+2\epsilon^2\overline{\mu}\frac{\partial \overline{h}_2}{\partial \overline{z}}\frac{\partial \overline{u}}{\partial \overline{z}}$$
$$-2\overline{\mu}\frac{\partial \overline{u}}{\partial \overline{r}}\bigg|_{\overline{r}=\overline{h}_2}-\frac{\overline{\gamma}}{\overline{h}_2}\left(1+\epsilon^2\left(\frac{\partial \overline{h}_2}{\partial \overline{z}}\right)^2\right) \quad (24b)$$
$$=\frac{1}{\epsilon^2}(\overline{p}_a-\overline{p})\left(1+\epsilon^2\left(\frac{\partial \overline{h}_2}{\partial \overline{z}}\right)^2\right)$$

$$2\overline{\mu}\epsilon\frac{\partial \overline{w}}{\partial \overline{z}}\frac{\partial \overline{h}_1}{\partial \overline{z}}-2\overline{\mu}\epsilon\frac{\partial \overline{h}_1}{\partial \overline{z}}\frac{\partial \overline{u}}{\partial \overline{r}}\bigg|_{\overline{r}=\overline{h}_1}$$
$$+\overline{\mu}\left(\frac{\partial \epsilon \overline{h}_1}{\partial \overline{z}}\right)^2\left(\frac{\partial \overline{w}}{\partial \epsilon \overline{r}}+\frac{\partial \epsilon \overline{u}}{\partial \overline{z}}\right)-\overline{\mu}\left(\frac{\partial \overline{w}}{\partial \epsilon \overline{r}}+\frac{\partial \epsilon \overline{u}}{\partial \overline{z}}\right)=0 \quad (24c)$$

$$
\begin{aligned}
&2\overline{\mu}\epsilon\frac{\partial\overline{w}}{\partial\overline{z}}\frac{\partial\overline{h}_2}{\partial\overline{z}} - 2\overline{\mu}\epsilon\frac{\partial\overline{h}_2}{\partial\overline{z}}\frac{\partial\overline{u}}{\partial\overline{r}}\Big|_{\overline{r}=\overline{h}_2} \\
&+\overline{\mu}\left(\frac{\partial\epsilon\overline{h}_2}{\partial\overline{z}}\right)^2\left(\frac{\partial\overline{w}}{\partial\epsilon\overline{r}} + \frac{\partial\epsilon\overline{u}}{\partial\overline{z}}\right) - \overline{\mu}\left(\frac{\partial\overline{w}}{\partial\epsilon\overline{r}} + \frac{\partial\epsilon\overline{u}}{\partial\overline{z}}\right) = 0
\end{aligned} \tag{24d}
$$

Finally, the dimensionless thermal boundary conditions for the outer and the inner surface read

$$
-\left(1+4\gamma_R\overline{T}^3\frac{\partial\overline{T}}{\partial\overline{r}}\right) = \epsilon\alpha(\overline{T}^4 - \overline{T}_f^4) + \epsilon\beta(\overline{T} - \overline{T}_a) \tag{25a}
$$

$$
-\left(1+4\gamma_R\overline{T}^3\right)\frac{\partial\overline{T}}{\partial\overline{r}} = \epsilon\beta(\overline{T} - \overline{T}_a) \tag{25b}
$$

In Equation (25), $\alpha = \frac{\sigma\epsilon_r T_s^3 L}{k_c}$ represents the ratio between the radiative and the conductive heat exchange, while $\beta = \frac{k_h L}{k_c}$ is the Nusselt number, that is, the ratio between convective and conductive heat transfer [24].

2.1.5. Determination of the Furnace and Ambient Temperature Profiles

As our furnace is very similar to the one used by Voyce et al. [22], we begin the determination of the temperature profiles of the furnace by extracting their experimental values. Voyce et al. [22] measured the temperature profile of the oven by inserting a thermocouple into the air hole of a capillary tube that was slowly introduced into the oven, allowing the thermal equilibrium between the fibre and the furnace to be assumed. They measured the temperature profile of the furnace at three peak temperatures, say, 1300 °C, 1600 °C, and 1760 °C, reporting that it is very challenging to measure temperature profiles at peak temperatures higher than 1700 °C as thermocouples become inaccurate and may melt. They measured the temperature distribution only in a restricted portion of the furnace, in the vicinity of the peak value where the temperature is high enough to permit the drawing process to occur. The length of this portion of the oven is approximately 12 mm. We extract the experimental data of Voyce et al. [22] and interpolate them with a functional form similar to the one proposed by Taroni et al. [24]:

$$
\overline{T}_f(\overline{z}) = \overline{T}_M\left(\frac{1}{5} + \frac{4}{5}exp\left(c(\overline{T}_M)(\overline{z}-0.5)^2\right)\right) \tag{26}
$$

see Figure 2, where $\overline{T}_M = \frac{T_M}{Ts}$ and

$$
c(\overline{T}_M) = 2.1053\overline{T}_M - 4.6316 \tag{27}
$$

With this selection of the parameters, we keep the temperature distribution centred in the middle of the "hot zone" and let the width of the distribution depend on the peak temperature. Our choice is motivated by the fact that the temperature peaks shift only slightly towards the left with respect to the furnace center when the peak temperature increases. On the other hand, the width of the temperature distribution is affected by the peak temperature. Using Equation (26), we always obtain interpolation curves with $R^2 > 0.9$ for the three temperature distributions measured by Taroni et al. [24]. We accept this result and do not try to obtain a closer match between interpolated and experimental data. In addition, we choose $\overline{T}_a(\overline{z}) = \frac{3}{4}\overline{T}_f$, in accordance with Taroni et al. [24].

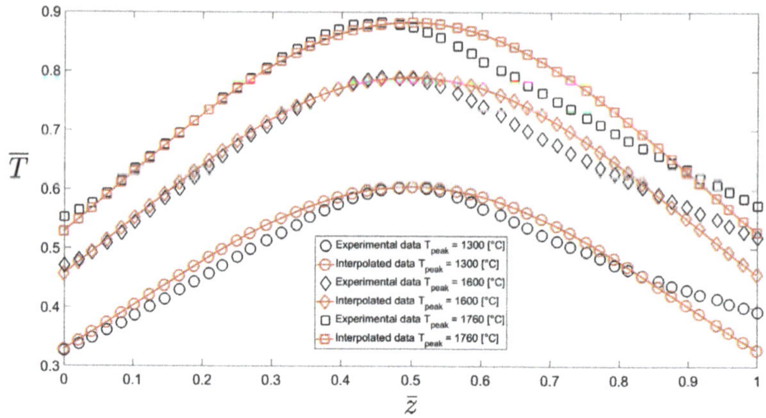

Figure 2. Furnace temperature profiles measured at three different peak temperatures by Voyce et al. [22] and interpolated temperature profiles according to the relation in (26).

2.1.6. The Functional Form of ϵ_r, k_h, α, and β

In general, values of the specific emissivity ϵ_r and the convective heat transfer coefficient k_h depend on fibre thickness, temperature, and properties of the material. Here, we follow the approach of Taroni et al. [24] by selecting

$$\epsilon_r = 1 - e^{-2.5\chi(h_2 - h_1)} \tag{28}$$

where we replace the fibre radius with the capillary thickness. Therefore, the specific emissivity ϵ_r decreases with decreasing capillary thickness. Concerning the convective heat transfer coefficient k_h, we utilize the form suggested by Geyling and Homsy [42]

$$k_h = \frac{w^{\frac{1}{3}}}{(h_2 - h_1)^{\frac{2}{3}}} \tag{29}$$

for the case of slow drawing ratios, while in the case of high drawing ratios we employ the correlation from Patel et al. [43] in the form suggested by Xue et al. [25]

$$k_h = 128.27 w^{0.574} \tag{30}$$

In Equation (29), we consider the capillary thickness instead of the radius. Finally, following the work of Taroni et al. [24], we let α vary with the capillary thickness:

$$\alpha = \frac{\sigma \epsilon_r T_s^3 L}{k_c}\left(1 - e^{-2.5\chi(h_2 - h_1)}\right) \tag{31}$$

However, in contrast to Taroni et al. [24] we utilize variable Nusselt numbers β, with k_h defined by Equations (29) and (30):

$$\beta = \frac{w^{\frac{1}{3}} L}{k_c (h_2 - h_1)^{\frac{2}{3}}} \tag{32}$$

in the case of slow drawing ratios and

$$\beta = \frac{128.27 w^{0.574} L}{k_c} \tag{33}$$

in the case of high drawing ratios.

2.1.7. The Viscosity of Silica Glass

Following Voyce et al. [22], we choose the correlation by Urbain et al. [44] for the viscosity in the temperature range 1400 °C $\leq$ T $\leq$ 2500 °C and the one by Hetherington et al. [45] for the temperature range 1000 °C $\leq T \leq$ 1400 °C. Urbain et al. [44] utilized the rotating cup technique to measure the viscosity in the interval between 10^{-1} and 10^{5} Poise and the isothermal penetration method to measure the viscosity in the interval between 10^{8} and 10^{13} Poise. Because the isothermal penetration method does not allow a glass metastable equilibrium to be reached, we opt for the correlation provided by Hetherington et al. [45]. They used a fibre elongation technique to measure the viscosity of the silica in this temperature range, checking that the silica was in an equilibrium condition [46]. Therefore, we utilize the following correlation:

$$\mu = 5.8 \cdot 10^{-8} exp\left(\frac{515400}{8.3145T}\right) \tag{34}$$

for the temperature range 1400 °C $\leq T \leq$ 2500 °C. Instead, we use

$$\mu = 3.8 \cdot 10^{-14} exp\left(\frac{712000}{8.3145T}\right) \tag{35}$$

for the temperature range 1000 °C < T $\leq$ 1400 °C. In (34) and (35), the viscosity μ is in $Pa \cdot s$ and the temperature T is in K.

3. Final Asymptotic Equations

Non-Rotating Capillary

We proceed with a regular parameter expansion of the unknowns

$$\begin{aligned}
\overline{w} &= \overline{w}_0(\overline{t},\overline{z}) + \epsilon^2 \overline{w}_1(\overline{t},\overline{z},\overline{r}) \\
\overline{u} &= \overline{u}_0(\overline{t},\overline{z},\overline{r}) + \epsilon^2 \overline{u}_1(\overline{t},\overline{z},\overline{r}) \\
\overline{p} &= \overline{p}_a + \epsilon^2 \overline{P}(\overline{t},\overline{z},\overline{r}) \\
\overline{T} &= \overline{T}_0(\overline{t},\overline{z},\overline{r}) + \epsilon^2 \overline{T}_1(\overline{t},\overline{z},\overline{r})
\end{aligned}$$

and utilize it in the previously derived non-dimensional mass conservation (Equation (19)), momentum (Equations (20) and (21)), and energy equations (Equation (22)) and boundary conditions (Equations (23) and (24)). By retaining the terms of the expansion at most up to order ϵ^2, we can obtain the equations of Fitt et al. [19]:

$$\frac{\partial(\overline{h}_1^2 \overline{w}_0)}{\partial \overline{z}} = \frac{\overline{p}_0 \overline{h}_2^2 \overline{h}_1^2 - \overline{\gamma} \overline{h}_2 \overline{h}_1 (\overline{h}_2 + \overline{h}_1)}{\mu(\overline{h}_2^2 - \overline{h}_1^2)} \tag{36}$$

$$\frac{\partial(\overline{h}_2^2 \overline{w}_0)}{\partial \overline{z}} = \frac{\overline{p}_0 \overline{h}_2^2 \overline{h}_1^2 - \overline{\gamma} \overline{h}_2 \overline{h}_1 (\overline{h}_2 + \overline{h}_1)}{\mu(\overline{h}_2^2 - \overline{h}_1^2)} \tag{37}$$

$$\frac{\partial}{\partial \overline{z}}\left(3\overline{\mu}\frac{\partial \overline{w}_0}{\partial \overline{z}}(\overline{h}_2^2 - \overline{h}_1^2) + \overline{\gamma}(\overline{h}_2 + \overline{h}_1)\right) = 0 \tag{38}$$

These are evolution equations for the inner and outer fibre surface and an axial momentum equation; for more details see [19]. In Equation (38), we have neglected the inertia terms due to the small values that the Reynolds number Re typically assumes; as a matter of fact, $Re \approx 10^{-8}$ [29]. With regard to the energy equation, we consider the case at the leading order, where $Pe = \frac{\widetilde{P}}{\epsilon^2}$ and $Br = \frac{\widetilde{Br}}{\epsilon^2}$:

$$\widetilde{P}\left(\frac{\partial\overline{T}}{\partial\overline{t}}+\overline{u}\frac{\partial\overline{T}}{\partial\overline{r}}+\overline{w}\frac{\partial\overline{T}}{\partial\overline{z}}\right)=\frac{1}{\overline{r}}\frac{\partial}{\partial\overline{r}}\left(\overline{r}\frac{\partial\overline{T}}{\partial\overline{r}}\right)+\frac{1}{\overline{r}}\frac{\partial}{\partial\overline{r}}\left(\gamma_R\overline{r}\frac{\partial\overline{T}^4}{\partial\overline{r}}\right)$$
$$+2\widetilde{Br}\overline{\mu}\left[\left(\frac{\partial\overline{u}}{\partial\overline{r}}\right)^2+\left(\frac{\overline{u}}{\overline{r}}\right)^2+\left(\frac{\partial\overline{w}}{\partial\overline{z}}\right)^2\right]+\widetilde{Br}\overline{\mu}\left[\frac{\partial\epsilon\overline{u}}{\partial z}+\frac{\partial\overline{w}}{\partial\epsilon\overline{r}}\right]^2 \quad (39)$$

This case corresponds to the general situation, where the convective heat transport balances the heat transfer across the fibre and the viscous dissipation [24]. Following Taroni et al. [24], we change variables to a more convenient coordinate frame that suits the geometry of the capillary:

$$\overline{x}(\overline{r},\overline{z})=\frac{\overline{r}-\overline{h}_1(\overline{z})}{\overline{h}_2(\overline{z})-\overline{h}_1(\overline{z})}\rightarrow\overline{r}=\overline{x}(\overline{h}_2-\overline{h}_1)+\overline{h}_1,\qquad \overline{\zeta}=\overline{z} \quad (40)$$

The continuity in Equation (19) at the leading order becomes

$$\overline{u}_0=-\frac{\partial\overline{w}_0}{\partial\overline{\zeta}}\frac{[\overline{x}(\overline{h}_2-\overline{h}_1)+\overline{h}_1]}{2}+\frac{\overline{A}(\overline{\zeta})}{[\overline{x}(\overline{h}_2-\overline{h}_1)+\overline{h}_1]} \quad (41)$$

where $\overline{A}$ is

$$\overline{A}=\frac{\overline{p}_0\overline{h}_2^2\overline{h}_1^2-\overline{\gamma}\overline{h}_2\overline{h}_1(\overline{h}_2+\overline{h}_1)}{2\overline{\mu}(\overline{h}_2^2-\overline{h}_1^2)} \quad (42)$$

and the axial momentum equation after being integrated once with respect to $\overline{z}$ assumes the form

$$\frac{\partial\overline{w}_0}{\partial\overline{\zeta}}=\left(\frac{\overline{F}-\overline{\gamma}(\overline{h}_2+\overline{h}_1)}{3\overline{\mu}(\overline{h}_{20}^2-\overline{h}_{10}^2)\overline{W}_0}\right)\overline{w}_0 \quad (43)$$

where the constant $\overline{F}$ arising from the integration physically represents the tension needed to pull the fibre. In addition, the following identity

$$\overline{h}_2^2-\overline{h}_1^2=\frac{(\overline{h}_{20}^2-\overline{h}_{10}^2)\overline{W}_0}{\overline{w}_0} \quad (44)$$

has been used, and represents an equation of mass conservation. We now change the variables in Equation (39) according to Equation (40) and utilize the asymptotic expansions previously defined to obtain

$$\frac{\partial\overline{T}}{\partial\overline{\zeta}}=\overline{\Psi}\frac{\partial\overline{T}_0}{\partial\overline{x}}+\overline{\Omega}\left(\frac{\partial^2\overline{T}_0}{\partial\overline{x}^2}\right)+\overline{\Lambda}\left(\frac{\partial\overline{T}_0}{\partial\overline{x}}\right)^2$$
$$+\frac{\widetilde{Br}\overline{\mu}}{\overline{w}_0}\left[3\left(\frac{\partial\overline{w}_0}{\partial\overline{\zeta}}\right)^2+\left(\frac{2\overline{A}}{\overline{\Theta}^2}\right)^2\right] \quad (45)$$

where

$$\overline{\Pi} = [\overline{x}(\overline{h}_2 - \overline{h}_1) + \overline{h}_1](\overline{h}_2 - \overline{h}_1)$$
$$\overline{\Theta} = [\overline{x}(\overline{h}_2 - \overline{h}_1) + \overline{h}_1]$$
$$\overline{\Lambda} = \frac{12\gamma_R \overline{T}_0^2}{\overline{w}_0(\overline{h}_2 - \overline{h}_1)^2}$$
$$\overline{\Omega} = \frac{1 + 4\gamma_R \overline{T}_0^3}{\overline{w}_0(\overline{h}_2 - \overline{h}_1)^2}$$
$$\overline{\Psi} = \frac{1}{\overline{w}_0}\left[\frac{1 + 4\gamma_R \overline{T}_0^3}{\overline{\Pi}} - \frac{\overline{A}}{\overline{\overline{\Pi}}} + \tilde{P}\frac{\partial \overline{w}_0}{\partial \overline{\zeta}}\frac{\overline{\Theta}}{2(\overline{h}_2 - \overline{h}_1)}\right]$$

The thermal boundary conditions in Equation (25) at $\overline{r} = \overline{h}_2$ and at $\overline{r} = \overline{h}_1$ become

$$\left.\frac{\partial \overline{T}_0}{\partial \overline{x}}\right|_{\overline{x}=1} = \left[-\frac{\epsilon\alpha(\overline{T}_0^4 - \overline{T}_f^4) + \epsilon\beta(\overline{T}_0 - \overline{T}_a)}{1 + 4\gamma_R \overline{T}_0^3}\right](\overline{h}_2 - \overline{h}_1) \tag{46a}$$

$$\left.\frac{\partial \overline{T}_0}{\partial \overline{x}}\right|_{\overline{x}=0} = -\frac{\epsilon\beta(\overline{T}_0 - \overline{T}_a)}{1 + 4\gamma_R \overline{T}_0^3}(\overline{h}_2 - \overline{h}_1) \tag{46b}$$

4. Results

4.1. Solution Method

We numerically solve the system of differential Equations (36), (37), (43) and (45) using a fourth-order Runge–Kutta–Merson method to integrate them in the axial direction. To this end, we first integrate Equation (43) and iteratively solve the whole system of equations until we find a value of the tension that satisfies the condition

$$\overline{w}_0(\overline{\zeta} = 1) - 1 < 10^{-3}$$

In addition, we discretize Equation (45) and the boundary conditions (46) in the radial direction using second order central difference schemes for both the first and the second derivatives.

All the relevant parameters and initial values for the different cases utilized in the numerical simulations are listed in Table 1. Although the values of the material parameters vary with the temperature, we assumed them to be constant, as their dependence on the temperature is weak. We only let the viscosity vary with the temperature, as its value changes by different orders of magnitude in the temperature range of interest. The step sizes in the axial and radial directions were chosen after a grid study to ensure that the results were trustworthy. We set $\Delta\zeta = 6.25 \times 10^{-7}$ and $\Delta x = 0.2$ for the case of the slow drawing ratios DR36-1, DR54-15, and DR72-2. In the case of the high drawing ratio DR1-102, we utilized $\Delta\zeta = 1.5625 \times 10^{-7}$ and $\Delta x = 0.5$.

Table 1. Main parameters and initial values used for the simulations. [a] Voyce et al. [22], [b] Lee and Yaluria [36], [c] Paek and runk [16], [d] Taroni et al. [24], [e] Huang et al. [47], [f] Myers [17], [g] Fitt et al. [19], [h] Luzi et al. [20].

Parameter	Symbol	Value	Units
Hot zone length [a]	L	0.12	m
Stefan-Boltzmann constant [b]	σ	5.67×10^{-8}	$W\,m^{-2}\,K^{-4}$
Refractive index [b]	n_0	1.5	-
Absorption coefficient [c]	χ	200	m^{-1}
Density [c]	ρ	2200	$kg\,m^{-3}$
Specific heat [c]	c_p	1000	$J\,kg^{-1}\,K^{-1}$
Thermal conductivity [d,e]	k_c	1.1	$W\,m^{-1}\,K^{-1}$
Glass softening temperature [f]	T_s	1900	K
Surface tension [g]	γ	0.25	$N\,m^{-1}$
Initial external radius [h]	h_{20}	1×10^{-2}	m
Initial internal radius [h]	h_{10}	3.65×10^{-3}	m
Drawing ratio	DR36-1		
Feed speed [h]	W_0	6×10^{-5}	$m\,s^{-1}$
Draw speed [h]	W_1	1.67×10^{-2}	$m\,s^{-1}$
Drawing ratio	DR54-15		
Feed speed [h]	W_0	9×10^{-5}	$m\,s^{-1}$
Draw speed [h]	W_1	2.5×10^{-2}	$m\,s^{-1}$
Drawing ratio	DR72-2		
Feed speed [h]	W_0	1.2×10^{-4}	$m\,s^{-1}$
Draw speed [h]	W_1	3.33×10^{-2}	$m\,s^{-1}$
Drawing ratio	DR1-102		
Feed speed [h]	W_0	1.67×10^{-5}	$m\,s^{-1}$
Draw speed [h]	W_1	1.7×10^{-1}	$m\,s^{-1}$

4.2. *Slow Drawing Ratios*

In this section, we first show representative results obtained by solving the system of Equations (36), (37), (43) and (45) for the case DR54-15. Figure 3a shows that the axial velocity profile assumes values close to zero at the beginning of drawing until $\bar{z} \approx 0.4$. Afterwards, it varies abruptly and suddenly increases, reaching the final value of one at $\bar{z} \approx 0.8$. To satisfy the mass conservation equation, the inner and outer surfaces greatly reduce their size before reaching the final dimensions; see Figure 3b. The radii significantly change their sizes and the axial velocity steeply increases only in a small portion of the furnace length measured by Voyce et al. [22]. They termed this part of the oven "hot zone", where the viscosity is low enough to enable the glass to deform. In Figure 3c, we contrast the evolution of the temperature profiles of the inner and outer surfaces of the capillary with that of the furnace along the drawing direction. At the beginning of the drawing, where the speed of the internal and the external surfaces of the capillary are very close to the feed speed, the temperature profiles of the glass surfaces are very close to the temperature distribution of the furnace, indicating thermal equilibrium between the furnace and the capillary. Starting from approximately $\bar{z} \approx 0.5$, the temperature profiles of the capillary surfaces deviate from that of the furnace, resulting in lower values of the glass temperature. Two possible causes lie in the reduced radiative heat exchange between the furnace and the fibre due to the smaller surface and in the significant increase of the convective heat exchange due to the higher values the capillary speed achieves toward the end of the drawing. Furthermore, we observe that the differences between the temperature profiles of the inner and outer radii of the capillary are negligible (see Figure 3c), indicating strong heat diffusion inside the glass due to the radiative contribution to the thermal conductivity. This is clear from the temperature contour displayed in Figure 4, which shows a substantial axial temperature variation across the fibre length and an imperceptible temperature change in the whole radial direction. In Figure 5, we can compare the final external diameters of the capillaries computed by numerically solving Equations (36), (37), (43) and (45) with the ones obtained experimentally by Luzi et al. [20] for the case of three slow drawing ratios without internal pressurization, that is, DR 36-1, DR 54-15, and DR 72-2, as reported in Table 1. In Figure 6, we compare their final air-filling fractions, that is, the ratio of their final internal radii h_{1f} over their final external radii h_{2f}. By doing this, we vary the peak temperature from 1850 °C up to 2050 °C in steps of 25 °C. The values of both the outer and inner final radii at the end of the drawing process gradually increase with decreasing furnace peak temperature for all three drawing ratios, as the higher viscosity at lower temperatures values hinders the capillary from further decreasing its size. With regard to the final external diameter, the numerical and experimental results are in excellent agreement, with a difference ranging from a minimum of approximately 0.06% up to a maximum of approximately 2%. With regard to the final air-filling fractions of the capillaries, the agreement between the numerical simulations and experimental results is excellent only in the temperature range between 1850 °C and 1975 °C, with a maximum difference of approximately only 2.8%. In the temperature range between 2000 °C and 2050 °C, the discrepancy between the numerical and experimental results varies between approximately 5% and 48%. This maximum extreme difference may be due to too low a value of the surface tension used in the simulations, that is, γ = 0.25 N/m. Another reason for the discrepancy may rely on the possible partial collapse of the inner surface of the capillary. The temperature profile may not be accurately described by the functional form provided by Equation (26) in the case of high peak temperatures, as it has been determined by fitting experimental data obtained at lower peak temperatures. This has a direct impact on the collapse of the capillary, as a simple order of magnitude analysis demonstrates that the collapse time of the inner capillary surface depends on both the viscosity and the surface tension of the molten glass [48]. Because the viscosity greatly varies with the temperature, a different temperature correlation may be required to approximate the furnace temperature profile at high peak temperatures. In addition, the insufficient information concerning the dependency of the surface tension with the temperature makes the estimation of the time required for collapse a very challenging task. In Figures 7 and 8 we examine cases with internal pressurization. First, we keep the pressure constant and vary the peak temperature, then we maintain the temperature and vary the internal pressure. In Figure 7a, we compare the final external diameter of the capillary obtained numerically with the experimental data of Luzi et al. [20], while in Figure 7b we compare the final air-filling fractions obtained numerically and experimentally. In both cases, we keep the internal pressure constant at $p_o = 9$ mbar and vary the peak temperature of the furnace from 1850 °C up to 2050 °C in steps of 25 °C. The numerical data match very well with the experimental data, with the maximum discrepancy between the final external diameters computed numerically and experimentally being approximately 11% in the case of T_{peak} = 2050 °C. We obtain similar agreement for the final external diameters and the final air-filling fractions in the case where we fix the peak temperature at 1950 °C and vary the internal pressure between 0 and 25 mbar; see Figure 8a,b. In this case, the maximum difference between the experimental values and the numerical computations is approximately 8%. The internal pressure counteracts the effects of the surface tension by promoting an enlargement of the inner hole, thereby increasing the size of the internal and external radii. The maximum deviations between the experimental and numerical results are again obtained for high values of the peak temperatures, where major uncertainties concerning the furnace temperature profile and the value of the surface tension and viscosity prevail.

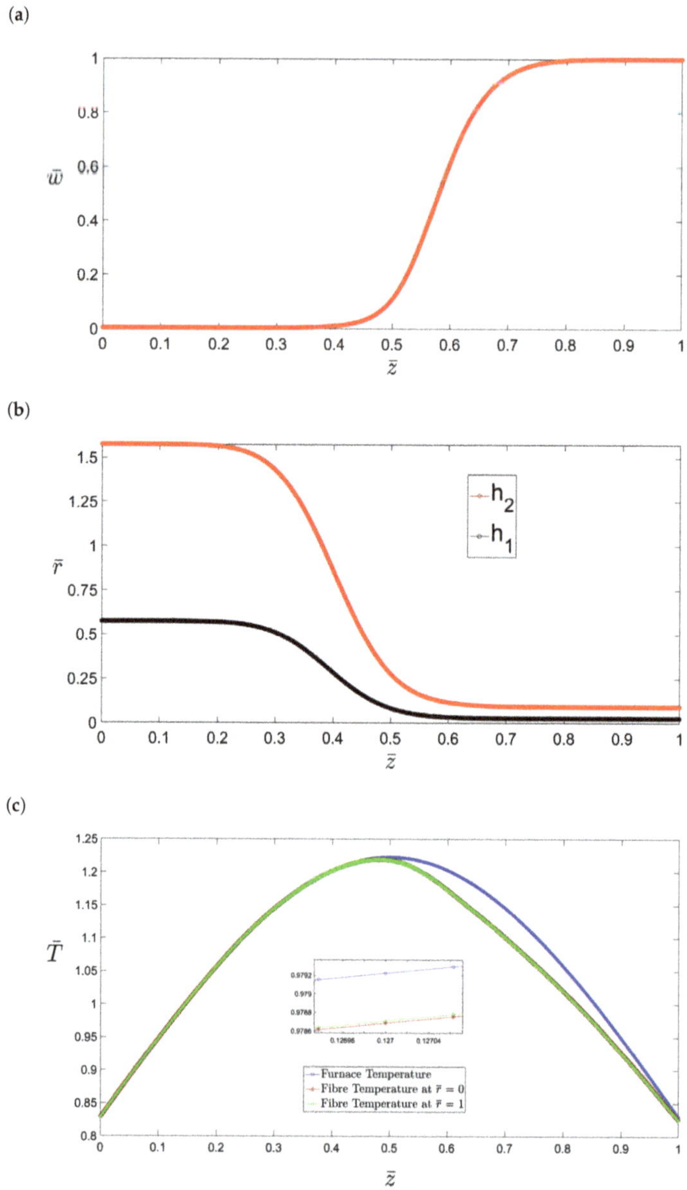

Figure 3. (**a**) Evolution of the axial component of the velocity $\overline{w}$ against the axial distance $\overline{z}$. The inner and outer surfaces of the capillary, $\overline{h}_1$ and $\overline{h}_2$, respectively, are plotted against the axial distance $\overline{z}$ in (**b**). The temperature profiles of the inner and outer capillary surfaces are plotted against the axial distance $\overline{z}$ together with the temperature profile of the furnace in (**c**). The inset in (**c**) magnifies the difference among the plotted temperature profiles of capillary surfaces and the temperature profile of the furnace.

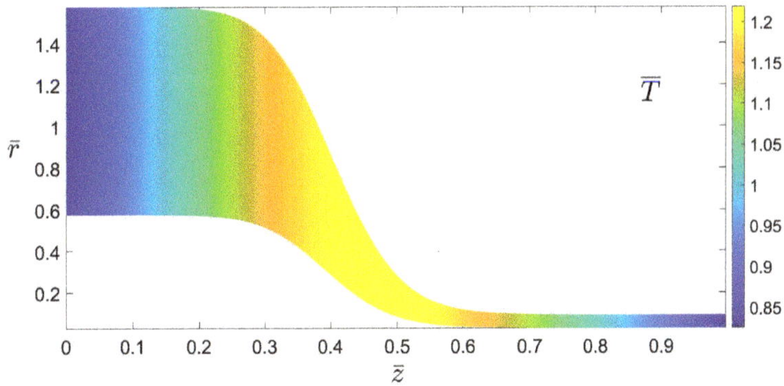

Figure 4. Fibre temperature contour in the case of DR54-15, T_{peak} = 2050 °C and no internal pressurization.

(**a**)

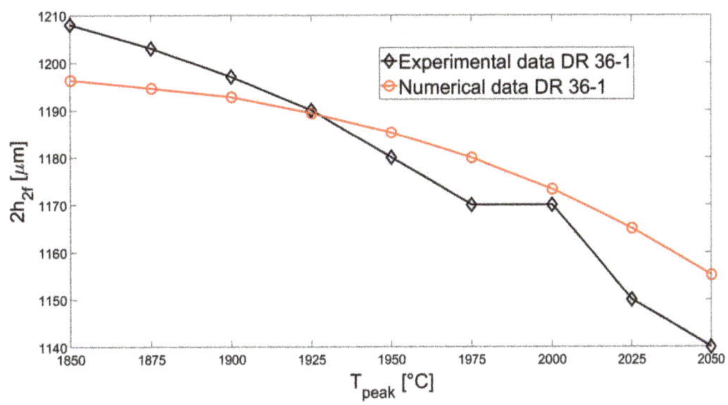

(**b**)

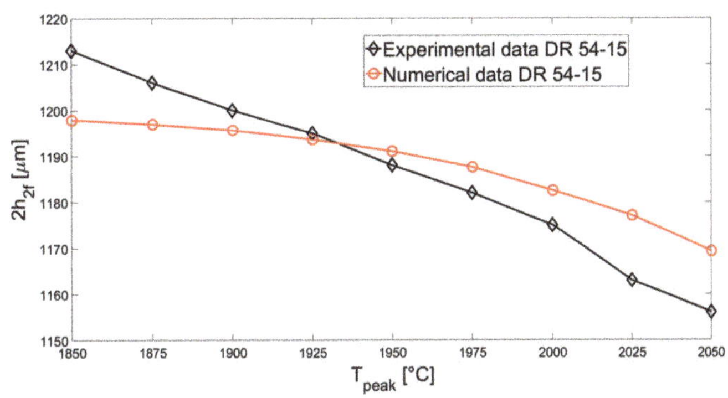

Figure 5. *Cont.*

(c)

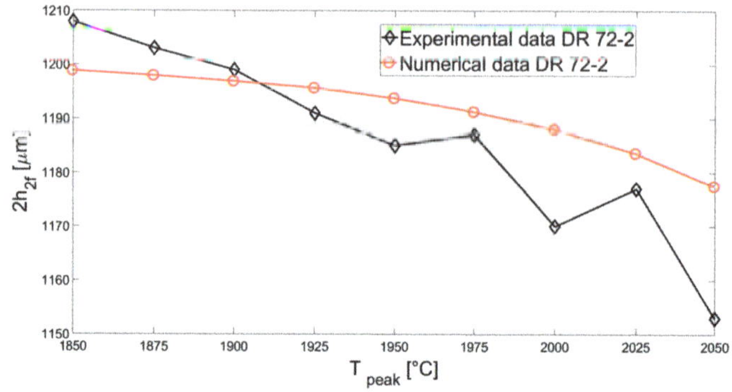

Figure 5. Comparison between the final external diameter of the capillary obtained numerically and experimentally by Luzi et al. [20] for the three different drawing ratios (**a**) DR36-1, (**b**) 54-15, and (**c**) 72-2 at different furnace peak temperatures.

(a)

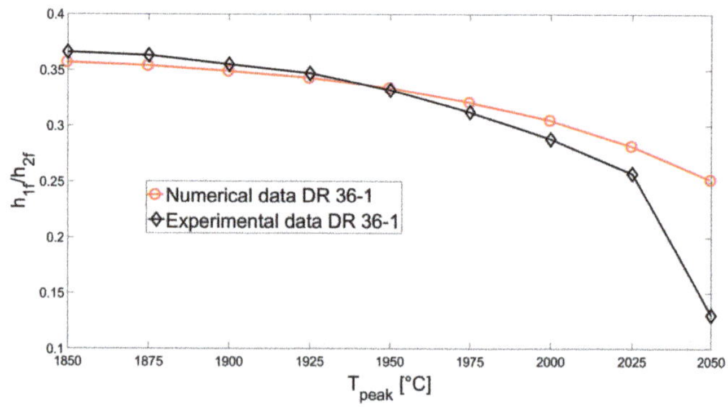

(b)

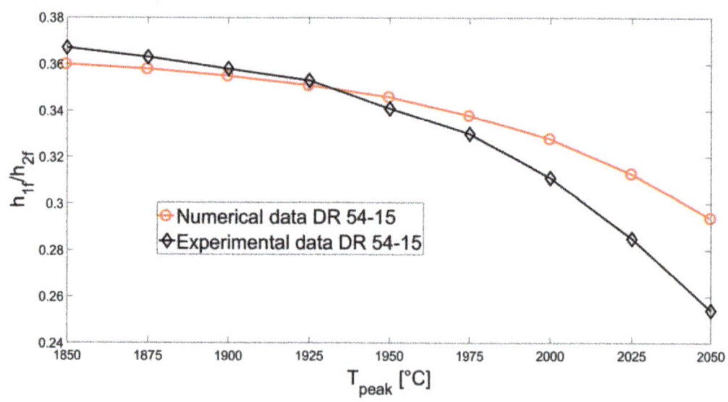

Figure 6. *Cont.*

(c)

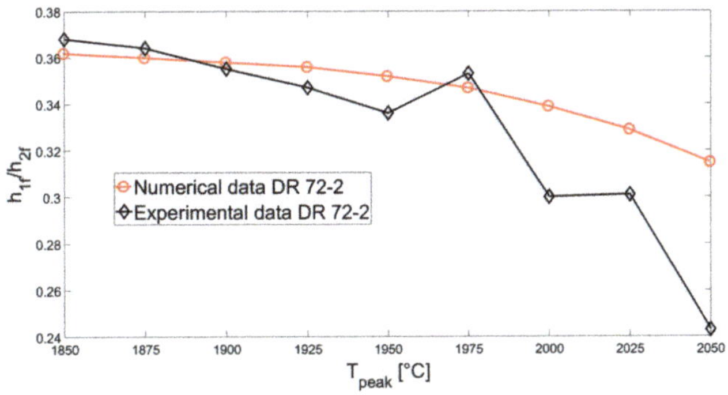

Figure 6. Comparison between the final air-filling fraction of the capillary obtained numerically and experimentally by Luzi et al. [20] for the three different drawing ratios (**a**) DR36-1, (**b**) 54-15, and (**c**) 72-2 at different furnace peak temperatures.

(a)

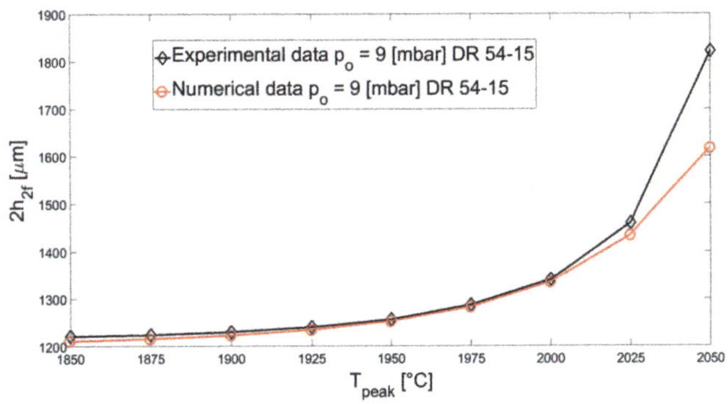

(b)

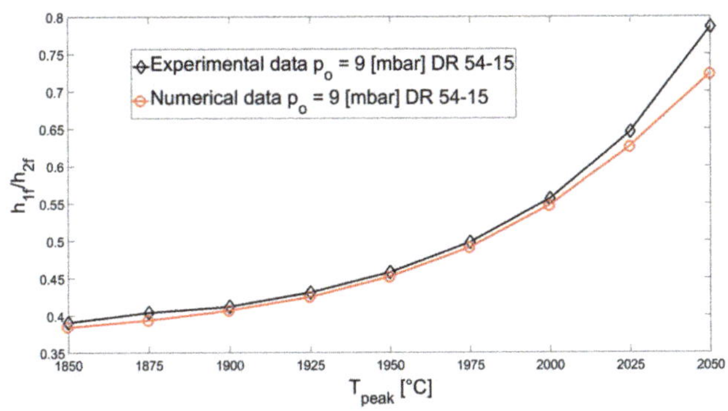

Figure 7. Comparison between the final outer diameter (**a**) and air-filling fraction (**b**) of the capillary obtained numerically and experimentally by Luzi et al. [20] for different furnace peak temperatures with a fixed value of internal pressurization.

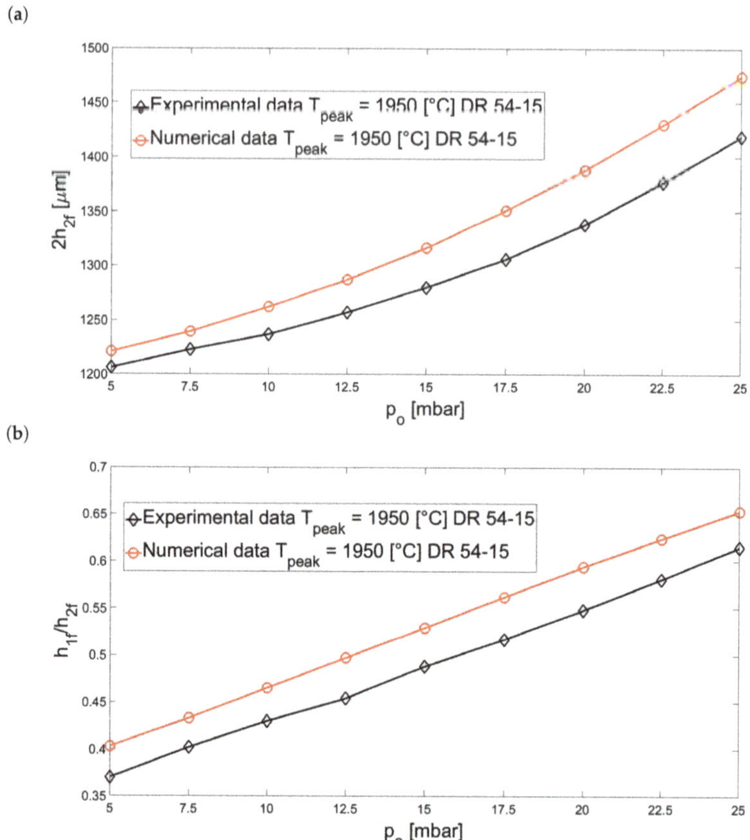

Figure 8. Comparison between the final outer diameter (**a**) and air-filling fraction (**b**) of the capillary obtained numerically and experimentally by Luzi et al. [20] for different values of internal pressurization and a fixed value of the furnace peak temperature.

4.3. High Drawing Ratios

In this section, we present exemplary results for the case of the high drawing ratio DR 1-102. This is relevant for fibre drawing, as the final dimensions of the capillary are of the same order of magnitude as commercially available optical fibres. Because the drawing speed is approximately ten times higher compared to the cases of the slow drawing ratios analyzed before, a smaller step size in the axial direction is required. Furthermore, because we realized that the temperature variations in the radial direction were not significant for the cases of the slow drawing ratios, we utilized a larger step size in the radial direction in the case of this high drawing ratio. This allowed us to choose a step size in the axial direction smaller than the one used with the slow drawing ratios, while it is not excessively small due to the constraint imposed by the Courant number for the stability of an explicit method. Moreover, we utilize Equation (33) to compute the Nusselt number β in the case of the high drawing ratio, as Equation (32) delivers values that are extremely high, promoting an unreasonable convective heat exchange that cools down the capillary too much and makes the numerical computations unstable. The high values of β are due to the high values of the axial velocity and the low values of the capillary size attained at the end of the drawing stage. Figure 9a shows that the axial velocity profile abruptly increases from a value approximately close to zero at $\bar{z} \approx 0.4$ up to about one at $\bar{z} \approx 0.6$. The increase of the axial velocity profile is steeper and the final value is attained at a shorter distance compared to the slow drawing ratios. In addition, the inner and outer surfaces of the capillary shrink considerably during the drawing process, and the final size of the capillary is very thin; see Figure 9b and inset. With regard to the temperature distribution within the capillary, we notice the absence of a significant variation of the temperature with the radial direction; see the inset of Figure 9c. Furthermore, in this case the temperature distribution in the axial direction is very close to that of the furnace until $\bar{z} \approx 0.45$. Afterwards, the temperature profile of the capillary significantly departs from that of the furnace, assuming lower values until the exit of the furnace is reached; see Figure 9c. The dimensionless values of the temperature of the capillary at different radial positions at the exit of the oven are approximately equal to 0.7, while the dimensionless value of the furnace temperature at the exit of the furnace is approximately equal to 0.82. In Figure 10, we compare the values of the final external diameters and those of the air volume fractions obtained numerically and experimentally for different values of the peak temperature without internal pressurization. The deviations between experiments and numerical simulations reach a maximum of approximately 12 % (see Figure 10a) and is again achieved at a high peak temperatures, that is, $T_{peak} = 2050$ °C.

(a)

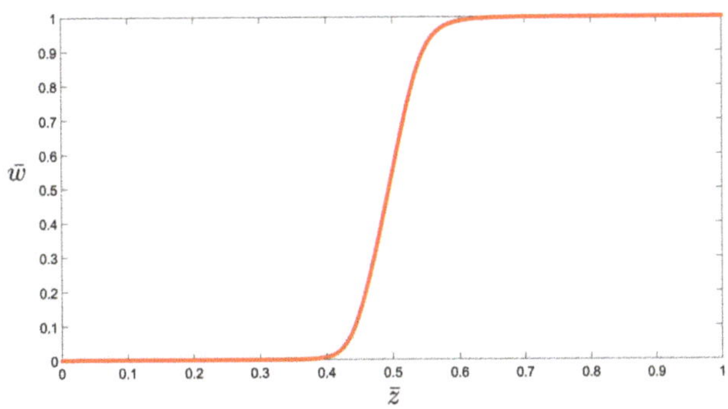

(b)

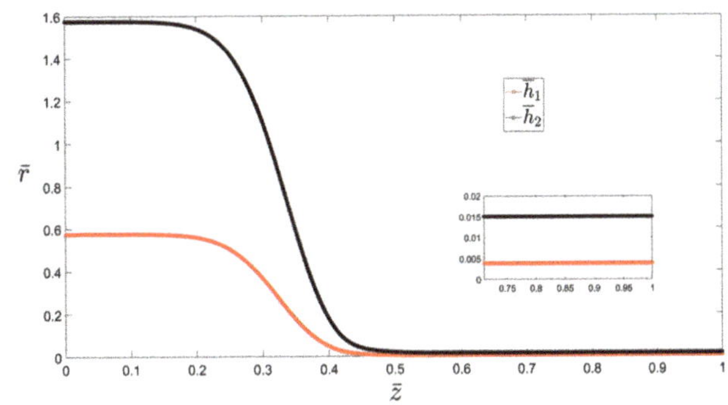

Figure 9. *Cont.*

(c)

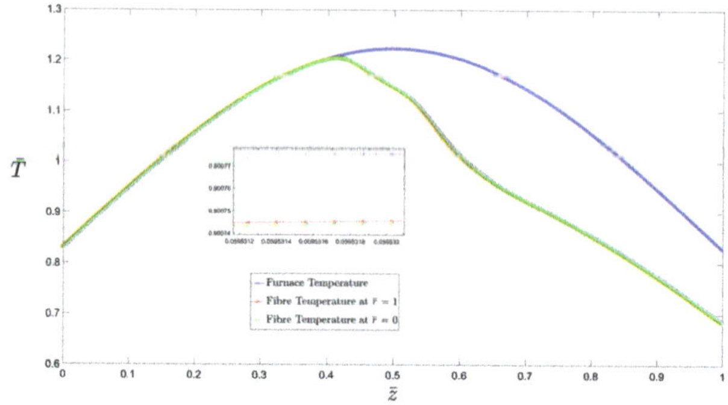

Figure 9. (**a**) Axial component of the velocity $\overline{w}$ against the axial distance $\overline{z}$, with $\overline{h}_1$ and $\overline{h}_2$ plotted against the axial distance $\overline{z}$ in (**b**) and the temperature profiles of the inner and outer capillary surfaces and the temperature profile of the furnace plotted against the axial distance $\overline{z}$ in (**c**). The inset in (**c**) shows the difference among the temperature profiles.

(a)

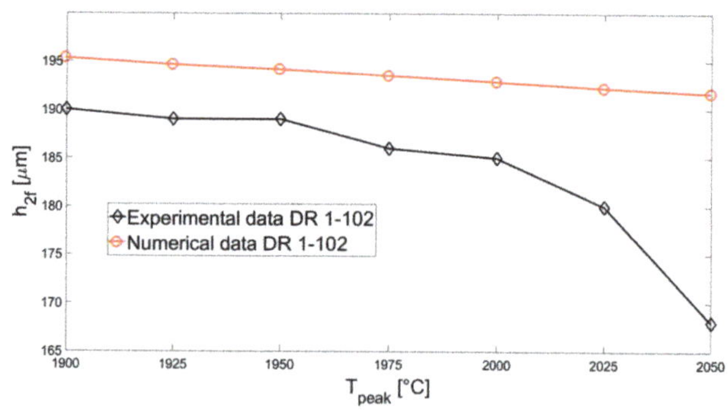

(b)

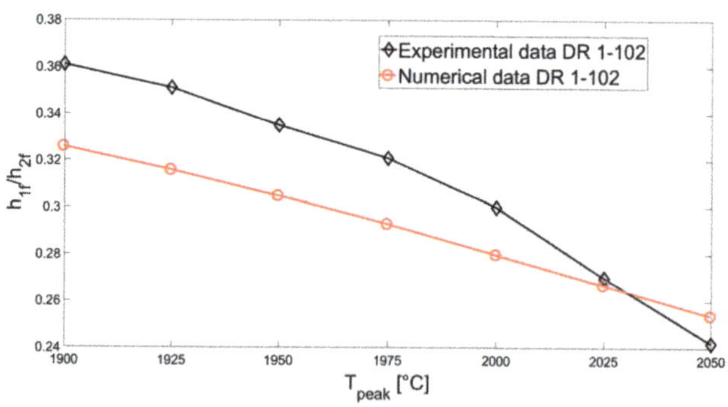

Figure 10. Comparison between the final outer diameter (**a**) and air-filling fraction (**b**) of the capillary obtained numerically and experimentally by Luzi et al. [20] for different furnace peak temperatures.

5. Discussion

In this work, we built a full asymptotic extensional-flow model to describe the evolution of a capillary during the drawing process. To this end, we analyzed and revisited the work of Taroni et al. [24] and integrated a new asymptotic energy equation into the momentum and mass conservation equations of Fitt et al. [19]. To check the validity of the model, we compared the numerical outcomes with the experimental data of Luzi et al. [20] for the case of low and high drawing ratios with and without internal pressurization. In the derivation of the energy equation, we considered conductive, convective, and radiative heat exchange, assuming that the capillary is optically thick. In addition, we included the effects of viscous dissipation in our model, as it may significantly affect the reduction of the size of the capillary [49]. Yin and Yaluria [39] pointed out that viscous dissipation must not be neglected, as it becomes important in the stage of the drawing where the size of the capillary shrinks due to the viscosity assuming large values and the velocity gradients being significant. To consider a general situation, we allow the temperature to depend on the axial and radial position of the capillary, thus choosing an approach where the convection is balanced by the transport across the capillary and by viscous dissipation effects. In the case of slow drawing ratios, the asymptotic model forecasts a slight reduction of the fibre temperature profile once the capillary size starts to decrease. In the case of high-speed ratios, however, the model predicts a sharp decrease of the capillary temperature profile toward the end of the drawing process compared to the initial phase, where thermal equilibrium prevails and the temperature of the capillary closely matches that of the furnace. The trend of this temperature profile of the capillary resembles the asymmetric fibre temperature profile utilized by Luzi et al. [28] to model the drawing process of a six-hole optical fibre. This fibre temperature profile sharply decreases when the size of the preform contracts and the axial velocity increases. In the contribution of Luzi et al. [28], it is shown that the use of an asymmetric temperature profile produces numerical results that are in good accordance with the experimental outcomes. On the other hand, the choice of a symmetric Gaussian temperature profile with a large width may produce completely incorrect results, even leading to the explosion of a capillary [20]. The asymmetric capillary temperature profile is produced by the selected functional forms of the Nusselt number β, that is, Equations (32) and (33). They result from the choice of the functional forms of the convective heat exchange k_h, i.e., Equations (29) and (30). Yin and Jaluria [39] remark that the heat transfer due to the radiative heat exchange is small at the end of the drawing, as it is directly proportional to the surface area of the fibre. This is reduced by approximately 1000 times from the beginning to the end of the drawing process. Thereby, the convective heat transfer may become the prevailing heat transfer mechanism because of the increase of the velocity at the end of the drawing. To compute k_h, we utilized the correlation by Patel et al. [43] in the case of high drawing ratios, and the one suggested by Geyling and Homsy [42] in the case of slow drawing ratios. The correlation suggested by Geyling and Homsy [42] produces extremely high values of the maximum Nusselt number β_{max} in the case of the high drawing ratio. In fact, $\beta_{max} \approx 28.04$ if the correlation of Geyling and Homsy is utilized, while $\beta_{max} \approx 5.02$ when the correlation by Patel et al. [43] is applied. This implies that $k_{hmax} \approx 257 \left[\frac{W}{m^2K}\right]$ using the correlation of Geyling and Homsy, and $k_{hmax} \approx 46 \left[\frac{W}{m^2K}\right]$ employing the correlation by Patel et al. [43]. This result is close to the values obtained by Choudhury and Yaluria [50], who numerically computed the heat transfer coefficient while investigating the drawing process of solid glass fibres. The results of our simulations reveal that the temperature profile of the capillary does not vary significantly in the radial direction due to the small size of the preform. A similar result was obtained by Taroni et al. [24] within approximately 1 cm of a solid fibre preform employing the surface radiation parameter $\gamma_R = 5$. In our case, $\gamma_R = 5.3$ indicates that the bulk radiation is predominant and the temperature distribution is approximately constant across the capillary. The results of our numerical simulations match very well overall with the experiments by Luzi et al. [20], except for a few isolated cases where the discrepancies are very high. This happens for high values of the peak temperature, where the temperature profile may not be well approximated by the relation (26). As a result, the partial collapse of the inner radius of the capillary resulting from the competition between viscosity and surface tension may not be captured. The agreement between experiments and simulations could be improved by utilizing a different correlation to model the furnace temperature profiles at high peak temperatures and an adequate correlation for the surface tension dependency on the temperature. Measurements of the furnace temperature profile are very difficult and prohibitive at high values of the furnace peak temperature [22], as thermocouples lose their accuracy and are inclined to melt. In addition, experimental and numerical investigations could be carried out to find different correlations for the convective heat transfer coefficient, as the models utilized to date have been developed for a solid thread [43] and for flow past a cylinder [42]. In addition, the present mathematical framework serves as a basis for modeling the drawing process of polymer optical fibres (POFs). As a first approximation, if the assumption of a Newtonian fluid is retained [25] the present mathematical formulation can be used in a straightforward way to model the drawing process of polymer axis-symmetric capillaries, as only a few process parameters need to be adjusted. However, because an adequate nonlinear viscoelastic model is necessary to accurately model the viscosity variations during the drawing of POFs, the governing equations need to be modified in a later stage of the research. Although the present asymptotic energy equation is only valid for annular capillaries and not for the general case of PCFs with arbitrary cross-sections, it provides a basis to improve simple models developed for more complicated geometries. For instance, the effects of viscous diffusion and the optically thick approximation can be included in the one-dimensional thermal model by Stokes et al. [34]. Moreover, this model can be extended to three dimensions, and work in this direction is currently ongoing.

6. Conclusions

In this work, we developed a novel asymptotic energy equation for the drawing process of an annular capillary. By coupling it with the mass, momentum, and evolution equations for the inner and outer surfaces by Fitt et al. [19], we built a complete asymptotic model for the capillary drawing process. The asymptotic system of equations is considerably simpler to solve numerically than the original full three-dimensional system of equations while being able to consider the main effects of heat transport across the capillary by convection, diffusion, and viscous dissipation. Heat is exchanged between the furnace and the capillary by convective and radiative mechanisms. To consider radiation within the capillary, we utilized the optically thick approximation and added an extra term to the conductive heat transfer coefficient. Although the validity of an optically thick capillary is likely to break down when the size of the capillary becomes very small, it nonetheless allows for quick and reliable computations of the geometry of the capillary compared to more involved approaches. Comparisons with experimental results for slow and high drawing ratios comprehensively show very good agreement for both pressurized and unpressurized capillaries. Discrepancies between experiments and simulations may be alleviated by a better choice of the value of the surface tension and a different correlation for the heat transfer coefficient.

Author Contributions: Supervision, B.G. and A.D.; Writing—original draft, S.L.; Writing—review & editing, G.L. All authors have read and agreed to the published version of the manuscript.

Funding: This work was supported by a National Research Foundation of Korea (NRF) grant funded by the Korean government (MSIT) (NRF-2021R1F1A1050103).

Institutional Review Board Statement: Not applicable.

Informed Consent Statement: Not applicable.

Data Availability Statement: Not applicable.

Conflicts of Interest: The authors declare no conflict of interest.

Appendix A

Appendix A.1. Two-Dimensional Mass and Momentum Equations for the Case of a Rotating Capillary

We assume axis-symmetry and an incompressible flow, i.e., $\rho = const$ and $\frac{\partial}{\partial \phi} = 0$, with a finite and non-zero v. The mass conservation equation assumes the same form as Equation (6), and many terms of the components of the stress tensor are identical to those defined in Section 2.1.2 for the case of a non-rotating capillary. The only components of the stress tensor that differ are

$$\tau_{r\phi} = \mu\left[r\frac{\partial}{\partial r}\left(\frac{v}{r}\right]\right)$$

and

$$\tau_{\phi z} = \mu\left(\frac{\partial v}{\partial z}\right)$$

The momentum equations along the radial and the azimuthal direction assume the form

$$\begin{aligned}\rho\left(\frac{\partial u}{\partial t} + u\frac{\partial u}{\partial r} - \frac{v^2}{r} + w\frac{\partial u}{\partial z}\right) = f_r - \frac{\partial p}{\partial r}\\ +\frac{2}{r}\frac{\partial}{\partial r}\left(r\mu\left(\frac{\partial u}{\partial r}\right)\right) + \frac{\partial}{\partial z}\left(\mu\left(\frac{\partial w}{\partial r} + \frac{\partial u}{\partial z}\right)\right) - 2\mu\frac{u}{r^2}\end{aligned} \tag{A1}$$

$$\begin{aligned}\rho\left(\frac{\partial v}{\partial t} + u\frac{\partial v}{\partial r} + \frac{uv}{r} + w\frac{\partial v}{\partial z}\right) = f_\phi + \frac{1}{r^2}\frac{\partial}{\partial r}\left[r^2\mu\frac{\partial}{\partial r}\left(\frac{v}{r}\right)\right]\\ +\frac{\partial}{\partial z}\left(\mu\frac{\partial v}{\partial z}\right)\end{aligned} \tag{A2}$$

while the one along the axial direction is identical to Equation (8). The kinematic and dynamic boundary conditions defined in the previous section for the case of a non-rotating capillary hold for the present case as well. In addition, we have two more tangential boundary conditions

$$\boldsymbol{s}_1^T \cdot \boldsymbol{\tau} \cdot \boldsymbol{n}_1 = 0 \tag{A3a}$$

$$\boldsymbol{s}_2^T \cdot \boldsymbol{\tau} \cdot \boldsymbol{n}_2 = 0 \tag{A3b}$$

in the tangential azimuthal direction. Now, the stress tensor $\boldsymbol{\tau}$ reads

$$\boldsymbol{\tau} = \begin{bmatrix} 2\mu\left(\frac{\partial w}{\partial z}\right) - p & \mu\left(\frac{\partial v}{\partial z}\right) & \mu\left(\frac{\partial w}{\partial r} + \frac{\partial u}{\partial z}\right) \\ \mu\left(\frac{\partial v}{\partial z}\right) & 2\mu\left(\frac{\partial v}{\partial r}\right) - p & \mu\left[r\frac{\partial}{\partial r}\left(\frac{v}{r}\right)\right] \\ \mu\left(\frac{\partial w}{\partial r} + \frac{\partial u}{\partial z}\right) & \mu\left[r\frac{\partial}{\partial r}\left(\frac{v}{r}\right)\right] & 2\mu\left(\frac{\partial u}{\partial r}\right) - p \end{bmatrix} \tag{A4}$$

and

$$\boldsymbol{s}_1^T = \boldsymbol{s}_2^T = 1\boldsymbol{e}_\phi$$

while $\boldsymbol{n}_i$ and $\boldsymbol{\tau}_i$ take the same form as Equations (11c) and (11b) for the case of a non-rotating capillary.

Appendix A.2. Energy Equation for the Case of a Rotating Capillary

In the case of a rotating capillary, the energy equation reads

$$
\begin{aligned}
\rho c_p\left(\frac{\partial T}{\partial t}+u\frac{\partial T}{\partial r}+w\frac{\partial T}{\partial z}\right) &= \frac{1}{r}\frac{\partial}{\partial r}\left(kr\frac{\partial T}{\partial r}\right)+\frac{\partial}{\partial z}\left(k\frac{\partial T}{\partial z}\right)\\
&+2\mu\left[\left(\frac{\partial u}{\partial r}\right)^2+\left(\frac{u}{r}\right)^2+\left(\frac{\partial w}{\partial z}\right)^2\right]+\mu\left[r\frac{\partial}{\partial r}\left(\frac{v}{r}\right)\right]^2\\
&+\mu\left(\frac{\partial v}{\partial z}\right)^2+\mu\left[\frac{\partial u}{\partial z}+\frac{\partial w}{\partial r}\right]^2
\end{aligned}
\tag{A5}
$$

Employing the Rosseland approximation and using Equation (15), we obtain

$$
\begin{aligned}
\rho c_p\left(\frac{\partial T}{\partial t}+u\frac{\partial T}{\partial r}+w\frac{\partial T}{\partial z}\right) &= \frac{1}{r}\frac{\partial}{\partial r}\left(k_c r\frac{\partial T}{\partial r}\right)\\
&+\frac{1}{r}\frac{\partial}{\partial r}\left(\tilde{k}_r r\frac{\partial T^4}{\partial r}\right)+\frac{\partial}{\partial z}\left(k_c\frac{\partial T}{\partial z}\right)+\frac{\partial}{\partial z}\left(\tilde{k}_r\frac{\partial T^4}{\partial z}\right)\\
&+2\mu\left[\left(\frac{\partial u}{\partial r}\right)^2+\left(\frac{u}{r}\right)^2+\left(\frac{\partial w}{\partial z}\right)^2\right]+\mu\left[\frac{\partial u}{\partial z}+\frac{\partial w}{\partial r}\right]^2\\
&\mu\left[r\frac{\partial}{\partial r}\left(\frac{v}{r}\right)\right]^2+\mu\left(\frac{\partial v}{\partial z}\right)^2
\end{aligned}
\tag{A6}
$$

Appendix A.3. Non-Dimensionalization of the Governing Equations in the Case of a Rotating Capillary

We non-dimensionalize Equations (6), (8), (A1), (A2) and (A6) by setting

$$
\begin{array}{lll}
r=\epsilon L\overline{r} & z=L\overline{z} & u=\epsilon W_1\overline{u}\\
w=W_1\overline{w} & h_1=\epsilon L\overline{h}_1 & h_2=\epsilon L\overline{h}_2\\
p=\dfrac{\mu_s W_1}{\epsilon^2 L}\overline{p} & T=T_s\overline{T} & t=\dfrac{L}{W_1}\overline{t}\\
v=\Omega L\overline{v} & \mu=\mu_s\overline{\mu} & \phi=\overline{\phi}
\end{array}
$$

where overbars indicate non-dimensional quantities, ϕ is the azimuthal angle, and Ω is the angular frequency. The continuity and momentum equations in the z direction are equal to Equation (19) and Equation (20), respectively. In turn, the equations in the r and in the ϕ direction assume the form

$$
\begin{aligned}
\epsilon^2 Re\left(\frac{\partial\overline{u}}{\partial\overline{t}}+\overline{u}\frac{\partial\overline{u}}{\partial\overline{r}}+\overline{w}\frac{\partial\overline{u}}{\partial\overline{z}}-\frac{S^2\overline{v}^2}{\epsilon^2\overline{r}}\right) &= \epsilon\frac{Re}{Fr^2}-\frac{\partial\overline{p}}{\partial\overline{r}}\\
+\frac{2}{\overline{r}}\frac{\partial}{\partial\overline{r}}\left[\overline{r}\,\overline{\mu}\left(\frac{\partial\overline{u}}{\partial\overline{r}}\right)\right]+\epsilon^2\frac{\partial}{\partial\overline{z}}\left[\overline{\mu}\left(\frac{\partial\overline{u}}{\partial\overline{z}}\right)\right]&+\frac{\partial}{\partial\overline{z}}\left[\overline{\mu}\left(\frac{\partial\overline{w}}{\partial\overline{r}}\right)\right]-2\overline{\mu}\left(\frac{\overline{u}}{\overline{r}^2}\right)
\end{aligned}
\tag{A7}
$$

$$
\begin{aligned}
Re\left(\frac{\partial\overline{v}}{\partial\overline{t}}+\overline{u}\frac{\partial\overline{v}}{\partial\overline{r}}+\overline{w}\frac{\partial\overline{v}}{\partial\overline{z}}+\frac{\overline{u}\,\overline{v}}{\overline{r}}\right) &=\\
\frac{1}{\epsilon^2\overline{r}^2}\frac{\partial}{\partial\overline{r}}\left[\overline{r}^2\,\overline{\mu}\left(\overline{r}\frac{\partial}{\partial\overline{r}}\left(\frac{\overline{v}}{\overline{r}}\right)\right)\right]&+\frac{\partial}{\partial\overline{z}}\left[\overline{\mu}\left(\frac{\partial\overline{v}}{\partial\overline{z}}\right)\right]
\end{aligned}
\tag{A8}
$$

The energy equation now reads

$$\epsilon^2 Pe\left(\frac{\partial \overline{T}}{\partial \overline{t}}+\overline{u}\frac{\partial \overline{T}}{\partial \overline{r}}+\overline{w}\frac{\partial \overline{T}}{\partial \overline{z}}\right)=\frac{1}{\overline{r}}\frac{\partial}{\partial \overline{r}}\left(\overline{r}\frac{\partial \overline{T}}{\partial \overline{r}}\right)$$
$$+\frac{1}{\overline{r}}\frac{\partial}{\partial \overline{r}}\left(\gamma_R \overline{r}\frac{\partial \overline{T}^4}{\partial \overline{r}}\right)+\epsilon^2\frac{\partial}{\partial \overline{z}}\left(\frac{\partial \overline{T}}{\partial \overline{z}}\right)+\epsilon^2\frac{\partial}{\partial \overline{z}}\left(\gamma_R\frac{\partial \overline{T}^4}{\partial \overline{z}}\right)$$
$$+2\epsilon^2 Br\overline{\mu}\left[\left(\frac{\partial \overline{u}}{\partial \overline{r}}\right)^2+\left(\frac{\overline{u}}{\overline{r}}\right)^2+\left(\frac{\partial \overline{w}}{\partial \overline{z}}\right)^2\right]+\epsilon^2 Br\overline{\mu}\left[\frac{\partial \epsilon \overline{u}}{\partial z}+\frac{\partial \overline{w}}{\partial \epsilon \overline{r}}\right]^2 \tag{A9}$$
$$+\epsilon^2 Br_R\overline{\mu}\left(\frac{\partial \overline{v}}{\partial \overline{z}}\right)^2+\epsilon^2 Br_R\overline{\mu}\left[\overline{r}\frac{\partial}{\partial \overline{r}}\left(\frac{\overline{v}}{\epsilon \overline{r}}\right)\right]^2$$

where

$$Br_R=\frac{\mu_s\Omega^2L^2}{T_s k_c}$$

is the Brinkman number associated with the rotational speed and $S=\frac{\Omega L}{W_1}$. The dimensionless kinematic, normal, and axial tangential boundary conditions are the same as in Equations (23) and (24). The tangential boundary conditions in the azimuthal direction for the inner and outer surface read

$$\left.\frac{\partial \overline{v}}{\partial \overline{r}}\right|_{\overline{r}=\overline{h}_1}-\left.\frac{\overline{v}}{\overline{r}}\right|_{\overline{r}=\overline{h}_1}-\epsilon^2\frac{\partial \overline{h}_1}{\partial \overline{z}}\frac{\partial \overline{v}}{\partial \overline{z}}=0 \tag{A10a}$$

$$\left.\frac{\partial \overline{v}}{\partial \overline{r}}\right|_{\overline{r}=\overline{h}_2}-\left.\frac{\overline{v}}{\overline{r}}\right|_{\overline{r}=\overline{h}_2}-\epsilon^2\frac{\partial \overline{h}_2}{\partial \overline{z}}\frac{\partial \overline{v}}{\partial \overline{z}}=0 \tag{A10b}$$

Finally, the dimensionless thermal boundary conditions for the outer and the inner surface are equal to Equations (25).

Appendix A.4. Final Asymptotic Equations for the Case of a Rotating Capillary

In this case, we expand the unknowns as follows:

$$\begin{aligned}
\overline{w}&=\overline{w}_0(\overline{t},\overline{z})+\epsilon^2\overline{w}_1(\overline{t},\overline{z},\overline{r})\\
\overline{u}&=\overline{u}_0(\overline{t},\overline{z},\overline{r})+\epsilon^2\overline{u}_1(\overline{t},\overline{z},\overline{r})\\
\overline{v}&=\overline{v}_0(\overline{t},\overline{z},\overline{r})+\epsilon^2\overline{v}_1(\overline{t},\overline{z},\overline{r})\\
\overline{p}&=\overline{p}_a+\epsilon^2\overline{P}(\overline{t},\overline{z},\overline{r})\\
\overline{T}&=\overline{T}_0(\overline{t},\overline{z},\overline{r})+\epsilon^2\overline{T}_1(\overline{t},\overline{z},\overline{r})
\end{aligned}$$

including the azimuthal component of the velocity field $\overline{v}$, then use them in the non-dimensional continuity (Equation (19)), momentum (Equations (20), (A7) and (A8)), and energy (Equation (A9)) equations and boundary conditions (Equations (23), (24) and (A10)). We obtain the equations of Voyce et al. [21] by expanding the unknowns up to order ϵ^2, as follows:

$$\frac{\partial(\overline{h}_1^2\overline{w}_0)}{\partial \overline{z}}=\frac{2\overline{p}_0\overline{h}_2^2\overline{h}_1^2-2\overline{\gamma}\overline{h}_2\overline{h}_1(\overline{h}_2+\overline{h}_1)+ReS^2\overline{B}^2\,\overline{h}_1^2\overline{h}_2^2(\overline{h}_2^2-\overline{h}_1^2)}{2\overline{\mu}(\overline{h}_2^2-\overline{h}_1^2)} \tag{A11}$$

$$\frac{\partial(\overline{h}_2^2\overline{w}_0)}{\partial \overline{z}}=\frac{2\overline{p}_0\overline{h}_2^2\overline{h}_1^2-2\overline{\gamma}\overline{h}_2\overline{h}_1(\overline{h}_2+\overline{h}_1)+ReS^2\overline{B}^2\,\overline{h}_1^2\overline{h}_2^2(\overline{h}_2^2-\overline{h}_1^2)}{2\overline{\mu}(\overline{h}_2^2-\overline{h}_1^2)} \tag{A12}$$

$$\frac{\partial}{\partial \overline{z}}\left(3\overline{\mu}\frac{\partial \overline{w}_0}{\partial \overline{z}}(\overline{h}_2^2-\overline{h}_1^2)+\overline{\gamma}(\overline{h}_2+\overline{h}_1)+\frac{1}{4}ReS^2\overline{B}^2(\overline{h}_2^4-\overline{h}_1^4))\right)=0 \tag{A13}$$

$$\frac{\partial}{\partial \overline{z}}\left(\overline{\mu}\frac{\partial \overline{B}}{\partial \overline{z}}(\overline{h}_2^4-\overline{h}_1^4))\right) = Re\overline{w}_0\left[\overline{h}_2^2\frac{\partial}{\partial \overline{z}}\left(\overline{h}_2^2\overline{B}\right)-\overline{h}_1^2\frac{\partial}{\partial \overline{z}}\left(\overline{h}_1^2\overline{B}\right)\right]+\overline{h}_1^2\overline{h}_2^2\overline{B}\frac{Re\overline{p}_0}{\overline{\mu}} + \frac{Re^2S^2\overline{h}_1^2\overline{h}_2^2\overline{B}^3\left(\overline{h}_2^2-\overline{h}_1^2\right)}{2\overline{\mu}} - \left(\overline{h}_1+\overline{h}_2\right)\frac{Re\overline{\gamma}\overline{B}\overline{h}_1\overline{h}_2}{\overline{\mu}} \tag{A14}$$

Equations (A11)–(A14) are the equations for the inner and outer fibre radii, axial momentum, and non-dimensional angular frequency equation, respectively; for more details, see [21]. Additionally, in this case we consider the steady-state situation and neglect the convective inertial terms in Equation (A13). As far as the energy equation is concerned we rescale the Péclet and Brinkman numbers as follows:

$$Pe = \frac{\widetilde{P}}{\epsilon^2}, \qquad Br = \frac{\widetilde{Br}}{\epsilon^2} \qquad Br_R = \frac{\widetilde{Br}_R}{\epsilon^2} \tag{A15}$$

obtaining

$$\widetilde{P}\left(\frac{\partial \overline{T}}{\partial \overline{t}}+\overline{u}\frac{\partial \overline{T}}{\partial \overline{r}}+\overline{w}\frac{\partial \overline{T}}{\partial \overline{z}}\right) = \frac{1}{\overline{r}}\frac{\partial}{\partial \overline{r}}\left(\overline{r}\frac{\partial \overline{T}}{\partial \overline{r}}\right)+\frac{1}{\overline{r}}\frac{\partial}{\partial \overline{r}}\left(\gamma_R\overline{r}\frac{\partial \overline{T}^4}{\partial \overline{r}}\right) + 2\widetilde{Br}\overline{\mu}\left[\left(\frac{\partial \overline{u}}{\partial \overline{r}}\right)^2+\left(\frac{\overline{u}}{\overline{r}}\right)^2+\left(\frac{\partial \overline{w}}{\partial \overline{z}}\right)^2\right]+\widetilde{Br}\overline{\mu}\left[\frac{\partial \epsilon\overline{u}}{\partial z}+\frac{\partial \overline{w}}{\partial \epsilon\overline{r}}\right]^2 + \widetilde{Br}_R\overline{\mu}\left(\frac{\partial \overline{v}}{\partial \overline{z}}\right)^2+\widetilde{Br}_R\overline{\mu}\left[\overline{r}\frac{\partial}{\partial \overline{r}}\left(\frac{\overline{v}}{\epsilon\overline{r}}\right)\right]^2 \tag{A16}$$

Using the coordinate system defined in Equation (40), the axial momentum equation becomes

$$\frac{\partial \overline{w}_0}{\partial \overline{\zeta}} = \left(\frac{\overline{F}-\overline{\gamma}(\overline{h}_2+\overline{h}_1)-\frac{1}{4}ReS^2\overline{B}^2(\overline{h}_2^4-\overline{h}_1^4)}{3\overline{\mu}(\overline{h}_{20}^2-\overline{h}_{10}^2)\overline{W}_0}\right)\overline{w}_0 \tag{A17}$$

where again the constant $\overline{F}$ is the tension required to pull the fibre. Utilizing the asymptotic expansions defined in this section and changing the variables in Equation (A16) according to Equation (40), we obtain

$$\frac{\partial \overline{T}}{\partial \overline{\zeta}} = \overline{\Psi}\frac{\partial \overline{T}_0}{\partial \overline{x}}+\overline{\Omega}\left(\frac{\partial^2 \overline{T}_0}{\partial \overline{x}^2}\right)+\overline{\Lambda}\left(\frac{\partial \overline{T}_0}{\partial \overline{x}}\right)^2 + \frac{\widetilde{Br}\overline{\mu}}{\overline{w}_0}\left[3\left(\frac{\partial \overline{w}_0}{\partial \overline{\zeta}}\right)^2+\left(\frac{2\overline{A}_R}{\overline{\Theta}^2}\right)^2\right]+\frac{\widetilde{Br}_R\overline{\mu}}{\overline{w}_0}\left(\frac{\partial \overline{B}}{\partial \overline{\zeta}}\overline{\Theta}\right)^2 \tag{A18}$$

where

$$\overline{\Pi} = [\overline{x}(\overline{h}_2-\overline{h}_1)+\overline{h}_1](\overline{h}_2-\overline{h}_1)$$

$$\overline{\Theta} = [\overline{x}(\overline{h}_2-\overline{h}_1)+\overline{h}_1]$$

$$\overline{\Lambda} = \frac{12\gamma_R\overline{T}_0^2}{\overline{w}_0(\overline{h}_2-\overline{h}_1)^2}$$

$$\overline{\Omega} = \frac{1+4\gamma_R\overline{T}_0^3}{\overline{w}_0(\overline{h}_2-\overline{h}_1)^2}$$

$$\overline{\Psi} = \frac{1}{\overline{w}_0}\left[\frac{1+4\gamma_R\overline{T}_0^3}{\overline{\Pi}}-\frac{\overline{A}_R}{\overline{\Pi}}+\widetilde{P}\frac{\partial \overline{w}_0}{\partial \overline{\zeta}}\frac{\overline{\Theta}}{2(\overline{h}_2-\overline{h}_1)}\right]$$

$$\overline{A}_R = \frac{2\overline{p}_0\overline{h}_2^2\overline{h}_1^2-2\overline{\gamma}\overline{h}_2\overline{h}_1(\overline{h}_2+\overline{h}_1)-ReS^2\overline{B}^2(\overline{h}_2^2\overline{h}_1^4-\overline{h}_1^2\overline{h}_2^4)}{4\overline{\mu}(\overline{h}_2^2-\overline{h}_1^2)}$$

Finally, the thermal boundary conditions assume the same form as in Equation (46).

References

1. Russell, P.S. Photonic-Crystal Fibers. *J. Light. Technol.* **2006**, *24*, 4729–4749. [CrossRef]
2. Buczynski, R. Photonic Crystal Fibers. *Acta Phys. Pol. A* **2004**, *106*, 141–167. [CrossRef]
3. Humbert, G.; Knight, J.; Bouwmans, G.; Russell, P.; Williams, D.; Roberts, P.; Mangan, B. Hollow core photonic crystal fibers for beam delivery. *Opt. Express* **2004**, *12*, 1477–1484. [CrossRef] [PubMed]
4. Wadsworth, W.J.; Percival, R.M.; Bouwmans, G.; Knight, J.C.; Birks, T.A.; Hedley, T.D.; Russell, P.S.T. Very High Numerical Aperture Fibers. *IEEE Photonics Technol. Lett.* **2004**, *16*, 843–845. [CrossRef]
5. Di Teodoro, F.; Brooks, C.D. 1.1 MW peak-power, 7 W average-power, high-spectral-brightness, diffraction-limited pulses from a photonic crystal fiber amplifier. *Opt. Lett.* **2005**, *30*, 2694–2696. [CrossRef]
6. Chernikov, S.V.; Zhu, Y.; Taylor, J.R.; Gapontsev, V.P. Supercontinuum self-Q-switched ytterbium fiber laser. *Opt. Lett.* **1997**, *22*, 298–300. [CrossRef]
7. Dainese, P.; Russell, P.J.S.; Joly, N.; Knight, J.C.; Wiederhecker, G.S.; Fragnito, H.L.; Laude, V.; Khelif, A. Stimulated Brillouin scattering from multi-GHz-guided acoustic phonons in nanostructured photonic crystal fibres. *Nat. Phys.* **2006**, *2*, 388–392. [CrossRef]
8. Monro, T.M.; Belardi, W.; Furusawa, K.; Baggett, J.C.; Broderick, N.G.R.; Richardson, D.J. Sensing with microstructured optical fibres. *Meas. Sci. Technol.* **2001**, *12*, 854–858. [CrossRef]
9. Marques, C.A.F.; Pospori, A.; Demirci, G.; Çetinkaya, O.; Gawdzik, B.; Antunes, P.; Bang, O.; Mergo, P.; André, P.; Webb, D.J. Fast Bragg Grating Inscription in PMMA Polymer Optical Fibres: Impact of Thermal Pre-Treatment of Preforms. *Sensors* **2017**, *17*, 891. [CrossRef]
10. Emiliyanov, G.; Jensen, J.B.; Bang, O.; Hoiby, P.E.; Pedersen, L.H.; Kjaer, E.M.; Lindvold, L. Localized biosensing with Topas microstructured polymer optical fiber. *Opt. Lett.* **2007**, *32*, 460–462. [CrossRef]
11. Hu, X.; Chen, Y.; Gao, S.; Min, R.; Woyessa, G.; Bang, O.; Qu, H.; Wang, H.; Caucheteur, C. Direct Bragg Grating Inscription in Single Mode Step-Index TOPAS/ZEONEX Polymer Optical Fiber Using 520 nm Femtosecond Pulses. *Polymers* **2022**, *14*, 1350. [CrossRef] [PubMed]
12. Hu, X.; Woyessa, G.; Kinet, D.; Janting, J.; Nielsen, K.; Bang, O.; Caucheteur, C. BDK-doped core microstructured PMMA optical fiber for effective Bragg grating photo-inscription. *Opt. Lett.* **2017**, *42*, 2209–2212. [CrossRef]
13. van Eijkelenborg, M.A.; Argyros, A.; Barton, G.; Bassett, I.M.; Fellew, M.; Henry, G.; Issa, N.A.; Large, M.C.; Manos, S.; Padden, W.; et al. Recent progress in microstructured polymer optical fibre fabrication and characterisation. *Opt. Fiber Technol.* **2003**, *9*, 199–209. [CrossRef]
14. van Eijkelenborg, M.; Large, M.; Argyros, A.; Zagari, J.; Manos, S.; Issa, N.; Bassett, I.; Fleming, S.; McPhedran, R.; de Sterke, C.M.; et al. Microstructured polymer optical fibre. *Opt. Express* **2001**, *9*, 319–327. [CrossRef] [PubMed]
15. Argyros, A. Applications of microstructured polymer optical fibers. In Proceedings of the 2010 Photonics Global Conference, Singapore, 14–16 December 2010; pp. 1–4. [CrossRef]
16. Paek, U.C.; Runk, R.B. Physical behavior of the neck–down region during furnace drawing of silica fibers. *J. Appl. Phys.* **1978**, *49*, 4417–4422. [CrossRef]
17. Myers, M.R. A model for unsteady analysis of preform drawing. *AIChE J.* **1989**, *35*, 592–602. [CrossRef]
18. Glicksman, L.R. The Dynamics of a Heated Free Jet of Variable Viscosity Liquid at Low Reynolds Numbers. *J. Basic Eng.* **1968**, *90*, 343–354. [CrossRef]
19. Fitt, A.D.; Furusawa, K.; Monro, T.M.; Please, C.P.; Richardson, D.J. The mathematical modelling of capillary drawing for holey fibre manufacture. *J. Eng. Math.* **2002**, *43*, 201–227. [CrossRef]
20. Luzi, G.; Epple, P.; Scharrer, M.; Fujimoto, K.; Rauh, C.; Delgado, A. Influence of Surface Tension and Inner Pressure on the Process of Fibre Drawing. *J. Light. Technol.* **2010**, *28*, 1882–1888. [CrossRef]
21. Voyce, C.J.; Fitt, A.D.; Monro, T.M. The mathematical modelling of rotating capillary tubes for holey-fibre manufacture. *J. Eng. Math.* **2008**, *60*, 69–87. [CrossRef]
22. Voyce, C.J.; Fitt, A.D.; Monro, T.M. Mathematical Modeling as an Accurate Predictive Tool in Capillary and Microstructured Fiber Manufacture: The Effects of Preform Rotation. *J. Light. Technol.* **2008**, *26*, 791–798. [CrossRef]
23. Voyce, C.J.; Fitt, A.D.; Hayes, J.R.; Monro, T.M. Mathematical Modeling of the Self-Pressurizing Mechanism for Microstructured Fiber Drawing. *J. Light. Technol.* **2009**, *27*, 871–878. [CrossRef]
24. Taroni, M.; Breward, C.J.W.; Cummings, L.J.; Griffiths, I.M. Asymptotic solutions of glass temperature profiles during steady optical fibre drawing. *J. Eng. Math.* **2013**, *80*, 1–20. [CrossRef]
25. Xue, S.C.; Tanner, R.I.; Barton, G.W.; Lwin, R.; Large, M.; Poladian, L. Fabrication of microstructured optical fibers-part I: Problem formulation and numerical modeling of transient draw process. *J. Light. Technol.* **2005**, *23*, 2245–2254. [CrossRef]
26. Xue, S.C.; Tanner, R.I.; Barton, G.W.; Lwin, R.; Large, M.; Poladian, L. Fabrication of microstructured optical fibers-part II: Numerical modeling of steady-state draw process. *J. Light. Technol.* **2005**, *23*, 2255–2266. [CrossRef]
27. Xue, S.C.; Large, M.; Barton, G.W.; Tanner, R.I.; Poladian, L.; Lwin, R. Role of material properties and drawing conditions in the fabrication of microstructured optical fibers. *J. Light. Technol.* **2006**, *24*, 853–860. [CrossRef]

28. Luzi, G.; Epple, P.; Scharrer, M.; Fujimoto, K.; Rauh, C.; Delgado, A. Numerical Solution and Experimental Validation of the Drawing Process of Six-Hole Optical Fibers Including the Effects of Inner Pressure and Surface Tension. *J. Light. Technol.* **2012**, *30*, 1306–1311. [CrossRef]
29. Stokes, Y.M.; Buchak, P.; Crowdy, D.G.; Ebendorff-Heidepriem, H. Drawing of micro-structured fibres: circular and non-circular tubes. *J. Fluid Mech.* **2014**, *755*, 176–203. [CrossRef]
30. Chen, M.J.; Stokes, Y.M.; Buchak, P.; Crowdy, D.G.; Ebendorff-Heidepriem, H. Microstructured optical fibre drawing with active channel pressurisation. *J. Fluid Mech.* **2015**, *783*, 137–165. [CrossRef]
31. Buchak, P.; Crowdy, D.G.; Stokes, Y.M.; Ebendorff-Heidepriem, H. Elliptical pore regularisation of the inverse problem for microstructured optical fibre fabrication. *J. Fluid Mech.* **2015**, *778*, 5–38. [CrossRef]
32. Buchak, P.; Crowdy, D.G. Surface-tension-driven Stokes flow: A numerical method based on conformal geometry. *J. Comput. Phys.* **2016**, *317*, 347–361. [CrossRef]
33. Chen, M.J.; Stokes, Y.M.; Buchak, P.; Crowdy, D.G.; Ebendorff-Heidepriem, H. Asymptotic Modelling of a Six-Hole MOF. *J. Light. Technol.* **2016**, *34*, 5651–5656. [CrossRef]
34. Stokes, Y.M.; Wylie, J.J.; Chen, M.J. Coupled fluid and energy flow in fabrication of microstructured optical fibres. *J. Fluid Mech.* **2019**, *874*, 548–572. [CrossRef]
35. Jasion, G.T.; Shrimpton, J.S.; Chen, Y.; Bradley, T.; Richardson, D.J.; Poletti, F. MicroStructure Element Method (MSEM): Viscous flow model for the virtual draw of microstructured optical fibers. *Opt. Express* **2015**, *23*, 312–329. [CrossRef]
36. Lee, S.; Jaluria, Y. Simulation of the transport processes in the neck-down region of a furnace drawn optical fiber. *Int. J. Heat Mass Transf.* **1997**, *40*, 843–856. [CrossRef]
37. Roy Choudhury, S.; Jaluria, Y. Thermal transport due to material and gas flow in a furnace for drawing an optical fiber. *J. Mater. Res.* **1998**, *13*, 494–503. [CrossRef]
38. Yin, Z.; Jaluria, Y. Zonal Method to Model Radiative Transport in an Optical Fiber Drawing Furnace. *J. Heat Transf.* **1997**, *119*, 597–603. [CrossRef]
39. Yin, Z.; Jaluria, Y. Neck Down and Thermally Induced Defects in High-Speed Optical Fiber Drawing. *J. Heat Transf.* **2000**, *122*, 351–362. [CrossRef]
40. Xue, S.C.; Poladian, L.; Barton, G.W.; Large, M. Radiative heat transfer in preforms for microstructured optical fibres. *Int. J. Heat Mass Transf.* **2007**, *50*, 1569–1576. [CrossRef]
41. Schlichting, H.; Gersten, K. *Boundary-Layer Theory*, 9th ed.; Springer: Berlin/Heidelberg, Germany, 2017.
42. Geyling, F.; Homsy, G.M. Extensional instabilities of the glass fibre drawing process. *Glass Technol.* **1980**, *21*, 95–102.
43. Patel, R.M.; Bheda, J.H.; Spruiell, J.E. Dynamics and structure development during high-speed melt spinning of nylon 6. II Mathematical modeling. *J. Appl. Polym. Sci.* **1991**, *42*, 1671–1682. [CrossRef]
44. Urbain, G.; Bottinga, Y.; Richet, P. Viscosity of liquid silica, silicates and alumino-silicates. *Geochim. Cosmochim. Acta* **1982**, *46*, 1061–1072. [CrossRef]
45. Hetherington, G.; Jack, K.H.; Kennedy, J.C. The viscosity of vitreous silica. *Phys. Chem. Glas.* **1964**, *5*, 130–136.
46. Doremus, R.H. Viscosity of silica. *J. Appl. Phys.* **2002**, *92*, 7619–7629. [CrossRef]
47. Huang, H.; Miura, R.M.; Wylie, J.J. Optical Fiber Drawing and Dopant Transport. *SIAM J. Appl. Math.* **2008**, *69*, 330–347. [CrossRef]
48. Lewis, J.A. The collapse of a viscous tube. *J. Fluid Mech.* **1977**, *81*, 129. [CrossRef]
49. Wylie, J.J.; HUANG, H. Extensional flows with viscous heating. *J. Fluid Mech.* **2007**, *571*, 359–370. [CrossRef]
50. Roy Choudhury, S.; Jaluria, Y. Practical aspects in the drawing of an optical fiber. *J. Mater. Res.* **1998**, *13*, 483–493. [CrossRef]

energies MDPI

Article

The Transient Thermal Disturbance in Surrounding Formation during Drilling Circulation

Minsoo Jang [1], Troy S. Chun [1] and Jaewoo An [2,*]

1 Department of Petroleum Engineering, Texas A&M University, College Station, TX 77843, USA
2 Department of Energy Science and Engineering, Stanford University, Stanford, CA 94305, USA
* Correspondence: jaewooan@stanford.edu; Tel.: +1-650-353-8071

Abstract: The injecting drilling mud is typically at the ambient temperature, relatively much colder than the deep formation, inducing a cooling effect in the formation. Although the cooled formation temperature gradually returns to its original temperature after drilling circulation, the recovery speed is slow due to low thermal diffusivity. Considering that any well tests begin in a short period after drilling ends, temperature recovery is not fully achieved before the tests. It means that the measured temperature of producing fluid is not that of the actual formation, significantly impairing the robustness of the subsequent thermal applications. Furthermore, there has been no quantified concept of thermal disturbance in the formation and its analysis. In this work, a proposed numerical transient heat transfer model computes the radial temperature in the drill pipe, annulus, and formation. The concept of quantifying thermal disturbance, named thermally disturbed radius (TDR), indicates how long the thermal disturbance occurs radially in the formation. The TDR increases with the more significant temperature difference between circulating fluid and formation. Thus, the TDR appears to be the largest at the bottom-hole depth. In the sensitivity of TDR of various operational parameters, circulation time (i.e., drilling time) is the most influential factor. Meanwhile, the other parameters do not significantly affect TDR: circulation rate, injecting mud temperature, and mud density. The sensitivity analysis concludes that as long as the operators control the drilling time, the uncertainty of the measured temperature after drilling can be manageable without limiting any other operational parameters.

Keywords: drilling circulation cooling; near-wellbore cooling; thermally disturbed radius; uncertainty of temperature

Citation: Jang, M.; Chun, T.S.; An, J. The Transient Thermal Disturbance in Surrounding Formation during Drilling Circulation. *Energies* **2022**, *15*, 8052. https://doi.org/10.3390/en15218052

Academic Editor: Mofazzal Hossain

Received: 16 October 2022
Accepted: 27 October 2022
Published: 29 October 2022

Publisher's Note: MDPI stays neutral with regard to jurisdictional claims in published maps and institutional affiliations.

Copyright: © 2022 by the authors. Licensee MDPI, Basel, Switzerland. This article is an open access article distributed under the terms and conditions of the Creative Commons Attribution (CC BY) license (https://creativecommons.org/licenses/by/4.0/).

1. Introduction

As hydrocarbon resource consumption increases, oil and gas development is inevitable in deep-depth onshore and offshore territories. The deep-depth development poses significant challenges during the drilling process because high formation temperature causes failures in current operation designs [1]. First, all drilling facilities need to secure the operational durability of devices in a high-temperature environment, primarily electrical equipment such as logging-while-drilling and measurement-while-drilling tools, motors, and turbines [2]. Additionally, this hostile environment can also impair the performance of drilling fluids. For the appropriate operational design, there have been studies to estimate the temperature of the flowing fluids accurately in the drilling pipe and annulus [3].

The pioneering work to estimate the radial temperature distribution in the reservoir was conducted by Bullard in 1946 [4]. His work was based on a similar diffusivity equation used in a transient well test for obtaining radial pressure, and this approach became one of the major categories to compute the temperature of a reservoir analytically [5]. Edwardson et al. implemented a similar method to Bullard's but modified it to be more practical to use by applying extrapolation in the calculation process [6]. He pointed out that the circulating mud cools down the near-wellbore formation and hypothesized around 10 ft of thermal disturbance.

Later, there were works estimating the temperature distribution in consideration of more realistic conditions based on numerical methods [2,7–10]. This research illustrated a more sophisticated temperature estimation by dividing domains more and applying each domain's heat and physical properties. At first, the reservoir system was divided into two domains: wellbore and formation [7]. Later, the wellbore was subdivided into the drill pipe and annulus [8], and then it was further divided into multi-grid in the radial direction in the drill pipe [2].

In addition to considering more dividing simulation domains, there have been attempts to improve the accuracy of the temperature calculation. Gorman et al. showed sophisticated radial temperature distributions by implementing the RANS turbulent model requiring great computing resources [11,12]. In spite of their valuable results, their work did not reveal the temperature at the payzone depth (i.e., near bottom-hole), the most important area in oil drilling. One noteworthy observation in their work is that temperature change by rotating drill pipe is minor because the axial velocity of the drilling fluid is much faster than that of the rotating drill pipe, suppressing the secondary motion induced by the rotating drill pipe [11].

In the transient well test, the temperature has been regarded as a complementary component to identifying reservoir parameters and production rate. However, with the advanced sensing platforms, several thermal applications have been developed using the measured temperature. Not only estimation techniques of production rate, but rate allocation in the multi-payzone system are introduced [13]. Furthermore, flow pattern identification through the real-time temperature data and identification of liquid condensate bank behavior was recently proposed [14].

Drilling mud is injected from the wellhead at an ambient temperature much lower than the deep formation temperature at the initial stage. Although the temperature of the circulating mud increases while it circulates through the deep reservoir, it still stays much lower compared to the deep formation temperature. Cold circulating mud takes heat from the adjacent formation, inducing a cooling effect during drilling operation [8,11]. At the formation surface facing the circulating mud (hereafter, "wellbore wall"), forced convection with a significant heat transfer rate is the dominant heat exchange component between the circulating mud and wellbore wall. On the other hand, although convection can take place in some formation parts containing a porous medium, most heat transfer inside the formation occurs only by conduction [7].

Compared to hydraulic diffusivity, thermal diffusivity is typically so low that thermal transient lasts several years even in a small reservoir with a radius of 50 m [15,16]. As conduction has the lowest heat transfer rate among the heat transfer mechanisms, it is not enough to offset the rapid temperature drop by the circulating mud [17]. Due to the imbalance of heat transfer, a thermal disturbance occurs in the radial direction inside the formation. As a result, the formation starts to cool from the near-wellbore area during the drilling operation. When the drilling operation ends, perforation is initiated after several hours of removing the remaining debris in the wellbore. During the perforation operation the temperature sensor measures the temperature of reservoir fluids (i.e., hydrocarbon and water) coming into the wellbore. However, the measurement deviates from the actual formation temperature cooled down by circulating mud during drilling and subsequent clean-up operations.

This discrepancy in temperature can significantly degrade the accuracy of the measured temperature as well as the robustness of the thermal applications. Estimation of production rate, one of the applications using the temperature data, requires a high degree of accuracy [13]. Even a slight error in the measured temperature may make the estimated results meaningless. While the cooling effect in the formation by drilling has been mentioned in some previous research works, few papers pay attention to the risk potential of inaccurate temperature measurement [5,7,17]. Furthermore, there has been no specific concept of this thermal disturbance by drilling circulation and its analysis.

In this work, a numerical transient heat transfer model is developed to compute the radial temperature distribution in the drill pipe, annulus, and formation. Using this tool, we introduce the quantified concept of the thermal disturbance, named thermally disturbed radius (TDR), which indicates how long the thermal disturbance occurs radially in the formation (Figure 1). Lastly, through the sensitivity analysis of TDR with various conditions during drilling, we propose operational guidelines to reduce the uncertainty of temperature measurement.

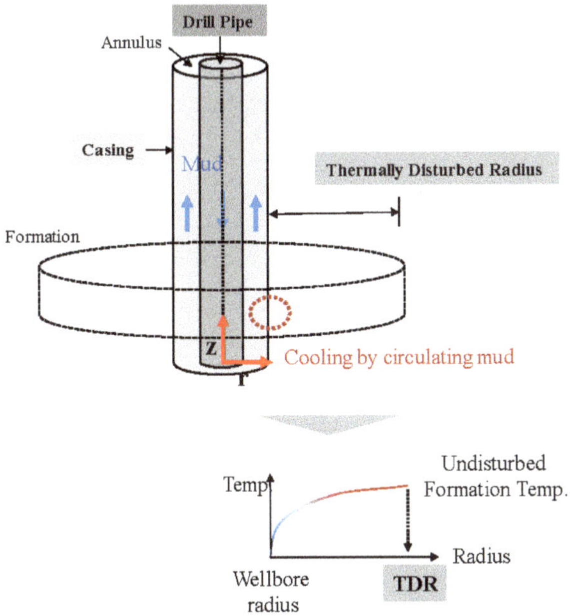

Figure 1. Schematic diagram of cooling effect by the circulating mud during drilling.

2. Methodology

2.1. Governing Equations

The operational system during drilling consists of five subsections: drilling hole (h), drill pipe (p), annulus (a), casing (c), and reservoir formation (f). To obtain the temperature profiles of the system, we deploy the governing equations of heat transfer balance in each section (Equation (1)). The imposed assumptions in this work are the following:

(1) Incompressible single-phase fluid (i.e., oil, no free gas);
(2) Constant heat properties;
(3) No reservoir flow;
(4) Negligible Joule–Thomson effect;
(5) The heated mud through deep formation is sufficiently cooled to the injecting temperature. The circulating mud is often observed to be heated by the circulation through the deep formation in the drilling practice; this phenomenon becomes more distinct when the circulation time gets longer.

The governing equations of heat transfer in the five sections are represented as

$$\begin{aligned}
A_h h_{h,p}\frac{\partial(\rho_m c_m T_h)}{\partial t} &= L_{h,p}h_{h,p}(T_p - T_h) + A_h h_{h,p}\frac{\partial}{\partial z}\left(\lambda_m \frac{\partial T_h}{\partial z}\right) + \frac{\partial(\rho_m c_m T_h q)}{\partial z} - Q_{h,a},\\
A_p\frac{\partial(\rho_p c_p T_p)}{\partial t} &= L_{h,p}h_{h,p}(T_h - T_p) + L_{p,a}h_{p,a}(T_a - T_p) + A_p\frac{\partial}{\partial z}\left(\lambda_p \frac{\partial T_p}{\partial z}\right),\\
A_a\frac{\partial(\rho_m c_m T_a)}{\partial t} &= L_{p,a}h_{p,a}(T_p - T_a) + L_{a,c}h_{a,c}(T_c - T_a) + A_a\frac{\partial}{\partial z}\left(\lambda_m \frac{\partial T_m}{\partial z}\right)\\
&\quad - \frac{\partial(\rho_m c_m T_h q)}{\partial z} + Q_{h,a},\\
A_c\frac{\partial(\rho_c c_c T_c)}{\partial t} &= L_{a,c}h_{a,c}(T_a - T_c) + Q_{f,c} + A_c\frac{\partial}{\partial z}\left(\lambda_c \frac{\partial T_c}{\partial z}\right),\\
\rho_f c_f\frac{\partial T_f}{\partial t} &= \frac{1}{r}\frac{\partial}{\partial r}\left(\lambda_f r\frac{\partial T_f}{\partial r}\right) + \frac{\partial}{\partial z}\left(\lambda_f \frac{\partial T_f}{\partial z}\right).
\end{aligned} \tag{1}$$

As the pipe flow is turbulent, the convection heat coefficient h is evaluated by the Dittus–Boelter equation, as shown in Equation (2).

$$\begin{aligned}
h &= Nu\lambda_m/d_x\\
Nu &= 0.023Re^{0.8}Pr^{0.4},\\
Re &= \rho_m q d_x/\mu A_x\ (x = h, a),\\
Pr &= \mu c_m/\lambda_m,\\
d_h &= 2r_h,\\
d_a &= 2(r_a - r_p),
\end{aligned} \tag{2}$$

where Nu is the Nusselt number, Re is the Reynolds number, Pr is the Prandtl number, μ is the viscosity, and d is the hydraulic diameter. $Q_{f,c}$ is the heat flux from the reservoir formation to the casing:

$$Q_{f,c} = -L_{c,r}\frac{\partial}{\partial r}\left(\lambda_f r\frac{\partial T_r}{\partial r}\right)\bigg|_{r=r_c} = -L_{c,r}\lambda_f\frac{\partial T_r}{\partial r}\bigg|_{r=r_c}. \tag{3}$$

The radial heat transfer from the drilling hole to the annulus exists at the bottom-hole due to the mass transport. $Q_{h,a}$ is the heat flux from the drilling fluid to the annulus approximated as

$$Q_{h,a} \approx -\frac{\partial(\rho_m c_m T q)}{\partial r}\chi_{z_{bth}}, \tag{4}$$

where $\chi_{z_{bth}}$ is the step function defined with the TDR and end locations of the bottom-hole $z_{bth,\ bot}$ and $z_{bth,\ top}$ as

$$\chi_{z_{bot}} = \begin{cases} 1;\ z_{bth,bot} \le z \le\ z_{bth,\ top} \\ \quad 0;\ otherwise \end{cases}. \tag{5}$$

To simplify the equations, the assumptions in the drill pipe and the casing sections are the following: the negligible vertical heat conduction and negligible heat accumulation by drilling rigs in relation to a large mass of soil. Those two assumptions lead to the same amount of heat flux inwards and outwards in those two sections. The simplified governing equation of the drill pipe is represented as

$$L_{h,p}h_{h,p}(T_h - T_p) + L_{p,a}h_{p,a}(T_a - T_p) = 0. \tag{6}$$

From this equation, we can derive the mathematical expression of T_p as a function of T_a and T_h:

$$T_p = \frac{L_{h,p}h_{h,p}T_h + L_{p,a}h_{p,a}T_a}{L_{h,p}h_{h,p} + L_{p,a}h_{p,a}}. \tag{7}$$

In the same way, the mathematical expression of T_c as a function of T_a and T_r reads

$$T_c = T_a - \frac{Q_{r,c}}{L_{a,c}h_{a,c}} = T_a - \frac{L_{c,r}\lambda_f}{L_{a,c}h_{a,c}}\frac{\partial}{\partial r}\left(r\frac{\partial T_r}{\partial r}\right)\bigg|_{r=r_c} = T_a - \frac{L_{c,r}\lambda_f}{L_{a,c}h_{a,c}}\beta. \tag{8}$$

Then, by inserting Equations (7) and (8) into Equation (1), the simplified system of the governing equations is given as

$$\begin{aligned} A_h\frac{\partial(\rho_m c_m T_h)}{\partial t} &= h_{h,a}(T_a - T_h) + A_h\frac{\partial}{\partial z}\left(\lambda_m\frac{\partial T_h}{\partial z}\right) + \frac{\partial(\rho_m c_m T_h q)}{\partial z} - Q_{h,a}, \\ A_a\frac{\partial(\rho_m c_m T_a)}{\partial t} &= h_{h,a}(T_h - T_a) - L_{c,r}\lambda_f\beta + A_a\frac{\partial}{\partial z}\left(\lambda_m\frac{\partial T_m}{\partial z}\right) - \frac{\partial(\rho_m c_m T_h q)}{\partial z} + Q_{h,a}, \\ \rho_f c_f\frac{\partial T_f}{\partial t} &= \frac{1}{r}\frac{\partial}{\partial r}\left(\lambda_f r\frac{\partial T_f}{\partial r}\right) + \frac{\partial}{\partial z}\left(\lambda_f\frac{\partial T_f}{\partial z}\right), \end{aligned} \tag{9}$$

where $h_{h,a}$ is the heat transfer coefficient between the drilling hole and the annulus defined as the length-weighted harmonic average of two convection heat transfer coefficients:

$$h_{h,a} = \frac{L_{h,p}h_{h,p}L_{p,a}h_{p,a}}{L_{h,p}h_{h,p} + L_{p,a}h_{p,a}}. \tag{10}$$

The initial conditions for T_h, T_a, and T_r are assumed with the geothermal gradient g_T and the surface temperature T_{surf} as

$$T_x(r,z,t=0) = T_{surf} - g_T z(x = h,\ a,\ f). \tag{11}$$

The boundary conditions are represented as

$$\begin{aligned} T_h(r,z=0,t>0) &= T_{inj}, \\ \frac{\partial T_h}{\partial z}(r,z=z_{bth,bot},\ t) &= 0, \\ T_a(r,z=0,t) &= T_{surf}, \\ \frac{\partial T_a}{\partial z}(r,z=z_{bth,bot},\ t) &= 0, \\ T_f(r,z=0,t) &= T_{surf}, \\ \frac{\partial T_f}{\partial z}(r,z=z_{bth,bot},\ t) &= 0, \\ T_f(r=r_e,z,t) = T_f(r=r_e,z,t=0) &= T_e, \end{aligned} \tag{12}$$

where r_e is the reservoir boundary radius.

2.2. Numerical Scheme

In this study, we derive the numerical transient heat transfer model in a two-dimensional r-z domain to derive the TDR (Figure 2). The finite volume method is implemented for the discretization, which is known for satisfying the heat energy balance of each element. The backward difference approximation for temporal discretization is used. For spatial discretization, we calculate the first derivatives with the forward difference approximation and the second derivatives with the central difference approximation. The governing equations in Equation (9) are discretized as

$$
\begin{aligned}
A_h \rho_m c_m \frac{T_{1,j}^{n+1} - T_{1,j}^{n}}{\Delta t} &= h_{h,a}\left(T_{2,j}^{n+1} - T_{1,j}^{n+1}\right) + A_h \lambda_m \frac{T_{1,j+1}^{n+1} - 2T_{1,j}^{n+1} + T_{1,j-1}^{n+1}}{\Delta z^2} + \\
&\quad \rho_m c_m q \frac{T_{1,j+1}^{n+1} - T_{1,j}^{n+1}}{\Delta z} - Q_{h,a,j}^{n+1} \text{ (drilling hole)}, \\
A_a \rho_m c_m \frac{T_{2,j}^{n+1} - T_{2,j}^{n}}{\Delta t} &= h_{h,a}\left(T_{1,j}^{n+1} - T_{2,j}^{n+1}\right) - L_{c,r} \lambda_f \beta_j^{n+1} + A_a \lambda_m \frac{T_{2,j+1}^{n+1} - 2T_{2,j}^{n+1} + T_{2,j-1}^{n+1}}{\Delta z^2} - \\
&\quad \rho_m c_m q \frac{T_{2,j+1}^{n+1} - T_{2,j}^{n+1}}{\Delta z} + Q_{h,a,j}^{n+1} \text{ (annulus)}, \\
\rho_f c_f \frac{T_{i,j}^{n+1} - T_{i,j}^{n}}{\Delta t} &= \frac{\lambda_f}{(r_{i-1/2} + r_{i+1/2})/2} \frac{r_{i+1/2} T_{i+1,j}^{n+1} - (r_{i+1/2} + r_{i-1/2}) T_{i,j}^{n+1} + r_{i-1/2} T_{i-1,j}^{n+1}}{\Delta r^2} + \\
&\quad \lambda_f \frac{T_{i,j+1}^{n+1} - 2T_{i,j}^{n+1} + T_{i,j-1}^{n+1}}{\Delta z^2} \text{ (reservoir : } i \geq 3). \qquad (13)
\end{aligned}
$$

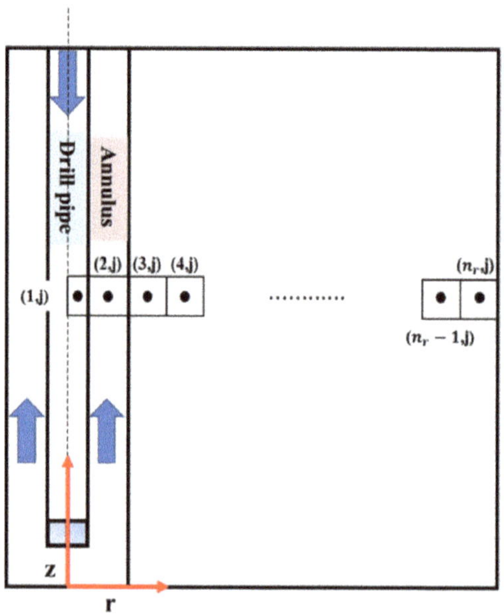

Figure 2. Scheme for numerical simulation of drill pipe, annulus, and formation.

Note that $T_h = T_{i=1}$, $T_a = T_{i=2}$, and $T_f = T_{i\geq 3}$. The subscripts i and j and the superscript n indicate indices for r, z, and t, respectively. The discretization sizes Δr, Δz and Δt are defined as

$$
\begin{aligned}
\Delta r &= \frac{r_e - r_c}{n_r - 2}, \\
\Delta z &= \frac{|z_{bth,bot}|}{n_z}, \qquad (14) \\
\Delta t &= \frac{t_{tot}}{n_t},
\end{aligned}
$$

where n_t is the total number of time steps, n_r and n_z are the total number of grid cells in r and z directions, and r_i and z_j are represented as

$$r_i = \begin{cases} r_h/2, & , i = 1 \\ (r_p + r_a)/2 & , i = 2, \\ r_c + (i - 5/2)\Delta r & , i > 2 \end{cases}$$
$$r_{i+1/2} = (r_i + r_{i+1})/2, z_j = z_{bth,bot} + (j - 1/2)\Delta z, \tag{15}$$

β_j^{n+1} and $Q_{h,a,j}^{n+1}$ in Equation (13) are approximated as

$$\beta_j^{n+1} \approx \frac{1}{r_{7/2} - r_{5/2}} \left(r_{7/2} \frac{T_{4,j}^{n+1} - T_{3,j}^{n+1}}{\Delta r} - r_{5/2} \frac{T_{3,j}^{n+1} - T_{2,j}^{n+1}}{r_3 - r_2} \right),$$
$$Q_{h,a,j}^{n+1} \approx \rho_m c_m q \frac{T_{2,j}^{n+1} - T_{1,j}^{n+1}}{\Delta r} \chi_{z_{bth}}(z_j). \tag{16}$$

We developed a MATLAB-based in-house simulator to solve the numerical systems with Equations (13) and (16). With the simulator, we can derive the temperature profile at each time step. In this study, we estimate TDR from the numerical transient simulation, where

$$\frac{|T(r = TDR) - T(r = r_e)|}{T(r = r_e)} < 0.001. \tag{17}$$

Figure 3 indicates a simplified flowchart of the calculation process obtaining the temperature distribution and TDR. The temperature distribution and TDR are computed for each time step.

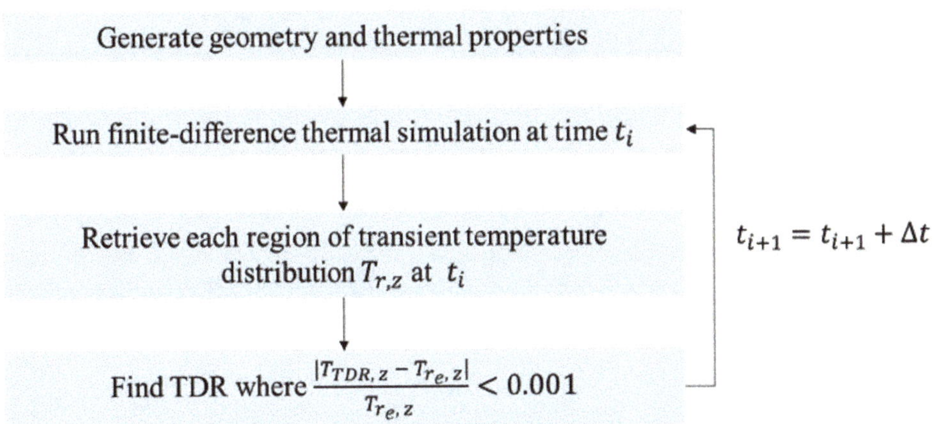

Figure 3. The flow chart of calculation.

3. Results

3.1. Validation

To validate the proposed transient heat transfer model, we compared the calculated temperature with the data from Yang et al.'s paper [18]. The necessary input parameters used are the same as those of Yang et al.'s work, as presented in Table 1. The model well has a total depth of 4900 m and 124 °C at the bottom-hole depth, a relatively deep high-temperature well in an operational manner. In the high-temperature well, thermal disturbance in the formation can be distinctively identified because heat exchange between the cold drilling fluid and formation remarkably occurs. The circulation time is 24 h, and the circulation rate is 18 L/s (approximately 285 gallons/min), which is the typical range of a drilling operational condition. Considering the high formation pressure in the deep well, we use slightly heavy mud with 1.2 kg/m^3 of mud density. We assume zero casing thickness with the negligible thermal resistance of the casing.

Table 1. Parameters of drilling operation and formation.

Parameter	Value
Circulation rate, q, L/s	18
Circulation time, t_{circ}, h	24
Inlet temperature, T_{inj}, °C	25
Surface temperature, T_{surf}, °C	15
Density of the drilling mud, ρ_m, kg/m3	1200
Density of the formation, ρ_f, kg/m3	2640
Thermal conductivity of the drilling mud, λ_m, W/m. °C	1.25
Thermal conductivity of the formation, λ_f, W/m. °C	1.75
Specific heat of the drilling fluid, c_m, J/kg. °C	1600
Specific heat of the formation, c_f, J/kg. °C	800
Well depth, $\lvert z_{bth,bot} \rvert$, m	4900
Geothermal gradient, g_T, °C /100 m	2.23
Wellbore diameter, $d_a (= d_c)$, mm	210
Inner diameter of the drill pipe, d_h, mm	121.4
Outer diameter of the drill pipe, d_p, mm	139.7
Discretization levels (n_r, n_z, n_t)	(200, 100, 7)

This work describes thermal disturbance in the radial direction. We validated our radial temperature estimation at the drilling hole, the annulus, and the wellbore wall (Figure 4). Though there is a slight difference in the lower depth near the bottom hole, the overall temperature distributions in all three cases are consistent with the reference data.

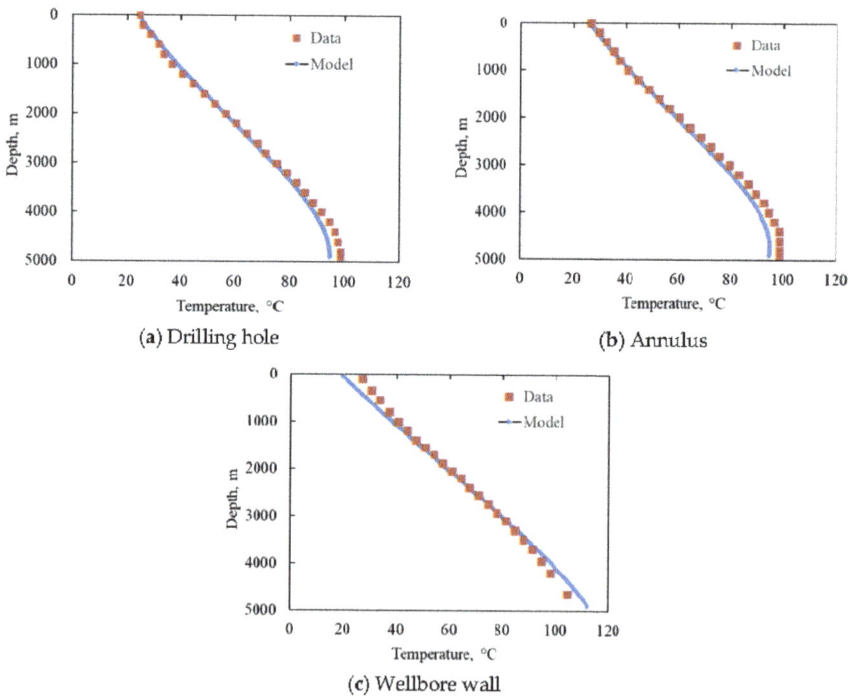

Figure 4. Comparison of calculated temperature with reference data.

3.2. Identification of Thermal Disturbance

We identified thermal disturbance by circulating mud from the radial temperature distribution in Figure 5. The solid red and blue curves are the radial temperature at the bottom-hole depth (i.e., 4900 m) and 2000 m, respectively. The dotted gray lines are thermally undisturbed temperature (i.e., formation temperature before drilling circulation) for each depth. In Figure 5a, the temperature difference between the solid curve and the dotted line indicates how much cooling occurs in the formation by cold circulating mud. As the formation radius is much longer than the wellbore radius, the inconsistency of radial temperature varies according to the radial distance from the wellbore. The near-wellbore formation is cooled by the circulating mud. The formation temperature far from the wellbore does not change, that is, the thermal disturbance does not reach the outer boundary of the formation at this moment. The radial temperature change at the bottom hole appears to be larger than that at a depth of 2000 m.

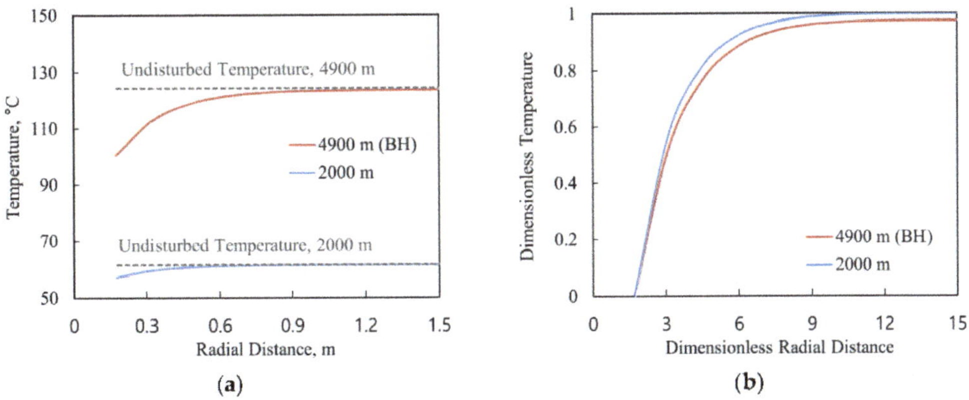

Figure 5. Radial temperature distributions of different depths: (**a**) radial temperature distribution and (**b**) dimensionless radial temperature distribution.

The radial distance and temperature were non-dimensionalized for clarity, following Equations (18) and (19), respectively. The profiles of the two curves in Figure 5b are almost identical. However, the required radial distance for converging to the undisturbed temperature (i.e., the dimensionless temperature is one) is different for each depth. When the depth is 2000 m, the curve converges to the undisturbed temperature when the radial distance is approaching 10. Meanwhile, the red curve at the bottom-hole depth does not reach even when the dimensionless radial distance is 15. The result in Figure 5b concludes that the thermal disturbance in the formation by circulating mud (i.e., TDR) is more extensive at the bottom hole than at 2000 m.

$$r_D = \frac{r}{r_c}, \tag{18}$$

$$T_D = \frac{T - T_e}{T_c - T_e}. \tag{19}$$

Figure 6 describes the thermal disturbance throughout the whole depth graphically. In this simulation, the outer reservoir boundary is 30 m, approximately 300 times the wellbore radius, which is sufficient length to describe thermal disturbance occurring over a few days. In the color map over the entire reservoir boundary (Figure 6a), the thermal disturbance is localized in the near-wellbore formation. Therefore, we represent Figure 6b as an enlarged version over $r_D \leq 30$ to visualize thermal disturbance effectively. Cooling and heating effects are determined by the temperature difference between the circulating mud and the formation. In the upper section of the depth up to 1000 m, the formation temperature near the wellbore increases because the formation temperature in this depth range is lower than the injection temperature. As the depth rises from 1000 m, the cooling effect in the near-wellbore formation begins to appear. Thus, the thermal disturbance by the circulating inducing cooling increases with deeper depth, and its degree reaches the maximum at the bottom hole. From these observations, TDR is inferred to be increasing as the depth approaches the bottom-hole depth.

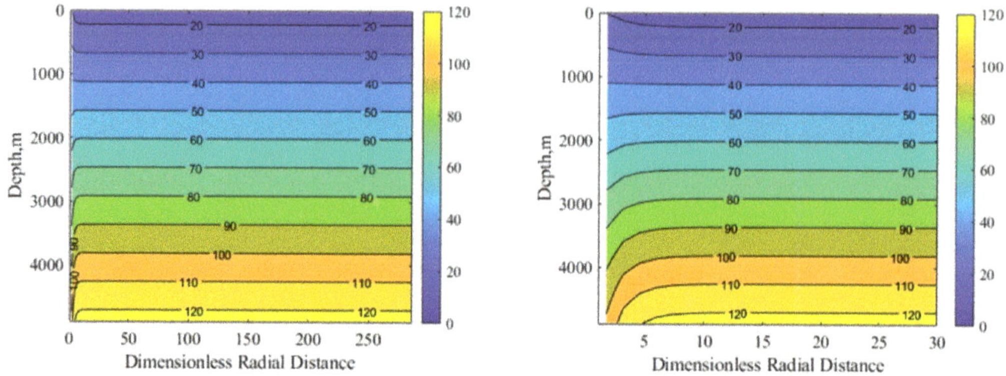

(a) (b)

Figure 6. Temperature profiles: (**a**) a whole radial range and (**b**) an enlarged version.

4. Sensitivity Analysis

In this section, we calculate TDR along the whole depth using Equation (17) and conduct its sensitivity analysis for various operating conditions. We selected four controllable factors in operation: circulation rate, circulation time, injecting temperature, and mud density. Table 2 presents the four operational parameters for the sensitivity.

Table 2. Operational parameters for sensitivity analysis at each section.

Section	4.1	4.2	4.3	4.4
Circulation rate, q, L/s	20, 40, 80	80	80	80
Circulation time, t_{circ}, h	168	24, 72, 168	168	168
Injecting temperature, T_{inj}, °C	25	25	25, 50, 75	25
Mud density, ρ_m, 1000 kg/m3	1.2	1.2	1.2	1.0, 1.2, 1.6

4.1. Circulation Rate

Circulation rate is an indicator of the carrying capacity that can remove drilling cuttings. Therefore, it is a major factor in determining the rate of penetration (ROP) [19]. Figure 7a,b describe the TDR and fluid temperature in the annulus along with the whole well depth. The TDR is non-dimensionalized as Equation (20) for efficient comparison.

$$\mathrm{TDR_D} = \frac{\mathrm{TDR}}{r_c}. \tag{20}$$

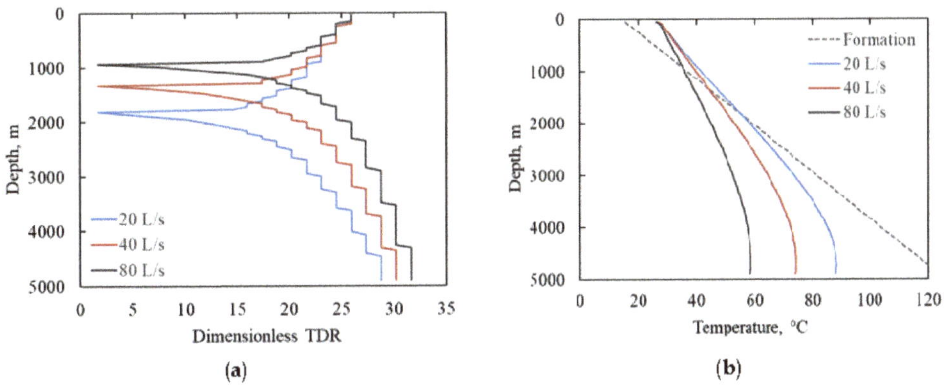

Figure 7. The sensitivity analysis of thermal disturbance with various circulation rates: (**a**) TDR and (**b**) fluid temperature in the annulus.

The circulation rate for comparison was determined to be 20 L/s, 40 L/s, and 80 L/s, shown in solid blue, red, and gray, respectively. In Figure 7a, all three TDR curves have similar shapes regardless of circulation rate. The TDR curve shape forms the minimum point at a certain depth, and below the depth, the TDR gradually increases as the depth increases. Despite the similarity of the curve shape, the minimum TDR depth of each case depends on the circulation rate. The smaller the circulation rate, the deeper the minimum TDR depth. The annulus fluid temperature can be evidence for the different minimum TDR depths, shown in Figure 7b. Because the annulus fluid is in direct contact with the formation, heat transfer occurs between the annulus fluid and the formation.

The circulation rate indicates the amount of heat capacity; the increasing fluid circulation requires more heat to change the fluid temperature. Thus, as the circulation rate increases, the temperature gradient in the annulus becomes smaller. In Figure 7b, the curve with the faster circulation deviates farther away from the formation temperature distribution (dotted gray line). Each temperature curve demonstrates a different slope intersecting the formation temperature line at different depths. The intersect depth is observed to be around 1000 m for 80 L/s, 1,400 m for 40 L/s, and 2000 m for 20 L/s, respectively. The intersect depth corresponds to the minimum TDR depth. At this depth, the heat transfer and, thus, the thermal disturbance occur minimally due to the slightest temperature difference between the circulating mud and the surrounding formation.

As observed in the previous chapter, the TDR is maximum at the bottom-hole depth in all three cases, where the temperature difference is the largest in the whole depth. The dimensionless TDR is around 28 for 20 L/s, 30.5 for 40 L/s, and 32 for 80 L/s, respectively. However, as seen from the estimated values, the increasing rate of TDR slows down compared to the rising circulation rate. Additionally, the TDR is expected to have an upper limit due to the limit of the temperature difference between the annulus fluid and the formation because the fluid temperature in the annulus cannot be lower than the injecting temperature. In conclusion, the larger circulation rate increases the TDR, but its effect is limited above a certain circulation rate.

4.2. Circulation Time

Figure 8 displays the TDR and annulus temperature distribution with various circulation times. The circulation time refers to the period of drilling and subsequent clean-up operations just before the perforation. Although circulation time can be varied significantly depending on the geologic and operational issues, one, three, and seven days of circulation are in the reasonable range in typical onshore fields. In Figure 8a, the minimum TDR depths indicate around 1000 m regardless of the different circulation times. The temperature distribution justifies this observation; the three temperature curves intersect the formation temperature line around 1000 m (Figure 8b).

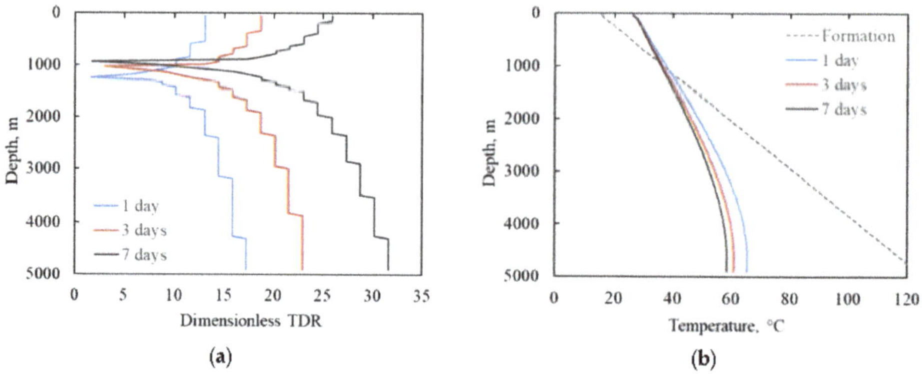

Figure 8. The sensitivity analysis of thermal disturbance with varying circulation times: (**a**) TDR and (**b**) fluid temperature in the annulus.

A noteworthy finding in Figure 8 is that the temperature variation with the various circulation times is somewhat trivial, while the TDR variation is significant. The TDR is around 17 for one day of circulation. The TDR increases almost double when the circulation lasts one week. This observation can be illustrated from the perspective of heat transfer rate in the two different domains: the formation and the wellbore wall. At the wellbore wall, the circulating mud is powered by pumps at the surface. The dominant forced convection by the pumping takes away heat from the formation. The heat removal triggers heat transfer from the outer formation boundary to the wellbore. As the fluid in the reservoir porous medium is static before the perforation, the flowability is insufficient to induce convection Thus, conduction exists as the only heat transfer mechanism in the formation. The heat conduction in the ground is less effective than convection heat transfer in fluid. The cooling effect appears to be moving radially towards the outer boundary. The thermal disturbance continuously propagates further from the wellbore, keeping the increase in the TDR.

A lengthy circulation time leads to a large TDR, which may impair temperature measurement accuracy after drilling. The circulation time is not easily manageable because it heavily depends on the total depth and ROP. Moreover, when operational problems take place, managing the circulation time becomes even more challenging. When the circulation time becomes longer, any alternative is required to compensate for the accuracy of the measured temperature.

4.3. Injecting Temperature

Steam injection has been most widely used and effective in thermal technology [20]. The steam is injected into the wellbore to decrease the hydrocarbon viscosity. Then, it displaces the hydrocarbon towards the wellbore for production [21]. Instead of the steam injection, hot water injection can be an attractive alternative option. Though hot water injection is less efficient than the steam injection, it is more versatile even in clay-bearing reservoirs, where the steam injection cannot be implemented. The hot water injection is also more suitable for deep reservoirs with high pressure [22]. During the water injection process, controlling injecting temperature has been the primary concern in heavy oil fields.

Figure 9 shows the change in TDR and annulus temperature, the injecting temperature being at 25 °C, 50 °C, and 75 °C lower than the bottom-hole temperature (i.e., 120 °C) of this reservoir. The injecting temperature can significantly alter the minimum TDR depth, as shown in Figure 9a. When the injecting temperature increases from 25 °C to 75 °C, the minimum TDR depth changes from 1000 m to 3000 m. This result is consistent with the depth where the temperature curves intersect the formation temperature line. Unlike the significant change in minimum TDR depth, the TDR at the bottom-hole change does not show much difference. These observations indicate that injecting temperature control can minimize the thermal disturbance at the bottom hole. If the injecting temperature is around the bottom-hole temperature with a sufficient flow rate, the minimum TDR depth can be near the bottom hole, thereby reducing temperature measurement uncertainty after drilling.

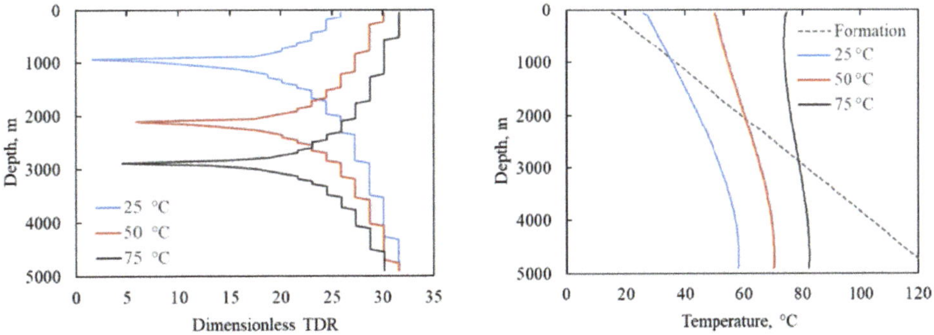

(a) (b)

Figure 9. The sensitivity analysis of thermal disturbance with varying injection temperatures: (**a**) TDR and (**b**) fluid temperature in the annulus.

4.4. Mud Density

Mud density is one of the key operating parameters for maintaining a stable wellbore condition during the drilling operation. Using an inappropriate mud density can lead to the risk of formation fracture or/and kick [23]. Figure 10 illustrates the changes in the TDR and fluid temperature in the annulus with various mud densities. The minimum TDR depth in all three cases is around 1000 m (Figure 10a). The minimum depth is evidenced by the similar intersection points between the fluid temperature curves and the formation temperature line (Figure 10b). As the mud density becomes heavier, the TDR increases, but the increasing interval is insignificant. Mud density is the minimal influential factor among operational parameters investigated previously: circulation rate, time, and injecting temperature. Changing mud density does not impact thermal disturbance; thus, the uncertainty of temperature measurement after drilling seldom increases.

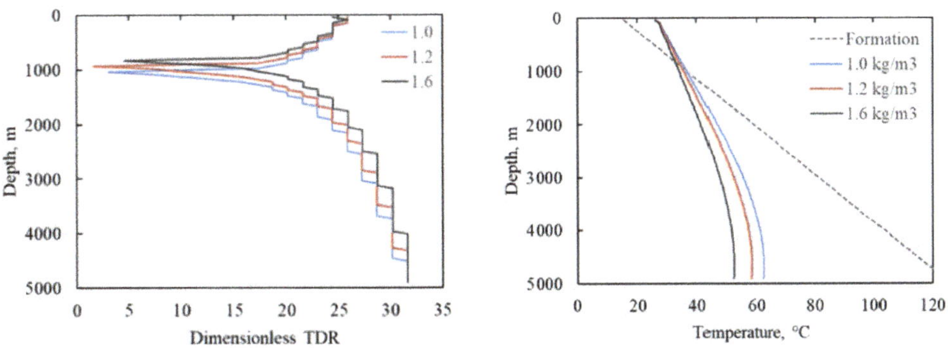

(a) (b)

Figure 10. The sensitivity analysis of thermal disturbance with various mud densities: (**a**) TDR and (**b**) fluid temperature in the annulus.

5. Conclusions

The proposed transient heat transfer model allows us to obtain the temperature of both wellbore and formation throughout the whole well depth. A quantified concept, TDR is introduced for quantitatively analyzing thermal disturbance in the formation. The TDR a newly introduced concept, can be an index indicating how uncertainty can exist in the temperature measurement in any well test after a drilling operation. The following are the delivered significant insights into understanding thermal disturbance:

1. The circulating mud cools down the formation near the wellbore during drilling causing the thermal disturbance in the formation. The TDR is the radial distance where the thermal disturbance occurs. Thus, the TDR acts as a quantified indicator of thermal disturbance.
2. The TDR tends to increase with the temperature difference between the fluids in the annulus and the formation, evidenced by the largest TDR being observed at the bottom hole.
3. Among the operational parameters, circulation time is the most influential factor in changing TDR, meaning measured temperature data may be inaccurate if the drilling operation time lasts longer. Meanwhile, other parameters (i.e., circulation rate, injecting temperature, and mud density) have insignificant effects.

Author Contributions: Conceptualization, M.J.; methodology, M.J. and J.A.; formal analysis, M.J. T.S.C., and J.A.; investigation, M.J. and J.A.; writing—original draft preparation, M.J. and J.A. writing—review and editing, T.S.C. and J.A.; supervision, J.A. All authors have read and agreed to the published version of the manuscript.

Funding: This research received no external funding.

Institutional Review Board Statement: Not applicable.

Informed Consent Statement: Not applicable.

Data Availability Statement: The data presented in this study are available on request from the corresponding author.

Conflicts of Interest: The authors declare no conflict of interest.

Nomenclature

A	horizontal cross-sectional area, m^2
c	specific heat capacity, $J/kg - °K$
d	hydraulic diameter, m
g_T	geothermal gradient, $°K/m$
$h_{x,y}$	convection heat transfer coefficient between sections x and y, $W/m^2 - °K$
$L_{x,y}$	intersection length between sections x and y, m
	total number of grid cells in r direction
n_z	total number of grid cells in z direction
n_t	total number of time steps
Nu	Nusselt number
Pr	Prandtl number
q	well flow rate, m^3/s
$Q_{x,y}$	heat flux per unit depth from section x to section y, W/m
r	radius, m
Re	Reynolds number
t	time, s
T	temperature, °K
z	depth, m
λ	thermal conductivity, $W/m - °K$
Δr	radial discretization size, m
Δt	temporal discretization size, s
Δz	vertical discretization size, m

β	average temperature gradient, °K/m
μ	viscosity, pa − s
ρ	density, kg/m^3
Subscript	
a	annulus
bot	bottom
bth	bottom hole
c	casing
circ	circulation
D	dimensionless
f	formation
e	reservoir boundary
h	drilling hole
i	r-index
j	z-index
inj	injection
m	drilling mud
p	drill pipe
surf	surface
top	top
Superscript	
n	*t*-index

References

1. Wu, B.; Zhang, X.; Jeffrey, R.G. A Model for Downhole fluid and Rock Temperature Prediction during Circulation. *Geothermics* **2014**, *50*, 202–212. [CrossRef]
2. Yang, M.; Luo, D.; Chen, Y.; Li, G.; Tang, D.; Meng, Y. Establishing a Practical Method to Accurately Determine and Manage Wellbore Thermal Behavior in High-Temperature Drilling. *Appl. Energy* **2019**, *238*, 1471–1483. [CrossRef]
3. Dirksen, R. Upgrading Formation-Evaluation Electronics for High-Temperature Drilling Environments. *J. Pet. Technol.* **2011**, *63*, 24–26. [CrossRef]
4. Bullard, E.C. The Time Necessary for a Bore Hole to Attain Temperature Equilibrium. *Geophys. J. Int.* **1947**, *5*, 127–130. [CrossRef]
5. Shen, P.Y.; Beck, A.E. Stabilization of Bottom Hole Temperature with Finite Circulation Time and Fluid Flow. *Geophys. J. Int.* **1986**, *86*, 63–90. [CrossRef]
6. Edwardson, M.J.; Girner, H.M.; Parkison, H.R.; Williams, C.D.; Matthews, C.S. Calculation of Formation Temperature Disturbances Caused by Mud Circulation. *J. Pet. Technol.* **1962**, *14*, 416–426. [CrossRef]
7. Luheshi, M.N. Estimation of Formation Temperature from Borehole Measurements. *Geophys. J. R. Astron. Soc.* **1983**, *74*, 747–776. [CrossRef]
8. Espinosa-Paredes, G.; Espinosa-Martínez, E.G. A Feedback-Based Inverse Heat Transfer Method to Estimate Unperturbed Temperatures in Wellbores. *Energy Convers. Manag.* **2009**, *50*, 140–148. [CrossRef]
9. Abdelhafiz, M.M.; Hegele, L.A.; Oppelt, J.F. Numerical Transient and Steady State Analytical Modeling of the Wellbore Temperature during Drilling Fluid Circulation. *J. Pet. Sci. Eng.* **2020**, *186*, 106775. [CrossRef]
10. Kutasov, I.M.; Eppelbaum, L.V. Wellbore and Formation Temperatures During Drilling, Cementing of Casing and Shut-In. In Proceedings of the World Geothermal Congress, Melbourne, Australia, 19–25 April 2015; International Geothermal Association: Den Haag, The Netherlands, 2015.
11. Gorman, J.M.; Abraham, J.P.; Sparrow, E.M. A novel, Comprehensive Numerical Simulation for Predicting Temperatures within Boreholes and the Adjoining Rock Bed. *Geothermics* **2014**, *50*, 213–219. [CrossRef]
12. Yang, H.; Li, J.; Liu, G.; Wang, J.; Luo, K.; Wang, B. Development of Transient Heat Transfer Model for Controlled Gradient Drilling. *Appl. Therm. Eng.* **2019**, *148*, 331–339. [CrossRef]
13. Ouyang, L.B.; Belanger, D. Flow Profiling by Distributed Temperature Sensor (DTS) System—Expectation and Reality. *SPE Prod. Oper.* **2006**, *21*, 269–281. [CrossRef]
14. Jang, M. Wellbore and Near-Wellbore Heat Transfer: General Theory and Practical Application. Ph.D. Thesis, Texas A&M University, College Station, TX, USA, 2021.
15. Palabiyik, Y.; Onur, M.; Tureyen, O.I.; Cinar, M. Transient Temperature Behavior and Analysis of Single-Phase Liquid-Water Geothermal Reservoirs during Drawdown and Buildup Tests: Part I. Theory, New Analytical and Approximate Solutions. *J. Pet. Sci. Eng.* **2016**, *146*, 637–656. [CrossRef]
16. Jang, M.; Chun, T.S.; An, J. An Analytical Heat Transfer Model in Oil Reservoir during Long-Term Production. *Energies* **2022**, *15*, 2544. [CrossRef]
17. Eppelbaum, L.V.; Kutasov, I.M. Pressure and Temperature Drawdown Well Testing: Similarities and Differences. *J. Geophys. Eng.* **2006**, *3*, 12–20. [CrossRef]

18. Yang, M.; Tang, D.; Chen, Y.; Li, G.; Zhang, X.; Meng, Y. Determining Initial Formation Temperature Considering Radial Temperature Gradient and Axial Thermal Conduction of the Wellbore Fluid. *Appl. Therm. Eng.* **2019**, *147*, 876–885. [CrossRef]
19. Bridges, S.; Robinson, L. *A Practical Handbook for Drilling Fluids Processing*, 1st ed.; Gulf Professional Publishing: Houston, TX USA, 2020.
20. Pwaga, S.; Iluore, C.; Hundseth, Ø.; Perales, F.J.; Idrees, M.U. *Comparative Study of Different EOR Methods*; Technical report Norwegian University of Science & Technology: Trondheim, Norway, 2010.
21. Alajmi, A.F. Heat Loss Effect on Oil Bank Formation during Steam Flood. *J. Pet. Sci. Eng.* **2021**, *199*, 108262. [CrossRef]
22. Askarova, A.; Turakhanov, A.; Markovic, S.; Popov, E.; Maksakov, K.; Usachev, G.; Karpov, V.; Cheremisin, A. Thermal Enhanced Oil Recovery in Deep Heavy Oil Carbonates: Experimental and Numerical Study on a Hot Water Injection Performance. *J. Pet Sci. Eng.* **2020**, *194*, 107456. [CrossRef]
23. Schubert, J.J.; Juvkam-Wold, H.C.; Choe, J. Well Control Procedures for Dual Gradient Drilling as Compared to Conventional Riser Drilling. *SPE Drill. Complet.* **2006**, *21*, 287–295. [CrossRef]

energies

Article

Aerodynamic Performance of VAWT Airfoils: Comparison between Wind Tunnel Testing Using a New Three-Component Strain Gauge Balance and CFD Modelling

Luis Santamaría, Mónica Galdo Vega, Adrián Pandal, José González Pérez, Sandra Velarde-Suárez and Jesús Manuel Fernández Oro *

Fluid Mechanics Area, Department of Energy, University of Oviedo, C/Wifredo Ricart s/n, 33204 Gijon, Asturias, Spain
* Correspondence: jesusfo@uniovi.es

Abstract: Vertical axis wind turbines are an emerging and in-development wind energy technology which are characterized by their complicated aerodynamics. Detached flow conditions, which are typically developed at operational tip speed ratios, demand a rigorous characterization of the airfoils for an accurate prediction of the turbine performance. In this work, a custom-built, three-component external strain gauge balance, specifically developed for airfoil testing, is validated. The physical reasons responsible for discrepancies with reference data are also analyzed. Two- and three-dimensional flat plates, as well as the DU06-W-200 airfoil, are tested in a wind tunnel. Lift and drag coefficients and pitching moments are obtained for a wide angular range at Re = 200,000. The results are compared with data from the bibliography and CFD simulations, performed with the recently developed GEKO (generalized k-omega) turbulence model, achieving remarkable agreement. Instantaneous forces are also analyzed with both experimental and CFD techniques, providing interesting results of the unsteady fluid dynamics. Finally, critical factors affecting the measurements are identified and enhancements are proposed for future works. In summary, a thorough evaluation of this new balance design is provided, showing its valuable potential for VAWT applications.

Keywords: airfoil testing; strain gauge balance; wind tunnel; GEKO turbulence model; vertical axis wind turbine; VAWT

Citation: Santamaría, L.; Galdo Vega, M.; Pandal, A.; González Pérez, J.; Velarde-Suárez, S.; Fernández Oro, J.M. Aerodynamic Performance of VAWT Airfoils: Comparison between Wind Tunnel Testing Using a New Three-Component Strain Gauge Balance and CFD Modelling. *Energies* **2022**, *15*, 9351. https://doi.org/10.3390/en15249351

Academic Editors: Artur Bartosik and Dariusz Asendrych

Received: 28 October 2022
Accepted: 8 December 2022
Published: 10 December 2022

Publisher's Note: MDPI stays neutral with regard to jurisdictional claims in published maps and institutional affiliations.

Copyright: © 2022 by the authors. Licensee MDPI, Basel, Switzerland. This article is an open access article distributed under the terms and conditions of the Creative Commons Attribution (CC BY) license (https://creativecommons.org/licenses/by/4.0/).

1. Introduction

At the present time, humanity is focused on the achievement of a reliable, affordable, and decarbonized energy system. The accomplishment of this goal involves undoubtedly the use of renewable energy, with wind energy harvested through horizontal axis wind turbines (HAWTs) leading the way due to their mature development.

However, such objectives start to require rapid actions and deeper strategies to fulfill the established deadlines (such as a net-zero-emissions scenario by 2050 [1]). These may even include the consideration of the real-time demand curves and on-site production for self-consumption. In this regard, wind energy production in urban environments is gaining much attention. In such restrictive placements, the lift-type vertical axis wind turbines (VAWT) seem to be the best candidates for wind energy extraction [2,3] due to their omnidirectionality, avoiding the need for orientation mechanisms; their ability to work better in variable wind conditions; and their lower noise emission [4].

Nevertheless, in contrast with the well-established HAWTs, which present higher efficiencies and superior rated power, the VAWT turbines are machines still requiring a vast amount of research to overcome the crucial issues that prevent them from achieving a profitable and efficient development status [5]. First of all, their aerodynamics are far more complicated than conventional turbines and there is no agreement yet on the best reference rotor design [6]; additionally, the application on urban environments means facing very-poor-quality wind resources [2,7]. As a result, recent efforts in this field have been directed towards increasing performance and overall energy production through the development of both flow augmentation and control devices [8]; although, the latter are even more attractive for higher-size turbines such as those for deep-water offshore environments. Therefore, a necessary first stage comprises the study of new airfoil designs (which include these characteristics) or existing airfoils in new set-ups, both requiring a precise determination of the aerodynamic properties.

To that extent, CFD methods are a highly valuable tool to improve understanding of the airflow around the turbine blades, the interactions with the flow control devices, and the effect of power augmentation devices. Moreover, they allow the analysis of different types of geometries at a lower cost. However, the accuracy of CFD simulations depends significantly on the selection of the appropriate turbulence model, computational grid construction, and numerical characteristics (temporal and spatial schemes). For that reason experimental validation is always required. Wind tunnel testing is widely employed although this technique faces some intrinsic problems such as prototype–tunnel interaction (blockage) which usually forces downscaling. This in turn derives from other issues, as an intensified relevance of surface roughness [9] and the increased difficulty of measuring airfoil drag at low incidence angles [10]. Furthermore, unsteady phenomena may arise due to complex aerodynamics in cases of highly loaded airfoils. This is especially relevant for VAWT turbines where high angles of attack occur, even when flow control devices are used [11]. During the regular operation of a VAWT, i.e., during a complete rotor turn, the blade angle of attack varies continuously going from positive to negative incidences of the relative incoming flow. Thus, in the pursuit of VAWT performance enhancement, or to properly design passive flow control devices, the understanding of airfoil behavior at different angles of attack is essential. To this effect, the development of accurate, reliable and affordable equipment, useful for this purpose, is inherently interesting.

Aerodynamic performance can be estimated from the integration of the pressure distributions measured with pressure taps [12] or directly with an aerodynamic balance. The first method provides more information but limits the number of geometries that can be tested, as every prototype has to be complexly manufactured to include the pressure tabs and tubes. On the other hand, while there are a wide variety of balance designs, external balances (placed outside the test section) are the most common for airfoil testing. Within external balances, two types are distinguished, single-piece (with multi-component load cells) and multi-piece (with several load cells) [13]. Single-piece aerodynamic balances are usually expensive and not commercial, as each application usually requires a specific range distribution; thus, they are custom-manufactured [14]. Meanwhile, multi-piece balances typically need more space, although in external balances, that is not commonly problematic and, thus, they are widely used [13,14].

Recently, a new design of a three-component external multi-part strain gauge balance, intended for studying the galloping of solar trackers, has been proposed [15,16]. Due to its relevant characteristics, as different load ranges in different directions and high frequency response, this balance has been identified as potentially attractive to test airfoils. Furthermore, its reduced size, scalability and ease of manufacturing make it even more interesting for this application.

This work presents a brand-new application of the aforementioned balance, including the testing of its capabilities, and the validation of its use for the evaluation of aerodynamic performance of VAWT airfoils. For that purpose, several prototypes have been tested in a wind tunnel using the balance, including a typical airfoil (DU06-W-200) developed for VAWT applications. Moreover, CFD simulations have been performed with recently developed turbulence models for complementary analysis and comparison.

The paper is structured as follows. Firstly, the experimental set-up used is presented in Section 2, including the description of the wind tunnel, the aerodynamic balance, and measurement procedures. Then, Section 3 describes the main characteristics of the CFD numerical modeling. Afterwards, the validation procedure is presented in Section 4. Two and three-dimensional flat plates and the DU06-W-200 airfoil are used for comparison, taking advantage of the available data in the open literature. In Section 5, the results are provided: the validation of the balance against bibliographic results is firstly presented for both the flat plates and the airfoil. In the following section, a deeper analysis is carried out with the help of CFD modelling, including the unsteady phenomena with respect to the pitching angle. Finally, after the presentation of the results, relevant conclusions and future works are provided in Section 6.

2. Experimental Methodology

2.1. Set-Up

This research was conducted in the facilities of the Energy Department of the University of Oviedo (Viesques University Campus at Gijón, Spain). In particular, a subsonic open-loop wind tunnel of 13.75 m in length and powered by a 30 kW axial fan with a diameter of 1.2 m was used for this research. It has a nozzle with a 1:12 area ratio, which provides a squared test section of 0.68×0.68 m^2 and allows wind velocities up to 35.5 m/s. A characteristic turbulence intensity of 0.7% for an averaged integral length scale of 0.1 m was obtained at the nozzle discharge. Although its typical configuration is arranged in an atmospheric, fully opened test section, an additional enlargement of the nozzle sidewalls was made to guarantee planar flow over the tested prototypes (discussed later in detail). A sketch of the wind tunnel is shown in Figure 1a. The test wind velocity is measured from the pressure difference in the nozzle, with a 1 in the $H_2O \pm 0.25\%$ differential pressure sensor. Note that, given the contraction ratio, velocities in the settling chamber were considered negligible.

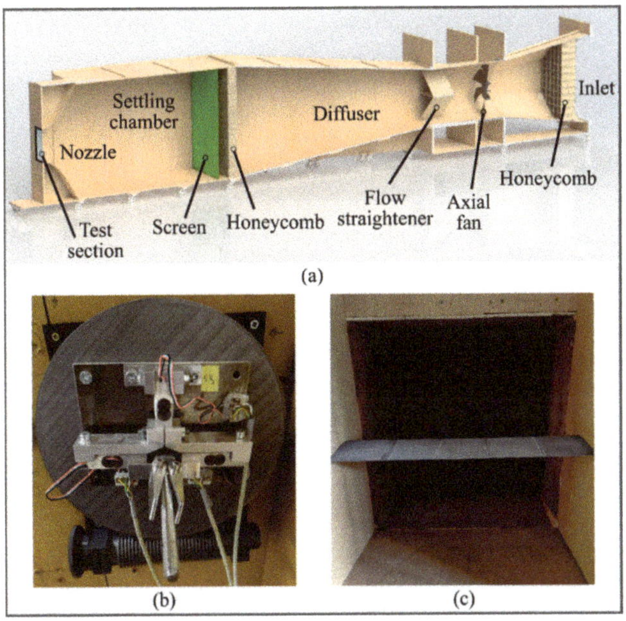

Figure 1. (**a**) Sketch of the wind tunnel (Courtesy of Angie L. Ramírez Celis). (**b**) Aerodynamic balance mounted on the mechanical orientation system. (**c**) DU06-W-200 airfoil prototype in the test section.

The custom-built aerodynamic balance under study is a strain gauge force balance with 3 components, which provides lift and drag forces and pitching moment. This balance was originally conceived to measure unsteady forces and torques on small-scale single-axis solar trackers, and successfully employed for recent aeroelastic investigations in our research group [15,16]. Precisely, the balance was designed to provide a larger range in one direction with respect to the other, which is also a very useful feature for testing airfoils where lift forces are much greater than drag forces. The balance, which rotates with the prototype to be tested, is composed of a floating axis supported by two symmetrically placed load cells which are attached to a frame supported by the third load cell. The assembly is designed so that forces and moments outside of the measuring plane are minimized. Two different balances were built for measuring ranges within 0.75 kg and 5 kg, respectively (ranges of the single load cell direction), although the design can be easily scaled to any other quantity. Each load cell has two strain gages that are connected to the same Wheatstone bridge circuit to provide an amplified output. The voltage from the bridges is measured with a data acquisition card, which allows a measuring frequency up to 20 kHz. Note that this is especially relevant when unsteady phenomena are to be studied with this kind of device. The rotation of the balance was performed with a mechanical orientation system which granted the variation in the pitching angle using a worm gear pair (Figure 1b). The system was manufactured by fused deposition modelling (FDM) and allowed a minimum angular step of 0.5°. Finally, measurement data analysis and calculations were performed with custom MATLAB codes in a computer.

Three different prototypes were tested in the aerodynamic balance for this work: two flat plates (of different dimensions) and the DU06-W-200 airfoil (Figure 1c). One of the flat plates was designed to perform as a theoretical 3D plate, featuring an aspect ratio (L/c, where L is the span and c is the chord or width) of 3.2, which was proved to be sufficient for the purpose. On the other hand, the 2D plate had a span as wide as the wind tunnel test section (a clearance of tenths of a millimeter was left so that there is no contact with the walls) and the same width of the 3D flat plate. Note that this allowed testing both at the same wind velocity with an equal Reynolds number. Hence, the 2D flat plate had an aspect ratio of 7.2. The airfoil prototype also had a span as wide as the wind tunnel, but the chord was chosen so that, at the objective Reynolds, the obtained forces were coherent with the range of the aerodynamic balance used. Thus, the resultant aspect ratio was 3.8, which, given the results obtained, proved to be sufficient to obtain 2D airfoil coefficients over this wall-to-wall prototype. The dimensions of the tested prototypes are included in Table 1.

Table 1. Tested prototypes and dimensions.

Prototype	Span [mm]	Chord/Width [mm]	Aspect Ratio [-]
Flat plate 3D	300	95	3.2
Flat plate 2D	680	95	7.2
DU06-W-200	680	180	3.8

The three models were made of PLA and manufactured with FDM, requiring subsequent sanding and polishing to achieve an adequate final roughness. The prototypes have in their core a steel rod to increase their stiffness. This rod has a fixed support connection to the balance and cylindrical joint in the wall of the other end, avoiding movements in the measurement plane and prototype bending. The balance calibration procedure already accounts for the effect of the second support.

2.2. Balance Calibration

In order to provide quality measurements, the aerodynamic balance was calibrated before each experiment. The calibration routine employed for this work assumed a linear response of the load cells, so a direct (exact solution) method was chosen to determine the correlation coefficients. The calibration, which implied a two-stage procedure to determine the calibration matrix, was completed under "no wind" conditions. In the first stage, the prototype was just mounted in the balance and a measurement was performed at free load, defining the zero-loading state. In the second stage, several measurements were performed with the balance loaded with known weights. Specifically, 3 load cases (LC) were carried out:

- Single load, F_{x_1}, in the horizontal direction (LC1).
- Single load, F_{y_2}, in the vertical direction (LC2).
- Combined load, obtained through a vertical load, F_{y_3}, displaced a known distance b from the axis (LC3).

From each load case, three outputs (one for each load cell) were obtained providing a 9 equation and 9 unknowns system to represent the direct correlation between loads and measured components. Matrix algebra can be applied to streamline this process in the following way.

First, a force matrix F_{LC} is defined with the three load cases, with one column for each component and one row for each load case.

$$F_{LC} = \begin{pmatrix} F_{x_1} & 0 & 0 \\ 0 & F_{y_2} & 0 \\ 0 & F_{y_3} & F_{y_3}b \end{pmatrix} \quad (1)$$

where F is the applied load, x and y are the horizontal and vertical directions, respectively, in the balance coordinate system, and b is the horizontal distance to the axis in load case 3.

With the output of the single load cell being associated with the measurements in the horizontal direction in these experiments, referred to as "signal 1" (s_1), and the other two, associated with the measurements in the vertical direction, referred to as "signal 2" (s_2) and "signal 3" (s_3), the output of the balance in the zero-loading state can be posed as a vector S_{LC_0} containing the values recorded in each load cell.

$$S_{LC_0} = \begin{pmatrix} S_{1_0} & S_{2_0} & S_{3_0} \end{pmatrix} \quad (2)$$

Following this, the matrix S_{LC} is defined with the load cells' output for the three load cases, yielding:

$$S_{LC} = \begin{pmatrix} s_{1,1} & s_{1,2} & s_{1,3} \\ s_{2,1} & s_{2,2} & s_{2,3} \\ s_{3,1} & s_{3,2} & s_{3,3} \end{pmatrix} \quad (3)$$

Then, the calibration matrix K with the coefficients that relate the output of the three load cells with the forces and moment is:

$$K = \begin{pmatrix} k_{1,1} & k_{1,2} & k_{1,3} \\ k_{2,1} & k_{2,2} & k_{2,3} \\ k_{3,1} & k_{3,2} & k_{3,3} \end{pmatrix} \quad (4)$$

Finally, applying the linear response assumption, the equation system is thus stated as:

$$F_{LC} = \left[S_{LC} - S_{LC_0}\right] \cdot K \quad (5)$$

where the no-load signals are discounted as the system offset. From this matrix system, matrix K can be directly deduced as $K = [S - S_0]^{-1} \cdot F_{xyz}$, thus obtaining the direct relation between load cell outputs and measured forces.

Once the calibration matrix is determined, it can be employed to obtain the forces acting on the models from the signals measured during the operation of the wind tunnel using a generalization of Equation (5) for a single-point measurement:

$$F_{xyz} = [S - S_0] \cdot K \quad (6)$$

where F_{xyz} and S are now row vectors with three columns.

As the calibration used is a linear, two-point method, the balance was additionally tested before the aerodynamic measurements' campaign to evaluate its accuracy. Figure 2 shows the response of the calibrated balance (y-axis) to 5 different known weights in the lower part of the balance range (x-axis), where the linearity of these types of sensors is mostly critical.

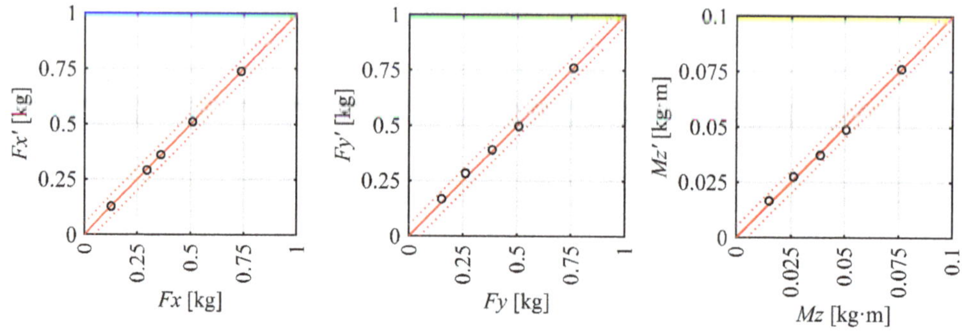

Figure 2. Balance response to a range of known weights using a two-point, linear calibration method

The figure shows that the calibration method used provides a sufficiently accurate linear response, even for the lower part of the balance range. The response in this range is better in the horizontal direction because its range is half than the others, so it is better prepared to measure small forces. This is especially interesting to the case of airfoil testing as drag forces are much lower than lift ones.

During the aerodynamic measurements campaign, the balance was calibrated before each experiment and tested after with known weights to validate the balance calibration. This reduces the influence of random errors produced by differences in the testing environment temperature, differences in the set-up assembly, etc. Up to 13 calibrations were performed during the campaign, providing useful statistical data of the balance performance. Table 2 shows the mean, standard deviation, maximum and median value of the relative errors between the known weights and the measured weights those 13 calibrations.

Table 2. Relative error statistics of 13 calibrations performed during the measurements campaign.

Component	Fx	Fy	Mz
$\bar{e}$ [%]	1.31	0.83	1.11
$s(e)$ [%]	1.13	0.86	1.41
e_{max} [%]	3.48	2.74	4.31
$\tilde{e}$ [%]	0.81	0.40	0.29

The mean relative error obtained was around 1%, slightly higher in the horizontal direction and slightly lower in the vertical direction. However, as the standard deviation reveals, there was some variability in the quality of the calibrations; hence, the mean is not very representative of the real performance of the balance. Note that although all 13 calibrations have been included in this analysis, a quality requirement was established in 1%. Thus, calibrations with errors above this, such as the one that achieved the maximum error shown in Table 2, were discarded and repeated. Nevertheless, as the median indicates these discarded cases were not common and the typical balance calibration errors were about 0.8% for the horizontal direction, 0.4% for the vertical direction, and 0.3% for the moment. In these calibrations, the loads were adjusted to the expected measured forces hence, in contrast with the previous figure, the vertical and moment errors are lower. The higher error in the horizontal force is probably due to the higher difficulty to produce a pure horizontal load, as opposed to the simplicity of vertical loading.

2.3. Aerodynamic Measurements

With the balance already calibrated, the procedure for the aerodynamic measurements is as follows. The offset signals S_0 can be equal to S_{LC_0} or not, depending on the chosen zero-load state of reference. This aspect is especially relevant when a given pitching angle α is fixed for the prototype–balance assembly because S_0 must be measured for every particular pitch. In addition, to obtain the forces F_{DLz} in the wind coordinate system (drag and lift, see Figure 3), a base–change matrix, M_{BC}, must be applied in the following way:

$$M_{BC} = \begin{pmatrix} \cos\alpha & -\sin\alpha & 0 \\ \sin\alpha & \cos\alpha & 0 \\ 0 & 0 & 1 \end{pmatrix} \tag{7}$$

$$F_{DLz} = F_{xyz} \cdot M_{BC} \tag{8}$$

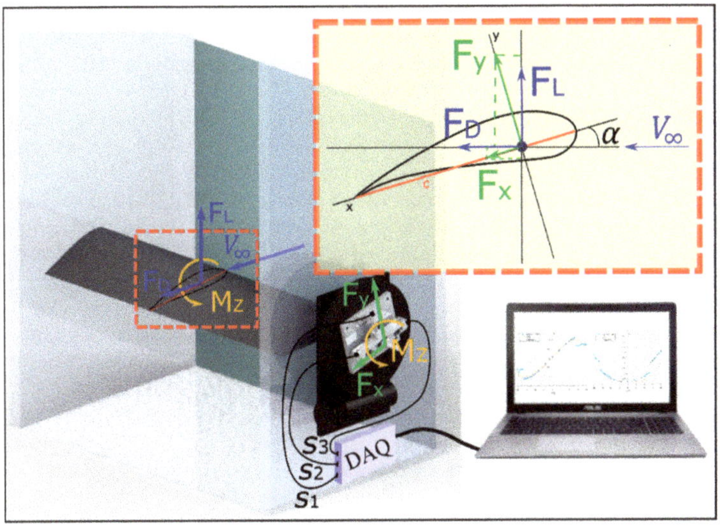

Figure 3. Diagram of the set-up with the variables and the different coordinate systems involved in the measuring procedure.

The aerodynamic coefficients can be finally obtained if the measured forces are made non-dimensional with the upstream dynamic pressure expressed as a force exerted on the prototype:

$$F_\infty = \frac{1}{2}\rho v_\infty^2 c L \tag{9}$$

leading to:

$$C_{DLM} = F_{DL} \cdot \begin{pmatrix} 1/F_\infty & 0 & 0 \\ 0 & 1/F_\infty & 0 \\ 0 & 0 & 1/F_\infty c \end{pmatrix} \tag{10}$$

where ρ is the air density, v_∞ the reference wind velocity, c is the chord/width of the prototype and L its span. The moment component must also be divided by the chord/width again to produce the non-dimensional moment coefficient. Hence, from Equation (10), the drag, lift and moment coefficients are retrieved directly.

The three aforementioned prototypes were tested following this procedure. Particularly, both flat plates were tested at a Reynolds number ($Re_c = v_\infty c/\nu$, where ν is the air kinematic viscosity) of 130,000 from 0° to 90° of the pitching angle using a constant angular step of 10°. Complementarily, the airfoil was tested at a Reynolds number of 200,000, going from −20° to 20° with a variable angular step, for a better characterization of the aerodynamic forces during the airfoil stall. The three gauge signals were recorded during almost 15 s at a typical acquiring rate of 20 kHz, which assured a sufficient number of points to guarantee correct repeatability and accuracy in the results.

3. Numerical Modelling

A 2D numerical model of the DU-06-W-200 airfoil was implemented in Ansys-FLUENT® v2020 to obtain the aerodynamic coefficients numerically. The Reynolds-averaged Navier–Stokes equations were resolved in an incompressible approach using eddy–viscosity turbulence modelling. Both Spalart–Allmaras (S-A) and new generalized k-ω (GEKO) models were used. The one-equation S-A model [17] is widely used for external aerodynamic applications. Although it is known to provide reasonable solutions for flows with adverse pressure gradients and separation, its accuracy to predict separation is lower than optimal two-equation models such as k-ω omega SST and GEKO. In addition, all k-ω models in Ansys are implemented with a y+-insensitive wall treatment, avoiding the discussion concerning the optimal selection of wall formulations in k-ε models [18].

GEKO is a recent turbulence model framework (based on the ω-equation) which introduces free parameters into the equations. The main advantage is that relevant parameters can be decided and tuned by the user for given operative ranges, and without a negative impact on the basic model calibration. The main tuning parameter for the GEKO model is the coefficient C_{SEP}, which controls the boundary layer separation, predicting a more aggressive detachment if its value is increased. In the case of airfoils, it is highly recommended to use a value between 2.0 and 2.5 [19]. Furthermore, the GEKO model has been executed also with the option for scale-adaptive simulation (SAS) activated, which deploys an improved URANS formulation for the resolution of the turbulent spectrum in unstable flow conditions. The SAS concept is based on the introduction of the von Kármán length scale into the turbulence scale equation, allowing the model to dynamically adjust to resolved structures in a URANS simulation, which results in an LES-like behavior in unsteady regions of the flow field (those with flow separation).

An extended domain, with a distance to the inlet of 12.5c and a distance to the outlet equal to 20c (domain size 32.5c × 25c), in line with typical values found in the literature, was considered accurate to avoid the effect of the boundaries on the development of the flow inside the domain region (see Figure 4). A C-mesh distribution has been employed around the airfoil, resulting in a [350 × 75] cell size for both pressure and suction sides of the airfoil. An averaged value of y += 1.7 (at Re_c = 200,000) has been achieved with the first mesh point located at roughly 0.05 mm from the wall. At the wake region, a structured mesh of [300 × 150] cells was also employed, resulting in 97,500 cells for the complete 2D model. Furthermore, an additional refined mesh with [525 × 150] nodes on the airfoil walls and 247,500 cells for the whole domain was also employed to check the solution sensitivity to the grid resolution.

The boundary conditions of the simulation domain are given in Figure 4 which includes details of the adopted mesh. A velocity inlet condition of 16.4 m/s was set at the domain inlet to match the Reynolds number (200,000) of the experimental measurements. Furthermore, up to 21 different angles of attack (AoA) were simulated to complete a detailed evolution of the aerodynamic coefficients, including negative and positive incidences: ±[0, 2, 4, 6, 8, 10, 12, 14, 16, 18, 20]. According to previous measurements, a turbulence intensity of 0.7% was fixed for a length scale one order of magnitude lower than the characteristic size of the test section in the wind tunnel. Both steady and unsteady simulations have been conducted, the latter necessary at high AoA for partially and fully detached flow conditions. A time-step size of 3×10^{-4} s was fixed in order to track the evolution of the vortex shedding with sufficient resolution. A time-averaged value of the airfoil coefficients was finally computed after periodically fluctuating regimes were achieved (typically, 50 times the airfoil chord flow-time).

The flow equations were discretized using the finite volume method with a second-order scheme for momentum and turbulent variables. Second-order accuracy was also selected in the transport equation for the pressure correction. The discretization of the temporal terms (when necessary) corresponds to a bounded second-order implicit formulation. The SIMPLE algorithm was used for the pressure–velocity coupling for all studied cases. Spatial discretization regarding gradient terms was selected to be the least-squares cell-based discretization. Finally, a convergence criterion of 10^{-6} was fixed for the velocity components of the momentum equation, while a minimum threshold of 10^{-5} was at least required for the rest of the equations. Simulations were performed using a four-node Intel Core i7-52820K at 3.3 GHz and 64 Gb RAM, with characteristic CPU times of 75 min for every execution (1 day of CPU time to complete the whole angular range) in the case of the refined mesh.

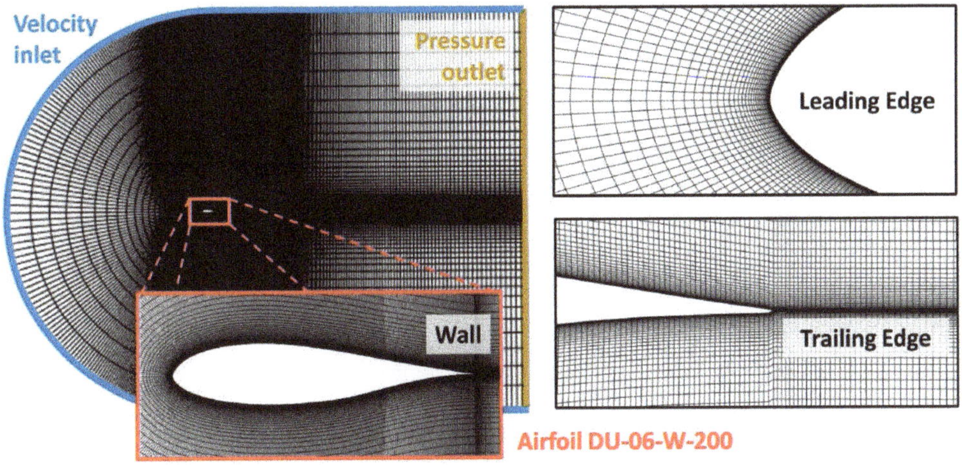

Figure 4. CFD simulation mesh detail views and boundary conditions.

4. Validation Data Sources

4.1. Flat Plate Data

The available data in the bibliography for the lift and drag coefficients in 2D and 3D flat plates have been obtained from [20]. This reference shows how the force on a flat plate can be calculated from the aerodynamic coefficients. The different data are provided for Reynolds numbers greater than 10,000 with an estimated precision of ±5%. Specifically, the aerodynamic coefficients on a flat plate of infinite span (2D) were obtained by applying the experimentally modified Kirchhoff and Rayleigh theory of discontinuous motion shown in the works of [21] and [22]. The coefficients for flat plates in three dimensions, presented in the equations shown by Blevins, were originally extracted from the experimental measurements presented by [23].

4.2. DU06-W-200 Airfoil

This airfoil was specifically designed for vertical axis wind turbine applications, with the objective of improving the self-starting abilities of this type of turbine. The available aerodynamic data of the DU06-W-200 airfoil in the literature, used to contrast the obtained results in the present database, are found in [24]. This source provides experimental work performed in the Low Turbulence Tunnel (LTT) at the Technical University of Delft. This wind tunnel has a 1.25 × 1.8 m test section and can achieve a maximum wind speed of 120 m/s. The tested model was made of solid aluminum with a 1.8 m span (the whole section width) and a chord of 0.25 m, and the aerodynamic coefficients were obtained with a six-component external balance. Two types of results are presented: "clean", corresponding to the bare airfoil in the wind tunnel with around 0.02% turbulence, and "dirty", which is the same testing conditions, but the airfoil features a zig-zag tape at 5% of the chord, simulating a much more turbulent test environment. From the available results, those corresponding to a Reynolds number of 300,000 have been chosen as a reference, as they are the closest to our experimental dataset.

5. Results

In this section, the results of the different tests are presented. Particularly, the results have been divided depending on the type of prototype used in the test: flat plate or DU06-W-200 airfoil.

5.1. Flat Plate

The aerodynamic coefficients (C_D, C_L) obtained with the flat plates are shown in Figure 5, compared with the data from the bibliography. Experimental curves are plotted with red discontinuous lines, using triangles for the 2D flat plate and squares for the 3D flat plate. On the other hand, 2D and 3D flat plate data from the bibliography are plotted with green and blue dotted lines, respectively.

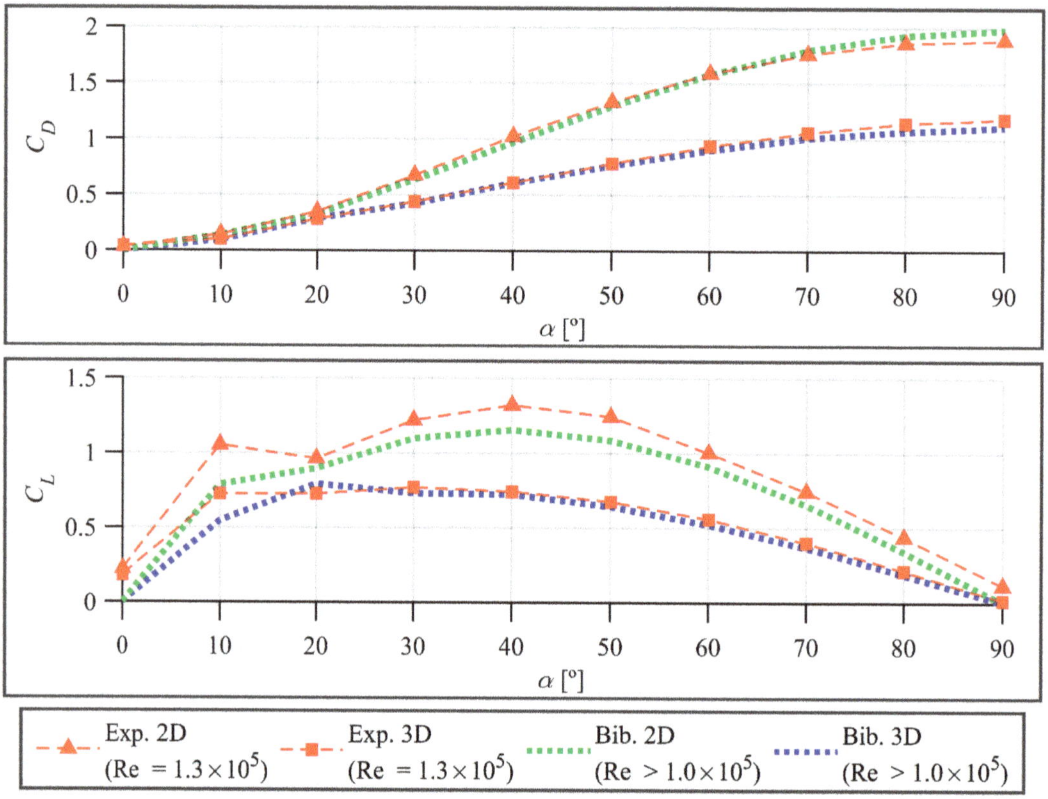

Figure 5. Experimental drag and lift coefficients from 2D and 3D flat plates compared with data from the bibliography.

The drag coefficient (top plot) exhibits a remarkable agreement for both flat plates in the whole angular range, with a very slight drift at high pitching angles. Complementarily, the lift coefficients (bottom plot) also show a good overall agreement, although with some overestimation for angles smaller than 20 degrees. This can be attributed to the lack of a complete symmetry between both pressure and suction sides of the plates. In fact, only the pressure sides are completely flat, because of a slight engrossment of the suction side at the mid-chord to accommodate a sufficiently robust shaft. Thus, at low pitching angles, when the flow is still attached, this geometrical defect raises the pressure difference between both sides leading to an increase in the lift force. However, at higher pitching angles, the flow in the suction side is completely detached and the experimental data matches the reference data with especially accurate results for the 3D flat plate. Despite the experimental lift coefficients for the 2D flat plate being slightly above the bibliography, the global trend is perfectly reproduced. The overall result suggests an accurate and precise performance of the balance, thus postulating it as a good candidate for airfoil testing.

5.2. DU06-W-200 Airfoil

Figure 6 shows the comparison of the experimental coefficients (C_D, C_L) of the airfoil measured with the aerodynamic balance and the data from the bibliography. Experimental data from this work are plotted with a red discontinuous line and triangle markers. Up to six complete tests were repeated in an effort to properly characterize the hysteresis zone related to the flow separation. Thus, in this figure, the markers and discontinuous line show the averaged coefficients from all the tests, while the light-red area bounds the maximum and minimum dispersion in the results.

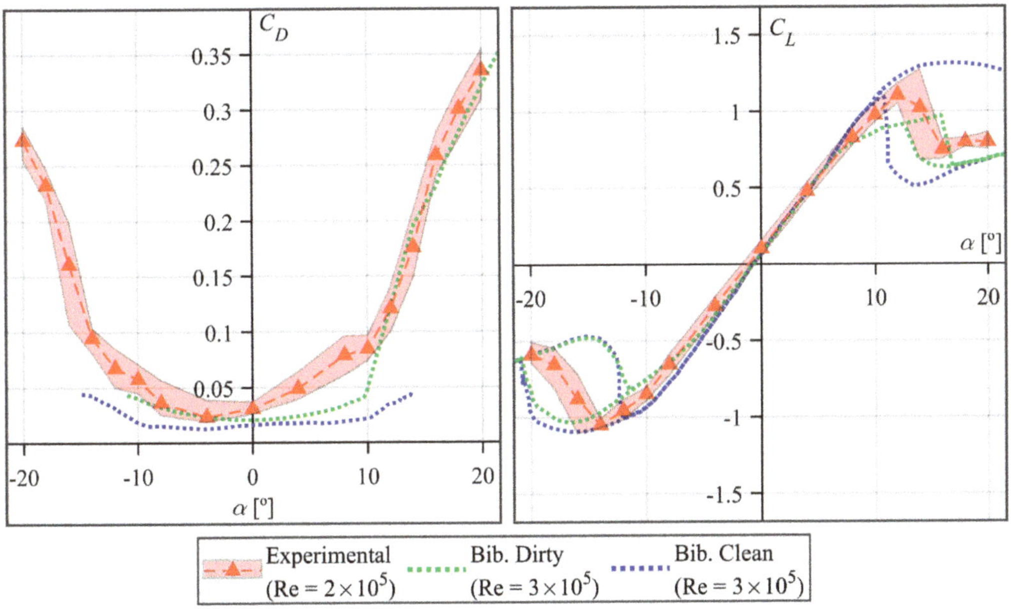

Figure 6. Experimental drag and lift coefficients from airfoil DU06-W-200 compared with data from the bibliography.

Regarding the drag coefficient (left plot), the obtained results are significantly higher than those from the "clean" dataset. This can be easily related to the big difference in the mean turbulence level (about 35 times) between both wind tunnels. Nevertheless, the obtained results match remarkably well for the "dirty" dataset, with the exception of the range of low positive angles, in which the obtained coefficients are higher. The reason of this discrepancy may probably lie in the presence of a light dimple in the airfoil shape, close to the leading edge at the pressure side. This defect is a consequence of the deburring of the seam scar produced in the layer shift as the airfoil is 3D-printed. That irregularity may be triggering turbulence transition on the airfoil (precisely in the stagnation point) and, thus, increasing the drag artificially. Furthermore, additional polishing of the area has also slightly modified the local slope of the airfoil, leading to a mismatch with respect to the original geometry.

Meanwhile, the lift coefficient curves (right plot) overlap perfectly for all the datasets at low pitching angles ($-6°$ to $6°$), where the flow is completely attached, and the incoming flow turbulence is not relevant. However, at higher angles ($-6°$ to $-11°$ and $6°$ to $10°$), the "clean" dataset bounds the maximum magnitude of the lift, with the obtained experimental results slightly below and the "dirty" dataset starting to decay due to the early flow separation. At negative angles of attack, both "clean" and "dirty" datasets maintain a slow and progressive detachment when the pitching angle is increased, until they finally drop at −20°. On the other hand, the experimental data from this work drops earlier at −16°, after achieving the maximum magnitude of the negative lift. Although our experiments have in fact not been performed to describe the hysteresis cycles, it is significant that the dispersion of the results resemble that phenomenon to some extent. Hence, the width of the hysteresis loop in the reference data is much higher than in the experiments, which practically crosses through the middle, dividing the others in half. This also occurs with the positive side of the curve. However, here, the difference between the lift drop of the "clean" dataset and the two others is much higher with the first one dropping outside the shown range (~22°) and the other two around 14°~15°. Furthermore, despite the experimental data achieving almost the same maximum lift coefficient as the "clean" dataset, the drop zone and hysteresis loop width match better the "dirty" dataset. The existence of a boundary layer on the side walls of the tunnel generating 3D effects at the ends of the tested wing may affect the hysteresis of flow separation on the wing. This could be the reason for the observed large differences between the experiment and CFD calculations at large angles of attack.

Considering the overall results, the aerodynamic balance used in this experiment clearly exhibits a notable accuracy, being able to reproduce the reference dataset both in drag and lift coefficients, and clearly characterizing the flow separation.

A further analysis of the experimental data obtained is discussed with the help of the CFD simulations performed. In Figure 7, the experimental coefficients are compared with the results from CFD simulations for different turbulence models and simulating conditions. Specifically, results from the simulation with the Spalart–Allmaras (S-A) model executed in a steady fashion are represented in dark gray, those computed steadily but for a generalized k-ω (GEKO) model are shown in green, whereas the GEKO unsteady simulations are shown in blue, distinguishing between the coarse mesh (light-blue discontinuous line) and the extra refined mesh (dark-blue continuous line).

Despite the simplifications of the S-A steady simulation, it performs remarkably well in reproducing the lift curve, with only a slight underestimation of the maximum. However, it is clearly unable to predict an accurate drag overshoot. On the other hand, the set of GEKO simulations produce subtle different results among them. The GEKO steady simulation significantly improves the results compared with the S-A, accurately characterizing the drag overshoot as well as the lift curve. Nevertheless, it overestimates both drag and lift magnitude at negative wide angles (−12° to −20°). The results from the GEKO unsteady simulation and with the coarse mesh are enhanced, but show that the initial mesh is not sufficiently accurate to reproduce the lift curve when flow separation starts to be significant In fact, it is still poorly predicting important flow features when the airfoil stalls, such as the instabilities of the boundary layers and the shedding of trailing vortexes. This is clearly improved with the extra-refined mesh, which produces the best results, especially for negative angles of attack. Yet, it still fails to predict accurate lift drops in the case of fully detached flow.

Since no data were found in the bibliography for the pitching moment, the experimental results have been directly compared with the GEKO unsteady simulation for the refined mesh in Figure 8. Furthermore, as a preliminary approach to evaluate unsteady capabilities of the aerodynamic balance, the RMS value of the fluctuations in the moment coefficient is also represented (dispersion bars) and compared in the figure. In this case, instead of presenting the averaged statistics of the whole dataset as before, only a single measurement has been used to ensure that these fluctuating results are consistent.

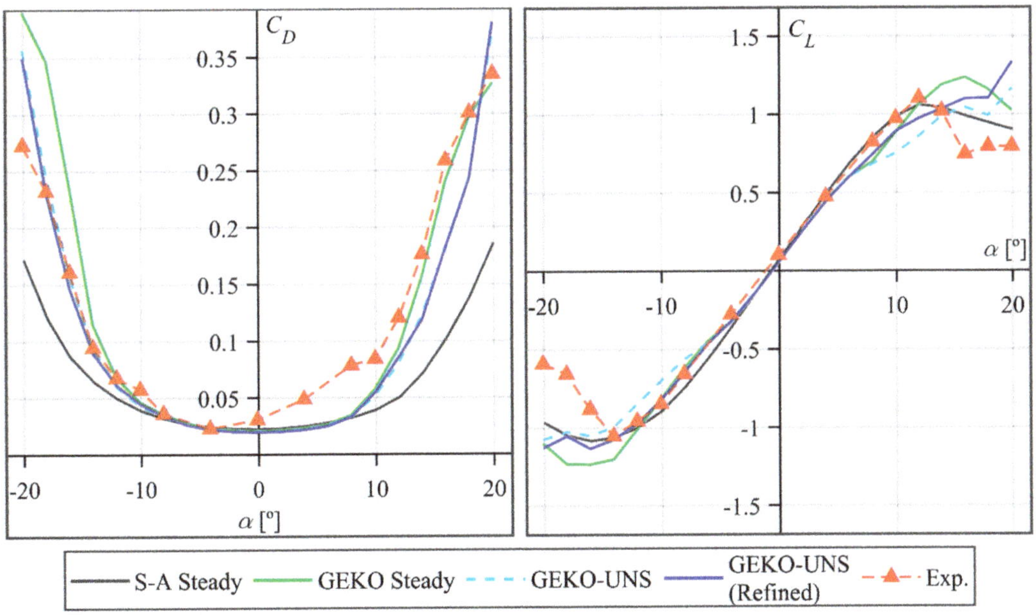

Figure 7. Experimental drag and lift coefficients from airfoil DU06-W-200 compared with results from CFD simulations for different turbulence models.

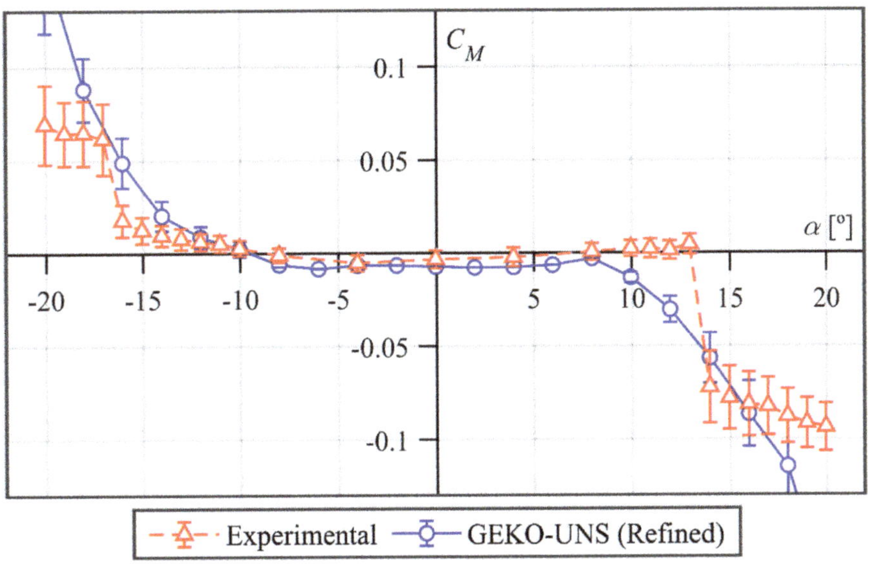

Figure 8. Experimental moment coefficient and fluctuating moment coefficient from airfoil DU06-W-200 compared with results from the best unsteady CFD model.

As expected, the moment coefficient for low pitching angles, which it is practically zero, matches perfectly between experiments and numerical results (solid and dashed lines). This is coherent with the hypothesis of the airfoil having the center of pressure approximately at 25% of the chord (the same location for the origin of coordinates in the

CFD and for the center of the shaft in the experimental prototype). However, as pitching angles increase (both in negative and positive directions), the CFD model predicts a smooth exponential-like rise in the coefficient magnitude, while the experimental results show a more drastic drop at −16° and 13°, followed by a moderate, linear-slope increase.

In the case of the RMS values, a very low level can be appreciated in the experimental data at low pitching angles. Likewise, the CFD model converges to a unique solution as there is no unsteadiness in the simulations. The flow separation can be easily identified in the figure by the sudden increase in the fluctuations in the experimental dataset, although in the CFD, there is a more progressive increase. Precisely, it was necessary to activate unsteady computations in the CFD model beyond ±8° of *AoA* to account for the inherent unsteadiness of the detached flow. This comparison also reveals that the dynamic sensitivity of the balance is enough to perceive the amplitude of the fluctuating forces, despite the structural damping of the wing model. Conversely, accurate frequency values are not feasible due to the high stiffness of the set-up, thus avoiding a complete fast response of the measurements.

For a deeper understanding of the unsteady phenomena involved in these fluctuating forces, the velocity field, pressure coefficient (defined as $C_p = 2(p - p_\infty)/\rho v_\infty^2$), and spectra of the fluctuating moment have been analyzed in detail for four positive pitching angles (8°, 12°, 16° and 20°) using the data from the refined GEKO unsteady simulation. The results are shown in Figure 9. The instantaneous velocity field at a particular instant in the simulation is represented on the left part of the figure in non-dimensional terms with respect to the upstream velocity. Meanwhile, on the upper-right plot, the pressure coefficient along the airfoil chord is represented for both suction and pressure sides at that same instant. In addition, shadowed areas have been introduced to illustrate how the coefficient is oscillating during a complete shedding cycle. Finally, on the right lower part the amplitude and oscillating frequency of the moment coefficient are shown, identifying the peak values.

The maps with the representation of the velocity magnitude allow the identification of the stagnation points in the lower part of the leading edge and show an evident trend towards an early flow separation as the pitching angle is progressively increased, with a remarkable thickening of the boundary layer. A counter-rotating pair of vortices is shed from the airfoil, growing in size as the pitching angle is more pronounced. This vortex shedding is coherent with the frequencies of the fluctuating moment, which show high frequency but low size of the vortex shedding at the lower pitching angles. Conversely, lower frequencies and higher sizes of the vortices are observed at higher angles, once the flow is fully detached. As a consequence, the aerodynamic coefficients are intensively fluctuating with amplitudes up to three times larger than those formed at the separation onset. Regarding the pressure coefficient, there is also a notable increase in the oscillations with the pitching angle, revealed as a progressive build-up of the C_p value in the pressure side, and a shift towards the trailing edge of the airfoil in the suction side. Note that from 12° onwards, a wide fluctuation can be observed in the trailing edge due to an oscillatory partial reattachment, which it is also responsible for the periodic variations on both drag and lift coefficients.

Previous assertions are validated by means of the Strouhal number, $St = fL/v_\infty$, which relates the vortex shedding of the large turbulent scales and the frequencies of the fluctuating moment. The frequency values (f) correspond to the first (fundamental) harmonic in the power spectrum of the fluctuations for the torque coefficient (see plot in the bottom right in Figure 9). The characteristic length (L) has been adopted as the maximum value of the integral length scale on the airfoil suction side (see Figure 10 below). The integral scale is estimated from the instantaneous values of the turbulent kinetic energy (k) and the turbulent dissipation rate (ε) according to $L = k^{3/2}/\varepsilon$ [25]. Using the convective inlet velocity (v_∞ = 16.4 m/s), typical values around 0.2 are found (see Table 3) for all the situations considered between 10 deg (partial detachment) and 20 deg (fully detached flow).

which is a characteristic value observed in separation of bluff-bodies at moderate-to-high Reynolds numbers.

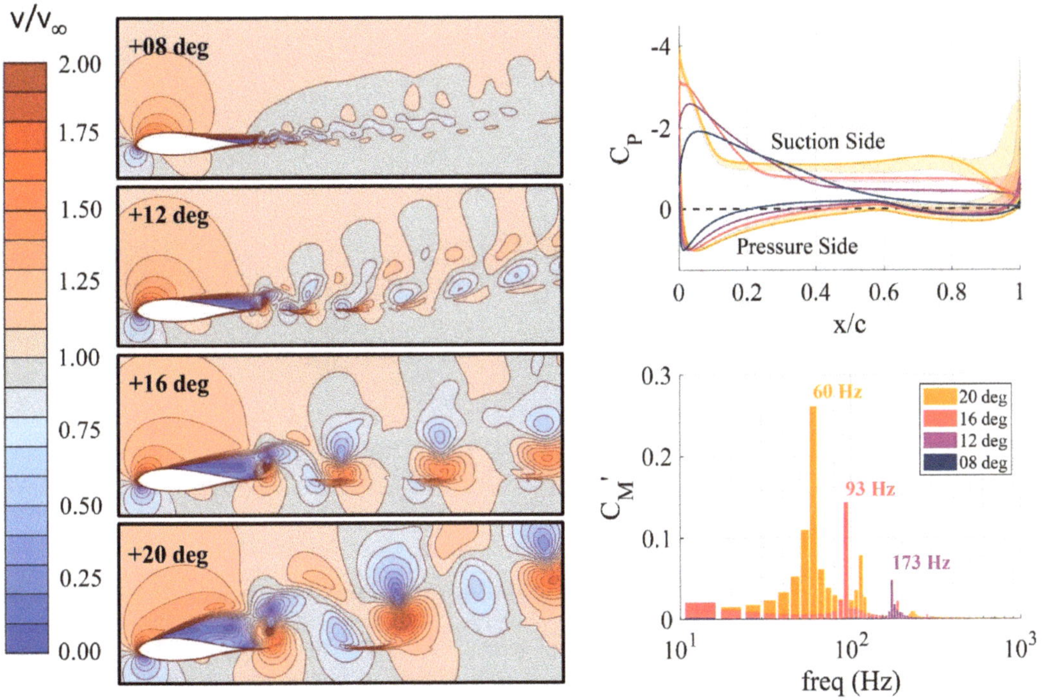

Figure 9. Velocity field, pressure coefficients and spectra of the fluctuating moment at *AoA* = 8°, 12°, 16° and 20°, obtained from the refined GEKO unsteady CFD simulation.

Table 3. Strouhal numbers of the detached flow for different angles of attack.

AoA	*f* [Hz]	*L* [m]	*St*
10 deg	233	0.014	0.197
12 deg	173	0.019	0.199
14 deg	126	0.024	0.185
16 deg	93	0.032	0.180
18 deg	73	0.036	0.160
20 deg	60	0.039	0.143

Figure 10 shows the computed values of integral length scales in the waked regions of the airfoil for different angles of attack. The figure reveals the vortical motion of the largest vortices, identified in a dark-blue color for an instantaneous snapshot, which illustrates the typical turn-out time of the vortices. At a low *AoA*, the size of the vortices is roughly in the order of magnitude of the airfoil thickness, with an intense vortex shedding (high frequency) revealed through the advection transport of the vortices street. For a high *AoA*, the vortices are progressively enlarged, now with a size in the order of magnitude of the airfoil chord, but with a lower shedding frequency (the generation rate of these large flow structures is significantly reduced as shown again in the convective transport along the airfoil wake).

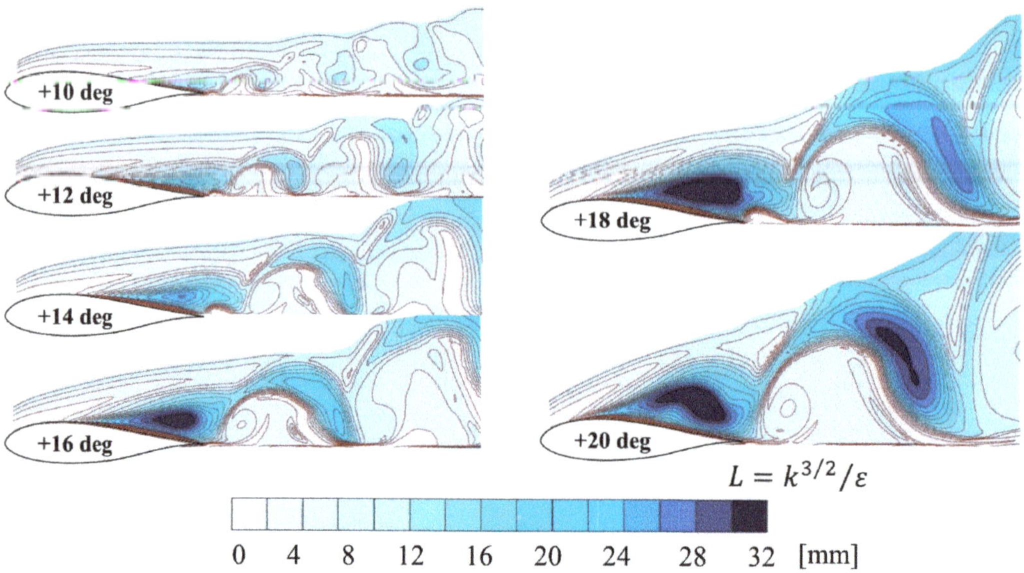

Figure 10. Integral length scales in the airfoil wake for positive *AoA* = 10°, 12°, 14°, 16°, 18° and 20°.

Further insight is now provided with a closer look to the detached regions of the airfoil suction side during a complete oscillation cycle. For that purpose, the longitudinal distribution of the mean pressure coefficients on the suction side of the airfoil is shown in a contour plot in Figure 11, for all the angles-of-attack simulated. A black dashed line identifies the averaged position of the separation point, revealing the severe engrossment of the detached region towards the leading edge for high pitching angles. Moreover, the contour map is complemented by a comprehensive view of the averaged detached regions over the airfoil, for *AoA* going from 6° to 20° (right plot), in order to illustrate the recirculation zones (identified with negative streamwise velocities).

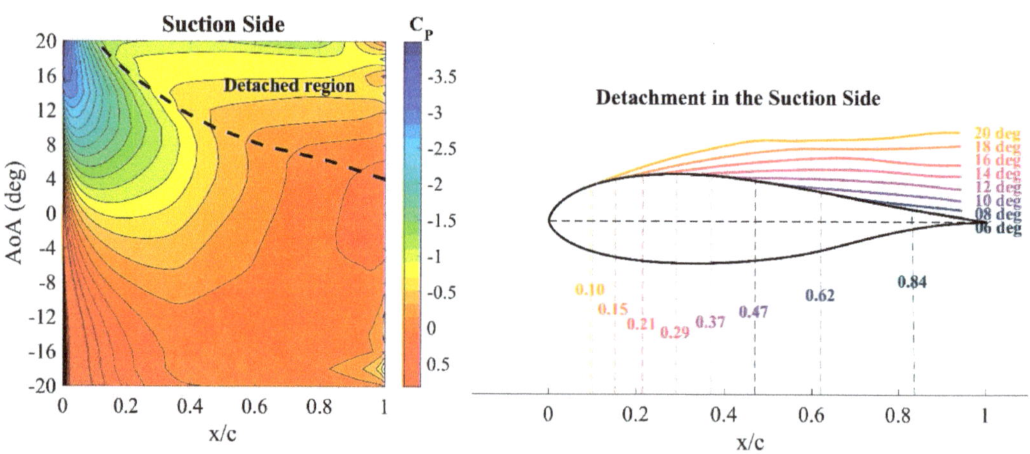

Figure 11. Pressure coefficient in the suction side, boundary layer and flow detachment point, for a wide set of pitching angles simulated with the refined GEKO unsteady model.

As expected, the position of the detachment point (where the wall shear stress equals zero) moves towards the leading edge, leaving a growing detached region. Note that at 20°, more than 80% of the suction face is exposed to fully detached flow. Although not shown here for brevity, a similar contour map is obtained for the pressure side, but symmetrically flipped with respect to the zero *AoA* and with a small shift, as detachment occurs at slightly higher angles for this side.

6. Conclusions

2D and 3D flat plates, in addition to the DU06-W-200 airfoil, have been tested in a wind tunnel equipped with a new external balance, designed for the measurement of aerodynamic forces. The inherent features of the balance make it suitable for an accurate characterization of VAWT airfoils, although specific validation has been required to ensure its performance. Furthermore, CFD simulations of the airfoil have been performed with different turbulence models and flow conditions, complementing the experimental results, and illustrating vividly the unsteady phenomena involved.

The experimental drag and lift coefficients obtained for both the flat plates and DU airfoil match the data available in the literature remarkably well, even at large angles of attack, when unsteady flow is rather relevant. The studied balance has proved its ability to capture fluctuating forces, although noise-filtering procedures must be implemented to obtain more reliable instantaneous data.

The new GEKO turbulence model used for this work offers accurate predictions of forces and moment, providing the best results when running unsteadily in the case of detached conditions, although the mesh requirements are significantly higher with respect to the steady simulations.

Finally, in light of the obtained results, the balance design as well as the CFD models have been successfully validated, and their integrated use has proved to be a highly recommendable and useful approach to test new airfoil geometries or airfoils featuring flow augmentation devices for VAWT applications.

Author Contributions: L.S., methodology, investigation, data curation, writing—original draft preparation; M.G.V., methodology, investigation; A.P., writing—original draft preparation; J.G.P., funding acquisition, resources; S.V.-S., supervision, resources; J.M.F.O., conceptualization, visualization, writing—review and editing, supervision. All authors have read and agreed to the published version of the manuscript.

Funding: This research was funded by the "Agencia Estatal de Investigación" (AEI) of the Spanish Ministry of Science and Innovation, in the context of the State Program to Promote Scientific-Technical Research and its Transfer, through the Project "Optimization through flow control techniques of a vertical axis wind turbine for urban environments" (ref. TED2021-131307B-I00), included in the NextGenerationEU funds of the European Community. Additionally, the support given by the University Institute for Industrial Technology of Asturias (IUTA) and the City Hall of Gijón, through the financed project SV-22-GIJON-1-04, is also recognized.

Acknowledgments: The authors would like to acknowledge the contribution of Katia Argüelles for her supervision and her administrative and technical support, as well as Bruno Pereiras for his advice and support during the measurement campaign.

Conflicts of Interest: The authors declare no conflict of interest. The funders had no role in the design of the study; in the collection, analyses, or interpretation of data; in the writing of the manuscript; or in the decision to publish the results.

References

1. IEA. *Net Zero by 2050—A Roadmap for the Global Energy Sector*; IEA: Paris, France, 2021.
2. Kumar, R.; Raahemifar, K.; Fung, A.S. A Critical Review of Vertical Axis Wind Turbines for Urban Applications. *Renew. Sustain. Energy Rev.* **2018**, *89*, 281–291. [CrossRef]
3. Hand, B.; Kelly, G.; Cashman, A. Aerodynamic Design and Performance Parameters of a Lift-Type Vertical Axis Wind Turbine: A Comprehensive Review. *Renew. Sustain. Energy Rev.* **2021**, *139*, 110699. [CrossRef]

4. Balduzzi, F.; Bianchini, A.; Carnevale, E.A.; Ferrari, L.; Magnani, S. Feasibility Analysis of a Darrieus Verti-cal-Axis Wind Turbine Installation in the Rooftop of a Building. *Appl. Energy* **2012**, *97*, 921–929. [CrossRef]
5. Du, L.; Ingram, G.; Dominy, R.G. A Review of H-Darrieus Wind Turbine Aerodynamic Research. *Proc. Inst. Mech. Eng. C J. Mech. Eng. Sci.* **2019**, *233*, 7590–7616. [CrossRef]
6. Barnes, A.; Marshall-Cross, D.; Hughes, B.R. Towards a Standard Approach for Future Vertical Axis Wind Tur-bine Aerodynamics Research and Development. *Renew. Sustain. Energy Rev.* **2021**, *148*, 111221. [CrossRef]
7. Micallef, D.; van Bussel, G. A Review of Urban Wind Energy Research: Aerodynamics and Other Challenges. *Energies* **2018**, *11*, 2204. [CrossRef]
8. Zhao, Z.; Wang, D.; Wang, T.; Shen, W.; Liu, H.; Chen, M. A Review: Approaches for Aerodynamic Performance Improvement of Lift-Type Vertical Axis Wind Turbine. *Sustain. Energy Technol. Assess.* **2022**, *49*, 101789. [CrossRef]
9. Howell, R.; Qin, N.; Edwards, J.; Durrani, N. Wind Tunnel and Numerical Study of a Small Vertical Axis Wind Turbine. *Renew. Energy* **2010**, *35*, 412–422. [CrossRef]
10. Selig, M.S.; McGranahan, B.D. Wind Tunnel Aerodynamic Tests of Six Airfoils for Use on Small Wind Turbines. *J. Sol. Energy Eng. Trans. ASME* **2004**, *126*, 986–1001. [CrossRef]
11. Wang, Z.; Zhuang, M. Leading-Edge Serrations for Performance Improvement on a Vertical-Axis Wind Turbine at Low Tip-Speed Ratios. *Appl. Energy* **2017**, *208*, 1184–1197. [CrossRef]
12. Kamliya Jawahar, H.; Alihan Showkat Ali, S.; Azarpeyvand, M.; Ilário da Silva, C.R. Aerodynamic and Aeroa-coustic Performance of High-Lift Airfoil Fitted with Slat Cove Fillers. *J. Sound Vib.* **2020**, *479*, 115347. [CrossRef]
13. Tropea, C.; Yarin, A.L.; Foss, J.F. *Handbook of Experimental Fluid Mechanics*; Springer: Berlin, Germany, 2007.
14. Tavakolpour-Saleh, A.R.; Setoodeh, A.R.; Gholamzadeh, M. A Novel Multi-Component Strain-Gauge External Balance for Wind Tunnel Tests: Simulation and Experiment. *Sens. Actuators A Phys.* **2016**, *247*, 172–186. [CrossRef]
15. Martínez-García, E.; Marigorta, E.B.; Gayo, J.P.; Navarro-Manso, A. Experimental Determination of the Resistance of a Single-Axis Solar Tracker to Torsional Galloping. *Struct. Eng. Mech.* **2021**, *78*, 519–528. [CrossRef]
16. Martínez-García, E.; Blanco-Marigorta, E.; Parrondo Gayo, J.; Navarro-Manso, A. Influence of Inertia and Aspect Ratio on the Torsional Galloping of Single-Axis Solar Trackers. *Eng. Struct.* **2021**, *243*, 112682. [CrossRef]
17. Spalart, P.R.; Allmaras, S.R. One-Equation Turbulence Model for Aerodynamic Flows. *Rech. Aerosp.* **1994**, 5–21. [CrossRef]
18. Menter, F.; Sechner, R.; Germany GmbH Matyushenko, A.A.; Petersburg, S. Best Practice: RANS Turbulence Modeling in Ansys CFD. 2019. Available online: https://fluidcodes.ir/rans-turbulence-modeling/ (accessed on 27 July 2022).
19. Menter, F.R.; Matyushenko, A.; Lechner, R. Development of a Generalized K-ω Two-Equation Turbulence Model. *Notes Numer. Fluid Mech. Multidiscip. Des.* **2020**, *142*, 101–109. [CrossRef]
20. Blevins, R.D. *Applied Fluid Dynamics Handbook*; Van Nostrand Reinhold Co.: New York, NY, USA, 1984.
21. Bradford Wick, B.H.; June, W. *Study of the Subsonic Forces and Moments on an Inclined Plate of Infinite Span*; UNT Libraries Government Documents Department: Denton, TX, USA, 1954.
22. Fage, A.; Johansen, F.C. On the Flow of Air behind an Inclined Flat Plate of Infinite Span. *Proc. R. Soc. Lond. Ser. A Contain. Pap. A Math. Phys. Character* **1927**, *116*, 170–197. [CrossRef]
23. Goldstein, S. (Ed.) *Modern Developments in Fluid Mechanics*; Oxford University Press: London, UK, 1938.
24. Claessens, M.C. The Design and Testing of Airfoils for Application in Small Vertical Axis Wind Turbines. Master's Thesis, Delft University, Delft, The Netherlands, 2006.
25. Pope, S.B. *Turbulent Flows*; Cambridge University Press: Cambridge, UK, 2000.

energies

Article

How to Improve an Offshore Wind Station

João Paulo N. Torres [1,2,*], Ana Sofia De Jesus [3] and Ricardo A. Marques Lameirinhas [1,3,*]

1 Instituto de Telecomunicações, 1049-001 Lisbon, Portugal
2 Academia Militar/CINAMIL, Av. Conde Castro Guimarães, 2720-113 Amadora, Portugal
3 Department of Electrical and Computer Engineering, Instituto Superior Técnico, 1049-001 Lisbon, Portugal; ist186940@tecnico.ulisboa.pt
* Correspondence: joaoptorres@hotmail.com (J.P.N.T.); ricardo.lameirinhas@tecnico.ulisboa.pt (R.A.M.L.)

Abstract: The ocean is approximately 71% of the Earth's surface and has a lot of resources available. Nowadays, human beings are looking for renewable ways to obtain energy. Offshore power can be obtained in several different ways. Offshore wind power is the most used renewable offshore energy. Since 2017, offshore wind power has a competitive price in comparison with conventional sources. In the 2010s, offshore wind power grew at over 30% per year. Although it has remained less than one percent of the overall world electricity generation, offshore wind power becomes quite relevant on the northern European countries from 2020. However, there are other ways to obtain energy offshore such as using tides and the sun. These types of farms are expensive and difficult to install and, therefore, we propose a combination of several renewable energies in one farm. The main ambition of this work is to try to reduce the installation and maintenance costs of the two types of offshore renewable energies by creating a structure capable of supporting the two types of turbines. To accomplish it, a theoretical study will be made, a brief state-of-the-art will be presented, the chosen items and the environment chosen for installation will be referred to, a prototype will be simulated using a multiphysics software and, finally, the results and conclusions will be presented, based on a Portuguese case study. How piezoelectric materials can enter offshore farms to increase efficiency is also referred to. The project proved to be possible of producing approximately 12.5 GWh of energy annually, more or less enough to supply 10 thousand homes. However, the installation of the piezoelectric materials did not prove to be viable as it is an expensive technology and does not produce a large amount of energy.

Keywords: co-located offshore renewable energies; piezoelectricity; tidal power; wind power

Citation: Torres, J.P.N.; De Jesus, A.S.; Marques Lameirinhas, R.A. How to Improve an Offshore Wind Station. *Energies* **2022**, *15*, 4873. https://doi.org/10.3390/en15134873

Academic Editors: Artur Bartosik and Dariusz Asendrych

Received: 7 June 2022
Accepted: 28 June 2022
Published: 2 July 2022

Publisher's Note: MDPI stays neutral with regard to jurisdictional claims in published maps and institutional affiliations.

Copyright: © 2022 by the authors. Licensee MDPI, Basel, Switzerland. This article is an open access article distributed under the terms and conditions of the Creative Commons Attribution (CC BY) license (https://creativecommons.org/licenses/by/4.0/).

1. Introduction

All over the world, nations are facing challenges in finding the best way to obtain energy. There are a global need for energy and mineral resources, which has been increasing year after year caused by the technological advances. The study reveals that the natural reserve of fossil fuels represents 150 years (coal), 58 years (natural gas) and almost 46 years (oil) of consumption at current rates. Fossil fuels are by definition a finite source (or at least, Humans consume that energy faster the it is renewed). Meanwhile, the pollution they cause (from climate-damaging greenhouse gases to health-endangering particles) has been setting negatively records, leading to dramatic consequences [1–3].

The world has an abundant source of natural, clean power derived from the wind, waves, tides, sun and others. Ocean, which covers approximately 71% of the Earth's surface, is seen as a huge opportunity. Offshore renewable are worldly recognised as reliable and stable electricity sources. Also, they are seen as support for water desalination and aquaculture [1–4].

The development of renewable technologies, associated to the global energy transition, promises to spur new industries, leading to jobs creation. Wind is the most explored and investigated offshore source. In 1991 was installed in Denmark the world's first offshore wind farm. Nowadays, millions of people consumes electricity from large-scale offshore wind farm projects [1–4].

The growth of European offshore wind, since the early 2000s, is supported by a steadily fallen over the years. In 2016, 10 European countries had already more than 12 MW grid connected production from 81 offshore wind farms, with a total of 3589 turbines installed. Nowadays, based on several signed commitments to reduce greenhouse gases emissions as well as the targets to reach carbon neutrality led to promising prediction regarding these renewable technologies [1–4].

It is estimated that the objective to have an installed capacity of at least 60 GW of offshore wind and at least 1 GW of ocean energy by 2030, with a view of reaching 300 GW and 40 GW of installed capacity, respectively, by 2050 is realistic and achievable. In addition to offshore wind energy, there are other ways to produce energy offshore [2,3].

Market forces, technological advances and price developments will continue to drive offshore renewable energy growth over the coming years. However, there are some obstacles. Offshore farms can be expensive and difficult to build and maintain. Therefore we have to take advantage and get the best out of this installation. Wind, sun and tides are unlimited resources that we can gather offshore. Therefore, the analysis and study of possible farms that can generate all of these types of energy together is an interesting and attractive research theme [2,3].

2. Overview and State-of-the-Art

Wind power or wind energy is an energy type that uses the kinetic energy of the wind and converts it to mechanical energy through wind turbines to turn electric generators for electrical power. The power content of the column of air is expressed by [2,3]:

$$P_{wind} = \frac{1}{2}\rho A v^3 \quad (1)$$

where ρ is the fluid density, A the cross-sectional area and v the fluid velocity. Figure 1 illustrates this.

Betz's law indicates the maximum power that can be extracted from the wind, independent of the design of a wind turbine in open flow.

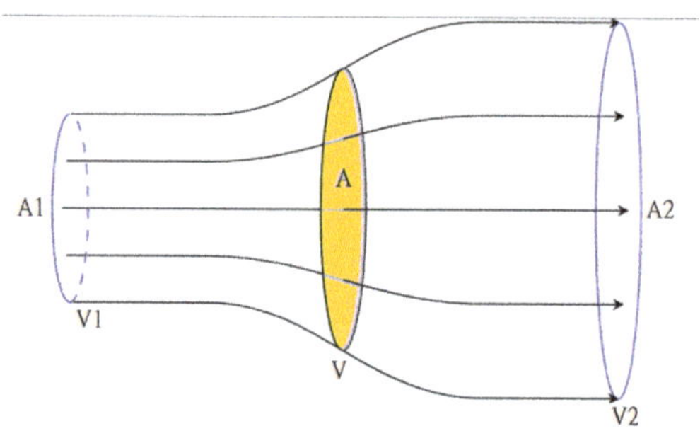

Figure 1. Schematic of fluid flow through a disk-shaped actuator. For a constant density fluid, the cross-sectional area varies inversely with speed.

The actual mechanical power P extracted by the rotor blades in watts is the difference between the upstream and the downstream wind powers:

$$P = \frac{1}{2}\rho A v(v_1^2 - v_2^2) \quad (2)$$

where $v = \frac{v_1+v_2}{2}$. We can calculate the power obtainable from a cylinder of fluid with cross-sectional area S and velocity v_1:

$$P = C_p \frac{1}{2} \rho A v_1^3 \tag{3}$$

where C_p is the power coefficient.

An offshore and an onshore wind turbine have similar technology. They diverge on the fact that offshore ones might produce more energy: higher wind speeds are available offshore comparing to on land, so offshore wind power's electricity generation is higher per amount of capacity installed.

Portugal only has 2 MW of offshore capacity in the floating wind turbine WindFloat near the Aguçadoura Wave Farm in Povoa de Varzim.

Tidal energy is power produced by the surge of ocean waters during the rise and fall of tides, where the intensity of the water from the rise and fall of tides is a form of kinetic energy.

Tidal stream generators draw energy from water currents in the same way wind turbines draw energy from air currents. The water density is about 800 times the density of air. This means that a single generator can provide significant power at low tidal flow velocities compared with similar wind speeds [5].

Piezoelectricity is the process of using crystals to convert mechanical energy into electrical energy, or vice versa, so there are two types of piezoelectric effect, direct piezoelectric effect and inverse piezoelectric effect. Since wind pressure is exerted on the wind turbine blades, the blades can be filled with piezoelectric materials to maximize the power obtained by wind force. When the wind impacts the piezoelectric panel, the change of pressure caused by the wind power is output as the voltage through the measuring meter and converted into the amount of energy [6].

The main objective of the co-location of offshore wind and tidal stream turbines is to reduce the cost of electricity generation from either technology separately [7–9].

Thus far, there is no yield that combines these two types of energy. However, there are some studies that simulate this installation.

A case-study site in the Pentland Firth that uses an eddy viscosity wake model in OpenWind to assess Wind energy, with a 3 MW rated power curve and thrust coefficient from a Vestas V90 turbine and to assess tidal energy, "is modeled using a semi-empirical superposition of self-similar plane wakes with a generic 1 MW rated power curve and thrust based on a full-scale, fixed-pitch turbine" [10].

The support structure loads due to wind, waves and current on a combined support structure featuring a single 3 MW wind turbine and 1 MW tidal turbine have been modeled for the same co-located farm case study of the Inner Sound of the Pentland Firth.

In this study, it has shown that the potential to share support structures by adding a tidal turbine to a wind turbine support looks promising.

This co-location results in a 70% increase in energy yield compared to operating the tidal turbines alone. It is found that, "within the space required around a single 3 MW wind turbine, co-location provides a 10–16% cost saving compared to operating the same size tidal-only array without a wind turbine. Furthermore, for the same cost of electricity, a co-located farm could generate 20% more yield than a tidal-only array" [7]. This also could help tidal stream technology move from being commercially uncompetitive alone to competitive when co-located with the wind [7,10].

There is other research that focuses on proposing and evaluating an optimized hybrid wind system and tidal turbines operating as a renewable energy generating unit in New Zealand.

It is known that using the capacities of wind and tidal power in renewable technologies would be a suitable alternative for fossil fuels and would help to prevent their detrimental effects on the environment. It is a cost-effective procedure for the power generation sector to maximize these renewables as a hybrid system.

This research indicates that Kaipara harbor has good potential for energy generatio from a hybrid system (wind plus tidal) with good wind energy yield and additional energy from tidal energy.

It is concluded that the installation of a hybrid system with wind and tidal energy in a certain place in New Zealand is beneficial. However, prototypes would have to be installed to complete this study [11].

Another study, which uses the coast of the United Kingdom, is considered. This co-location of wind and tidal power will offer synergies of shared infrastructures that will help to reduce both capital and operational costs. With this purpose, a simulation and modeling of the system in terms of generation, structural forces, integration to the grid and economics is performed.

Although having Levelized Cost of Energy (LCOE) higher than a normal offshore wind farm, the co-location of both technologies are demonstrating feasible. The advantages of the use of tidal are that we can produce a lot of reliable energy with smaller turbines and seize more energy from the same location [1,12].

This study proposes a single structured tower with hybrid renewable energy cultivation on the southwest coast of Yemen.

3. Environment Chosen for Installation

The proposed installation will be simulated in the atmospheric conditions of Povoa de Varzim in Viana do Castelo, where the WindFloat Atlantic project is installed. The average wind speed in this location is presented in Figure 2.

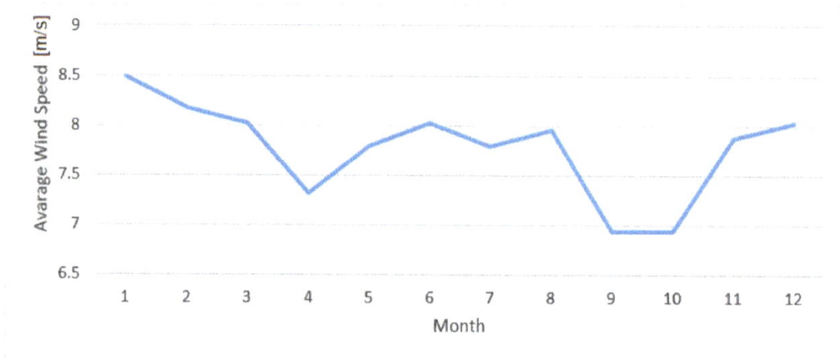

Figure 2. Wind Speed (m/s) Variability at 100 m high.

In the following figure, we can see how the average wind speed varies during a year at the height of 100 m. This information was provided by the Global Wind Atlas website.

The data collected is for a height of 100 m, and our rotor is about 190 m high. The following equation, the logarithmic profile law, can determine the wind speed at a certain height, which is used to compute the values on Table 1.

$$v_h = v_{an} \times \left| \frac{ln(Z_h/z_0)}{ln(Z_{an}/z_0)} \right| \quad (4)$$

where v_h is the e velocity at the desired height Z_h, v_{an} is the velocity at the anemometer height Z_{an} and z_0 is the surface roughness length, which is 0.005 m/s for the blown sea [13].

Unlike what happens with the wind, obtaining information about the currents' speed is very difficult. This information is scarce and hard to find. This information was provided by the Instituto Hidrografico (IH), which provided data on current speeds in 2015 at the chosen location.

Table 1. Wind Speed at 100 m and 190 m.

	Wind Speed at 100 m	Wind Speed at 190 m
Minimum	5.61 m/s	5.97 m/s
Average	7.78 m/s	8.28 m/s
Maximum	10.52 m/s	11.20 m/s

Only the maximum and average values, on Table 2 of the current speed will be considered as the minimum value is null.

Table 2. Current speed.

	Current Speed at 40 m Deep
Average	1.512 m/s
Maximum	1.920 m/s

4. Modelation

The wind turbines used in the simulation will be identical to those used in the WindFloat Atlantic project located in Povoa de Varzim because it is the only project of offshore wind energy in Portugal. The WindFloat Atlantic project has three V164-8.4 MW wind turbines mounted on the semi-submersible WindFloat platforms that are anchored to the seabed at a water depth of 100 m.

In the proposed project, we will use a turbine with the same specifications that an AR1500 tidal turbine. After analyzing the characteristics of PVDF and PZT, the piezoelectric materials chosen to apply in the structure were the PVDF. The PVDF model chosen was PROWAVE FS-2513P SENSOR, PIEZO FILM [14].

5. Calculations

To know the power generated by the turbines, we need to know the wind and the current speed, the surface area of the blades, the fluid density and the power coefficient, which varies with wind and current speed.

The equation that allows us to obtain the power has already been mentioned previously (3). For our wind speed values, results are presented on Table 3.

Table 3. Values of C_p according to chosen wind speeds.

Wind Speed (m/s)	Power Coefficient
5.97	0.41
8.28	0.44
11.20	0.44

The blade's surface area is given by πr^2, 21,124 m^2, where the radius of the surface is 82 m. The normally considered value of air density will be used in these calculations; 1.2225 Kg/m^3. The power estimated for different wind speeds is presented on Table 4.

Table 4. Values of Power according to wind speed.

Wind Speed (m/s)	Power (MW)
5.97	1.129
8.28	3.232
11.20	7.970

Calculating the energy produced during a year, a maintenance period of 15 days was considered when the turbines are not working. At rated power (8 MW), the energy produced would be 16,800 MWh a year. At average power (3.232 MW), the energy produced would be 11,946 MWh a year.

To know the power generated by the tidal turbines we use the same equation but with different data.

The power coefficient for the chosen tidal turbine is not known. Through a study of some different tidal turbines, it was estimated that the power coefficient would be approximately 0.4, and that is the value used in the calculations.

Tidal turbines are also significantly smaller than wind turbines, and because of that the blade's surface area is also smaller, with a radius of 9 m and a surface area of 254,469 m^2

For tidal turbines, the fluid considered is water, and the fluid density is 999 kg/m^3 The power estimated for different wind speeds is on Table 5.

Table 5. Values of power according to current speed.

Current Speed (m/s)	Power (kW)
1.512	175.746
1.902	349.834

Calculating the energy produced during a year, a maintenance period of 15 days was considered when the turbines are not working. At rated power (1.5 MW), the energy produced would be 5040 MWh a year. At average power (175.746 kW), the energy produced would be 591 MWh a year.

5.1. Simulations

To simulate the proposed structure, where tidal and wind turbines will be installed as well as piezoelectric materials, it is necessary to simulate some conditions such as wind flow and pressure around the structure.

A 3D model of the structure was constructed using Geometry tool provided by the software

To make the process easier, the structure was built in parts, and then all the parts were joined as presented on Figure 3. The wind blade has been modeled in 3D and is only a geometric approximation of a real blade due to 3D design software limitations. The blade is 82 m long (actual length of the model Vestas V164-8.0). The tidal turbine was designed using the same base as the wind turbine blade. The tidal turbine is significantly shorter with each blade just 9 m long, but it is wider and stronger. The floating platform used in the WindFloat project was also designed. Each pole is 30 m high, being 50 m apart. The main tower is 190 m high and joins all of the designed components.

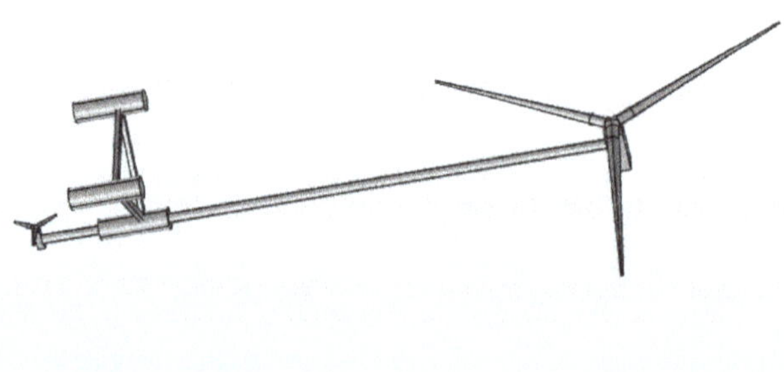

Figure 3. Structure drawn using the software.

In order to proceed with the simulations, it is necessary to attribute materials to the designed structures. The structure is made of different materials, including being coated by anti-corrosion materials due to the corrosive effect of water and rain, but it is mainly composed of iron, so the material chosen for all of the structure was iron.

5.2. Wind Tunnel Simulation

The wind flow around the structure will be simulated using the software turbulent flow. This module aims to solve the Navier–Stokes equations presented in (5) by modeling the fluid domain around the airflow as a mesh of discrete elements. The set of Equations (5) counts with the fluid density ρ, the fluid velocity $\mathbf{u}$, the fluid pressure p and the fluid dynamic viscosity μ. The finer the mesh, the more precise the results are. However, it is always a trade off, as thinner mesh elements imply more computation time. Due to this, the smallest geometry details were removed using the "Cleanup" tool. The equations will be evaluated using a computational assessment technique. Up-to-date, there are three different techniques: Direct Numerical Simulation (DNS), which solves all the eddies, from the largest to the smallest, Large Eddy Simulation (LES), where only the large-scale eddies are resolved and Reynolds-Averaged Navier–Stokes (RANS), a completely different time-averaged method that does not resolve eddies explicitly, choosing to instead model its effect using the concept of turbulent viscosity. RANS is not an explicit method and, therefore, is less computationally expensive, with that being the primary reason for its use in this work [13].

The Navier–Stokes are moment0conservation equations, relating the inertial force $\rho\left(\frac{\partial \mathbf{u}}{\partial t} + \mathbf{u}\cdot\nabla\mathbf{u}\right)$, with the pressure force $-\nabla p$, the viscous forces $\nabla\cdot\left[\mu(\nabla\mathbf{u} + (\nabla\mathbf{u})^T - \frac{2}{3}(\nabla\cdot\mathbf{u})\mathbf{I})\right]$ as well as with an external force $\mathbf{F}$.

$$\begin{cases} \rho\left(\frac{\partial \mathbf{u}}{\partial t} + \mathbf{u}\cdot\nabla\mathbf{u}\right) = -\nabla p + \nabla\cdot\left[\mu(\nabla\mathbf{u} + (\nabla\mathbf{u})^T - \frac{2}{3}(\nabla\cdot\mathbf{u})\mathbf{I})\right] + \mathbf{F} \\ \frac{\partial \rho}{\partial t} + \nabla\cdot(\rho\mathbf{u}) = 0 \\ \rho = \rho(p,T) \end{cases} \tag{5}$$

After choosing the assessment technique or turbulence model type, a turbulence model needs to be chosen accordingly. For the RANS, the software makes nine different turbulence models available (which can be consulted at the software library). However, for the case study, the k-ϵ turbulence model, where k refers to turbulent kinetic energy and ϵ the rate of dissipation on turbulent kinetic energy, is selected. Of the reasons behind its election, its good performance for complex geometries, its stability and the possibility to use wall functions stood out. Wall functions are adopted to resolve the thin boundary layer near the wall, preventing the use of a very fine mesh. Essentially, they provide an offset so that the mesh does not need to go near the wall; moreover, being the straightforward solution, a relationship is used to characterize the flow [13].

An incompressible flow approximation is used, assuming the fluid's density as constant, implying that the divergence of the fluid velocity is zero, as suggested in Expression (6). The wind tunnel dimensions are 295 m in height, 200 m in width and 100 m in depth. The above-mentioned structure is placed inside the tunnel. Only the parts of the structure that are inside this wind tunnel are subject to the wind, as illustrated on Figure 4.

$$\nabla\cdot\mathbf{u} = 0 \Rightarrow -\frac{2}{3}(\nabla\cdot\mathbf{u}) = 0. \tag{6}$$

The main goal of the following simulation is to calculate the power generated by the wind turbine by Betz's law and through the equation described above.

As altitude increases, atmospheric pressure decreases. As altitude increases, the amount of gas molecules in the air decreases, and the air becomes less dense than air nearer to sea level. One can calculate the atmospheric pressure at a given altitude. Temperature and humidity also affect the atmospheric pressure. Pressure is proportional to temperature and inversely proportional to humidity. However, at the height of 190 m, this variation in air pressure is not significant. Therefore, the reference air pressure and temperature were considered, as they are values that do not significantly affect the results. Regarding the wind speed, three different simulations were performed, considering the minimum, average and maximum speed, as in Figure 5. This is the input inserted at the open boundary.

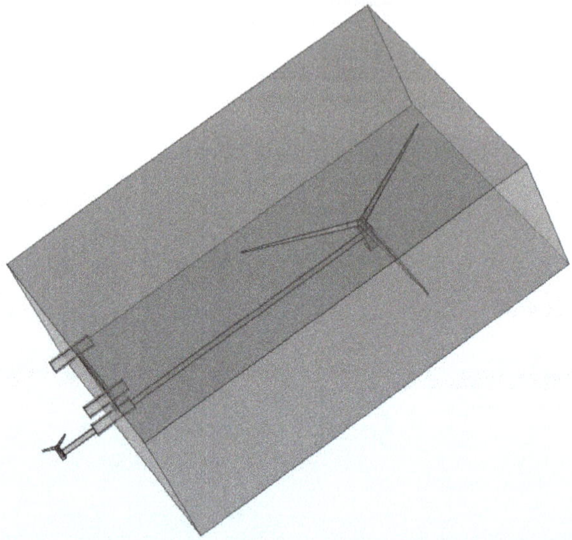

Figure 4. Wind tunnel using the software.

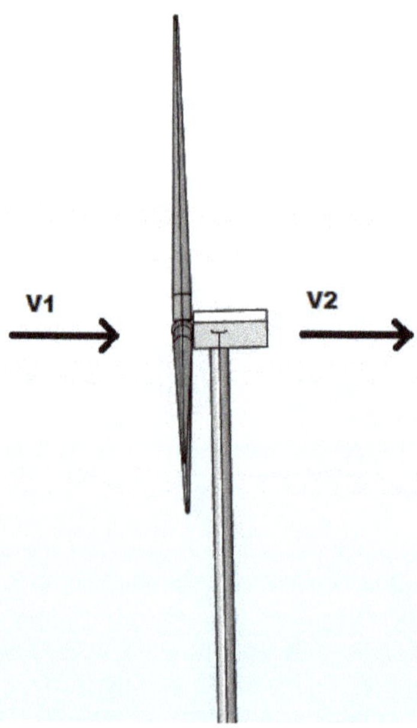

Figure 5. Betz Law.

In Figure 6, the resulting wind flow is illustrated, being characterized by velocity magnitude, orientation and direction. The velocity magnitude's values are differentiated by colors whose legend can be seen at the right border. This figure is for an input speed of 8.28 m/s, but several simulations were performed for different wind speeds.

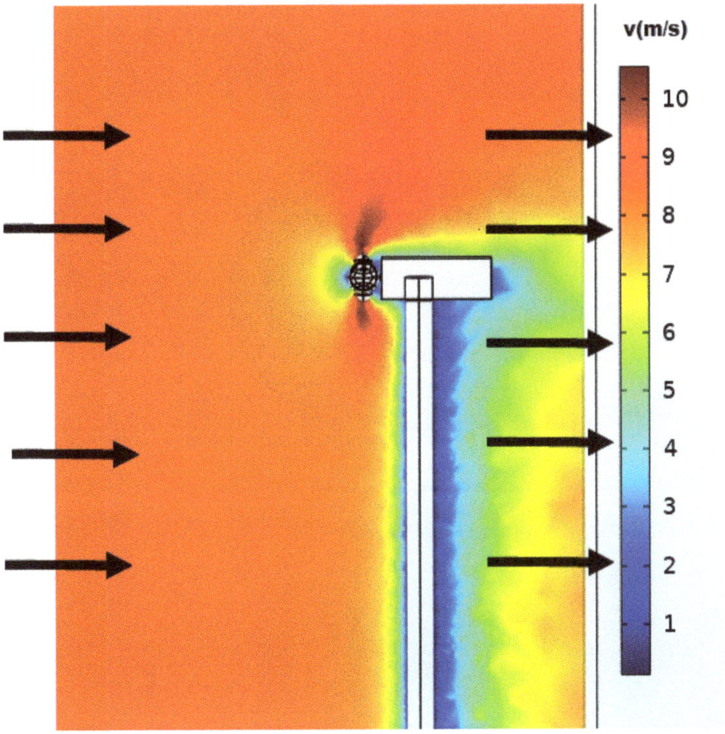

Figure 6. Wind speed using the software.

As we have seen before, the main goal is to take v_2. The values taken from the performed simulations are shown in Table 6.

Table 6. Values of wind speed before and after the rotor.

Wind Speed v_1 (m/s)	Wind Speed v_2 (m/s)
5.97	4.2
8.28	5.75
11.20	7.78

5.3. Water Tunnel Simulation

The water flow around the structure will be simulated using the software's turbulent flow. As mentioned before, this module aims to solve the Navier–Stokes equations but instead of air, we are using water.

The water tunnel dimensions are 70 m in height, 200 m in width and 100 m in depth. The above-mentioned structure is placed inside the tunnel. Only the part of the structure that is subject to the water is inside this water tunnel, as presented in Figure 7.

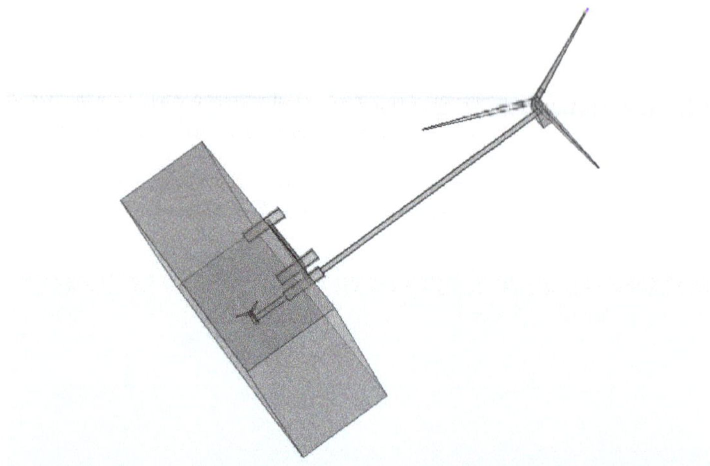

Figure 7. Water tunnel using the software.

The concept of the following simulation is the same as the wind turbine simulation. The goal is to obtain v_2 to calculate the power generated by the tidal turbine, as presented in Figure 8.

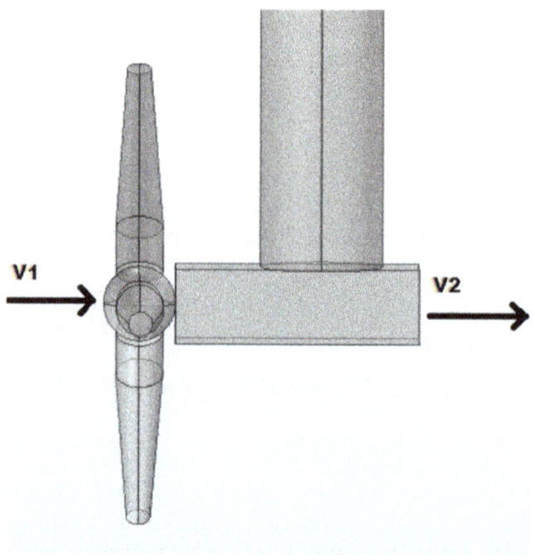

Figure 8. Betz Law.

As depth increases, atmospheric pressure also increases. The deeper you go under the sea, the greater the pressure of the water pushing down on you. For every 33 feet (10.06 m) you go down the pressure increases by one atmosphere. The atmospheric pressure at a depth of 40 m is about 5 atm. The water temperature considered will be 10 °C, as it is the average temperature of the water in the chosen place. However, the temperature does not have a considerable effect on the simulation. The average and maximum current speeds were chosen as inputs at the open boundary. Results are on Figure 9.

In the figure above, the resulting water flow is illustrated, being characterized by velocity magnitude, orientation and direction. The velocity magnitude's values are differentiated by colors whose legend can be seen at the right border. This figure is for an input speed of 1.902 m/s, but several simulations were performed for different current speeds as presented on Table 7.

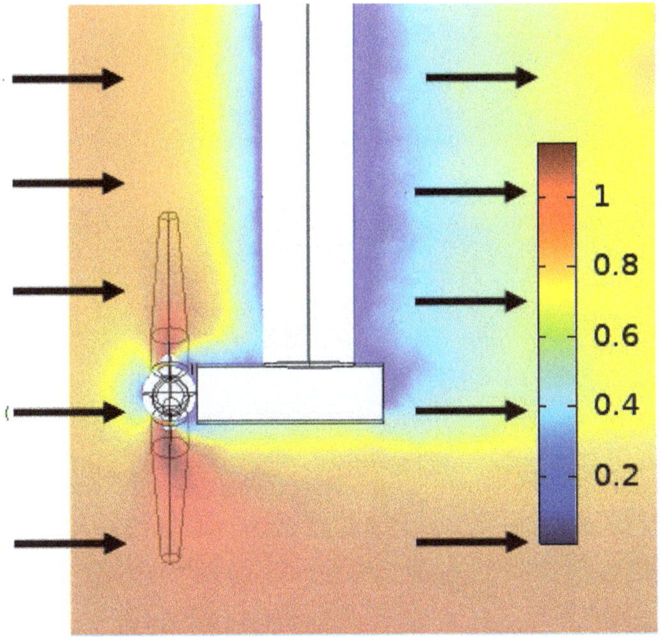

Figure 9. Current Speed (m/s) using the software.

Table 7. Values of current speed before and after the rotor.

Current Speed v_1 (m/s)	Current Speed v_2 (m/s)
1.512	1.12
1.902	1.35

6. Piezoelectric Materials Simulation

The piezoelectric effect will be simulated using the software. The piezoelectricity interface combines the solid mechanics and electrostatics interfaces with the constitutive relationships required to model the piezoelectric phenomena. Both direct and inverse piezoelectric effects can be modeled.

The piezoelectric coupling can be formulated using either the strain–charge or stress–charge form. There are three modules within the structural mechanics and acoustics modules branch that offer this feature for simulating piezoelectricity: the acoustics module, MEMS (microelectromechanical systems) and module and structural mechanics module.

The acoustics module includes dedicated tools for modeling wave generation and propagation in fluids, linear elastic materials, porous media and piezoelectric materials. It is used for piezoelectric transducers as transmitters to radiate sound to the surrounding fluids and as receivers to detect sound coming from the surrounding fluids.

The MEMS module includes a terminal feature that allows you to connect a piezoelectric device to an electrical circuit. The electrical circuit can be used to excite the transducer as well as receive detected signals. The terminal feature also enables the computation of the lumped parameters of the piezoelectric device, such as admittance and scattering parameters (S-parameters).

The structural mechanics module provides efficient modeling features such as the shell and membrane interfaces [15].

The simulations will be run by the last module with a geometry as illustrated on Figure 10. This is a finite element analysis (FEA) software package tailored for analyzing the mechanical behavior of solid structures. The structural mechanics module brings built-in multiphysics couplings that include thermal stress, fluid–structure interaction and piezoelectricity [15].

As previously mentioned, the piezoelectric material used is a polyvinylidene difluoride (PVDF) film with an external depth of 13 mm and an external width of 25 mm.

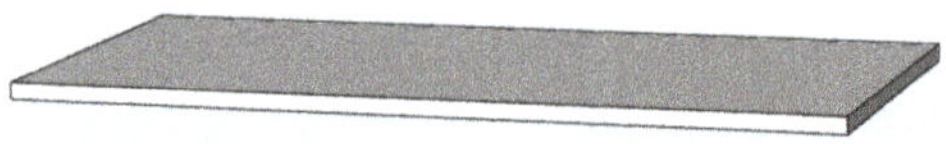

Figure 10. PVDF film drawn using the software.

Before any simulation is performed on the piezoelectric material, it is necessary to know the pressure that the wind exerts in the area where the piezoelectric materials will be applied.

The piezoelectric materials will be installed in the area shown in Figure 11.

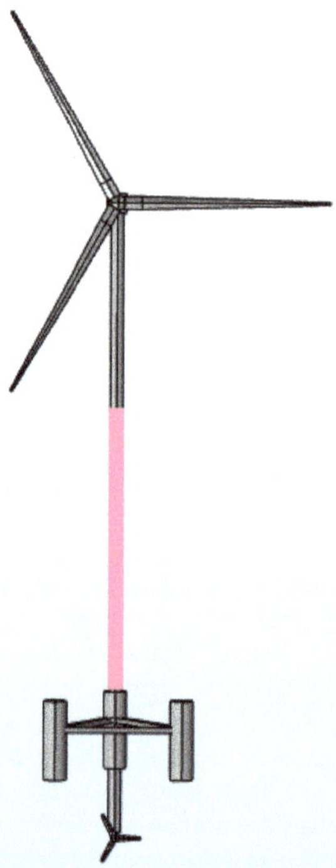

Figure 11. Area where piezoelectric materials will be applied.

The piezoelectric materials will be applied in the pink area shown in the figure because they receive the pressure of the wind, and their operation is not significantly affected by the wind turbines. This area is approximately 108 × 2.62 m.

The wind was calculated at the average height of 54 m using the formula mentioned before in Expression (3). Results are compiled on Table 8.

Table 8. Values of wind speed at 54 m.

Wind Speed at 100 m (m/s)	Wind Speed at 54 m (m/s)
5.61	5.26
7.78	7.30
10.52	9.87

Several simulations were performed to find out the pressure exerted by the wind on the structure. The pressure exerted on the tower with a wind speed of 7.30 m/s is shown on Figure 12 and all the results are compiled in Table 9.

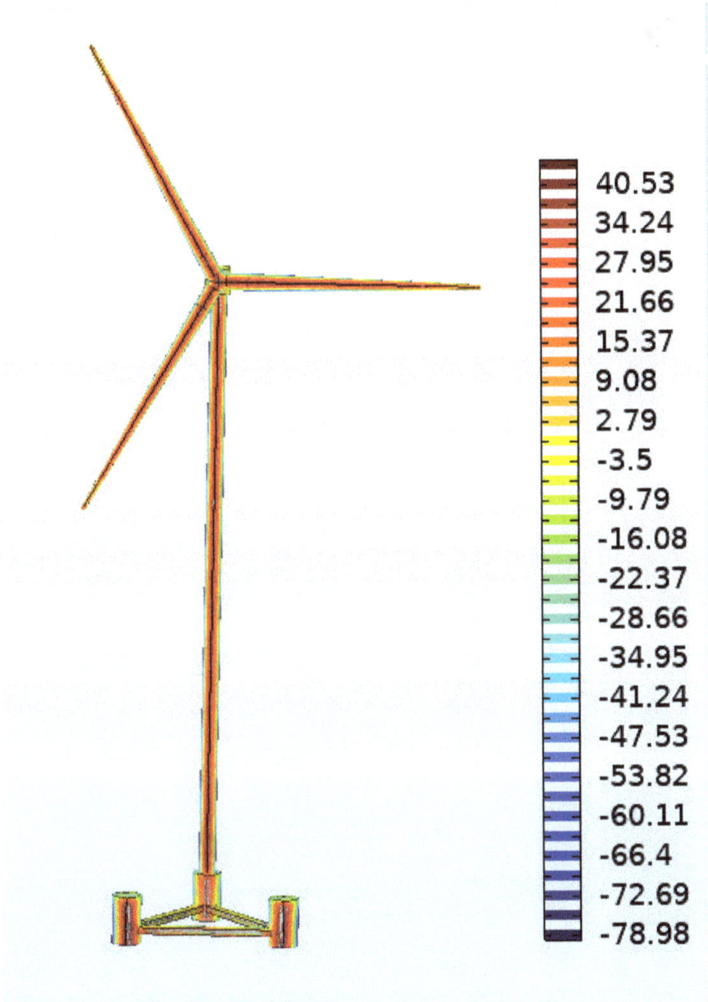

Figure 12. Pressure (Pa) using the software.

Table 9. Values of pressure according to wind speed.

Wind Speed at 54 m (m/s)	Pressure (Pa)
5.26	20.010
7.30	38.546
9.87	70.478

Knowing the pressure that the wind exerts on the structure, we were able to extract the current density and voltage of the piezoelectric material, presented on Table 10. Therefore the simulation input is the pressure that we received from the previous simulations.

Table 10. Values of voltage and current.

Pressure (Pa)	Voltage (V)	Current Density Norm (A/m^2)
20.010	1.412×10^{-3}	6.54×10^{-5}
38.546	1.520×10^{-3}	1.03×10^{-4}
70.478	2.781×10^{-3}	2.35×10^{-4}

Figure 13 shows the behavior of the piezoelectric material when the wind pressure is exerted.

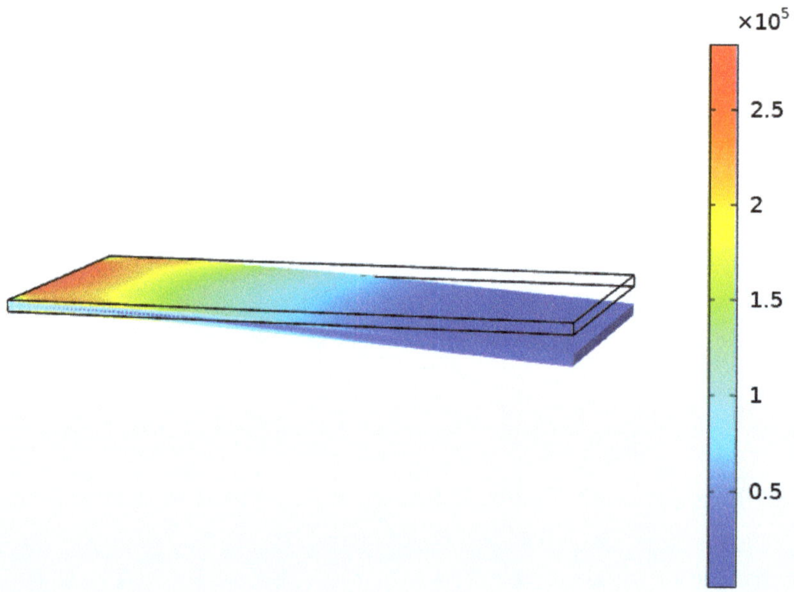

Figure 13. PVDF film stress using the software.

7. Results

7.1. Wind Turbines

The simulation's main goal is to carry out a comparison between the results calculated theoretically and the results calculated through the simulations performed.

After several simulations were made, where the main goal was to calculate the wind speed at the exit of the rotor, the calculations were carried out to calculate the power generated by the wind turbine.

Through the theoretical equations and with the data taken from the simulations performed, it is possible to calculate the power generated through Equation (2).

To calculate the power generated by the wind turbine, it is necessary to use the following data: $p = 1.225\ \mathrm{kg/m^3}$ and A = 21,124 m^2.

Using the data from Table 11, the generated power was calculated. A wind speed of 5.97 m/s corresponds to a power of 1.184 MW, a wind speed of 8.28 m/s corresponds to a power of 3.222 MW, and a wind speed of 11.20 m/s corresponds to a power of 7.970 MW.

Table 11. Values of wind speed before and after the rotor.

Wind Speed v_1 (m/s)	Wind Speed v_2 (m/s)
5.97	4.2
8.28	3.232
11.20	7.78

To find the power coefficient C_p at a given wind speed, all you have to do is divide the power produced by the total power available in the wind at that speed. Thus, through the simulations and the generated power calculations, we were able to calculate the power coefficient. A wind speed of 5.97 m/s corresponds to a power coefficient of 0.43, a wind speed of 8.28 m/s corresponds to a power coefficient of 0.44, and a wind speed of 11.20 m/s corresponds to a power coefficient of 0.44.

Calculating the energy produced during a year, a maintenance period of 15 days was considered when the turbines are not working. At average power (3.222 MW), the energy produced would be 11909 MWh.

Table 12 shows a comparison between the results obtained theoretically and the results obtained through simulations.

Table 12. Comparison of theoretical and simulated results.

Wind Speed v_1 (m/s)	Power Generated (MW) (Theoretical Calculation)	C_p (Theoretical)
5.97	1.129	0.41
8.28	3.232	0.44
11.20	7.453	0.41
Wind Speed v_1 (m/s)	**Power Generated (MW) (trough Simulation)**	**C_p (trough Simulation)**
5.97	1.184	0.42
8.28	3.222	0.44
11.20	7.970	0.44

Analyzing the results obtained, we can see that the theoretical results and those obtained through the simulations are quite similar, which allows us to say that the designed structure has a very close approximation to the real one.

There are several factors that influence the efficiency of a wind turbine. Many aerodynamic factors affect wind turbine power generation, such as wind speed, air density, temperature, air pressure, area swept, and height, etc.

The typical cut-in, rated and cut-out wind-speed values are in the range of 3–5, 10–15 and 25–30 m/s, respectively. The wind turbine chosen has a cut-in wind speed of 4 m/s, a rated wind speed of 13 m/s and a cut-out wind speed of 25 m/s. For the wind speed below the cut-in value, the turbine will produce worthless power. When this happens, the turbines usually enter a parking mode. The turbine is also shut down and kept in parking mode when wind speed is above the cut-out value or during emergency conditions due to security. For wind-speed values between the cut-in and rated, the power P curve maintains a cubic relationship with respect to wind speed [16]. Figure 14 presents a power curve of a Vestas V164-8.0 turbine.

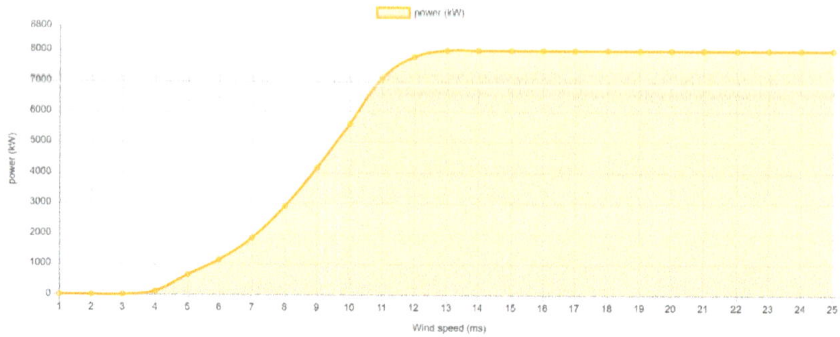

Figure 14. Power Curve of Vestas V164-8.0 (from [17]).

Therefore, the location chosen for the installation of these turbines is a good location since the wind speeds practiced in this location are suitable for a good operation of the turbines.

The wind turbine's characteristics are usually related to the degree of air density. The energy produced by the wind is directly commensurate to the degree of air density [16].

Air pressure and temperature affect the air density; they are directly proportional to the pressure and inversely proportional to the temperature. The scale of the pressure and temperature decreases with increasing elevation [16]. However, we saw that the air density does not have a significant change due to temperature and pressure and, therefore, in the simulations, the value of air density was always the same, 1.225 kg/m^3.

It is also concluded that the produced power is directly proportional to the rotor-swept area. When the swept area and the diameter of the rotor are large, an increase in energy produced will be earned from the wind.

The tower height is another important factor in the power generated by the turbine. The energy available in the wind is proportional to the cube of the wind speed so any small increase in wind speed will result in an impact on the economic factor. An efficient method to get the turbine in stronger winds is to mount them on taller towers [16].

Near the Earth's surface, friction reduces the wind speed. Frictionless surfaces, such as a quiet sea, offer small amounts of resistance, so the difference in the wind speed with the height is not high. On the other hand, wind undergoes a major change by irregular surfaces such as forests and buildings [16]. Obstacles such as structures and trees can significantly affect wind speed. They often create turbulence in their neighborhood. The slowdown effect on the wind from an obstacle increases with the height and length of the obstacle. This effect is more pronounced close to the obstacle and close to the ground. It is a good thing to have few major obstacles close to wind turbines, especially in the case they are upwind in the prevailing wind direction, i.e., "in front of" the turbine [16]. Thus, a similar turbine installed at sea or on land with buildings or trees around it will have different efficiency.

The results obtained for these turbines were as expected. Since these turbines are installed in the chosen location, the calculations and results obtained are only a confirmation of the results of the operation of these turbines.

Data provided to the public by EDP (Energias de Portugal) say that a turbine installed at the chosen location generates around 25 GWh during a year. Through our calculations this result is a little lower, about 12 GWh a year. This difference is due to the fact that we do not use results from real wind speeds and only approximations. The wind speeds used are annual averages of the wind speeds practiced annually and, therefore, there are wind speeds that we were not able to consider in the calculations performed.

7.2. Tidal Turbines

As with the simulations for the wind turbine, the main goal was to calculate the wind speed at the exit of the rotor to calculate the power generated by the currents.

The power equation is the same as for the wind turbine, only changing the data: $p = 999$ kg/m^3 and $A = 254.469$ m^2.

Using the data from Table 13, the generated power was calculated. A current speed of 0.912 corresponds to a power of 38.492 kW, and a current speed of 1.902 m/s corresponds to a power of 371.006 kW.

Table 13. Values of current speed before and after the rotor.

Wind Speed v_1 (m/s)	Wind Speed v_2 (m/s)
1.512	1.12
1.902	1.35

To find the power coefficient C_p at a given current speed, all you have to do is divide the power produced by the total power available in the current at that speed. Thus, through the simulations and the generated power calculations, we were able to calculate the power coefficient. A current speed of 1.512 corresponds to a power coefficient of 0.40 and a current speed of 1.902 m/s corresponds to a power coefficient of 0.42.

Calculating the energy produced during a year, a maintenance period of 15 days was considered when the turbines are not working. At average power (1.512 m/s), the energy produced would be 575.744 MWh.

Table 14 shows a comparison between the results obtained theoretically and the results obtained through simulations.

Table 14. Comparison of theoretical and simulated results. * This value of Cp is just an approximation due to lack of data.

Current Speed v_1 (m/s)	Power Generated (kW) (Theoretical Calculation)	C_p * (Theoretical)
1.512	175.746	0.40
1.902	349.834	0.40
Current Speed v_1 (m/s)	**Power Generated (kW) (trough Simulation)**	**C_p (trough Simulation)**
1.512	172.583	0.39
1.902	371.006	0.42

Analyzing the results obtained, we can see that the theoretical results and those obtained through the simulations are quite similar, like the wind turbine, which allows us to say that the designed structure has a very close approximation to the real one.

Unlike wind turbines, tidal turbines do not have much information. However, we know that its efficiency is also affected by some factors such as current speed, water density, temperature, water pressure, area swept and height, etc.

The tidal turbine chosen has a cut-in current speed of less than 1 m/s, a rated current speed of 3 m/s and a cut-out current speed of 5 m/s. Like the wind turbine, in a current speed below the cut-in value, the turbine will produce negligible power and is usually shut down and entered into parking mode. The turbine will also shut down and be kept in parking mode when current speed is above the cut-out value or during emergency conditions due to security. For current speed values between the cut-in and rated, the power P curve maintains a cubic relationship with respect to the current speed. In Figure 15 is a power curve of a AR1500 tidal turbine.

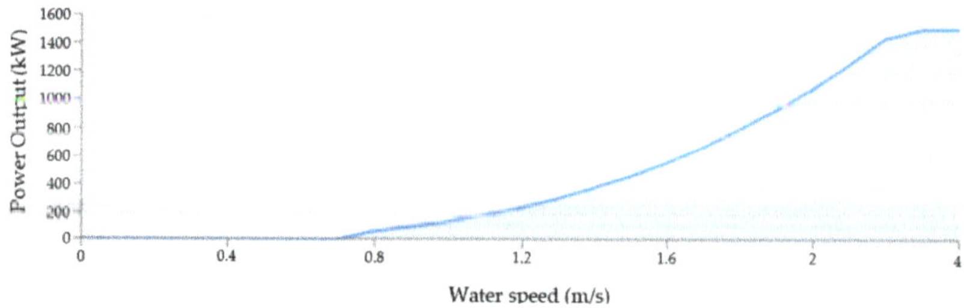

Figure 15. Power Curve of AR1500 (from [18]).

The location chosen for the installation of these turbines is not the best since the current speeds practiced in this location are usually lower than the rated current speed. These turbines are normally placed in places where the currents are stronger. However, the current speed in this location is enough for the turbines to properly work.

The energy produced by the current is directly commensurate to the degree of water density. The water density considered was 999 kg/m^3. The density does increase with depth, but only to a tiny extent. At the bottom of the deepest ocean, the density is only increased by about 5%, so the change can be ignored in most situations.

It is also concluded that the produced power is directly proportional to the rotor-swept area, just like the wind turbine. When the swept area and the diameter of the rotor are large, an increase in energy produced will be earned from the current.

The current speed changes with depth. Up to a 50 m depth, the current is influenced by the wind. The current speed is normally greatest near the surface.

The tidal turbine's operation is also influenced by the presence of obstacles, and the current speed will be less in the presence of them. However, at sea, the presence of obstacles that influence the operation of the turbines is less likely.

Unlike wind turbines, there is no data on the operation of these tidal turbines in the chosen location. Thus, the results obtained cannot be compared; they can only be analyzed.

7.3. Piezoelectric Materials

The purpose of the use of piezoelectric materials is different from the use of wind and tidal turbines. These objects produce a small amount of energy compared to the turbines. They are used to take advantage of the space that exists on the tower and to make the best use of the wind.

Knowing that the available area is approximately 108 × 2.62 m for the application of piezoelectric materials and considering that each piezoelectric has a height of 0.2 mm and a width of 13 mm, we can install a total of 540,200 materials in rows of 200 materials. The materials in each row are connected together in series, and the rows are connected in parallel. This allows that if a material goes bad, only the queue with the material that stops working is affected, making it affect the system as little as possible.

The power generated by piezoelectric materials is only an approximation since these materials produce energy due to pressure fluctuations exerted on them, and these fluctuations are not possible to simulate with the software used. Since this simulation is not possible, this calculation was performed using an approximation. The pressure exerted on the material during 1 s was calculated, and it was assumed that this pressure was the same exerted during the 24 h of a day.

The power calculation is performed through the results taken from the simulation (voltage and current density) and multiplied by the area where the wind pressure is exerted.

After performing these calculations, it is estimated that a generated power of 20 W and daily energy of 480 Wh and 168 kWh a year.

It was seen that the reduced generated power was capable of feeding a lighting signal for a turbine, but the flags found have powers between 30 W and 60 W.

The calculation of the power generated is a calculation based only on the average wind speed occurring at the chosen location and, therefore, it is possible to generate higher power and, in turn, feed an LED light (30 W).

8. Conclusions

The main goal of this research work is to understand if there are ways to produce more energy in an offshore wind power station. For this, we studied the possibility of co-localized wind and tidal energy. The possibility of placing piezoelectric materials in the structure was also studied. The costs of this type of technology are quite high and, therefore, the possibility of joining several technologies is a way to reduce the costs of installing these technologies separately and, consequently, we produce more energy. To achieve that, an intensive study of all technologies was made.

First, a viable location was chosen for the installation of the project. There is an offshore wind station on Póvoa de Varzim, and because of that, the location chosen was there. Then, the atmospheric conditions were studied. Subsequently, the necessary items for the proposed infrastructure were chosen. The proposed structure was designed in the software used. The structure designed and simulated to study the co-location of the two types of energy was composed of two types of turbines: AR 1500 Tidal turbine and Vestas V164-8.0 Wind Turbine. Finally, the results obtained after some simulations using the software were discussed.

Through the simulations, we managed to produce a high amount of energy through the installed turbines. As it was calculated, the wind turbine produces a greater amount of energy than the tidal turbine. This is due to the fact that the size of the wind turbine is relatively larger because, for the same turbine dimensions, the tidal turbine can produce a significantly greater amount of energy.

The project proved that it is possible to produce approximately 12.5 GWh of energy annually, more or less enough to supply 10 thousand homes. However, the installation of piezoelectric materials did not prove to be viable as it is an expensive technology and does not produce a large amount of energy. Furthermore, these materials would be exposed to critical atmospheric conditions, which would lead to regular maintenance that is not justifiable.

Green energies are quite expensive energies, and offshore energy is no exception. The biggest disadvantage of this project is undoubtedly the high costs of all components as well as installation and maintenance. The fact that we only performed the simulation for a wind turbine and a tidal turbine helps to increase the project costs. An offshore wind farm pays off if the number of installed turbines is higher because it greatly increases the energy produced, and costs do not increase as much with the increase in turbines.

In conclusion, the project presented is very expensive but a very promising project due to the fact that the technologies used are innovative and produce a large amount of clean energy.

In this study, it was assumed that the platform of the WindFloat project also supports a tidal turbine. The tidal turbine used is a turbine that is supposed to be on the bottom, supported by a structure of its own. A study on whether it would be possible to adapt this turbine to the proposed project and whether the floating platform would be able to support the weight of this turbine will have to be carried out.

A major problem with these facilities is the environmental impact. A study on the impact that the project would have at an environmental level also needs to be carried out.

As for piezoelectric materials, a study on these is also effective. This technology will be a good solution if a piezoelectric material is found that is capable of withstanding difficult atmospheric conditions and capable of being more efficient than the materials proposed in this project.

Lastly and most importantly, a detailed economic analysis would have to be carried out. Most of these projects are quite expensive and do not have detailed budgets available, which leads to great difficulty in projecting a correct budget.

Author Contributions: Conceptualisation: J.P.N.T. and R.A.M.L.; Software: A.S.D.J.; Methodology J.P.N.T. and A.S.D.J.; Investigation: J.P.N.T., A.S.D.J. and R.A.M.L.; Formal analysis: J.P.N.T.; writing A.S.D.J. All authors have read and agreed to the published version of the manuscript.

Funding: This work was supported in part by FCT/MCTES through national funds and in part by cofounded EU funds under Project UIDB/50008/2020. This work was also supported by FCT under the research grant UI/BD/151091/2021.

Institutional Review Board Statement: Not applicable.

Informed Consent Statement: Not applicable.

Data Availability Statement: Not applicable.

Conflicts of Interest: The authors declare no conflict of interest.

References

1. Marques Lameirinhas, R.A.; Torres, J.P.N.; de Melo Cunha, J.P. A Photovoltaic Technology Review: History, Fundamentals and Applications. *Energies* **2022**, *15*, 1823. [CrossRef]
2. Ragheb, M.; Ragheb, A.M. Wind Turbines Theory—The Betz Equation and Optimal Rotor Tip Speed Ratio. In *Fundamental and Advanced Topics in Wind Power*; Carriveau, R., Ed.; IntechOpen: Rijeka, Croatia, 2011; Chapter 2. [CrossRef]
3. Ragheb, M.; Ragheb, A.M. *Wind Energy Convertion Theory, Betz Equation*; Wind Energie: Berlin, Germany, 2017.
4. dos Santos Castilho, C.; Torres, J.P.N.; Ferreira Fernandes, C.A.; Marques Lameirinhas, R.A. Study on the Implementation of a Solar Photovoltaic System with Self-Consumption in an Educational Building. *Energies* **2021**, *14*, 2214. [CrossRef]
5. Nachtane, M.; Tarfaoui, M.; Goda, I.; Rouway, M. A review on the technologies, design considerations and numerical models of tidal current turbines. *Renew. Energy* **2020**, *157*, 1274–1288. [CrossRef]
6. Caliò, R.; Rongala, U.B.; Camboni, D.; Milazzo, M.; Stefanini, C.; De Petris, G.; Oddo, C.M. Piezoelectric Energy Harvesting Solutions. *Sensors* **2014**, *14*, 4755–4790. [CrossRef]
7. Lande-Sudall, D.; Stallard, T.; Stansby, P. Co-located offshore wind and tidal stream turbines: Assessment of energy yield and loading. *Renew. Energy* **2018**, *118*, 627–643. [CrossRef]
8. Alfakih, S.; De, T.; Ali, S.; Aneeq, A.; Hayat, K. *Simulation Model of Single Structured Tower Hybrid Wind and Tidal Energy Cultivation Based on Yemen's South West Coast*; EDP Sciences: Les Ulis, France, 2019; Volume 107. [CrossRef]
9. Lande-Sudall, D.; Stallard, T.; Stansby, P. Co-located deployment of offshore wind turbines with tidal stream turbine arrays for improved cost of electricity generation. *Renew. Sustain. Energy Rev.* **2019**, *104*, 492–503. [CrossRef]
10. Lande-Sudall, D. Co-Located Offshore Wind and Tidal Stream Turbines. Ph.D. Thesis, The University of Manchester (United Kingdom), Manchester, UK, 2017.
11. Majdi Nasab, N.; Kilby, J.; Bakhtiaryfard, L. The Potential for Integration of Wind and Tidal Power in New Zealand. *Sustainability* **2020**, *12*, 1807. [CrossRef]
12. Martínez, I.D. Offshore Wind and Tidal Stream Power Plant. Master's Thesis, Universitat Politècnica de Catalunya, Barcelona, Spain, 2019; p. 79.
13. Engana Carmo, J.; Neto Torres, J.P.; Cruz, G.; Marques Lameirinhas, R.A. Effect of the Inclusion of Photovoltaic Solar Panels in the Autonomy of UAV Time of Flight. *Energies* **2021**, *14*, 876. [CrossRef]
14. Riley, M.; Taheri, F.; Islam, M. A critical review of materials available for health monitoring and control of offshore structures. In Proceedings of the 51 st Canadian Chemical Engineering Conference, Halifax, NS, Canada, 14–17 October 2001; pp. 14–17.
15. Piezoelectric Simulations. Available online: https://cdn.comsol.com/wordpress/2014/12/Piezo_COMSOL_50.compressed.pdf (accessed on 20 May 2022).
16. El-Ahmar, M.; El-Sayed, A.H.M.; Hemeida, A. Evaluation of factors affecting wind turbine output power. In Proceedings of the 2017 Nineteenth International Middle East Power Systems Conference (MEPCON), Cairo, Egypt, 19–21 December 2017; IEEE: Piscataway, NJ, USA, 2017; pp. 1471–1476.
17. Models, W.T. Vestas V164-8.0. Available online: https://en.wind-turbine-models.com/turbines/318-vestas-v164-8.0 (accessed on 18 April 2022).
18. Newbold, C.; Akrami, M.; Dibaj, M. Scenarios, Financial Viability and Pathways of Localized Hybrid Energy Generation Systems around the United Kingdom. *Energies* **2021**, *14*, 5602. [CrossRef]

energies MDPI

Article

Effect of Deflocculant Addition on Energy Savings in Hydrotransport in the Lime Production Process

Beata Jaworska-Jóźwiak [1,*] and Marek Dziubiński [2,*]

1 Department of Management and Production Engineering, Kielce University of Technology, al. Tysiaclecia Panstwa Polskiego 7, 25-314 Kielce, Poland
2 Department of Chemical Engineering, Lodz University of Technology, ul. Wolczanska 213, 90-924 Lodz, Poland
* Correspondence: b.jaworska@tu.kielce.pl (B.J.-J.); marek.dziubinski@p.lodz.pl (M.D.)

Abstract: The subject of the research was limestone hydromixture consisting of particles of a mean size of 45.5 μm conveyed by water in a pipeline of a total length of 632 m. In the paper, the results of rheological measurements of tested hydromixtures after the application of deflocculant consisting of waste product from the lime production process in the form of mineral particles and commonly known dispersant were presented. Calculations of pressure drop including hydromixtures with volume concentrations in the range of 21.30–50.00%, and density ranging from 1140–1410 kg/m^3 in a pipeline of 200 mm diameter are presented. A decrease in friction losses in the flow in the pipeline of hydromixtures with different mass concentrations after the addition of deflocculant was observed. The study revealed that the addition of deflocculant resulted in a reduction of friction in the pipeline, enabling the pumping of hydromixtures with twice higher solids concentrations than originated from industrial installation, with a lower volumetric flow rate. This resulted in a decrease of the power consumption of the motor driving the pump, and obtained significant energy savings in the hydromixture transport process. The maximum energy saving achieved was equal to 58%.

Keywords: hydrotransport; energy savings; Bingham plastic fluids; deflocculant; pressure drop

Citation: Jaworska-Jóźwiak, B.; Dziubiński, M. Effect of Deflocculant Addition on Energy Savings in Hydrotransport in the Lime Production Process. *Energies* **2022**, *15*, 3869. https://doi.org/10.3390/en15113869

Academic Editors: Phillip Ligrani and Roberto Alonso González Lezcano

Received: 30 March 2022
Accepted: 20 May 2022
Published: 24 May 2022

Publisher's Note: MDPI stays neutral with regard to jurisdictional claims in published maps and institutional affiliations.

Copyright: © 2022 by the authors. Licensee MDPI, Basel, Switzerland. This article is an open access article distributed under the terms and conditions of the Creative Commons Attribution (CC BY) license (https://creativecommons.org/licenses/by/4.0/).

1. Introduction

Transportation of liquids or hydromixtures through conduits is an important element of technological solutions in many industries. The advantages of such transport include: low costs, maintaining high purity of the transported medium, the elimination of environmental pollution, as well as easy monitoring and automation of such a process.

The demand to pump hydromixtures consisting of increasingly finer solid particles with their increasing concentration in industrial installations has been growing. Pumping hydromixtures with high solids concentrations in pipelines requires the application of pumps with a high discharge pressure, which results in a significant energy consumption and leads to greater pump wear [1–3].

During the transportation process of hydromixture, a substantial reduction of frictional pressure drop in the pipeline may occur. It is therefore often necessary to consider the problems of calculating the power requirements for pumping through a given pipeline system, the selection of the optimum pipe diameter, the measurement and control of the flow rate, etc. The knowledge of these factors facilitates controlling the total plant cost installation operation [4]. Several studies for pressure drop prediction in slurry flow are available in the literature, for instance [5–7].

It is clear that the transportation of utilities through pipelines involves significant energy expenditure. In recent years, with the growing difficulties with obtaining energy and the need to optimize transport costs, it has become necessary to develop new energy-saving technologies and methods of media transport that enable the reduction of energy consumption [8–10]. Such methods include: methods using the phenomena of the abnormal reduction of any pressure drop when macromolecular polymers are added to the transported medium (Toms effect), methods using surfactants (surface-active substances or deflocculants (a chemical or compound that prevents flocculation) as additives to reduce pressure drop during media transport, two-phase liquid-gas transport methods using compressed air (mainly used while concentrated suspensions exhibit complex rheological properties, especially when non-Newtonian shear thinning liquids are transported), and methods of modifying the rheological properties of the transported media. Each of these methods has found a wide application in practice and has been the subject of many research works [11–15].

A great deal of works have been published in the literature devoted to explanation of the mechanism of pressure drop reduction [16–22]. Despite the fact that many hypotheses of this phenomenon have been put forward, it has not been possible to propose universal model equations for predicting the frictional pressure drop.

The objective of this study is to develop a method of reduction of the electrical power consumption in the transportation of hydromixture in the pipeline. Such studies are part of the issues of industrial process intensification and multiphase fluid mechanics due to their wide applicability.

2. Materials and Methods

The hydromixture used in the study originated from an industrial installation of a Polish limestone mine. The solid fraction contained in the hydromixture consisted of dusty particles of quarried stone containing a high percentage of calcium carbonate. The grain composition of the solid fraction, determined using a laser grain size analyzer, is shown in Figure 1.

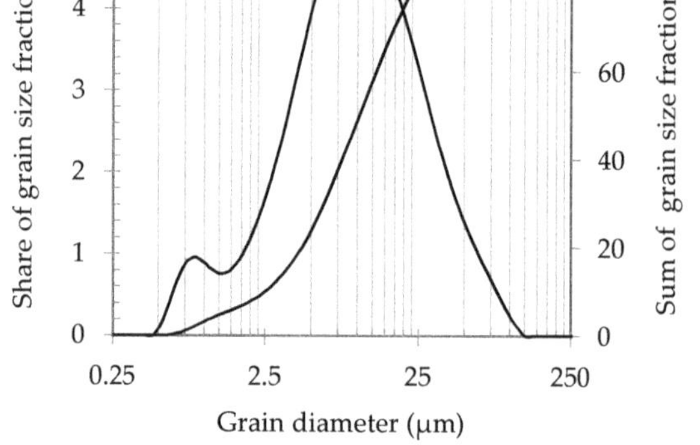

Figure 1. Grain size distribution curve of the tested hydromixture.

The grain size ranged from 0.5 μm to 163.5 μm, with an average grain size of 45.5 μm. The dust fraction with a mean grain diameter ranging from 2 μm to 50 μm had the largest share in the sample (about 65%). The sand fraction, characterizing particles with an average grain diameter larger than 50 μm, represented almost one third of all particles. The clay particles constituted only 3.38% of all the particles contained in the hydromixture. Chemically, the test hydromixture consisted mainly of calcium oxide CaO (73.6%) and silicon oxide (SiO_2—13%). The other chemical compounds in the hydromixture were magnesium oxide (MgO—0.6%), iron oxide (Fe_2O_3—0.3%), aluminum oxide (Al_2O_3—1.1%) and sulfur trioxide (SO_3—0.3%). Substances that could not be identified accounted for approximately 11.1% of the total.

Rheological studies of hydromixture were performed using an Anton Paar (Lyon, France) MCR 302 rotational viscometer at T = 20 °C for hydromixtures with a mass concentration of C_m = (21.30−50.00)% and a solid volume concentration of C_v = (10−29)% in a range of hydromixtures density from about 1140 kg/m^3 to 1410 kg/m^3.

The effect of frictional pressure drop reduction in the pipeline was obtained by introducing to the hydromixture an addition of substances with highly dispersive properties, in the form of fine mineral particles and chemical compounds, called deflocculants (drag reduction agents—DRA). Such an effect is called chemical treatment and, due to its potential practical use, has become the subject of much interest in various branches of industry and scientific works [23–28]. This effect should result in frictional pressure drop reduction in the flow of the hydromixture with an increase in the concentration of the solid phase contained in it, and result in a reduction of the energy consumption necessary to overcome the friction occurring in the pipeline. The reduction in the frictional pressure drop entails a reduction in the total cost of the hydrotransport process.

The composition of the proposed deflocculant was developed based on years of experimental research, knowledge derived from the literature, and cooperation with an industrial enterprise.

Sodium water glass and calcareous groats were selected as the deflocculant components based on their easy access in the product market, low manufacturing/acquisition cost and environmental neutrality. In the case of sodium water glass, the unquestionable advantage is its common occurrence and low purchase cost. Sodium water glass is formed by the reaction between silica and sodium hydroxide or silica and sodium carbonate. The addition of sodium water glass was intended to lower the viscosity, leading to a reduction in the shear stress created in the transported hydromixture. Calcareous groats, a residue from the lime slaking process that produces hydrated lime, consist of mineral particles with a grain diameter of less than 1.8 mm and an average grain diameter of 240 µm. Figure 2b presents an image of the calcareous groats grains surface made by a Phenom Pro X (Thermo Fisher Scientific, Waltham, MA, USA) scanning electron microscope (SEM) after applying a 10 kV electron accelerating voltage.

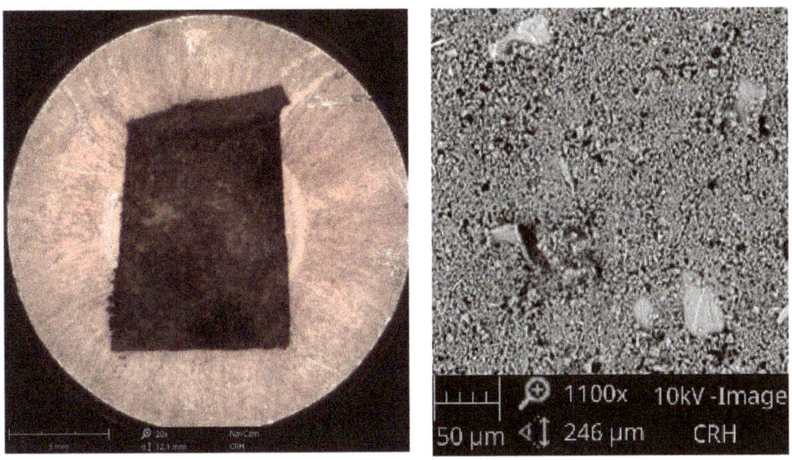

(a) (b)

Figure 2. (**a**) Picture of dried lime slurry sample. (**b**) Surface image of calcareous groats grains made by scanning electron microscopy (SEM) with 1100× magnifications.

This substance is inexpensive and easy to obtain, and is stored in large amounts in the production plant area. Chemically, calcareous groats are mostly composed of calcium oxide and magnesium oxide (70.81%), silicon dioxide (1.52%) and aluminum (0.84%).

The remaining substances are minor impurities formed in the process of hydration of a mixture of lime and water. Table 1 presents the results of the chemical composition analysi of calcareous groats performed with a scanning electron microscope integrated with an energy-dispersive X-ray spectrometer (EDS).

Table 1. Chemical composition of calcareous groats made using an energy-dispersive X-ray spectrometer (EDS).

Element Number	Element Symbol	Element Name	Atomic Concentration (%)	Error (%)
20	Ca	Calcium	13.1	0.0
8	O	Oxygen	79.7	0.1
6	C	Carbon	2.2	0.9
53	I	Iodine	0.2	0.0
7	N	Nitrogen	4.3	2.1
12	Mg	Magnesium	0.5	0.8

These two substances (sodium water glass and calcareous groats) combined together in appropriate proportions, depending on the mass concentration of the hydromixture were added to tested samples with a high solid phase concentration in order to reduce the viscosity of the hydromixture.

Method of Measurement

The rheological parameters of hydromixtures used in this study were performed by an Anton Paar MCR 302 rotational viscometer. A coaxial double-gap cylinder system was chosen to measure the shear stresses generated in the hydromixture. In this measuring system, the outer cylinder, which is also the vessel into which the sample is applied, is a stationary element that allows for easy temperature control and stabilization. The inner cylinder (rotor) is driven by a motor for which the speed can be determined and the torque measured. The torque is proportional to the power of the current supplying the motor. It is converted to shear stress, and the rotor speed to the shear rate. In the performed measurements, the value of the shear rate was set, while the value of the shear stress was measured by the device. The measurements were carried out at 20 °C in the range of a shear rate from about 8–10 s^{-1} to 100 s^{-1}.

3. Results

Hydraulic conveying belongs to the most reliable and economically justified modes of hydrotransport. Due to its ecological values, it is also the preferred solution applied in environmental engineering. Transport of hydromixture with significant amounts of solids requires determination of the hydromixture's rheological properties to adopt an appropriate rheological model of the hydromixture. Such a model, determined on the basis of experimental research, allows the calculation of the frictional pressure drop during the flow of a given mixture through industrial installation. It is also essential to determine a condition at which laminar flow transforms into turbulence.

3.1. Rheological Decsription of Tested Hydromixture

The flow curves of hydromixtures with mass concentrations of 21.30%, 28.14%, 35.00%, 42.75% and 50.00% without addition (denoted as 'pure') and with deflocculant addition (denoted as '+DFL') are presented in Figure 3. The rheometric measurements were compared with the results calculated for the Bingham plastics model where the solid line indicates the fit of the rheological model to the experimental data. The shape of the flow curves indicates that the tested hydromixtures belong to the group of non-Newtonian, pseudo-plastic fluids

Non-Newtonian fluids are generally highly viscous systems throughout the technically important range of the shear rate and are very difficult to transport in a pipeline.

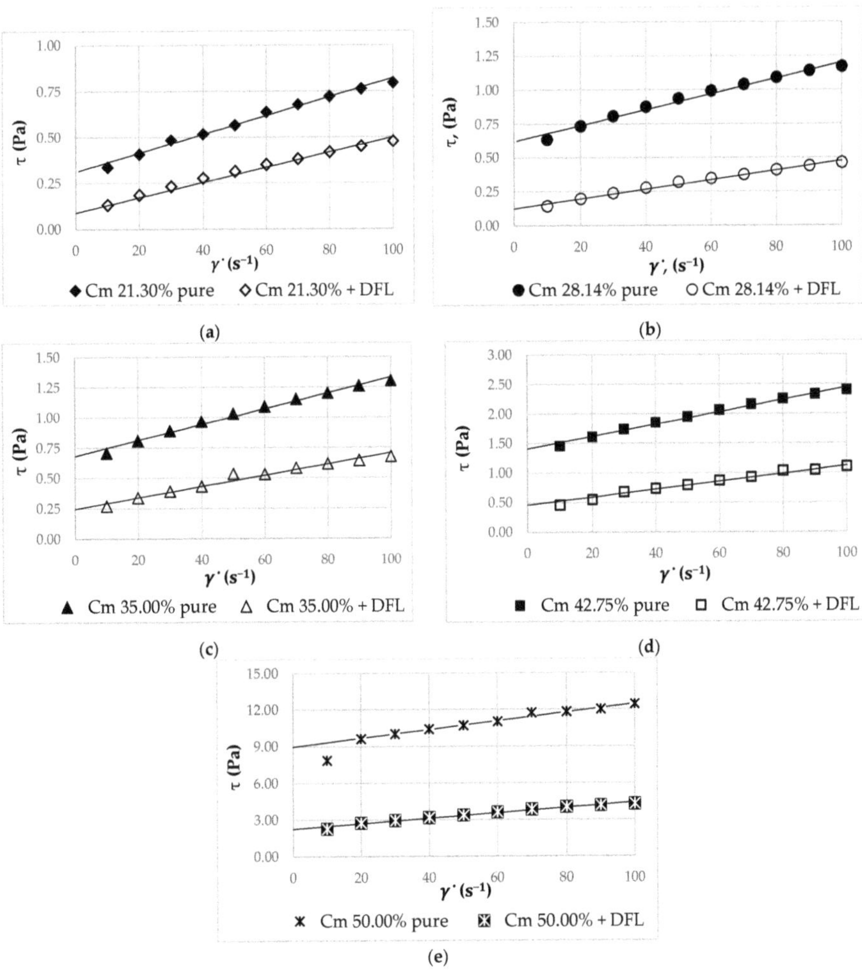

Figure 3. The experimental flow curves of tested hydromixtures with a mass concentration of: (**a**) 21.30%; (**b**) 28.14%; (**c**) 35.00%; (**d**) 42.75%; and (**e**) 50.00%.

The flow curves of hydromixtures were described by the Bingham model as follows:

$$\tau = \tau_y + \eta_p \cdot y \tag{1}$$

where η_p is plastic viscosity and τ_y is a yield stress.

The rheological properties of the tested hydromixtures can be described by the rheological Bingham plastics model in the shear rate range from about 8–10 s^{-1} to 100 s^{-1}. This range of shear rate occurs in the hydromixture flow in the analyzed pipeline.

In Table 2, the parameters of analyzed hydromixtures with deflocculant (+DFL) and without the addition of deflocculant (pure), such as density (ρ_m), yield stress value (τ_y), plastic viscosity (η_p) and the coefficient value of fitting the mathematical model to the experimental data (R^2) are shown.

Table 2. Rheological parameters of the tested hydromixtures.

Parameters		ρ_m (kg/m^3)	τ_y (Pa)	η_p (Pa·s)	R^2 (%)
C_m 21.30%	pure	1140	0.3109	0.0051	98.90
	+DFL		0.0887	0.0041	98.88
C_m 28.14%	pure	1195	0.6209	0.0058	98.26
	+DFL		0.1253	0.0035	99.51
C_m 35.00%	pure	1254	0.6812	0.0065	98.69
	+DFL		0.2461	0.0046	98.62
C_m 42.75%	pure	1330	1.4073	0.0104	98.96
	+DFL		0.4576	0.0066	98.55
C_m 50.00%	pure	1410	8.9667	0.0350	98.53
	+DFL		2.2654	0.0216	97.85

3.2. Mathematical Model

Calculation of the pressure drop of the tested hydromixture depends on the character of the flow (laminar or turbulent), determined by the Reynolds number for Bingham fluids. The critical value of the Reynolds number is described by an empirical equation of the form [4]:

$$(N_{Re})_{cr} = \frac{1 - \frac{4}{3}\phi_c + \frac{\phi_c^4}{3}}{8 \cdot \phi_c} \cdot N_{He} \quad (2)$$

where the parameter ϕ_c and the Hedström number are calculated as follows:

$$N_{He} = \frac{\rho \cdot d^2 \cdot \tau_y}{\eta_p^2} \quad (3)$$

and

$$\frac{\phi_c}{(1 - \phi_c)^3} = \frac{N_{He}}{16800} \quad (4)$$

Flow of the Bingham plastics fluid is laminar if the calculated Reynolds number (6) is below the critical value (2), above which the transition to turbulent flow occurs.

The critical value of the Reynolds number for Bingham plastic fluids describing transition from laminar to turbulent flow for tested hydromixtures were calculated from Equation (2), and is presented in Table 3.

Table 3. Parameters at the transition from a laminar to turbulent flow regime of tested hydromixtures.

Parameters		$(N_{Re})_{cr}$	N_{He}	ϕ_c
C_m 21.30%	pure	12,350	545,116	0.7191
	+DFL	9261	240,638	0.6443
C_m 28.14%	pure	14,620	881,926	0.7567
	+DFL	11,878	488,744	0.7100
C_m 35.00%	pure	14,222	809,050	0.7500
	+DFL	12,621	583,611	0.725
C_m 42.75%	pure	13,417	692,408	0.7385
	+DFL	12,478	559,037	0.721
C_m 50.00%	pure	11,218	412,770	0.695
	+DFL	9695	273,810	0.657

The friction factor for the laminar flow of Bingham plastic fluids (f_L) was calculated from the expression known as the Buckingham–Reiner equation [4]:

$$f_L = \frac{16}{N_{Re}} \cdot \left[1 + \frac{1}{6} \cdot \frac{N_{He}}{N_{Re}} - \frac{1}{3} \cdot \frac{N_{He}^4}{f_L^3 \cdot N_{Re}^7} \right] \approx \frac{16}{N_{Re}} \cdot \left[1 + \frac{N_{He}}{8 \cdot N_{Re}} \right] \quad (5)$$

where the Reynolds number is given by:

$$N_{Re} = \frac{U_s \cdot d \cdot \rho}{\eta_p} \quad (6)$$

For the Bingham plastics fluids there is a gradual deviation from purely laminar to fully turbulent flow. For turbulent flow, the friction factor (f_T) can be represented by the empirical expression of Darby and Melson [21], and modified by Darby [22]:

$$f_T = \frac{10^a}{N_{Re}^{0.193}} \quad (7)$$

where

$$a = -1.47 \cdot \left[1 + 0.146 exp\left(-2.9 \cdot 10^{-5} \cdot N_{He} \right) \right] \quad (8)$$

The friction factor for Bingham plastics fluids can be calculated for any Reynolds number, from laminar through turbulent, from the equation [22]:

$$f = \left(f_L^m + f_T^m \right)^{1/m} \quad (9)$$

where f_L is calculated from the Equation (5) and f_T is given by the Equation (7).

The m value is calculated from the expression:

$$m = 1.7 + \frac{40000}{N_{Re}} \quad (10)$$

Equation (9) was applied for calculations of pressure drop under turbulent condition for d < 335 mm, $N_{Re} \leq 3.4 \cdot 10^5$, and $1000 \leq N_{He} \leq 6.6 \cdot 10^7$, and were used in further calculations of the friction factor.

3.3. Physical Model of Hydromixture Transport

The subject of the research was the transport of finely dispersed hydromixture from a reservoir to the settling tank in a pipeline of a total length of 632 m and a diameter equaling 200 mm. The total height difference between the inlet and the outlet length of the pipeline was 11 m.

We assumed that the flow of the tested hydromixture is fully developed, axially-symmetrical and isothermal. The flowing medium consisted of a solid phase and water as a carrier liquid.

The calculations of the total pressure drop in the pipeline was carried out for the hydromixture transported under industrial conditions, i.e., with a solid phase concentration of C_m = 21.30% and a volumetric flow rate of Q_v = 110 m^3/h, which corresponds to a solid phase mass transport of 27 t/h. The results of the calculations performed for the base hydromixture were compared with the flow pressure drop values for hydromixtures with mass concentrations of 28.14%, 35.00%, 42.75% and 50.00%, to which the deflocculant additive was applied.

Based on the assumptions of the physical model, the pressure drop (Δp) of the hydromixture flow in the tested pipeline was calculated from the expression:

$$\Delta p = \frac{\rho_m \cdot U_s^2}{2} + \rho_m \cdot g \cdot \Delta h + \cdot p_{1-2} \quad (11)$$

where ρ_m is the density of the hydromixture, U_s is the mean flow velocity, and Δh is the difference in level between the pump inlet and the pipeline outlet. Δp_{1-2} is the linear losses due to friction in pipeline sections, which were calculated from the Darcy–Weisbach equation as follows:

$$\Delta p_{1-2} = \lambda \cdot \frac{\rho_m \cdot U_s^2}{2} \cdot \frac{L}{d} \quad (12)$$

where λ is the friction factor, L is the total length of the pipeline, and d is the pipeline diameter.

The values of the friction factor λ were determined from the equation:

$$\lambda = 4 \cdot f \quad (13)$$

3.4. Results of the Pressure Drop Calculation

The results of the pressure drop calculations for the example hydromixtures of mass concentrations of 21.30%, 42.75% and 50.00%, without the addition of deflocculant (pure) and with deflocculant (DFL), are presented in Figure 4.

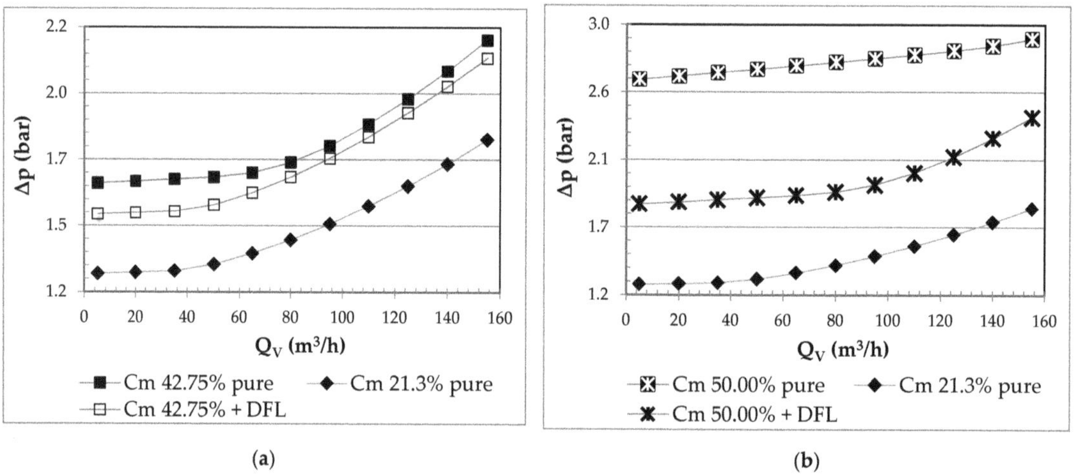

Figure 4. The results of frictional pressure drop calculations for hydromixtures of selected solid phase concentrations: (**a**) 42.75%, and (**b**) 50.00%.

The pressure drop values presented in the graphs above Figure 4 show an increasing trend of increasing volumetric flow rate and mass concentration of the hydromixture. However, the growth is smaller for the hydromixtures with the addition of deflocculant due to the viscosity decrease. This effect allows for the transportation of a constant amount of solid phase mass in hydromixtures of higher mass concentrations and a smaller pressure drop.

3.5. Electricity Consumption in the Hydromixture Transport

As a result of the deflocculant addition, an economic effect was achieved by the reduction of electricity consumption when pumping the hydromixture in a pipeline. In order to determine the obtained economic effect, the power at the pump shaft (P_w) required to pump the hydromixture was calculated using the equation:

$$P_w = \frac{Q_v \cdot \Delta p}{\eta} \quad (14)$$

where Q_v is the volumetric flow rate and η is the pump efficiency.

According to data acquired from the manufacturer of the equipment used in the production plant, the pump efficiency is 62%.

Knowing the pump motor efficiency of η_S = 93.6%, the power consumption of the motor driving the pump (P_S) can be expressed as:

$$P_s = \frac{P_w}{\eta_s} \tag{15}$$

Assuming the constancy of the transported solid particles equal 27 t in the transport of hydromixtures with higher than 21.30% mass concentrations, the volumetric flow rates decrease and equal 80 m^3/h for a hydromixture with a mass concentration of 28.14%, 61 m^3/h for a hydromixture with a mass concentration of 35.00%, 48 m^3/h for a hydromixture with a mass concentration of 42.75%, and 38 m^3/h for a hydromixture with a mass concentration of 50.00%, respectively.

A decrease in the viscosity of hydromixture was achieved as the result of deflocculant addition, which can be observed in the flow curves of hydromixtures shown in Figure 3. Transport of hydromixture with a smaller volumetric flow rate but the same amount of solids is possible. This results in a significant decrease of energy consumption during the transportation of such hydromixtures compared to transportation of base hydromixture, with the mass concentration of 21.30%, which is presented in Table 4.

Table 4. Power consumption of the electric motor driving the centrifugal pump.

C_m (%)	C_m 21.30% Pure	C_m 28.14% + DFL	C_m 35.00% + DFL	C_m 42.75% + DFL	C_m 50.00% + DFL
Q_v (m^3/h)	110	80	61	48	38
P_S (kW)	8.03	5.55	4.27	3.5	3.34

Figure 5 presents the electricity consumption during pumping of the hydromixture with the addition of deflocculant as a function of the mass concentration of the hydromixture. The maximum energy savings is 58% over the transport of basic hydromixture with the concentration of 21.30% by mass.

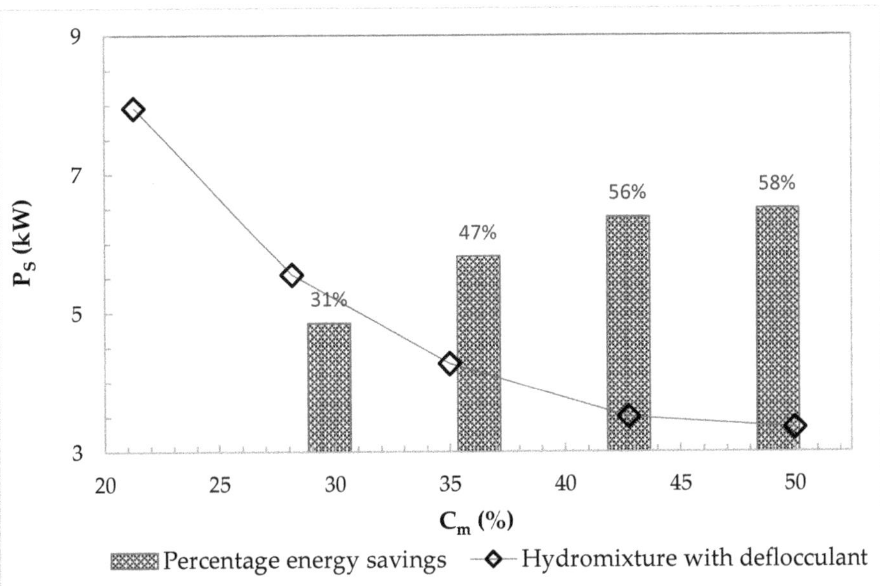

Figure 5. The electricity consumption during pumping of the hydromixture.

4. Discussion

The basis for the calculation of the pressure drop during the transportation of hydromixture in a pipeline is a determination of the rheological properties of tested hydromixtures.

The results of rheological behaviour measurements of hydromixture with different mass concentrations were described by the Bingham plastics model in the shear rate range from about 8–10 s^{-1} to 100 s^{-1} with high accuracy (98–99%), which confirms that the studied hydromixtures were shear-thinning non-Newtonian fluids with a yield stress. For all the tested hydromixtures, decreases in viscosity with increasing shear rate were observed.

Calculations of the friction factors of the analyzed hydromixtures in a 200 mm diameter pipeline was performed using Equations (5), (7) and (9). The friction factor values in the range of laminar flow calculated from Equations (5) and (9) were almost identical. Similarly, the friction factor values for the turbulent flow of non-Newtonian liquid calculated from Equations (7) and (9) were also very similar. In the paper, the possibility of friction factor calculating for non-Newtonian fluids in large diameter pipelines in turbulent flow using both Equations (7) and (9) is confirmed. The calculated friction factors of the hydromixtures in a pipeline of 200 mm diameter and 632 m length confirmed that the addition of deflocculant affects the reduction of friction factors for all the analyzed hydromixtures. Therefore, it was possible to pump hydromixtures with higher solids concentrations, resulting in significant energy savings in their hydrotransport and reduced water consumption in the limestone hydromixture transport process.

The positive effect of using a deflocculant in the transport of the hydromixture occurring in the analyzed pipeline has an influence on the increase in the efficiency of the whole production process. The additional effect that increases the efficiency of the hydromixture transport process is the reduction of the amount of water used in the pumping process, which improves environmental protection by reducing the amount of burdensome slurry. As a result, the estimated operating cost of the industrial installation of hydromixture transportation in a pipeline is significantly reduced.

5. Conclusions

The paper presents a new possible method of hydromixture transportation in pipelines enabling significant energy savings in such transportation. This effect was achieved by using an additive of deflocculant, whose composition was self-developed and contains environmentally neutral mineral particles. The proposed method, except for the energy savings, reduces the amount of water used in such transportation, thus reducing the amount of burdensome slurries, which is an additional ecological aspect of the proposed method.

Author Contributions: Conceptualization, B.J.-J. and M.D.; methodology, B.J.-J.; software, B.J.-J.; validation, M.D.; formal analysis, B.J.-J.; investigation, B.J.-J.; resources, B.J.-J.; data curation, B.J.-J.; writing—original draft preparation, B.J.-J.; writing—review and editing, M.D.; visualization, B.J.-J.; supervision, M.D.; project administration, B.J.-J.; funding acquisition, B.J.-J. All authors have read and agreed to the published version of the manuscript.

Funding: This research received no external funding.

Data Availability Statement: Not applicable.

Acknowledgments: The chemical composition of calcareous groats analysis was performed using a scanning electron microscope integrated with an energy-dispersive X-ray spectrometer (EDS) at the Institute of Physics at the Silesian University of Technology.

Conflicts of Interest: The authors declare no conflict of interest.

Nomenclature

a	parameter given by Equation (8)
C_m	mass concentration of hydromixture, (%)
C_v	volume concentration of hydromixture, (%)
d	pipe diameter, (m)
DFL	deflocculant
f	friction factor for Bingham plastic fluids
f_L	friction factor for laminar flow
f_T	friction factor for turbulent flow
L	total length of a pipeline, (m)
m	parameter given by Equation (10)
N_{He}	Hedström number
$(N_{Re})_{cr}$	critical value of Reynolds number
P_w	power at the pump shaft, (W)
P_S	power consumption of the motor driving the pump, (W)
Q_v	volumetric flow rate, (m^3/h)
U_s	mean flow velocity, (m/s)
Greek letters	
Δh	difference in level between the pump inlet and the pipeline outlet, (m)
Δp	pressure drop, (Pa)
Δp_{1-2}	flow pressure drop due to friction, (Pa)
η	pump efficiency, (%)
η_p	plastic viscosity, (Pa·s)
λ	friction factor
ρ_m	density of hydromixture, (kg/m^3)
τ	shear stress, (Pa)
τ_y	yield stress, (Pa)
$\dot{\gamma}$	shear rate, (s^{-1})
φ_c	parameter calculated from Equation (4)

References

1. Newitt, D.M.; Richardson, J.F.; Abbot, M.; Turtle, R.B. Hydraulic Conveying of Solids in Horizontal Pipes. *Trans. Inst. Chem. Eng.* **1955**, *33*, 93–110.
2. Parzonka, W.; Kenchington, J.M.; Charles, M.E. Hydrotransport of solids in horizontal pipes: Effects of solids concentration and particle size on the deposit velocity. *Can. J. Chem. Eng.* **1981**, *59*, 291–296. [CrossRef]
3. Shrikant Dhodapkar, K.J.; Shenggen, H. Fluid-Solid Transport in Ducts. In *Multiphase Flow Handbook*; Clayton, T.C., Ed.; Taylor and Francis Group: London, UK, 2006; Chapter 4; pp. 1–101.
4. Chhabra, R.P.; Richardson, J.F. *Non-Newtonian Flow and Applied Rheology*, 2nd ed.; Butterworth-Heinemann: Oxford, UK, 2008; pp. 111–123, 142–148.
5. Abd Al Aziz, A.I.; Mohamed, H.I. A Study of the Factors Affecting Transporting Solid—Liquid Suspension through Pipelines. *Open J. Fluid Dyn.* **2013**, *3*, 152–162. [CrossRef]
6. Matousek, V. Pressure drops and flow patterns in sand-mixture pipes. *Exp. Therm. Fluid Sci.* **2002**, *26*, 693–702. [CrossRef]
7. Hashemi, S.A.; Wilson, K.C.; Sanders, R.S. Specific energy consumption and optimum operating condition for coarse-particle slurries. *Powder Technol.* **2014**, *262*, 183–187. [CrossRef]
8. Wu, J.; Graham, L.; Wang, S.; Parthasarathy, R. Energy efficient slurry holding and transport. *Miner. Eng.* **2010**, *23*, 705–712. [CrossRef]
9. Edelin, D.; Czujko, P.C.; Castelain, C.; Josset, C.; Fayolle, F. Experimental determination of the energy optimum for the transport of floating particles in pipes. *Exp. Therm. Fluid Sci.* **2015**, *68*, 634–643. [CrossRef]
10. Ihle, C.F. The least energy and water cost condition for turbulent, homogeneous pipeline slurry transport. *Int. J. Miner. Process.* **2016**, *148*, 59–64. [CrossRef]
11. Shook, C.A.; Roco, M.C. *Slurry Flow. Principles and Practice*; Butterworth-Heinemann: Oxford, UK, 1991; p. 35.
12. Jurban, B.A.; Zurigat, Y.H.; Goosen, M.F.A. Drag Reducing Agents in Multiphase Flow Pipelines: Recent Trends and Future Needs. *Pet. Sci. Technol.* **2005**, *23*, 1403–1424.
13. He, M.; Wang, Y.; Forssberg, E. Parameter studies on the rheology of limestone slurries. *Int. J. Miner. Process.* **2006**, *78*, 63–77. [CrossRef]
14. Deckers, J.; Vleugels, J.; Kruth, J. Additive Manufacturing of Ceramics: A Review. *J. Ceram. Sci. Technol.* **2014**, *5*, 245–260.

15. Silva, R.; Garcia, F.A.P.; Faia, P.M.G.M.; Rasteiro, M.G. Settling Suspensions Flow Modelling: A Review. *Powder Part. J.* **2015**, *3*, 41–56. [CrossRef]
16. Dziubiński, M. A general correlation for two-phase pressure drop in intermittent flow of gas and non-Newtonian liquid mixture in a pipe. *Chem. Eng. Res. Des.* **1995**, *73*, 529–534. [CrossRef]
17. Chhabra, R.; Dziubiński, M. Predicting two phase pressure drop for the flow of gas/non-Newtonian liquid mixtures in horizonta pipes. *Int. J. Fluid Mech.* **1989**, *2*, 63–69.
18. Richardson, J.F.; Dziubiński, M. Two-phase flow of gas and non-Newtonian liquids in horizontal pipes. Superficial gas velocit for maximum power saving. *J. Pipelines* **1985**, *5*, 107–111.
19. Vlasak, P.; Chara, Z.; Konfrst, J.; Krupicka, J. Distribution of concentration of coarse particle-water mixture in horizontal smoot pipe. *Can. J. Chem. Eng.* **2016**, *94*, 1040–1047. [CrossRef]
20. Naveh, R.; Tripathi, N.M.; Kalman, H. Experimental pressure drop analysis for horizontal dilute phase particle-fluid flow *Powder Technol.* **2017**, *321*, 355–368. [CrossRef]
21. Darby, R.; Melson, J.M. A Friction Factor Equation for Bingham Plastic Slurries and Suspensions for All Flow Regimes. *Chem Eng.* **1981**, 59–61.
22. Darby, R. Predict Friction Loss in Slurry Pipes. In *Chemical Engineering*; ABI/I-NFORM Global: New York, NY, USA, 1992; p. 11 Available online: https://www.researchgate.net/publication/292650993_Predict_friction_loss_in_slurry_pipes (accessed on 2 March 2022).
23. Chhabra, R.P.; Farooqi, S.I.; Richardson, J.F. Isothermal two-phase flow of air and aqueous polymer solutions in a smoot horizontal pipe. *Chem. Eng. Res. Des.* **1983**, *62*, 22–32.
24. He, M.; Wang, Y.; Forssberg, E. Slurry rheology in wet ultrafine grinding of industrial minerals: A review. *Powder Technol.* **200** *147*, 94–112. [CrossRef]
25. Wang, Y.; Yu, B.; Zakin, J.L.; Shi, H. Review on Drag Reduction and Its Heat Transfer by Additives. *Adv. Mech. Eng.* **2011**, 874 [CrossRef]
26. Zhou, Y.B.; Xu, N.; Ma, N.; Li, F.C.; Wei, J.J.; Yu, B. On Relationships among the Aggregation Number, Rheological Property, an Turbulent Drag-Reducing Effect of Surfactant Solutions. *Adv. Mech. Eng.* **2011**, 345328. [CrossRef]
27. Matras, Z.; Kopiczak, B. Intensification of drag reduction effect by simultaneous addition of surfactants and high molecula polymer into the solvent. *Chem. Eng. Res. Des.* **2015**, *96*, 35–42. [CrossRef]
28. Sikorski, C.F.; Lehman, R.L.; Shepherd, J.A. The effects of viscosity reducing chemical additives on slurry rheology and pipelin transport performance for various mineral slurries. In Proceedings of the Seventh International Technical Conference on Slurr Transportation, Nevada, NV, USA, 23–26 March 1982.

energies

Article

Settling of Mesoplastics in an Open-Channel Flow

Luka Kevorkijan [1], Elvis Žic [2], Luka Lešnik [1] and Ignacijo Biluš [1,*]

[1] Chair of Power, Process and Environmental Engineering, Faculty of Mechanical Engineering, University of Maribor, 2000 Maribor, Slovenia
[2] Department of Hydraulic Engineering, Faculty of Civil Engineering, University of Rijeka, 51000 Rijeka, Croatia
* Correspondence: ignacijo.bilus@um.si

Abstract: Pollution of water by plastic contaminants has received increasing attention, owing to its negative effects on ecosystems. Small plastic particles propagate in water and can travel long distances from the source of pollution. In order to research the settling motion of particles in water flow, a small-scale experiment was conducted, whereby spherical plastic particles of varying diameters were released in an open-channel flow. Three approaches were investigated to numerically simulate the motion of particles. The numerical simulation results were compared and validated with experimental data. The presented methods allow for deeper insight into particle motion in fluid flow and could be extended to a larger scale to predict the propagation of mesoplastics in natural environments.

Keywords: dense discrete phase model; discrete element method; diameter; discrete phase model; mesoplastics; particle; settling

Citation: Kevorkijan, L.; Žic, E.; Lešnik, L.; Biluš, I. Settling of Mesoplastics in an Open-Channel Flow. *Energies* **2022**, *15*, 8786. https://doi.org/10.3390/en15238786

Academic Editors: Artur Bartosik and Dariusz Asendrych

Received: 20 October 2022
Accepted: 17 November 2022
Published: 22 November 2022

Publisher's Note: MDPI stays neutral with regard to jurisdictional claims in published maps and institutional affiliations.

Copyright: © 2022 by the authors. Licensee MDPI, Basel, Switzerland. This article is an open access article distributed under the terms and conditions of the Creative Commons Attribution (CC BY) license (https://creativecommons.org/licenses/by/4.0/).

1. Introduction

Research on the topic of particle motion is ongoing. Many pollutants that are present in the environment form a dispersed multiphase system, in which the continuous phase is the medium of the environment (gas or liquid) and the dispersed phase is made of pollutant particles. Plastic particles with diameters between 1 and 10 mm, termed mesoplastic particles or mesoplastics, represent a ubiquitous and concerning pollutant [1,2]. However, there is still a lack of consensus with respect to the definition of plastic debris size classification [1–3]. In the literature, plastic particles with diameters of less than 5 mm are often classified as microplastic particles or microplastics [3–5].

One source of pollution is the wastewater generated by various industries [6], for example, the textile industry [7]. Owing to the negative impact of plastic particles on the environment, it is of particular interest and importance to research how they spread in the environment [8–13]. The propagation of mesoplastics occurs in rivers and oceans; therefore, the topic of the motion of particles suspended in water should be the focus of further research [14,15]. Depending on their properties, mesoplastics can settle and propagate as sediment, flow freely suspended with water or float on the free surface of the water. In rivers, natural and man-made obstacles exist, which produce different flow features and regimes that can be replicated in a laboratory-scale physical model to investigate various scientific and engineering problems with respect to different modes of mesoplastic propagation.

The transport of particles on a laboratory scale has been studied in open-channel flows in the past. Cook et al. [16] experimentally studied the longitudinal dispersion of microplastic particles and suggested the use of a fluorescent tracer to study the movement of plastic particles in the environment. Yu et al. proposed an empirical formula to predict the incipient motion of microplastic particles with diameters up to 5 mm [17]. The general importance and applicability of studying particle settling was highlighted in an article by Yi and colleagues, who studied the settling of sturgeon eggs in an open channel [18]. In addition to experimental work, different numerical simulation approaches have been used to study the propagation of plastic particles. The dispersal of microplastic particles in coastal waters was studied by Fatahi and colleagues [15] using computational fluid dynamics (CFD); a discrete phase model (DPM) was used to track the motion of particles. Roy et al. [19] conducted a parametric study of plastic particle propagation using a Lagrangian tracking model in a lid-driven cavity flow. On a larger scale, an Eulerian modelling approach is sometimes adopted, as in a study of the transport and deposition of microplastic particles in the Baltic sea conducted by Schernewski et al. [14], who concluded that microplastic emissions resulted mostly from wastewater, as well as sewer and stormwater overflow. Another approach to Lagrangian tracking is the dense discrete phase model (DDPM), which is often used to simulate particle-laden flows in pipelines [20]. Recent attempts to model the motion of microplastic or mesoplastic particles include works by Holjević et al. [21] and Travaš et al. [22], who modelled the transport of different microplastic (or mesoplastic) particles in a laminar open-channel flow.

In this work, the settling of mesoplastic particles is investigated in a 3D turbulent open-channel flow using numerical simulations and experiments.

2. Materials and Methods

Two methods of Lagrangian tracking of particles in fluid flow, available as part of ANSYS Fluent, a commercial computational fluid dynamics (CFD) software, were used in this work. The first is the discrete phase model (DPM) [23], in which the particles are assumed to be point masses, and their volume is not accounted for in the equations of fluid flow. The second is the dense discrete phase model (DDPM) [23], in which the volume of particles are accounted for in the governing equations of fluid flow via their volume fraction. Within the DDPM framework, particle–particle interaction can be included. Multiple approaches are available to describe particle–particle interaction; in this work, this interaction is investigated via the discrete element method (DEM). To this end, an additional DEM software is used, Altair EDEM, and connected to ANSYS Fluent to exchange information about particles.

2.1. Discrete Phase Model

2.1.1. Liquid Flow

The governing equations of flow are continuity Equation (1) and the Navier–Stokes Equation (2)

$$\frac{\partial(\rho_c)}{\partial t} + \nabla\cdot(\rho_c \boldsymbol{u}) = 0, \tag{1}$$

$$\frac{\partial(\rho_c \boldsymbol{u})}{\partial t} + \nabla\cdot(\rho_c \boldsymbol{u}\boldsymbol{u}) = -\nabla p + \mu_c \nabla^2 \boldsymbol{u} + \rho_c \boldsymbol{g} + S_M, \tag{2}$$

where ρ_c is the continuous phase (liquid) density, $\boldsymbol{u}$ is the continuous phase (liquid) velocity, μ_c is the continuous phase (liquid) dynamic viscosity and p is the pressure in the continuous phase (liquid). The term $\rho_c \boldsymbol{g}$ represents the gravitational force, where $\boldsymbol{g}$ is the gravitational acceleration.

The presence of particles in the flow can affect the flow field, which is accounted for by the additional momentum term S_M. For turbulence, the RANS approach is adopted with the realizable $k-\varepsilon$ turbulence model, where two additional transport equations for turbulence kinetic energy (k) and dissipation of turbulence kinetic energy (ε) are used.

$$\frac{\partial(\rho_c k)}{\partial t} + \nabla\cdot(\rho_c k \boldsymbol{u}) = \nabla\cdot\left[\left(\mu_c + \frac{\mu_t}{\sigma_k}\right)\nabla k\right] + G_k + G_b - \rho_c \varepsilon - Y_M + S_k, \tag{3}$$

$$\frac{\partial(\rho_c \varepsilon)}{\partial t} + \nabla\cdot(\rho_c \varepsilon \boldsymbol{u}) = \nabla\cdot\left[\left(\mu_c + \frac{\mu_t}{\sigma_\varepsilon}\right)\nabla \varepsilon\right] + \rho_c C_1 S \varepsilon - \rho_c C_2 \frac{\varepsilon^2}{k + \sqrt{\nu\varepsilon}} + C_{1\varepsilon}\frac{\varepsilon}{k} C_{3\varepsilon} G_b + S_\varepsilon, \tag{4}$$

where the coefficients are

$$C_1 = max\left[0.43; \frac{\eta}{\eta+5}\right], \tag{5}$$

$$\eta = S\frac{k}{\varepsilon}, \tag{6}$$

and the strain rate magnitude is

$$S = \sqrt{2SS}. \tag{7}$$

The model constants are $C_{1\varepsilon} = 1.44$, $C_2 = 1.9$, $\sigma_k = 1.0$ and $\sigma_\varepsilon = 1.2$. The expression for eddy viscosity (μ_t) is

$$\mu_t = \rho_c C_\mu \frac{k^2}{\varepsilon} \tag{8}$$

In contrast to the standard $k - \varepsilon$ model, the coefficient (C_μ) is a function of mean strain, mean rotation rate, turbulence kinetic energy and dissipation of turbulence kinetic energy. In Equations (3) and (4), there are additional terms that describe the production of turbulence kinetic energy (G_k), buoyancy effects (G_b) and compressibility effects (Y_M). σ_k and σ_ε are the turbulent Prandtl numbers for the turbulent kinetic energy (k) and the dissipation of turbulent kinetic energy (ε), respectively. To account for the effects of particles on turbulence, source terms for the turbulence kinetic energy (S_k) and the dissipation of turbulence kinetic energy (S_ε) are also included.

2.1.2. Particle Motion

Particles are tracked in the Lagrangian frame; particle position and velocity are obtained by integration of the governing ordinary differential equations along the particle trajectory. The motion of a single particle is governed by the conservation of mass equation and Newton's second law.

$$\frac{dm_p}{dt} = 0, \tag{9}$$

$$m_p\frac{dv}{dt} = \mathbf{F}, \tag{10}$$

$$I_p\frac{d\omega}{dt} = \mathbf{M}, \tag{11}$$

where m_p is the particle mass, I_p is the particle moment of inertia, $\boldsymbol{v}$ is the particle velocity, $\boldsymbol{\omega}$ is the particle angular velocity, $\boldsymbol{F}$ is the resultant force on the particle and $\boldsymbol{M}$ is the resultant torque on the particle. The forces acting on the particle are the drag ($\boldsymbol{F_d}$), the buoyancy force ($\boldsymbol{F_b}$), the gravitational force ($\boldsymbol{F_g}$), the pressure gradient force ($\boldsymbol{F_{pg}}$), the virtual mass force ($\boldsymbol{F_{vm}}$) and the Magnus lift due to particle rotation ($\boldsymbol{F_{ml}}$):

$$m_p\frac{dv}{dt} = \mathbf{F}_d + \mathbf{F}_b + \mathbf{F}_g + \mathbf{F}_{pg} + \mathbf{F}_{vm} + \mathbf{F}_{ml}, \tag{12}$$

These forces are expressed as:

$$\mathbf{F}_d = m_p\frac{18\mu_c}{\rho_p d_p^2}\frac{C_D Re_p}{24}(\boldsymbol{u} - \boldsymbol{v}), \tag{13}$$

$$\mathbf{F}_b + \mathbf{F}_g = m_p\frac{\boldsymbol{g}(\rho_p - \rho_c)}{\rho_p}, \tag{14}$$

$$\mathbf{F}_{pg} = m_p\frac{\rho_c}{\rho_p}\frac{D\boldsymbol{u}}{Dt}, \tag{15}$$

$$\mathbf{F}_{vm} = \frac{1}{2}m_p\frac{\rho_c}{\rho_p}\left(\frac{D\boldsymbol{u}}{Dt} - \frac{d\boldsymbol{v}}{dt}\right), \tag{16}$$

$$F_{ml} = \frac{1}{2} A_p C_{RL} \rho_c \frac{|u - v|}{|\Omega|} ((u - v) \times \Omega), \quad (17)$$

where ρ_p is the particle density, d_p is the particle diameter, C_D is the drag coefficient, Re_p is the particle Reynolds number, A_p is the projected particle surface area, C_{RL} is the rotational lift coefficient and Ω is the relative angular velocity of the particle. For the drag coefficient (C_D), a correlation by Morsi and Alexander [24] is adopted, as it is applicable over a wide range of Reynolds numbers.

For the rotational lift coefficient (C_{RL}), an approach proposed by Tsuji et al. [25] is adopted, as it is applicable to high particle Reynolds numbers ($Re_p < 1600$). The torque applied to the particle (M) can be expressed in terms of rotational drag acting on the particle. Equation (11) is rewritten as

$$I_p \frac{d\omega}{dt} = \frac{\rho_c}{2} \left(\frac{d_p}{2} \right)^5 C_\omega |\Omega| \cdot \Omega, \quad (18)$$

where C_ω is the rotational drag coefficient. The correlation proposed by Dennis et al. [26] is used.

2.2. Dense Discrete Phase Model

2.2.1. Liquid Flow

The governing equations of flow presented in Section 2.1.1 are modified to account for the presence of the particles via their volume fraction

$$\frac{\partial(\alpha_c \rho_c)}{\partial t} + \nabla \cdot (\alpha_c \rho_c u) = 0, \quad (19)$$

$$\frac{\partial(\alpha_c \rho_c u)}{\partial t} + \nabla \cdot (\alpha_c \rho_c u u) = -\alpha_c \nabla p + \nabla \cdot \left[\alpha_c \mu_c \left(\nabla u + u^T \right) \right] + \alpha_c \rho_c g + K_M (v - u) + S_{M,explicit}, \quad (20)$$

where α_c is the continuous-phase volume fraction, K_M is the particle-averaged interphase momentum exchange coefficient (implicit part of the momentum exchange with discrete phase) and $S_{M,explicit}$ is the momentum source term resulting from the displacement of fluid in the presence of a discrete phase.

2.2.2. Particle Motion

The governing equations of particle motion are the same as those described in Section 2.1.2. Equation (12) can be extended to include particle–particle interaction force (F_{DEM})

$$m_p \frac{dv}{dt} = F_d + F_b + F_g + F_{pg} + F_{vm} + F_{ml} + F_{DEM}. \quad (21)$$

Particle–particle interaction is modelled according to DEM. Deformations of particles during collisions are accounted for by introducing the collision overlap between particles instead of collision deformations, as shown in Figure 1. Contact between colliding particles is modelled with the Hertz–Mindlin model [27,28], whereby the particle interaction is represented by a spring–dashpot system, as shown in Figure 2.

The particle interaction force is split into the normal and tangential components, where normal is defined at the point of contact between particles.

$$F_{DEM} = F_{DEM,n} n + F_{DEM,t} t. \quad (22)$$

The normal and tangential components of the interaction force are expressed in terms of the spring–dashpot system as the spring force and the damping force.

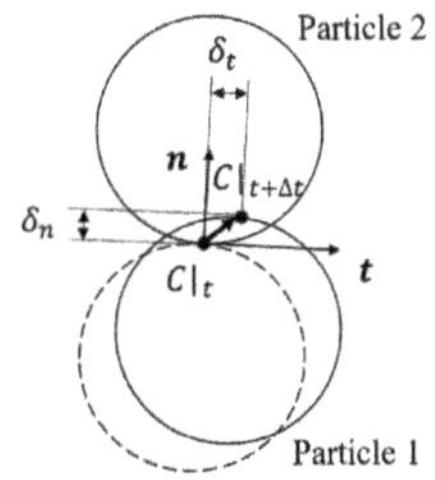

Figure 1. Particle overlap definition.

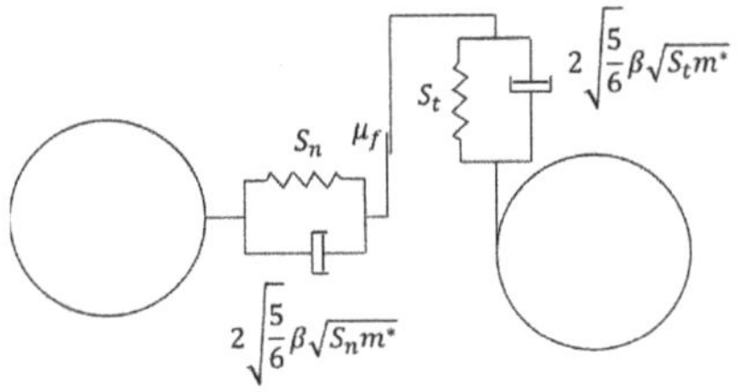

Figure 2. Hertz–Mindlin spring and dashpot system.

$$F_{DEM,n} = F_{n,spring} + F_{n,damping} = \frac{4}{3}E^*\sqrt{R^*}\delta_n^{\frac{3}{2}} - 2\sqrt{\frac{5}{6}}\beta\sqrt{S_n m^*}v_{rel,n}, \tag{23}$$

$$F_{DEM,t} = F_{t,spring} + F_{t,damping} = 8G^*\sqrt{R^*\delta_n}\delta_t - 2\sqrt{\frac{5}{6}}\beta\sqrt{S_t m^*}v_{rel,t}, \tag{24}$$

where E^* is the equivalent Young's modulus, G^* is the equivalent shear modulus, R^* is the equivalent radius, δ_n is the overlap in the normal direction, δ_t is the overlap in the tangential direction, β is a constant, S_n is the normal stiffness, S_t is the tangential stiffness, m^* is the equivalent mass, $v_{rel,n}$ is the normal component of the relative velocity and $v_{rel,t}$ is the tangential component of the relative velocity. These are defined as

$$\frac{1}{E^*} = \frac{1-\nu_1^2}{E_1} + \frac{1-\nu_2^2}{E_2}, \tag{25}$$

$$\frac{1}{G^*} = \frac{2(1+\nu_1)(1-\nu_1^2)}{E_1} + \frac{2(1+\nu_2)(1-\nu_2^2)}{E_2}, \tag{26}$$

$$\frac{1}{R^*} = \frac{R_1+R_2}{R_1R_2}, \tag{27}$$

$$\beta = \frac{-\ln e}{\sqrt{\ln^2 e + \pi^2}}, \tag{28}$$

$$S_n = 2E^*\sqrt{R^*\delta_n}, \tag{29}$$

$$S_t = 8G^*\sqrt{R^*\delta_n}, \tag{30}$$

$$m^* = \frac{m_1m_2}{m_1+m_2}, \tag{31}$$

$$v_{rel,n} = v_{1,n} + v_{2,n}, \tag{32}$$

$$v_{rel,t} = v_{1,t} + v_{2,t}, \tag{33}$$

where ν is the Poisson ratio, E is the Young's modulus, R is the particle radius, e is the coefficient of restitution, m is the particle mass, v_n is the particle normal velocity and v_t is the particle tangential velocity. Subscripts 1 and 2 refer to particle 1 and particle 2, respectively, in collision, as shown in Figure 1. Normal and tangential velocities of particles are observed according to the local collision coordinate system, as shown in Figure 1.

2.3. Experiment

The Faculty of Civil Engineering of the University of Rijeka is home to the GUNT HM 162 experimental flume (shown on Figure 3), which is an open-channel flow physical model with a test section with a length of 12,500 mm and a rectangular cross section measuring 309 mm in width and 450 mm in height. This experimental flume is also fitted with a separate particle inlet.

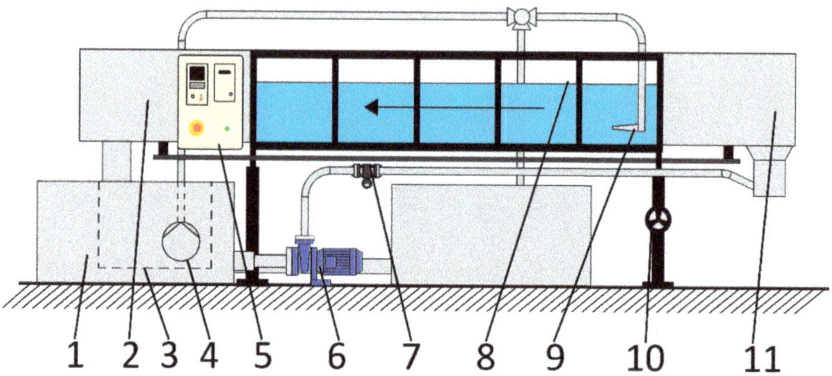

Figure 3. HM 162 experimental flume: 1—water tank, 2—water outlet element, 3—sediment screen basket, 4—sediment pump, 5—PLC (switch box), 6—centrifugal pump, 7—flow rate sensor, 8—open-channel test section, 9—particle (sediment) inlet, 10—inclination adjustment element, 11—water inlet element.

The HM 162 experimental flume consists of an open-channel test section that is a part of a closed water circuit. The water is circulated from a water tank into the pipe and through the water inlet element into the experimental section by a Lowara SHS4 centrifugal pump, which can provide a maximum head of 16.1 m and has an operating flow rate in the range of 5.4 to 130 $m^3\ h^{-1}$. The water flows back into the water tank through the outlet element at the end of the experimental section. The water pump is controlled by the PLC (switch box) to adjust the flow rate. The actual flow rate is reported by the Endress+Hauser Promag 10L electromagnetic flow rate sensor, which is built into the pipe between the water pump and inlet section and has a measuring range of 5.4 to 180 $m^3\ h^{-1}$. The water level and the tilt of the open-channel section are also adjustable via the adjustable overflow edge and the inclination adjustment element, respectively.

The particle inlet that is supplied by GUNT (as shown in Figure 3) is intended for dense sediment flows, with high sediment flow rates achieved by means of a separate GUNT-manufactured sediment pump. The pump is designed for fine-grained sand with a granulation of up to 2 mm and can achieve a maximum flow rate of 36 $m^3\ h^{-1}$. Because this work is focused on small batches of particles that are passively introduced into the water and are larger than 2 mm, a new particle inlet was constructed at the hydraulic laboratory of the Faculty of Civil Engineering, University of Rijeka.

A metal funnel was fitted with a 3D-printed opening mechanism and fixed inside a 3D-printed holding plate. The holding plate was attached to four steel wires, which were attached to the base plate. When the base plate was placed on top of the experimental section side walls, the funnel was positioned as shown in Figure 4.

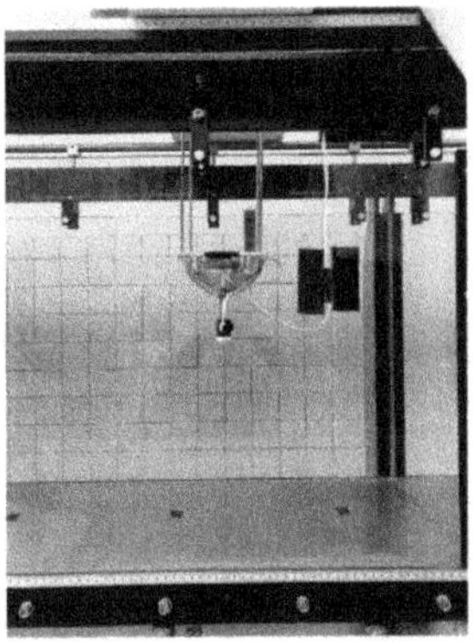

Figure 4. Detailed view of the particle inlet funnel.

We investigated the settling and particle size distribution of 3D-printed spherical plastic particles with a density of 1140 kg m^{-3}. The particle size distribution is presented in Table 1; each particle was colored according to its size for size identification.

Table 1. Particle sizes used in the settling experiment.

Particle Diameter (mm)	2	2.5	3.5	4
Number of particles	62	67	70	58
Color	Black	Red	Green	White

The volumetric flow rate was set to 27 m^3 h^{-1}, and the water level was set to 350 mm. The tilt of the channel was horizontal (no tilt). As shown in Figure 4, the particle inlet was positioned in the middle of the width of the test section and 5450 mm downstream of the beginning of the test section. The bottom of the funnel where particles enter the water was positioned 80 mm below the free surface.

Particle release and settling were filmed with two cameras: one obtaining a top view and one capturing a side view of the channel. Settling times and downstream settling distances were determined from the videos.

3. Computational Setup

For the purpose of conducting a mesh study, three different structured, hexahedral computational meshes of the whole channel geometry were prepared with the ANSYS (Canonsburg, PA, USA) Meshing module. To reduce the number of cells and therefore the computational time, the computational domain was reduced only to the area of interest, where the domain was shortened to 5250 mm by moving the inlet 3550 mm downstream and moving the outlet 3700 mm upstream of the experimental open-channel test section. Three different meshes of this reduced domain were generated, with element sizes between corresponding mesh densities kept constant. The final mesh statistics are presented in Table 2.

A velocity inlet that matches the volumetric flow rate was used in the experiment, relative pressure at the outlet was set to 0 Pa and the operating pressure was set to 101,325 Pa. A no-slip condition was imposed on the walls of the flume, and a free-slip wall condition was imposed on the free surface of the flow. All walls were treated as smooth walls.

The continuous phase was water with constant density of 998.2 kg m^{-3} and a constant viscosity of 0.001003 kg m^{-1} s^{-1}. The particles were made with VeroBlue RGD840 material with a density of 1140 kg m^{-3}, Poisson's ratio of 0.49 and a Young's modulus of 2650 MPa. For the DEM approach, properties of the walls, particle–particle interaction and particle–wall interaction properties were further prescribed. For the bottom steel wall, a density of 7800 kg m^{-3} was prescribed with a Poisson's ratio of 0.3 and a Young's modulus of 210,000 MPa. The glass side walls were prescribed a density of 2500 kg m^{-3}, a Poisson's ratio of 0.22 and a Young's modulus of 70,000 MPa. For both the particle–particle interaction and particle–wall interaction, the coefficient of restitution was 0.5, the coefficient of static friction was 0.5 and the coefficient of rolling friction was 0.01. Temperature dependence of material properties was not considered, and an isothermal simulation approach was adopted, with all material properties reported for 20 °C.

Table 2. Information about analyzed computational meshes: M1—coarse mesh, full domain; M2—medium mesh, full domain; M3—fine mesh, full domain; M1r—coarse mesh, reduced domain; M2r—medium mesh, reduced domain; M3r—fine mesh, reduced domain.

Label	M1	M2	M3	M1r	M2r	M3r
Number of elements	47,400	120,736	309,738	21,000	53,536	137,104
Min y^+	19.3	13.4	9.69	17.6	13.0	9.73
Max y^+	63.8	47.5	36.4	57.3	41.3	30.1
Ave. y^+	54.2	39.2	28.8	49.4	37.4	27.5
Min. element volume (m^3)	2.7×10^{-5}	1.06×10^{-5}	4.14×10^{-6}	2.7×10^{-5}	1.06×10^{-5}	4.14×10^{-6}
Max. element volume (m^3)	2.7×10^{-5}	1.06×10^{-5}	4.14×10^{-6}	2.7×10^{-5}	1.06×10^{-5}	4.14×10^{-6}
Min. orthogonality	1	1	1	1	1	1
Max. aspect ratio	1.78	1.74	1.75	1.78	1.74	1.75

Pressure–velocity coupling of equations was achieved via the Coupled scheme, gradients were evaluated using the least squares cell-based method, pressure at the cell faces was interpolated using the PRESTO! method and the third-order accurate QUICK spatial discretization scheme was adopted for all equations. Temporal discretization of equations was achieved by the bounded second-order implicit time integration scheme. A time step of 0.05 s was chosen, which was shorter than the particle response time, resulting in a maximum Courant number of less than 1. The iterative solution of equations within each time step was limited to 100 iterations; however, a scaled residuals convergence criterion of 10^{-4} was achieved before this limitation.

For the particles, an inlet velocity of 0 m s^{-1} was imposed, and a mass flow rate for each size fraction was determined to match the number of particles for a given size fraction within a time window of 1.1 s, which was determined in the experiment.

4. Results and Discussion

Vertical and horizontal fluid velocity profiles were plotted along the respective directions at the particle inlet location before particles were injected into domain. As shown in Figures 5 and 6, both in the case of a whole domain and in the case of a reduced domain, all three mesh densities predict a similar velocity profile. The main difference is observed near the channel walls, where the coarse mesh underpredicts both the velocity gradient and the velocity magnitude. In comparison to the whole domain, significantly different velocity profiles are obtained on the meshes of the reduced domain. However, this can be corrected by applying the calculated velocity profiles on the fine mesh from the whole-domain case as an inlet boundary condition in the reduced-domain case.

A comparison of calculated settling times obtained using three approaches and experimental values is presented in Figure 7. Satisfactory agreement was achieved for all three simulation approaches, except for the smallest particles with a diameter of 2 mm. A similar observation can be derived from the comparison of the downstream distance travelled by particles before settling on the bottom of the channel, as presented in Figure 8. Interaction between particles does not appear to be significant, as the DDPM approaches with and without inclusion of DEM produce similar results.

A more detailed comparison of the DPM and DDPM approaches is presented in Figures 9 and 10, showing a time evolution of settling velocity and horizontal particle velocity, respectively. Minor differences can be observed, indicating that the point particle assumption of DPM is sufficient for treatment of mesoplastic particles in an open-channel flow.

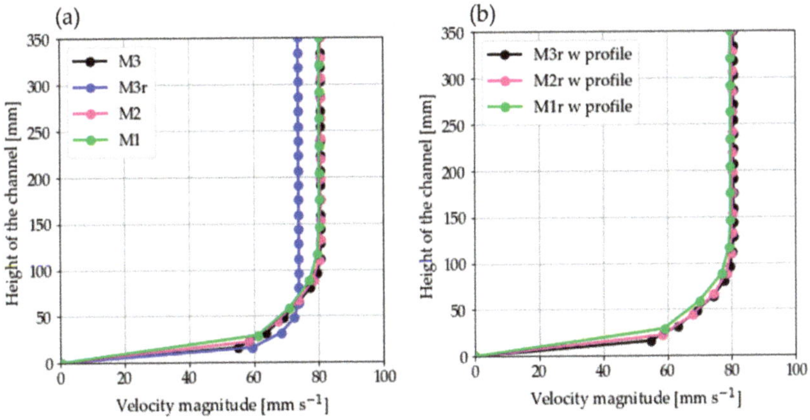

Figure 5. Vertical velocity profiles of water: (**a**) comparison between three mesh densities for the full domain and the fine mesh of the reduced domain; (**b**) comparison between three mesh densities of the reduced domain with a velocity profile boundary condition.

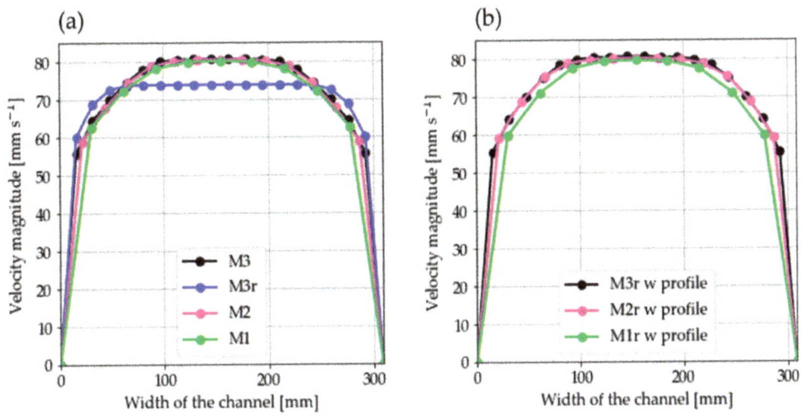

Figure 6. Horizontal velocity profiles of water: (**a**) comparison between three mesh densities for the full domain and the fine mesh of the reduced domain; (**b**) comparison between three mesh densities of the reduced domain with a velocity profile boundary condition.

Figure 11 presents the propagation of a particle cloud as observed in the DPM simulation and the experiment at three different times. The color of the largest 4 mm particles is changed from white to yellow in the visualization of the simulation results for improved visibility. Additional vertical lines are drawn in the photos from the experiment that correspond to vertical lines drawn in the pictures from the simulation due to the difference in perspective. As shown in Figure 11, general agreement can be observed between simulation and experimental results, although with a noticeably tighter grouping of particles by size in the case of the simulation.

From both the experiment and simulations, it is evident that larger plastic particles settle more quickly than smaller particles, as expected, with the larger particles reaching a higher settling velocity. As shown in Figure 9, all particles reach terminal velocity (flattening of all curves for time evolution of velocity). Similarly, all particles reach the maximum horizontal velocity of the flow; as expected, smaller particles reach this velocity quicker than larger particles. Because larger particles reach the bottom of the channel sooner than smaller particles, their horizontal velocity also drops sooner. As shown by the velocity profile presented in Figure 5, the velocity adheres to a no-slip law (boundary condition); therefore, when particles approach the bottom of the channel, their velocity reduces accordingly. Figure 10 shows that settled particles retain some horizontal velocity in the simulation; however, this behavior is not observed in the experiment because an accurate particle–wall interaction was outside the scope of this study, and only the settling motion in the fluid was analyzed.

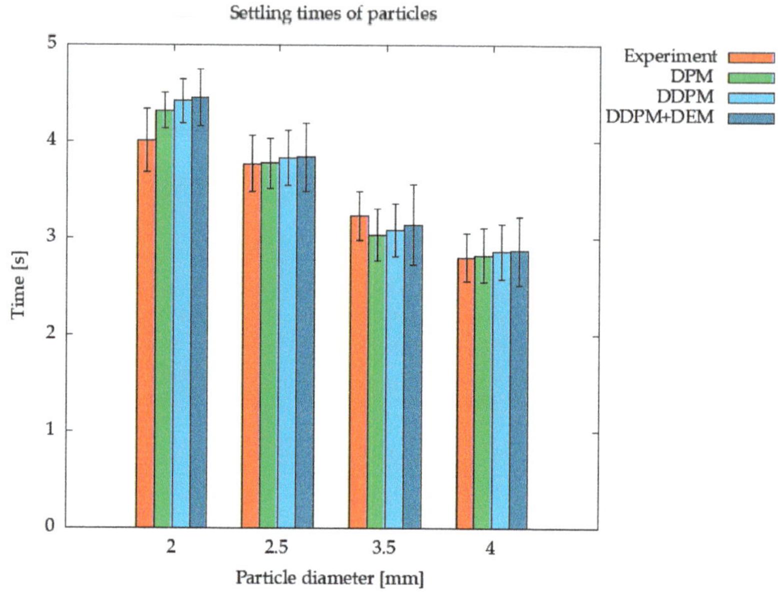

Figure 7. Average settling times of plastic particles; error bars represent one standard deviation.

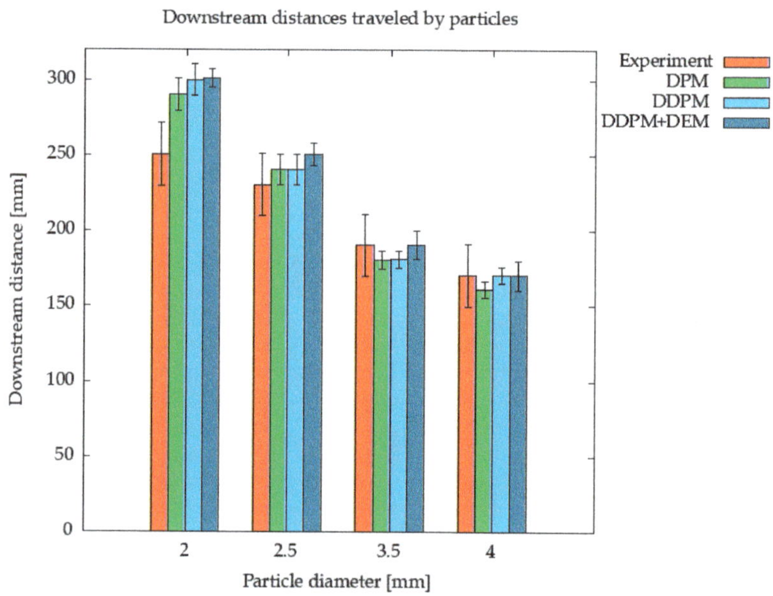

Figure 8. Average downstream settling distance of plastic particles; error bars represent one standard deviation.

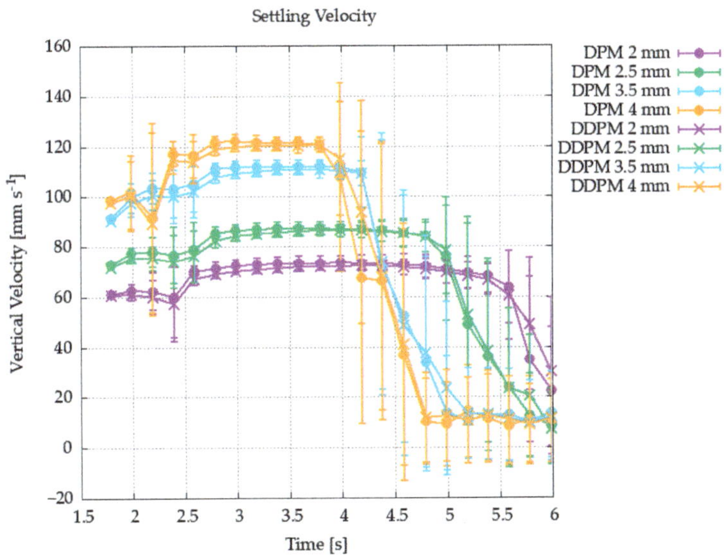

Figure 9. Settling velocity of plastic particles; points represent average values, and error bars represent one standard deviation.

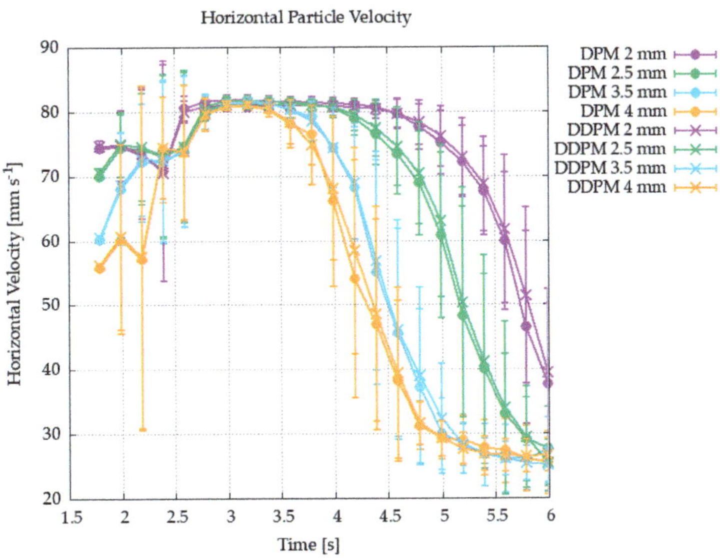

Figure 10. Horizontal velocity of plastic particles; points represent average values, and error bars represent one standard deviation.

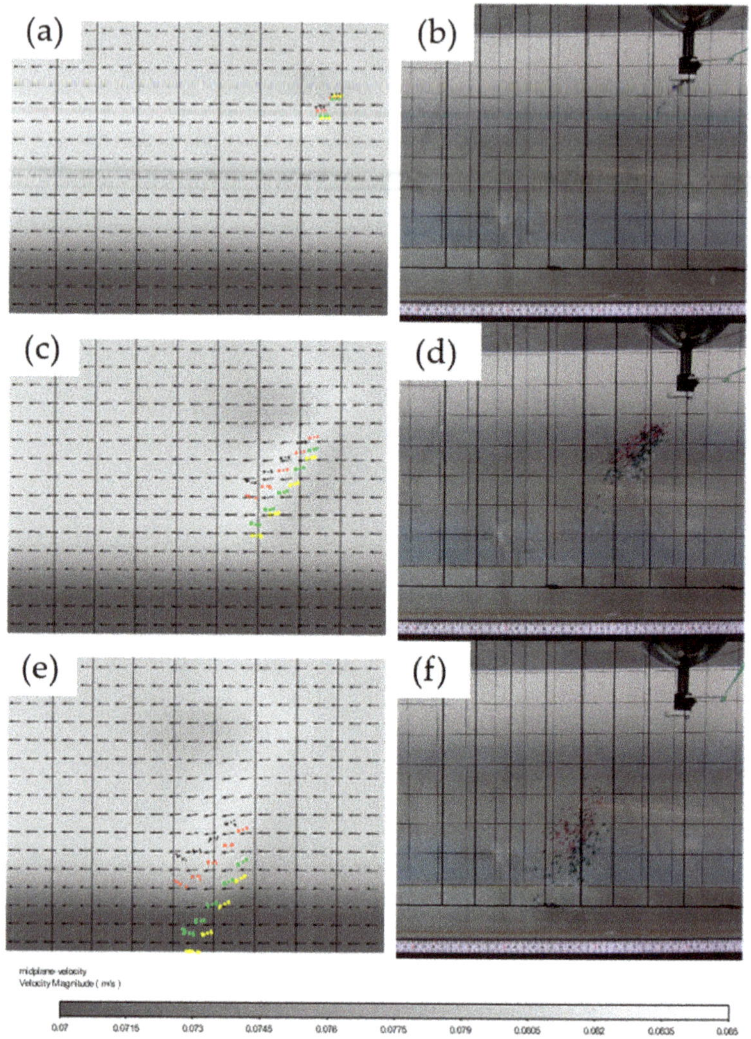

Figure 11. Comparison of the particle cloud as observed in the experiment and the simulation using the DPM approach: (**a**) and (**b**) 0.5 s after the release of particles; (**c**) and (**d**) 1.5 s after the release of particles; (**e**) and (**f**) 2.5 s after the release of particles. Particles are colored by size as described in Table 1. Yellow color is used instead of white in the visualization of the simulation results.

5. Conclusions

In this study the settling of mesoplastic particles of varying sizes in water was investigated. Settling times and downstream distance travelled by particles were measured experimentally. Numerical simulations were performed with three different approaches and compared to the experimental results. The computational effort was reduced by reducing the size of the domain (length of channel), and special care was taken to ensure the correct velocity profile was calculated at the position where particles were introduced into the domain, as the velocity profile of the water influences the velocity of the particles. The following conclusions can be drawn from this research:

(1) Larger particles travel a shorter distance downstream, settle quicker and reach a higher terminal velocity than smaller particles, as expected based on previous research on this topic.
(2) All investigated approaches to simulate settling of mesoplastic particles are appropriate and allow for a detailed investigation of particle motion. Interaction between particles is negligible and can be omitted, as it requires additional computational effort.
(3) All presented modelling approaches produced tighter spatial grouping of particles compared to the experiment, indicating the need for future research on the topic particle motion modelling and evaluation of existing models.

The investigated approaches for simulation of particle settling in an open-channel flow can be used on a larger scale to predict the propagation of mesoplastics in rivers.

Author Contributions: Conceptualization, I.B.; methodology, E.Ž.; software, L.K.; validation, L.L. and L.K.; writing—original draft preparation, L.K.; writing—review and editing, I.B., L.L. and E.Ž.; visualization, I.B. and E.Ž.; supervision, I.B.; project administration, L.K. and L.L. All authors have read and agreed to the published version of the manuscript.

Funding: The authors wish to thank the Slovenian Research Agency (ARRS) for their financial support within the framework of the Research Programme P2-0196 Research in Power, Process and Environmental Engineering. This paper is the result of a project on the Development of Research Infrastructure at the University Campus in Rijeka (RC.2.2.06-0001), cofunded by the European Regional Development Fund (ERDF) and the Ministry of Science and Education of the Republic of Croatia.

Data Availability Statement: All data are contained within the article.

Acknowledgments: The authors wish to thank Duje Kalajžić for his help in the experiments.

Conflicts of Interest: The authors declare no conflict of interest.

References

1. Bermúdez, J.R.; Swarzenski, P.W. A microplastic size classification scheme aligned with universal plankton survey methods. *MethodsX* **2021**, *8*, 10–15. [CrossRef] [PubMed]
2. Hartmann, N.B.; Hüffer, T.; Thompson, R.C.; Hassellöv, M.; Verschoor, A.; Daugaard, A.E. Are We Speaking the Same Language? Recommendations for a Definition and Categorization Framework for Plastic Debris. *Environ. Sci. Technol.* **2019**, *53*, 1039–1047. [CrossRef] [PubMed]
3. Frias, J.P.G.L.; Nash, R. Microplastics: Finding a consensus on the definition. *Mar. Pollut. Bull.* **2018**, *138*, 145–147. [CrossRef] [PubMed]
4. Wagner, M.; Lambert, S. (Eds.) *Freshwater Microplastics*; Springer International Publishing: Cham, Switzerland, 2018; Volume 58.
5. GESAMP. Sources, Fate and Effects of Microplastics in the Marine Environment: A Global Assessment. 2015. Available online: www.imo.org (accessed on 18 October 2022).
6. Sun, J.; Dai, X.; Wang, Q.; van Loosdrecht, M.C.M.; Ni, B.J. Microplastics in wastewater treatment plants: Detection, occurrence and removal. *Water Res.* **2019**, *152*, 21–37. [CrossRef]
7. Čurlin, M.; Pušić, T.; Vojnović, B.; Vinčić, A. STEM Approach in Assessment of Microplastic Particles in Textile Wastewater. *Teh. Vjesn.* **2022**, *29*, 1777–1781. [CrossRef]
8. Auta, H.S.; Emenike, C.U.; Fauziah, S.H. Distribution and importance of microplastics in the marine environment—A review of the sources, fate, effects, and potential solutions. *Environ. Int.* **2017**, *102*, 165–176. [CrossRef]
9. Danopoulos, E.; Twiddy, M.; Rotchell, J.M. Microplastic contamination of drinking water: A systematic review. *PLoS ONE* **2020**, *15*, e0236838. [CrossRef]
10. Eriksen, M.; Lebreton, L.C.; Carson, H.S.; Thiel, M.; Moore, C.J.; Borerro, J.C.; Reisser, J. Plastic Pollution in the World's Oceans: More than 5 Trillion Plastic Pieces Weighing over 250,000 Tons Afloat at Sea. *PLoS ONE* **2014**, *12*, e111913. [CrossRef]
11. Klein, S.; Dimzon, I.K.; Eubeler, J.; Knepper, T.P. *Analysis, Occurrence, and Degradation of Microplastics in the Aqueous Environment*; Springer: Cham, Switzerland, 2018; Volume 58.
12. Lebreton, L.C.M.; Van Der Zwet, J.; Damsteeg, J.W.; Slat, B.; Andrady, A.; Reisser, J. River plastic emissions to the world's oceans. *Nat. Commun.* **2017**, *8*, 15611. [CrossRef]
13. Ren, Z.; Gui, X.; Xu, X.; Zhao, L.; Qiu, H.; Cao, X. Microplastics in the soil-groundwater environment: Aging, migration, and co-transport of contaminants—A critical review. *J. Hazard. Mater.* **2021**, *419*, 126455. [CrossRef]
14. Schernewski, G.; Radtke, H.; Robbe, E.; Haseler, M.; Hauk, R.; Meyer, L.; Labrenz, M. Emission, Transport, and Deposition of visible Plastics in an Estuary and the Baltic Sea—A Monitoring and Modeling Approach. *Environ. Manag.* **2021**, *68*, 860–881. [CrossRef] [PubMed]

15. Fatahi, M.; Akdogan, G.; Dorfling, C.; Van Wyk, P. Numerical study of microplastic dispersal in simulated coastal waters using cfd approach. *Water* **2021**, *13*, 3432. [CrossRef]
16. Cook, S.; Chan, H.L.; Abolfathi, S.; Bending, G.D.; Schäfer, H.; Pearson, J.M. Longitudinal dispersion of microplastics in aquati flows using fluorometric techniques. *Water Res.* **2020**, *170*, 115337. [CrossRef] [PubMed]
17. Yu, Z.; Yao, W.; Loewen, M.; Li, X.; Zhang, W. Incipient Motion of Exposed Microplastics in an Open Channel Flow. *Environ. Sc Technol.* **2022**, *56*, 14498–14506. [CrossRef]
18. Yi, Y.; Jia, W.; Yang, Y.; Zhang, S. Effects of diameter, density, and adhesiveness on settling velocity and drag coefficient of tw sturgeon species eggs in flow. *J. Hydraul. Res.* **2022**, *60*, 229–239. [CrossRef]
19. Roy, N.; Wijaya, K.P.; Götz, T.; Sundar, S. A mathematical model governing the short-range transport of microplastic particles in lid-driven cavity with an obstacle. *Commun. Nonlinear Sci. Numer. Simul.* **2021**, *101*, 105893. [CrossRef]
20. Fu, S.; Guo, X.Y.; Dong, L.H.; Sheng, K.; Sun, A. Numerical Simulation of Migration Laws of Dense Particle Flow in Pipelines. *In J. Simul. Model.* **2022**, *21*, 89–100. [CrossRef]
21. Holjević, T.; Travaš, V.; Kranjčević, L.; Družeta, S. Analysis of Microplastic Particle Transmission. *J. Marit. Transp. Sci.* **2022**, 4 237–244. [CrossRef]
22. Travaš, V.; Kranjčević, L.; Družeta, S.; Holjević, T.; Lučin, I.; Alvir, M.; Sikirica, A. Model gibanja čestica mikroplastike u nehomogenom i laminarnom polju brzine. *Hrvatske Vode* **2021**, *29*, 201–213.
23. ANSYS. *Ansys Fluent Theory Guide*; ANSYS: Canonsburg, PA, USA, 2021.
24. Morsi, S.A.; Alexander, A.J. An investigation of particle trajectories in two-phase flow systems. *J. Fluid Mech.* **1972**, *55*, 193–20 [CrossRef]
25. Tsuji, Y.; Oshima, T.; Morikawa, Y. Numerical simulation of pneumatic conveying in the horizontal pipe. *KONA Powder Part.* **1985**, *3*, 38–51. [CrossRef]
26. Dennis, S.C.R.; Singh, S.N.; Ingham, D.B. The steady flow due to a rotating sphere at low and moderate Reynolds numbers. *Fluid Mech.* **1980**, *101*, 257–279. [CrossRef]
27. Mindlin, R.D.; Deresiewicz, H. Elastic Spheres in Contact Under Varying Oblique Forces. *J. Appl. Mech.* **1953**, *20*, 327–34 [CrossRef]
28. Hertz, H. Über die Berührung fester elastischer Körper. *J. Angew. Math.* **1882**, *92*, 156–171. [CrossRef]

MDPI
St. Alban-Anlage 66
4052 Basel
Switzerland
www.mdpi.com

Energies Editorial Office
E-mail: energies@mdpi.com
www.mdpi.com/journal/energies

Disclaimer/Publisher's Note: The statements, opinions and data contained in all publications are solely those of the individual author(s) and contributor(s) and not of MDPI and/or the editor(s). MDPI and/or the editor(s) disclaim responsibility for any injury to people or property resulting from any ideas, methods, instructions or products referred to in the content.

www.ingramcontent.com/pod-product-compliance
Lightning Source LLC
Chambersburg PA
CBHW061758210926
51850CB00006B/591
* 9 7 8 3 0 3 6 5 9 8 5 3 6 *